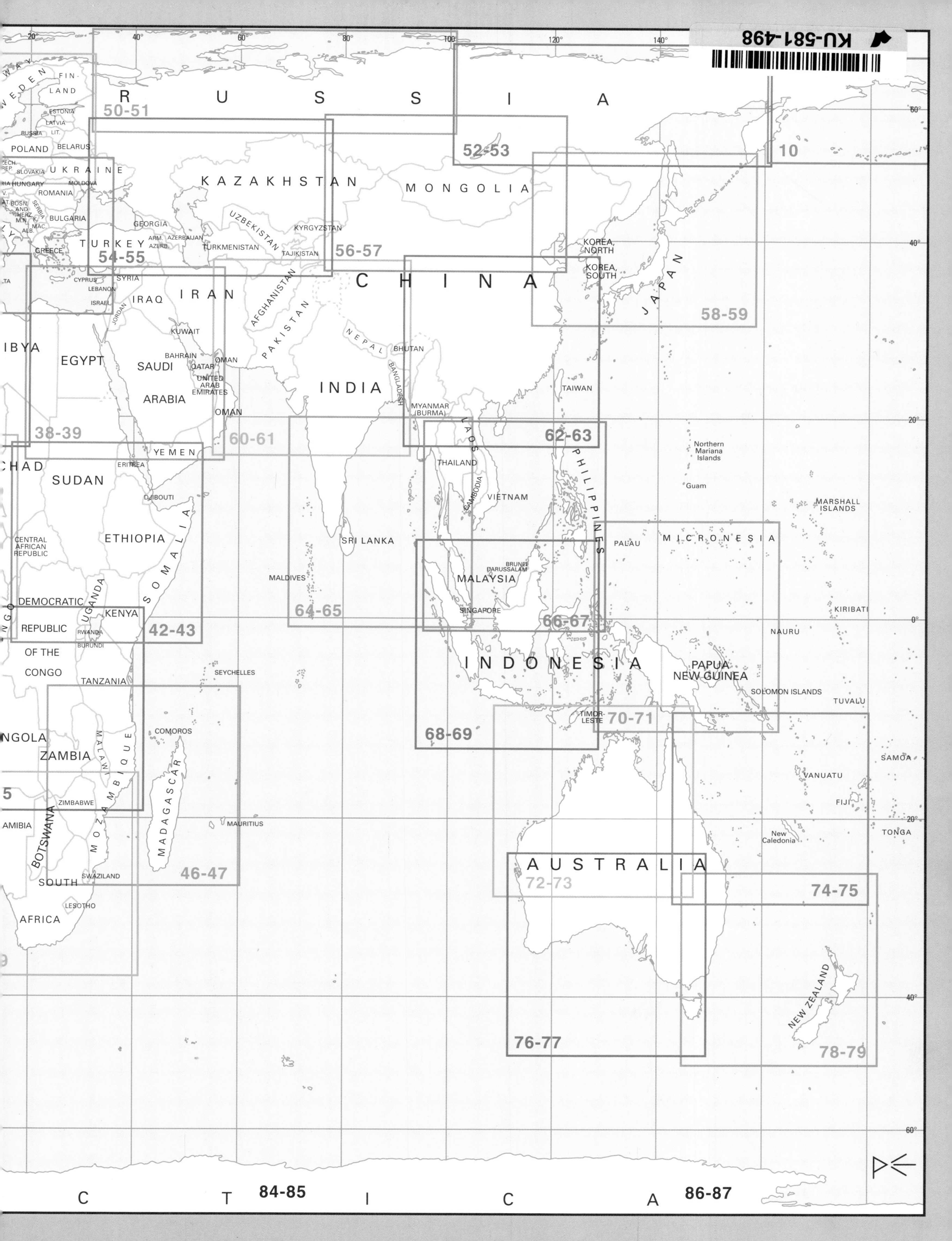

TOPOGRAPHIC MAPS

PETERS WORLD ATLAS

THE EARTH IN ITS TRUE PROPORTIONS

h.f.ullmann

THIS ATLAS IS PUBLISHED WITH THE SAME CONTENT IN ALL COUNTRIES OF THE WORLD

Original title: *Peters Weltatlas*
ISBN: 978-3-8331-5559-8

Project coordination for h.f.ullmann: Daniel Fischer, Lars Pietzschmann

Revised and updated edition.

Research assistants and advisors: Dr. E. C. Barrett, Professor Dr. Ulrich Bleil, Michael Benckert, Wolfgang Behr, Professor Heinz Bosse, Professor Walter Buchholz, Dr. Nicola Bradbear, Carol Claxton, Professor Dr. Heinrich Dathe, Hellmuth Färber, Jean Fernand-Laurent, Kurt Ficker, Professor Dr. Fritz Fischer, Karlheinz Gieseler, Professor Dr. Manfred Görlach, Professor Dr. Ulrich Grosse, Dipl.-Geogr. Arnulf Hader, Birgit Hahn, Dipl.-Ing. Max Hann, Dirk Hansom, Dr. Günther Heidmann, Professor Dr. Wolf Herre, Karl-Heinz Ingenhaag, Dr. Andreas Kaiser, Toni Kaufmann, Michael Kidron, Professor Dr. Gunther Krause, Dr. Manfred Kummer, Daniel Lloyd, Konrad G. Lohse, Dipl.-Ing. Wolfgang Mache, Dr. Udo Moll, Olive Pearson, Dr. Aribert Peters, Dipl.-Ing. Werner Peters, Thomas Plümer, Dr. Detlev Quintern, D.H. Reichstein, Luise Scherf, Hellmuth Schlien, Professor Dr. Herrmann Schulz, Ronald Segal, Professor Dr. Axel Sell, Eduard Spescha, Dr. Walter Stützle, Dr. Jürgen Wendler, Professor Adolf Witte, Professor Dr. Karl Wohlmuth, Siegfried Zademack, Madeleine Zeller

Cartography: KÜMMERLEY+FREY, BERN
(graticules, coastlines, borders, seas, rivers and lakes)
OXFORD CARTOGRAPHERS, www.oxfordcarto.com
(topographic and thematic maps)
KARTOGRAPHIE HUBER, MUNICH
(update)

Cartography editor: Terry Hardaker; Editorial coordination: Hazel Hand, Penny Watson; Editorial revision: Ann Leleu, Georg Möller, Anja Peters; Computer programming: H. Morelli; Map compilation: Katherine Armitage, Claire Carlton, John Hall, Sheila Hodson, Christine Johnston, Jean Kelly, Tanya Lillington, Angela Morrison, Kay Roberts, Fiona Sutcliffe; Technical coordination: John Dawson, Piet Summerfield; Map design: Mick Dyer, Gerhard Engel, Bob Hawkins, Sally Horn, Robert Hundley, Jeff Jones, David Lewis, Sue Lovell, Colin McCarthy, Michael Oakley, Judith Wood; Calculations: Peter Langran, John Williams; Preparation of reliefs: David Angus; Relief shading: Terry Hardaker; Computer cartography including typesetting and editing: Barbara Croucher, Duncan Croucher, Betty Döppl, Petra Faltermeier, Karin Geier, Franz Huber, Ingrid Kampfhenkel, Hermann Lechner, Lothar Meier, Rudolf Nowak, Margaret Prietzsch, Anton Sommer, Iris Sommer, Dr. Jürgen Wendler, Thorsten Wieland

Typesetting: Blenheim Colour

Cover design: Simone Sticker, Martin Wellner

Project coordination for the English edition for h.f.ullmann: Lars Pietzschmann

Overall responsibility for the production: h.f.ullmann publishing, Potsdam, Germany

Printed in China

ISBN 978-3-8331-5560-4

10 9 8 7 6 5 4 3 2 1
X IX VIII VII VI V IV III II I

If you would like to be informed about forthcoming h.f.ullmann titles, you can request our newsletter by visiting our website (**www.ullmann-publishing.com**) or by e-mailing us at: newsletter@ullmann-publishing.com.
h.f.ullmann, Birkenstraße 10, 14469 Potsdam, Germany

FOREWORD

A world in which all cultures have the same status, in which manual work rates as high as intellectual work and in which the economy no longer serves the enrichment of one person at the expense of others, but the satisfaction of everyone's needs and a fulfilled, free life for everybody: This objective is the central thread of Professor Arno Peters's life's work. The historian, economist, and cartographer (1916-2002) thus presents the various epochs, cultures, and fields, such as politics, art or science, as being of equal importance in his "Synchronoptic World History". History no longer appears only as the history of the rulers of this world. Similarly, the "Peters World Atlas", having been published in many countries since 1989 and now present in an updated form, is also committed to the thought of the equal rank of all peoples.

Transferring the spherical form of the earth to a two-dimensional map is not unlike pressing an orange peel onto a flat surface. Inevitably, this causes distortions. Which of these distortions will materialise on the printed map depends on the manner of calculation, the so-called projection. With the Peters Projection, Arno Peters found a way to realistically render the proportions: No country gives the impression of being bigger than it really is. This conformal map of the world, on which the "Peters World Atlas" is based, has been disseminated in many millions of copies by organisations of the United Nations, including the children's fund UNICEF, since it was first published in 1973.

The picture of the world that people can look at in the form of a map is co-decisive as to how they see themselves in relation to others. With this idea in mind, Arno Peters decided to depict all countries and continents on the same scale in his atlas. As he never tired of emphasising, the dominant position of Europe in the world corresponded to the practice of representing Europe on a larger scale than the other countries and continents in old maps. His atlas is meant to aid in overcoming this outdated, Eurocentric world view linked with colonialism and in promoting solidarity among the peoples of the world. The atlas is the work of a pioneer of globalisation, but of a globalisation which is truly worthy of this designation, that is to say which is more than the mere dismantling of trade barriers.

Jürgen Wendler, PhD, who collaborated with Arno Peters for many years

CONTENTS

Map Finder for the Topographic Maps . . . . . Front Endpaper
Foreword 3
Introduction 6-7
Page Finder for Countries 8
Key to the Topographic Section 9
Sources for Thematic Section 96
Introduction to the Thematic Section 97
Political World Map 112-113
Relief and Vegetation of the World . . . . . Rear Endpaper

THE WORLD IN 43 MAPS AT THE SAME SCALE

THE AMERICAS

Alaska, Canada (North-West) 11
Canada (North) 12-13
Canada (South-West), USA (West) 14-15
Canada (South-East), USA (East) 16-17
USA (South), Mexico, Hawaii 18-19
USA (South-East), Cuba, Caribbean 20-21
Central America, South America (North) 22-23
South America (West) 24-25
South America (East) 26-27
South America (Central) 28-29
South America (South) 30-31

EUROPE

Northern Europe 32-33
Southern Europe 34-35
Eastern Europe 50, 54

AFRICA

Africa (North-West) 36-37
Africa (North-East) 38
Africa (West) 40-41
Africa (East) 42-43
Africa (South-West) 44-45
Africa (South-East), Madagascar 46-47
Africa (Southern) 48-49

ASIA

Russia (North-Western Siberia) 51
Russia (Northern Siberia) 52-53
Russia (North-Eastern Siberia) 10
Asia (North-West) 54-55
Russia (South-East), Kazakhstan, Mongolia, China (North) 56-57
China (North-East), Korea, Japan 58-59
China (South-East), South-East Asia (North) . . . 62-63
Asia Minor 35, 54
Arabian Peninsula 38-39
Asia (South) with India (North) 60-61
India (South), South-East Asia 64-65
South-East Asia, Philippines, Borneo 66-67
Indonesia, Malaysia 68-69
New Guinea, South Pacific 70-71

AUSTRALASIA

Australia (West) 72-73
Australia (North-East), South Pacific 74-75
Australia (South) 76-77
Australia (South-East), New Zealand 78-79

ANTARCTICA

Antarctica (Byrd Land) 80-81
Antarctica (Graham Land) 82-83
Antarctica (Maud Land) 84-85
Antarctica (Wilkes Land) 86-87

THE ARCTIC

Arctic (Canada) 88-89
Arctic (Greenland) 90-91
Arctic (Novaya Zemlya) 92-93
Arctic (Siberia) 94-95

NATURE, HUMANKIND AND SOCIETY IN 212 THEMATIC WORLD MAPS

The Continents 98-99
560 Million Years Ago · 280 Million Years Ago · 180 Million Years Ago · 120 Million Years Ago · 60 Million Years Ago · Today

Mountains 100-101

Rivers and Seas 102-103

Sun and Climate 104-105
Hours of Sunshine · Incidence of Rays · Evaporation · Temperature · Temperature Variation · Precipitation

Natural Dangers 106-107
Volcanoes and Earthquakes · Maritime Earthquakes and Tsunami · Weather related and natural Disasters · Retreat of Glaciers

Vegetation 108-109
Woodland · Pastureland · Arable Land · Barren Land

People and Cities 110-111

States 112-113

Languages 114-115
French · Spanish · Portuguese · English · German · Russian · Italian · Bengali · Malay · Arabic · Chinese · Japanese · Hindi · Swahili

Religions 116-117
Natural Religions · Hinduism · Christianity · Islam · Buddhism · Confucianism · Shinto · No Religion

Education 118-119
Illiteracy · Primary Education · Secondary Education · University · Pupil/Teacher Ratio · Expenditure on Education

The Sciences 120-121
University Teachers · Publication of Scientific Books · Researchers and Engineers · Expenditure on the Sciences

Information 122-123
Newspapers · Radio · Television · Books

Communications 124-125
Telephones · Mobile Phones · Fax · Internet

Traffic Density 126-127
Railways · Cars · Air Traffic · Shipping

Population Structure 128-129
Population Density · Population Growth · Adults/Children Ratio · Men/Women Ratio

Life Expectancy 130-131

Health 132-133
Medical Attention · Available Hospital Beds · Drinking Water Supplies · Infant Mortality

HIV/AIDS-Malaria Risk Areas 134-135

Nutrition 136-137
Calorie Consumption · Undernourished Population · Adiposity, Men · Adiposity, Women

Soil Cultivation 138-139

Use of Chemical Fertilizers 140-141

Crop Yield 142-143

Staple Foodstuffs 144-145
Wheat · Barley · Rye · Oats · Millet · Soya · Maize Sorghum · Rice

Animal Husbandry 146-147
Cattle · Horses · Mules · Donkeys · Camels · Buffalo · Elephant · Reindeer · Sheep · Pigs · Goats · Bees · Chickens · Ducks · Turkeys · Silkworm

Fishing 148-149
Size of Catches · Herring · Cod · Tuna · Flatfish · Squid

Hunting 150-151
Hare, Rabbit · Roe Deer · Red Deer · Elk · Fox · Wolf · Bear · Jackal · Antelope · Gazelle · Kangaroo · Lynx · Opossum · Beaver · Pheasant · Partridge

Natural Products 152-153
Wood · Rubber · Silk · Cotton · Wool · Skins · Hemp · Flax · Honey · Eggs · Grain · Meat · Vegetables · Fruit · Sugar · Milk

Mineral Resources 154-155
Iron · Copper · Tin · Lead · Gold · Silver · Platinum · Diamonds · Aluminium · Zinc · Manganese · Nickel · Chrome · Sulphur · Potassium · Phosphate

Energy 156-157
Oil · Natural Gas · Water Power · Uranium · Coal · Total Consumption

Steel and Automobile Production 158-159

Urbanisation 160-161

World Trade 162-163
Natural Products (Import) · Natural Products (Export) · Industrial Products (Import) · Industrial Products (Export)

The Living Conditions of the Nations (Human Development Index) 164-165

Poor Nations - Rich Nations 166-167

Economic Growth 168-169

Employment Structure 170-171
Agriculture · Transport · Industry · Mining · Building · Trade · Energy · Social Professions · Banking

Economic Order 172-173

Public Finances
Gross Domestic Product (GDP) 174-175

National Debt 176-177

Unemployment 178-179

Inflation 180-181

Child Labour 182-183

Inequality 184-185

Prostitution 186-187

The Status of Women 188-189
Paid Employment · Housework · Number of Children · Monogamy/Polygamy · Participation in Legislation · Higher Education

Relative Military Strength 190-191
Armed Forces · Tanks · Fighter Planes · Naval Vessels · Nuclear Weapons · Reserve Forces

The Conquest of Space 192-193
Launching Sites for Space Flights · Earth Satellites · Manned Space Flights · Interplanetary Space Flights

Index 194-232

CARTOGRAPHIC INTRODUCTION

It may come as a shock to realise that all of the atlases we have known until now present a distorted picture of the world. The nature of this distortion, and the reason for it, are now so obvious that it seems hardly possible to have overlooked it for 400 years. The distortion caused by attempting to represent the spherical earth on flat paper is more or less unavoidable, but the distortion caused by the use of inconsistent scales, which has acquired the unquestioned sanction of habit, is not.

We have come to accept as „natural" a representation of the world that devotes disproportionate space to large-scale maps of areas perceived as important, while consigning other areas to small-scale general maps. And it is because our image of the world has become thus conditioned, that we have for so long failed to recognise the distortion for what it is - the equivalent of peering at Europe and North America through a magnifying glass and then surveying the rest of the world through the wrong end of a telescope.

There is nothing „natural" about such a view of the world. It is the remnant of a way of thinking born even before the age of colonialism and fired by that age. Few thinking people today would subscribe to a world-view of this kind. Yet, until now, no atlas has existed that provided a picture of the world undistorted by varying scales.

A single scale

All topographic maps in this atlas are at the same scale: each double-page map shows one-sixtieth of the earth's surface. This means that all the topographic maps can be directly compared with one another. Among the many surprises this unique feature offers may be, for some users, the relative sizes of Great Britain (page 32) and the island of Madagascar (page 47); or, perhaps, the areas respectively covered by Europe (pages 32-33) and North Africa (pages 36-37). For most people it will soon become apparent that their hazy and long-held notions of the sizes of different countries and regions are, in a lot of cases, quite drastically wrong.

But what do we mean by scale? The scale indicator that appears on reference maps only shows distance scale. It enables the user to calculate the factor needed to multiply distances so as to compare them with those on other maps. This is a complex and somewhat tedious exercise that the great majority of users understandably neglects to carry out. Moreover, the number of different scales in conventional atlases is surprisingly high, in general between twenty and fifty. Thus, the concept of relative scale must become increasingly vague in the user's mind. What is generally not mentioned is that, because it is impossible to transfer the curved surface of the globe correctly to a flat plane, the scale indicator on a map is only valid for a single part of the map, such as a line of latitude.

Distance is only one aspect of scale. Area has also to be considered. Whereas there can be no maps with absolute fidelity of distance, there can be maps with fidelity of area. Fidelity of area (or equal area) means that all countries are shown at the correct size in relation to others. The maps in this atlas preserve fidelity of area, a feature never previously achieved in an atlas. In the Peters Atlas all topographic maps have equal area scale: 1 square centimetre on the maps equals 6,000 square kilometres in reality.

While uniform scale for all parts of the world enables comparisons between all places on an equal footing, the local geography covered at larger scales in other world atlases is omitted from the Peters Atlas. To include enlargements of the more densly populated regions would be to defeat the purpose of this atlas.

A single symbology

This equality of scale offers further advantages besides direct comparability. The basis of any map compilation is the simplification of reality, which cartographers call „generalisation." This transfer of the real character of the earth's surface into a system of lines and symbols, which can be graphically represented, has to be adapted to the scale employed. Thus a river or road with all its turns and windings at a scale of 1:100,000 can be drawn nearer to reality (that is, with more detail) than at a scale of 1:1,000,000. Symbols also vary for different scales. Thus the same symbol can mean a town with 50-100,000 inhabitants at one scale, but a city with 1-5 million inhabitants at another. The same elevation may be differently coloured on maps of different scales. All such difficulties vanish in this atlas, which by way of its single scale has only one level of generalisation and a single set of symbols.

Topographic map colours

The green/brown colouring of most current atlases represents the topographic relief of the region; green stands for low-lying areas, brown for mountainous country, with different shades of the two colours for different elevations. Since, however, both colours (as also the blue of the sea and the white of snow-covered mountains) are borrowed from nature, the user of the atlas may be forgiven for assuming the green parts on the map to represent areas with vegetation and the brown parts to be the barren land. Although this is broadly true in Europe it may not be so elsewhere. For example, in North Africa the lower areas, even those below sea level, are usually deserts, and it is only above a certain height in the mountains that vegetation begins. The green/brown colouring is thus unsuitable for representing relief in an atlas dedicated to an accurate worldview. So in this atlas green represents vegetation, brown barren land, and a mixture of the two colours represents thin or scattered vegetation. Global vegetation data were obtained from 1985-86 satellite photography with the help of the Remote Sensing Unit at the Department of Geography of Bristol University. The resolution of this imagery down to individual units of 20 square kilometres, and its conversion to the Peters base maps by the Remote Sensing Unit, makes this the most up-to-date statement available of the distribution of world vegetation.

Topographic map relief

The tradition of showing relief by coloured layers running from green in lowlands, through brown to purple, mauve or white in high mountains, has one further serious disadvantage. The features of the land surface are only shown along the contours, or lines joining points of equal height. In between, we receive no information about the surface of the land. „Hillshading," or rendering the complete landscape with 3-D shadows to depict the relief, overcomes this. Cartographers have struggled with the best way to create hillshading for hundreds of years. In this atlas the 3-D relief comes from photographing specially made plaster relief models and blending these photos with hand-rendered colouring. In the blending, the relief shading has also been enhanced to eliminate awkward shadows, as for example when the angle of the light from the photo ran directly down a mountain chain thus reducing its impact. The addition of „spot heights" for selected peaks and other points on the map lends precision to the artwork.

The Peters Projection
Anyone who has ever tried to peel an orange and press the peel into a continuous flat piece without tearing will have grasped the fundamental impossibility that lies at the heart of all cartographic endeavour: that fidelity of shape, distance and angle are of necessity lost in flattening the surface of a sphere. On the other hand it is possible to retain three other qualities: fidelity of area, fidelity of axis and fidelity of position. Fidelity of area makes it possible to compare various parts of a map directly with one another, and fidelity of axis and position guarantee correct relationships of north-south and east-west axes by way of a rectangular grid.

In 1973 Arno Peters published his world map, which unites in a single flat map all three achievable qualities - fidelity of area, fidelity of axis and fidelity of position. In this way the real comparative sizes of all countries in the world are clearly visible. For this atlas Arno Peters has generalised the projection principle upon which his World Map was based, so that now each regional map represents the maximum possible freedom from distortion, of area as well as shape. Since map distortions through a projection decrease in proportion to the size of the area depicted - the smaller the area, the smaller the distortion - the forty-three topographical maps in this atlas are considerably closer to reality than is the Peters World Map. In particular these individual maps correct the distortions which are unavoidable on the Peters World Map in the equatorial and polar regions. An indication of how this has been applied can be seen from the shapes of the page areas on the Map Finder (front endpaper). In the north they are horizontal and thin while towards the Equator they are nearly square. The degree of departure from the normal page proportion is a guide to the amount of shape correction applied to the regional maps.

The eight polar maps on pages 80-95 have the same scale as all the other topographical maps. They also have fidelity of area, and represent one-sixtieth of the earth's surface on each double page. Thus the size of the countries and continents shown on them can be directly compared with all the other 35 topographical maps. The fidelity of position and axis which is necessarily lost on polar maps is also absent on these maps.

The thematic maps
The second part of this atlas directs attention to the whole earth. The author has collected data for 212 individual world thematic maps under 48 subject headings. Each of these subject headings is given a double page spread, but if more than one topic is covered under any subject, separate maps are given. Thus under the subject heading „Life Expectancy" only one topic is covered so the double spread comprises a single map, whereas the subject heading „Animal Husbandry" requires sixteen topics and therefore displays sixteen individual maps. The principle of one topic per map also enables all the maps to be represented by simple grades of colour, with, usually, a single hue chosen for each topic. Within this hue the range from light to dark colour represents low to high values of the topic. In this way all the thematic maps can be understood at a glance, without the necessitiy for complicated symbols or explanations.

The graticule
The traditional zero meridian running through Greenwich was adopted worldwide in 1884, when Britain was the strongest European colonial power and ruled over a quarter of the world. After the ending of colonialism and with the closure of Greenwich Observatory, there is no reason other than custom for retaining this zero meridian. The international dateline, which is dependent upon the zero meridian, also needs correction, since over its whole length it has been partially diverted where it cuts an inhabited area. The retention of the division into 360 degrees is also, it can be argued, an anomaly in the age of almost worldwide decimalisation.

Arno Peters has therefore proposed a new decimal grid in which the zero meridian and the international dateline would become a single line placed in the middle of the Bering Strait, and the earth is divided into 100 decimal degrees east-west and north-south. As the new graticule is however not yet generally established, in this Atlas the old graticule has been maintained.

The index
Someone consulting an index in search of a district, town or river has until now had to memorise, besides the page number at least two grid figures, two letters, or a letter and a number. There can be few users of an atlas who have not experienced the irritation of forgetting at least one item in this unwieldy string of digits by the time the relevant map has been located, and the time-consuming exercise of turning back to the index to recall this information. In the Peters Atlas there is, apart from the page number, only a single letter, which can be easily remembered. This innovatory and simple indexing system is explained on page 194.

Digital cartography
The thematic maps in this edition have been revised using the latest available data and completely redrawn digitally, using Apple Mac technology.

The original edition of the atlas used computer techniques to adapt the world map to the 43 individual double spreads in the topographic section, minimising shape distortion on each spread. This process used Scitex technology with geographic data from the Erdgenössische Technische Hochschule in Berne. The same process is employed in the present edition, which also continues to feature the hand-crafted workmanship of the land surface colouring. The atlas thus reconciles the traditional and digital approaches.

Chief Cartographer
Oxford Cartographers

PAGE FINDER FOR COUNTRIES

Afghanistan . . . 60
Albania . . . 35
Algeria . . . 36-37
Andorra . . . 34
Angola . . . 44
Antigua & Barbuda . . . 21
Argentina . . . 28-29, 30-31
Armenia . . . 54
Australia . . . 72-73, 74, 76-77, 78
Austria . . . 34-35
Azerbaijan . . . 54
Bahamas . . . 21
Bahrain . . . 39
Bangladesh . . . 61
Barbados . . . 23
Belarus . . . 33
Belgium . . . 32-33
Belize . . . 19
Benin . . . 41
Bhutan . . . 61
Bolivia . . . 25, 28
Bosnia & Herzegovina . . . 35
Botswana . . . 48-49
Brazil . . . 23, 24-25, 26-27, 28-29
Brunei Darussalam . . . 69
Bulgaria . . . 35
Burkina Faso . . . 40-41
Burma (Myanmar) . . . 62, 65
Burundi . . . 45
Cambodia . . . 65, 66
Cameroon . . . 41
Canada . . . 11, 12-13, 14-15
Cape Verde . . . 234
Central African Republic . . . 41, 42
Chad . . . 37, 38, 41, 42
Chile . . . 28, 30
China . . . 56-57, 58, 61, 62-63
Colombia . . . 22-23, 24-25
Comoros . . . 47
Congo . . . 41, 42, 44
Congo, Dem. Rep. of the . . . 42, 44-45, 46
Costa Rica . . . 22
Côte d'Ivoire . . . 40
Croatia . . . 35
Cuba . . . 20-21
Cyprus . . . 35
Czech Republic . . . 33, 34-35
Denmark . . . 33
Djibouti . . . 43
Dominica . . . 23
Dominican Republic . . . 21
Ecuador . . . 24
Egypt . . . 38
El Salvador . . . 22
Equatorial Guinea . . . 41
Eritrea . . . 38-39, 43
Estonia . . . 33
Ethiopia . . . 43
Fiji . . . 75
Finland . . . 33
France . . . 34
French Guiana . . . 26
Gabon . . . 41, 44
Gambia . . . 40
Georgia . . . 54
Germany . . . 33
Ghana . . . 40
Great Britain & Northern Ireland . . . 32
Greece . . . 35
Greenland . . . 90-91
Grenada . . . 23
Guatemala . . . 19, 22
Guinea . . . 40
Guinea-Bissau . . . 40
Guyana . . . 23
Haiti . . . 21
Honduras . . . 19, 22
Hungary . . . 35
Iceland . . . 32
India . . . 60-61, 64
Indonesia . . . 68-69, 70
Iran . . . 39, 54-55, 60
Iraq . . . 39
Ireland . . . 32
Israel . . . 38
Italy . . . 34-35
Jamaica . . . 21
Japan . . . 58-59
Jordan . . . 38
Kazakhstan . . . 54-55, 56
Kenya . . . 43, 45
Kiribati . . . 235
Korea, North . . . 58
Korea, South . . . 58
Kosovo . . . 35
Kuwait . . . 39
Kyrgyzstan . . . 55, 56
Laos . . . 62, 65, 66
Latvia . . . 33
Lebanon . . . 38
Lesotho . . . 49
Liberia . . . 40
Libya . . . 37, 38
Liechtenstein . . . 34
Lithuania . . . 33
Luxembourg . . . 33
Macedonia . . . 35
Madagascar . . . 47
Malawi . . . 45
Malaysia . . . 66-67
Maldives . . . 64
Mali . . . 36-37, 40-41
Malta . . . 35
Marshall Islands . . . 235
Mauritania . . . 36, 40
Mauritius . . . 47
Mexico . . . 18-19
Micronesia . . . 71
Moldova . . . 33, 35
Monaco . . . 34
Mongolia . . . 56-57
Montenegro . . . 35
Morocco . . . 36
Mozambique . . . 46
Myanmar (Burma) . . . 62, 65
Namibia . . . 48
Nauru . . . 235
Nepal . . . 61
Netherlands . . . 33
New Zealand . . . 79
Nicaragua . . . 22
Niger . . . 37, 41
Nigeria . . . 41
Norway . . . 33
Oman . . . 39, 60
Pakistan . . . 60
Palau . . . 71
Panama . . . 22
Papua New Guinea . . . 70-71
Paraguay . . . 28-29
Peru . . . 24-25
Philippines . . . 67
Poland . . . 33
Portugal . . . 34
Qatar . . . 39
Romania . . . 35
Russia . . . 10, 33, 50-51, 52-53, 54-55, 56-57, 58-59
Rwanda . . . 45
St. Kitts & Nevis . . . 21
St. Lucia . . . 23
St. Vincent & the Grenadines . . . 23
Samoa . . . 235
San Marino . . . 34
Sao Tomé & Principe . . . 41
Saudi Arabia . . . 38-39
Senegal . . . 40
Serbia . . . 35
Seychelles . . . 235
Sierra Leone . . . 40
Singapore . . . 66
Slovakia . . . 33, 35
Slovenia . . . 35
Solomon Islands . . . 71
Somalia . . . 43
South Africa . . . 48-49
Spain . . . 34
Sri Lanka . . . 64
Sudan . . . 38, 42
Suriname . . . 26
Swaziland . . . 49
Sweden . . . 33
Switzerland . . . 34
Syria . . . 38-39
Taiwan . . . 61
Tajikistan . . . 55
Tanzania . . . 45, 46
Thailand . . . 62, 66
Timor-Leste . . . 69, 72
Togo . . . 41
Tonga . . . 235
Trinidad & Tobago . . . 23
Tunisia . . . 34, 36-37
Turkey . . . 35, 54
Turkmenistan . . . 54-55, 60
Tuvalu . . . 235
Uganda . . . 42, 45
Ukraine . . . 33, 35, 54
United Arab Emirates . . . 39
Uruguay . . . 28-29
USA . . . 11, 14-15, 16-17, 18-19, 20
Uzbekistan . . . 55
Vanuatu . . . 75
Venezuela . . . 23
Vietnam . . . 62, 66
Western Sahara . . . 36
Yemen . . . 39, 43
Zambia . . . 44-45
Zimbabwe . . . 49

THE WORLD IN 43 MAPS AT THE SAME SCALE
EACH MAP SHOWS ONE-SIXTIETH OF THE EARTH'S SURFACE.

The colours used on the maps simulate those found in nature.

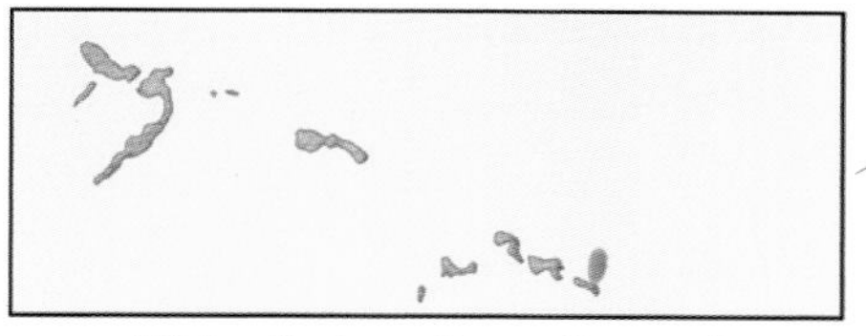
Water (Lakes, Seas, Oceans)

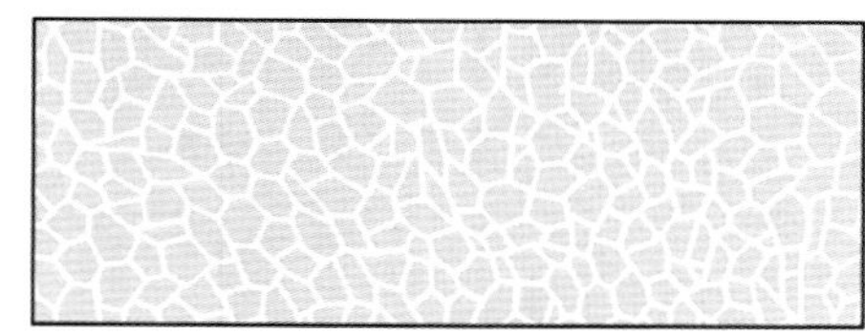
Ice Shelf

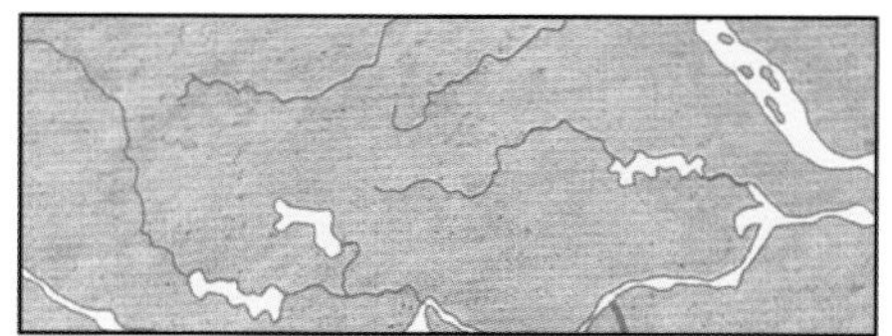
Vegetation (Plains)

Barren Land (Plains)

Continental Ice (Plains)

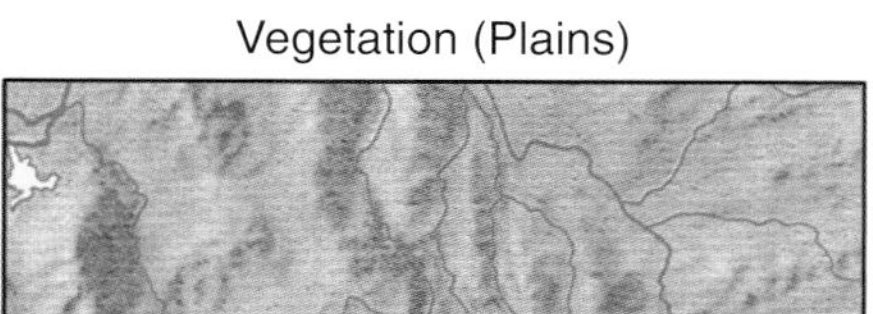
Vegetation (Hills)

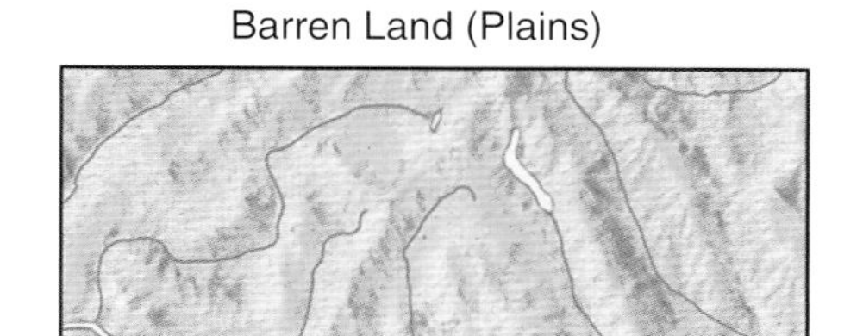
Barren Land (Hills)

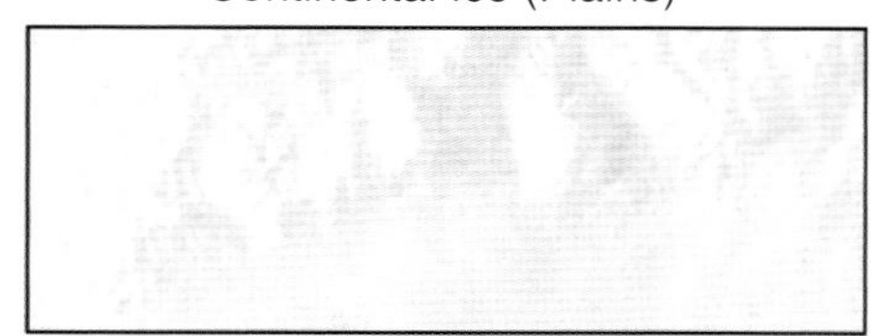
Continental Ice (Hills)

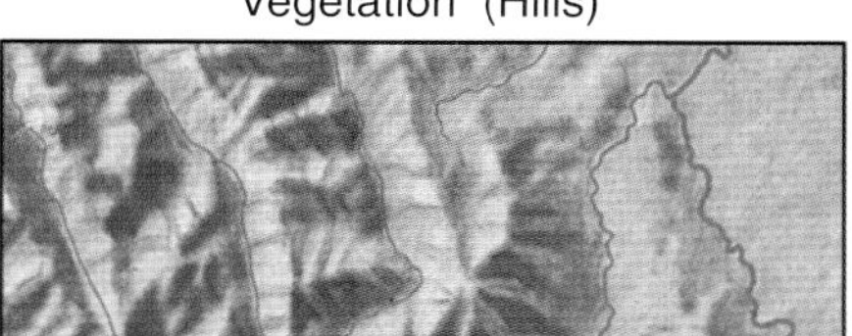
Vegetation (Mountains)

Barren Land (Mountains)

Continental Ice (Mountains)

Spot Heights:

•1236	1236 Metres above Sea Level
⊙1369	City 1369 Metres above Sea Level

Communications:

- Railway
- Road
- Motorway
- Motorway in Tunnel
- River
- Canal

Boundaries:

- International Boundary
- International Boundary on River
- Disputed International Boundary
- State Boundary

On each double page the 1000 largest and most important cities and towns are shown; if the double page shows sea as well as land, there are proportionately fewer:

- fewer than 100,000 inhabitants
- 100,000 - 1,000,000 inhabitants
- 1,000,000 - 5,000,000 inhabitants
- over 5,000,000 inhabitants

Adjoining map indicator:

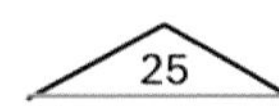

Map of adjoining area is on page 25.

Other physical features:

- Waterfall

- Swamp, Marsh

- Salt Lake
- Coral Reef

Other man-made features:

- ∴ Archaeological Site
- Great Wall of China

Latitude and Longitude:

25°E	25 degrees Longitude East
50°W	50 degrees Longitude West
30°N	30 degrees Latitude North
60°S	60 degrees Latitude South
	Tropics

Type styles:

Mato Grosso	Physical Features
Kolhapur	Cities and towns (capital cities underlined)
BELGIUM	Countries
TEXAS	States
INDIAN OCEAN	Oceans, Seas

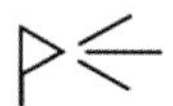
Peters Projection (fidelity of area, axis and position)

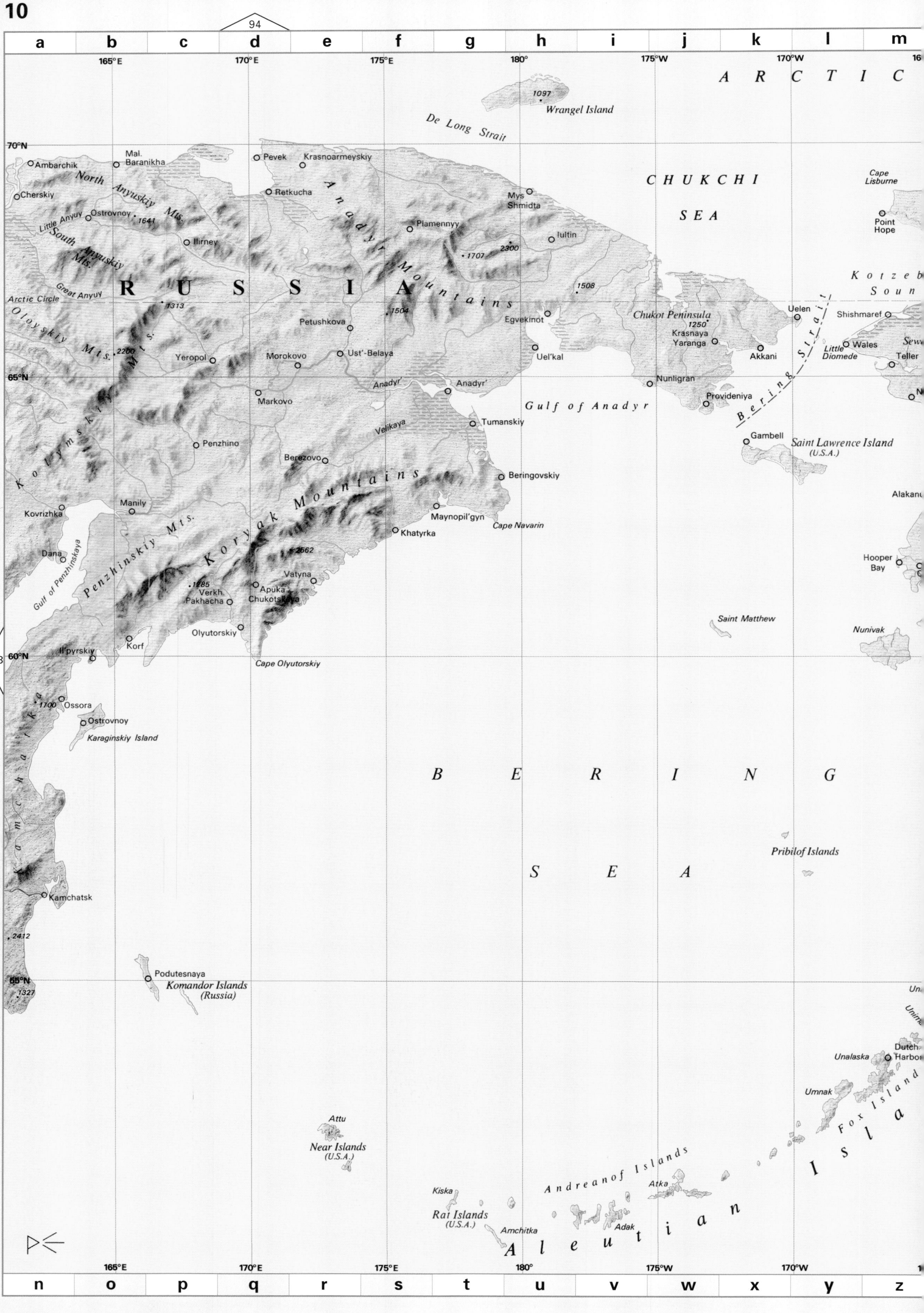

94

a b c d e f g h i j k l m

165°E 170°E 175°E 180° 175°W 170°W

ARCTIC

Wrangel Island
1097
De Long Strait
70°N
Ambarchik
Mal. Baranikha
Pevek
Krasnoarmeyskiy
North Anyuskiy Mts.
Cherskiy
Retkucha
Anadyr Mountains
Mys Shmidta
CHUKCHI SEA
Cape Lisburne
Point Hope
Little Anyuy
Ostrovnoy
1641
South Anyuskiy Mts.
Ilirney
Plamennyy
Iultin
2300
1707
1508
Kotzebue Sound
Great Anyuy
Arctic Circle
RUSSIA
1313
1504
Egvekinot
Chukot Peninsula
1250
Uelen
Shishmaref
Oloyskiy Mts.
Petushkova
Krasnaya Yaranga
Akkani
Little Diomede
Wales
Teller
Bering Strait
2200
Yeropol
Morokovo
Ust'-Belaya
Uel'kal
65°N
Anadyr'
Anadyr'
Nunligran
Markovo
Gulf of Anadyr
Provideniya
Kolymskiy Mts.
Tumanskiy
Velikaya
Gambell
Saint Lawrence Island (U.S.A.)
Penzhino
Berezovo
Beringovskiy
Koryak Mountains
Manily
Kovrizhka
Maynopil'gyn
Cape Navarin
Alakanuk
Penzhinskiy Mts.
Khatyrka
Dana
Gulf of Penzhinskaya
2562
Hooper Bay
Vatyna
1285
Verkh. Pakhacha
Apuka
Chukotskaya
Saint Matthew
Olyutorskiy
Nunivak
Korf
Il'pyrskiy
53
60°N
Cape Olyutorskiy
1700
Ossora
Ostrovnoy
Karaginskiy Island
Kamchatka
BERING SEA
Pribilof Islands
Kamchatsk
2412
Podutesnaya
Komandor Islands (Russia)
55°N
1327
Unimak
Dutch Harbor
Unalaska
Umnak
Fox Islands
Attu
Near Islands (U.S.A.)
Andreanof Islands
Atka
Kiska
Rat Islands (U.S.A.)
Amchitka
Adak
Aleutian Islands

165°E 170°E 175°E 180° 175°W 170°W

n o p q r s t u v w x y z

This map shows 1/60 of the earth's surface

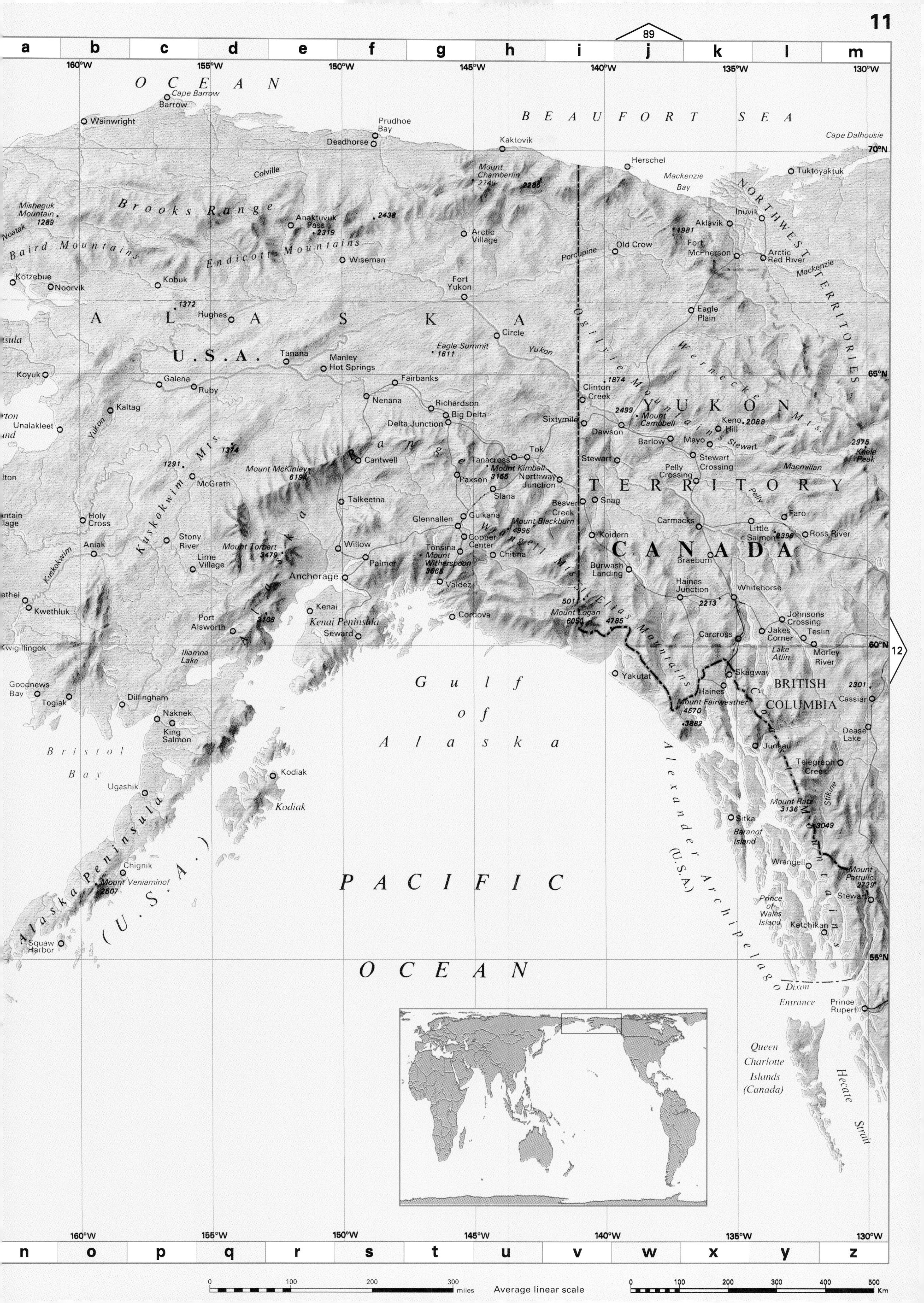
a b c d e f g h i j k l m
89
160°W 155°W 150°W 145°W 140°W 135°W 130°W
OCEAN
Cape Barrow
Barrow
Wainwright
Prudhoe Bay
Deadhorse
Kaktovik
BEAUFORT SEA
Cape Dalhousie
70°N
Herschel
Mackenzie Bay
Tuktoyaktuk
Colville
Mount Chamberlin 2749
2286
Misheguk Mountain 1289
Brooks Range
Anaktuvuk Pass 2319
2438
Noatak
Baird Mountains
Endicott Mountains
Arctic Village
Inuvik
Aklavik
1981
Old Crow
Porcupine
Fort McPherson
Arctic Red River
NORTHWEST TERRITORIES
Mackenzie
Kotzebue
Noorvik
Kobuk
Wiseman
Fort Yukon
Eagle Plain
1372
Hughes
ALASKA
Circle
U.S.A.
Tanana
Manley Hot Springs
Eagle Summit 1611
Yukon
Ogilvie Mountains
Wernecke Mts.
Koyuk
Galena
Ruby
Fairbanks
65°N
1874
Clinton Creek
Kaltag
Nenana
Richardson
Big Delta
Delta Junction
Sixtymile
2499
Mount Campbell
Dawson
YUKON
Keno Hill
2088
Unalakleet
Yukon
1374
Barlow
Mayo
Stewart
2975 Keele Peak
Tok
Tanacross
Stewart
Stewart Crossing
1291
Kuskokwim Mts.
Alaska Range
Cantwell
Mount McKinley 6194
Mount Kimball 3155
Paxson
Northway Junction
Pelly Crossing
Macmillan
McGrath
TERRITORY
Pelly
Slana
Talkeetna
Beaver Creek
Snag
Faro
Holy Cross
Glennallen
Gulkana
Mount Blackburn 4996
Carmacks
Little Salmon
2398
Ross River
Stony River
Aniak
Mount Torbert 3479
Willow
Copper Center
Wrangell Mts.
Koidern
CANADA
Kuskokwim
Lime Village
Tonsina
Mount Witherspoon 3666
Chitina
Palmer
Braeburn
Anchorage
Burwash Landing
Valdez
Haines Junction
Whitehorse
Bethel
Kwethluk
Kenai
5011
Mount Logan 6050
St. Elias Mountains
2213
Johnsons Crossing
Port Alsworth
3108
Kenai Peninsula
Seward
Cordova
4785
Jakes Corner
Teslin
Kwigillingok
Iliamna Lake
Carcross
Lake Atlin
Morley River
60°N
12
Yakutat
Skagway
BRITISH COLUMBIA
Goodnews Bay
Togiak
Dillingham
Gulf of Alaska
Haines
Mount Fairweather 4670
2301
Cassiar
Naknek
King Salmon
3882
Dease Lake
Bristol Bay
Juneau
Alexander Archipelago (U.S.A.)
Coast Mountains
Telegraph Creek
Ugashik
Kodiak
Kodiak
Stikine
Mount Ratz 3136
3049
Sitka
Baranof Island
Alaska Peninsula
(U.S.A.)
Chignik
Mount Veniaminof 2507
Wrangell
PACIFIC
Mount Pattullo 2729
Prince of Wales Island
Stewart
Ketchikan
Squaw Harbor
55°N
OCEAN
Dixon Entrance
Prince Rupert
Queen Charlotte Islands (Canada)
Hecate Strait
160°W 155°W 150°W 145°W 140°W 135°W 130°W
n o p q r s t u v w x y z
0 100 200 300 miles
Average linear scale
0 100 200 300 400 500 Km

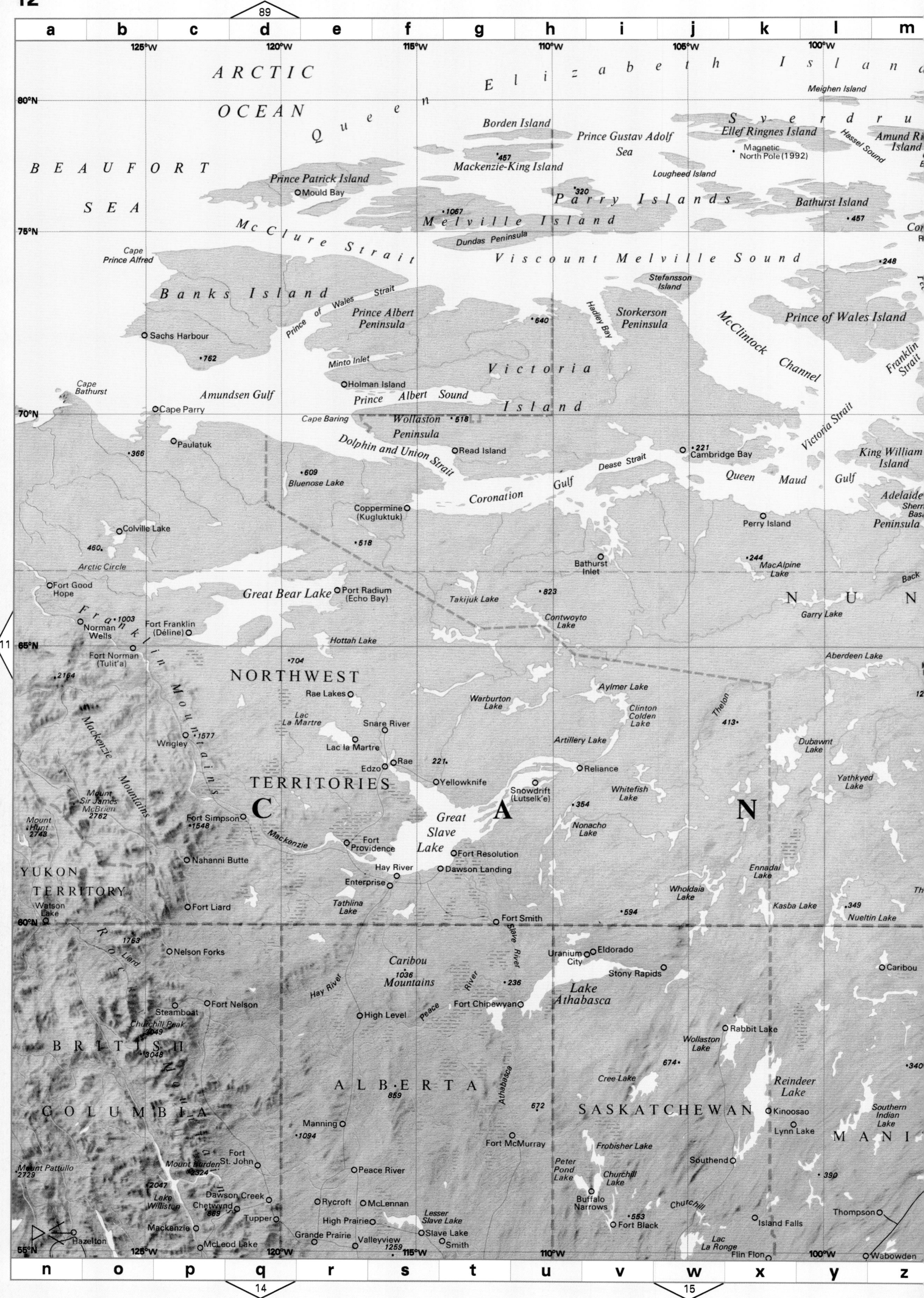
89
a b c d e f g h i j k l m
125°W 120°W 115°W 110°W 105°W 100°W
ARCTIC OCEAN
Queen Elizabeth Islands
Meighen Island
80°N
Borden Island
Prince Gustav Adolf Sea
Sverdrup
Ellef Ringnes Island
Hassel Sound
Amund Ringnes Island
Magnetic North Pole (1992)
BEAUFORT SEA
Mackenzie-King Island
Lougheed Island
Prince Patrick Island
Mould Bay
Parry Islands
Bathurst Island
Melville Island
McClure Strait
75°N
Dundas Peninsula
Cape Prince Alfred
Viscount Melville Sound
Stefansson Island
Banks Island
Prince of Wales Strait
Prince Albert Peninsula
Hadley Bay
Storkerson Peninsula
McClintock Channel
Prince of Wales Island
Franklin Strait
Sachs Harbour
Minto Inlet
Victoria Island
Cape Bathurst
Holman Island
Amundsen Gulf
Prince Albert Sound
Cape Parry
70°N
Cape Baring
Wollaston Peninsula
Victoria Strait
Paulatuk
Dolphin and Union Strait
Read Island
Cambridge Bay
King William Island
Dease Strait
Queen Maud Gulf
Bluenose Lake
Coronation Gulf
Adelaide Peninsula
Coppermine (Kugluktuk)
Perry Island
Colville Lake
Arctic Circle
Bathurst Inlet
MacAlpine Lake
Fort Good Hope
Great Bear Lake
Port Radium (Echo Bay)
Takijuk Lake
Contwoyto Lake
Garry Lake
Back
NUN
Norman Wells
Fort Franklin (Déline)
Franklin Mountains
65°N
Hottah Lake
Fort Norman (Tulit'a)
Aberdeen Lake
NORTHWEST
Rae Lakes
Aylmer Lake
Warburton Lake
Clinton Colden Lake
Thelon
Mackenzie Mountains
Lac La Martre
Snare River
Artillery Lake
Dubawnt Lake
Wrigley
Lac la Martre
Rae
Edzo
Reliance
Yathkyed Lake
TERRITORIES
Yellowknife
Snowdrift (Lutselk'e)
Whitefish Lake
Mount Sir James McBrien 2762
Mount Hunt 2743
Fort Simpson
CANADA
Nonacho Lake
Great Slave Lake
Mackenzie
Fort Providence
Fort Resolution
Nahanni Butte
Hay River
Dawson Landing
Ennadai Lake
YUKON TERRITORY
Enterprise
Wholdaia Lake
Watson Lake
Fort Liard
Tathlina Lake
Kasba Lake
Nueltin Lake
60°N
Fort Smith
Rocky Mountains
Liard
Slave River
Nelson Forks
Uranium City
Eldorado
Caribou Mountains
Stony Rapids
Caribou
Hay River
Peace River
Lake Athabasca
Steamboat
Fort Nelson
Fort Chipewyan
High Level
Peace
Rabbit Lake
Churchill Peak
BRITISH COLUMBIA
Wollaston Lake
ALBERTA
Athabasca
Cree Lake
Reindeer Lake
SASKATCHEWAN
Kinoosao
Southern Indian Lake
Manning
Lynn Lake
Fort McMurray
Frobisher Lake
MANI
Mount Pattullo
Mount Burden
Fort St. John
Peter Pond Lake
Southend
Peace River
Churchill Lake
Lake Williston
Dawson Creek
Chetwynd
Rycroft
McLennan
Buffalo Narrows
Churchill
Tupper
High Prairie
Lesser Slave Lake
Fort Black
Island Falls
Thompson
Mackenzie
Grande Prairie
Slave Lake
Hazelton
Valleyview
Smith
Lac La Ronge
55°N
McLeod Lake
Flin Flon
Wabowden
n o p q r s t u v w x y z
14
15
11

91

a b c d e f g h i j k l m

90°W 85°W 80°W 75°W 70°W 65°W

80°N 75°N 70°N 65°N 60°N 55°N

Nansen Sound
Axel Heiberg Island
Eureka
Agassiz Ice Cap
Ellesmere Island
Islands
Norwegian Bay
Bjorne Peninsula
Graham Island
Sydney Ice Cap • 1326
Smith Bay
North Lincoln Land
Grise Fjord
Jones Sound
Devon Island • 1887
Dundas Harbour
Baffin Bay
Strait
Cape Clarence
Lancaster Sound
Prince Regent Inlet
set Island
Bylot Island • 2134
Admiralty Inlet
Arctic Bay
• 1189
• 549
Borden Peninsula
Brodeur Peninsula
Eclipse Sound
Pond Inlet
Buchan Gulf
244 •
Bernier Bay
Gulf of Boothia
• 1554
518
Clyde
Baffin Island
Barnes Ice Cap • 1250
Fury and Hecla Strait
Spence Bay (Taloyoak)
Pelly Bay
• 558
Jens Munk Island
Rowley Island
Foley
Henry Kater Pen.
Home Bay
Davis Strait
Simpson Peninsula
Pelly Bay
Committee Bay
Hall Beach
Melville Peninsula
Wales Island
Prince Charles Island • 30
Kivitoo
Penny Ice Cap • 2591
Broughton Island
Cumberland Peninsula
Rae Isthmus
Repulse Bay
381 •
Foxe Basin
Koukdjuak
Nettiling Lake
Nunatak
• 2134
Cape Dyer
Exeter Sound
Pangnirtung
Lyon Inlet
Vansittart Island
Wager Bay
Roes Welcome Sound
Nabukjuak
Cumberland Sound
Hoare Bay
Kigisa
Southampton Island • 625
Foxe Channel
Foxe Peninsula • 411
Cape Dorset
Amadjuak Lake
Hall Peninsula • 1148
Coral Harbour
Bell Peninsula
Chesterfield Inlet
Chesterfield Inlet (Igluligaarjuk)
Salisbury Island
305
Iqaluit (Frobisher Bay)
Evans Strait
Fisher Strait
Nottingham Island
Hudson Strait
Meta Incognita Peninsula
Frobisher Bay
Labrador
Coats Island
Rankin Inlet
Big Island
Lake Harbour
Loks Land
Whale Cove
Mansel Island
Ivujivik • 540
Salluit
Purtuniq
• 661
Resolution Island
Kangiqsujuaq
Eskimo Point (Arviat)
Akulivik
Cape Hopes Advance
Labrador Sea
Ungava
Akpatok Island
Port Burwell
Cape Chidley
Hudson
Povungnituk
Ungava Bay
Torngat Mountains
ATLANTIC OCEAN
Ottawa Islands
Peninsula 390 •
1621
Ramah
NEWFOUNDLAND
Churchill
Inukjuak
Kangiqsualujjuaq
Bay
Rivière aux Feuilles
Kuujjuaq
Koksoak
Labrador
Clintock
Nutak
• 472
Lake Minto
QUEBEC
• 1076
Nelson
York Factory
• 451
Fraser
Nain
Belcher Islands
241 Lac à l'Eau-Claire
196 •
Fort Severn
Shamattawa
ONTARIO
Severn
Winisk
Cape Henrietta Maria
Kuujjuarapik
Caniapiscau
• 876
Hopedale
Lake Bienville
D A V U T

90

n o p q r s t u v w x y z

16 17

0 100 200 300 miles Average linear scale 0 100 200 300 400 500 Km

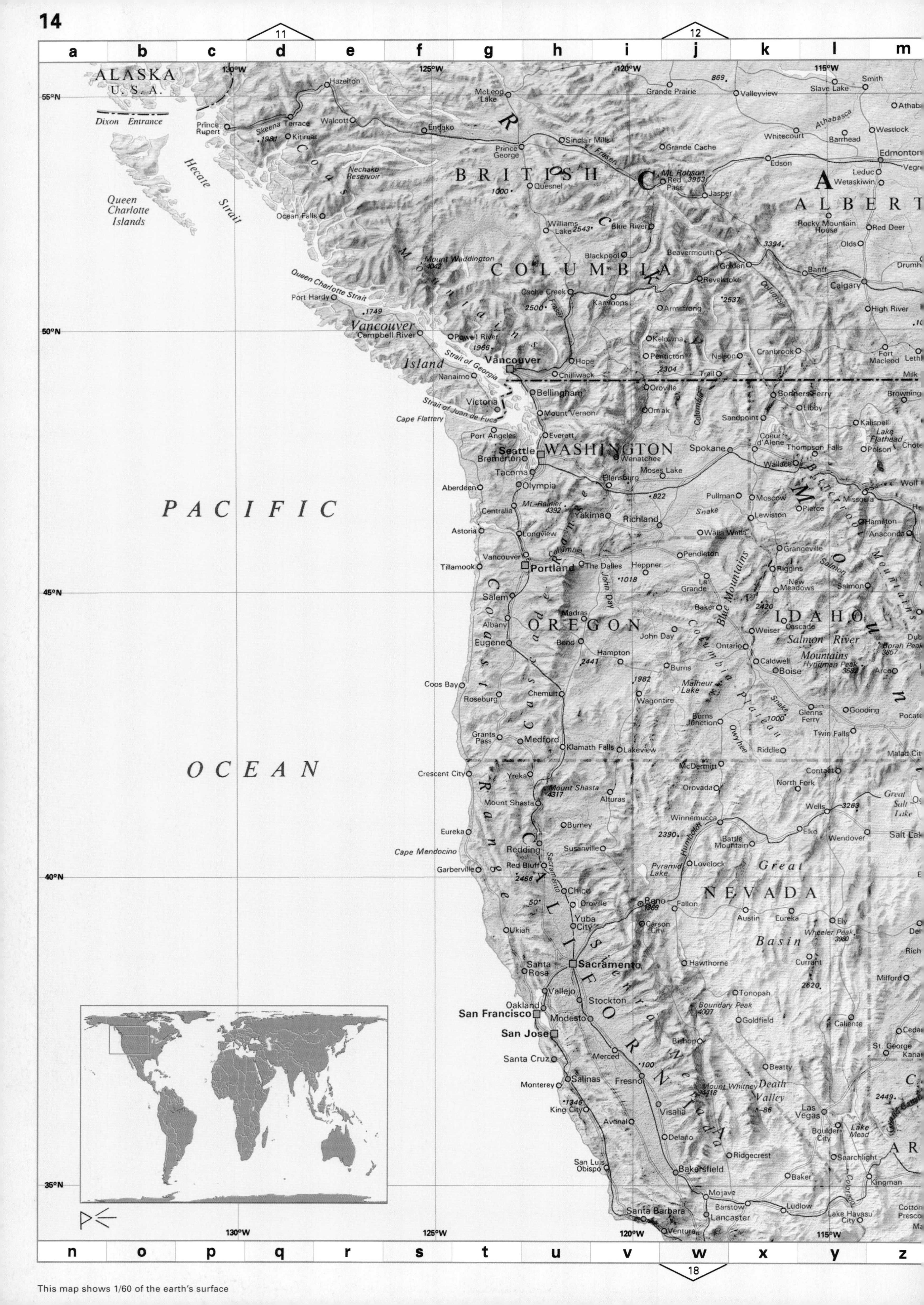

11
12
a b c d e f g h i j k l m
ALASKA
U. S. A.
Dixon Entrance
Queen Charlotte Islands
Hecate Strait
Prince Rupert
Skeena
Terrace
Kitimat
Hazelton
Walcott
Endako
McLeod Lake
Prince George
Sinclair Mills
Fraser
Grande Prairie
Valleyview
Slave Lake
Smith
Athabasca
Whitecourt
Barrhead
Westlock
Grande Cache
Edson
Edmonton
Leduc
Wetaskiwin
Nechako Reservoir
BRITISH COLUMBIA
Quesnel
Mt. Robson 3953
Red Pass
Jasper
ALBERTA
Rocky Mountain House
Red Deer
Olds
Ocean Falls
Williams Lake
Blue River
Blackpool
Beavermouth
Golden
Banff
Calgary
Revelstoke
Mount Waddington 4042
Coast Mountains
Queen Charlotte Strait
Port Hardy
Cache Creek
Kamloops
Armstrong
High River
Vancouver Island
Campbell River
Powell River
Vancouver
Hope
Kelowna
Penticton
Nelson
Cranbrook
Fort Macleod
Strait of Georgia
Nanaimo
Chilliwack
Trail
Victoria
Strait of Juan de Fuca
Cape Flattery
Bellingham
Mount Vernon
Oroville
Omak
Bonners Ferry
Libby
Sandpoint
Kalispell
Flathead Lake
Polson
Port Angeles
Everett
Seattle
Bremerton
WASHINGTON
Wenatchee
Spokane
Coeur d'Alene
Thompson Falls
Wallace
Tacoma
Moses Lake
Ellensburg
Aberdeen
Olympia
Pullman
Moscow
Missoula
Centralia
Mt. Rainier 4392
Yakima
Richland
Snake
Lewiston
Pierce
Hamilton
Anaconda
Astoria
Longview
Walla Walla
Columbia
Vancouver
Pendleton
Grangeville
PACIFIC OCEAN
Tillamook
Portland
The Dalles
Heppner
John Day
Riggins
New Meadows
Salmon
La Grande
Blue Mountains
Salem
Madras
OREGON
Baker
IDAHO
Weiser
Cascade
Salmon River Mountains
Albany
Eugene
Bend
John Day
Ontario
Borah Peak 3857
Hampton
Caldwell
Boise
Hyndman Peak 3682
Arco
Burns
Coos Bay
Malheur Lake
Roseburg
Chemult
Wagontire
Columbia Plateau
Glenns Ferry
Gooding
Burns Junction
Owyhee
Twin Falls
Grants Pass
Medford
Klamath Falls
Lakeview
Riddle
Cascade Range
Crescent City
Yreka
McDermitt
Contact
Mount Shasta 4317
Mount Shasta
Alturas
Orovada
North Fork
Wells
Great Salt Lake
Burney
Winnemucca
Eureka
Susanville
Humboldt
Battle Mountain
Elko
Wendover
Salt Lake
Cape Mendocino
Redding
Red Bluff
Sacramento
Pyramid Lake
Lovelock
Great Basin
Garberville
Chico
NEVADA
Oroville
Reno
Fallon
Austin
Eureka
Ely
Yuba City
Carson City
Wheeler Peak 3980
Ukiah
Sacramento
Hawthorne
Currant
Santa Rosa
Vallejo
Stockton
Tonopah
Milford
Oakland
San Francisco
Modesto
Boundary Peak 4007
Goldfield
Caliente
San Jose
Bishop
St. George
Santa Cruz
Merced
CALIFORNIA
Sierra Nevada
Beatty
Monterey
Salinas
Fresno
Mount Whitney 4418
Death Valley
King City
Visalia
Las Vegas
Boulder City
Lake Mead
Avenal
Delano
Ridgecrest
Searchlight
San Luis Obispo
Bakersfield
Baker
Kingman
Mojave
Barstow
Ludlow
Santa Barbara
Lancaster
Lake Havasu City
Ventura
130°W
125°W
120°W
115°W
55°N
50°N
45°N
40°N
35°N
n o p q r s t u v w x y z
18
This map shows 1/60 of the earth's surface

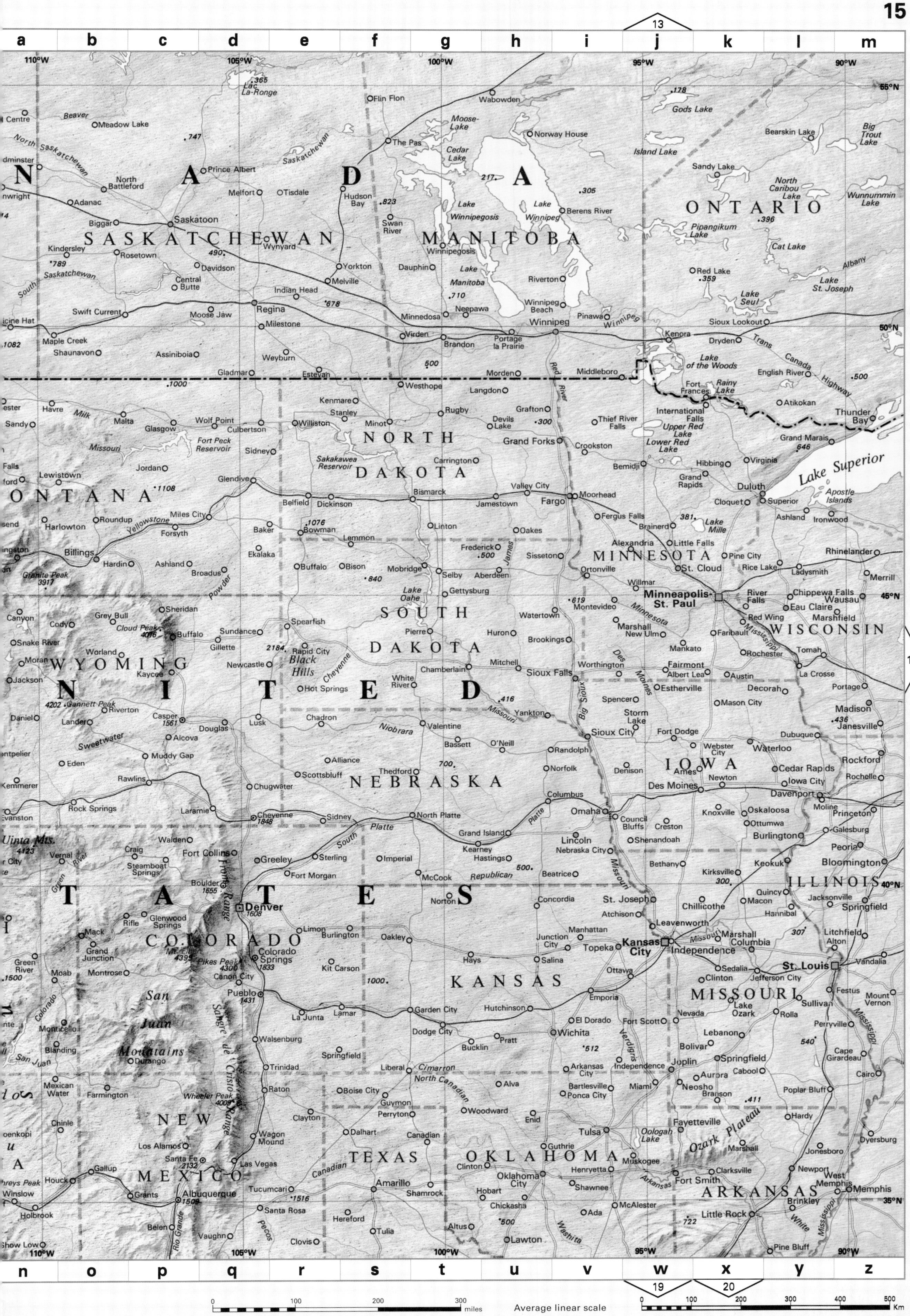

13

a b c d e f g h i j k l m

110°W 105°W 100°W 95°W 90°W

.365 Lac La-Ronge
Flin Flon
Wabowden
.178 Gods Lake
55°N
Centre
Beaver
Meadow Lake
Moose-Lake
Norway House
Big Trout Lake
Bearskin Lake
North Saskatchewan
.747
Saskatchewan
The Pas
Cedar Lake
Island Lake
dminster
Prince Albert
Sandy Lake
North Battleford
217
.305
North Caribou Lake
Wunnummin Lake
nwright
Adanac
Melfort
Tisdale
Hudson Bay
.823
Lake Winnipegosis
Lake Winnipeg
Berens River
Biggar
Saskatoon
Swan River
Pipangikum Lake
.396
CANADA
SASKATCHEWAN
MANITOBA
ONTARIO
Kindersley
Rosetown
Wynyard
.490
Winnipegosis
Cat Lake
.789
Davidson
Yorkton
Dauphin
Lake Manitoba
Red Lake
.359
Albany
Saskatchewan
South
Central Butte
Melville
Riverton
Lake St. Joseph
Indian Head
.678
.710
Lake Seul
Swift Current
Moose Jaw
Regina
Neepawa
Winnipeg Beach
Minnedosa
Pinawa
Winnipeg
icine Hat
Milestone
Winnipeg
Sioux Lookout
50°N
.1082
Maple Creek
Virden
Brandon
Portage la Prairie
Kenora
Dryden
Trans Canada Highway
Shaunavon
Assiniboia
Weyburn
Lake of the Woods
.500
English River
.500
Gladmar
Estevan
Morden
Middleboro
Red River
Rainy Lake
Fort Frances
.1000
Westhope
Langdon
International Falls
Atikokan
Thunder Bay
ester
Havre
Milk
Kenmare
Stanley
Rugby
Grafton
Sandy
Malta
Glasgow
Wolf Point
Culbertson
Williston
Minot
Devils Lake
.300
Thief River Falls
Upper Red Lake
Missouri
Fort Peck Reservoir
Sidney
NORTH DAKOTA
Grand Forks
Crookston
Lower Red Lake
Grand Marais
646
Lake Superior
Falls
Jordan
Sakakawea Reservoir
Carrington
Bemidji
Hibbing
Virginia
Lewistown
Grand Rapids
ford
MONTANA
.1108
Glendive
Bismarck
Valley City
Moorhead
Duluth
Apostle Islands
Belfield
Dickinson
Jamestown
Fargo
Cloquet
Superior
send
Harlowton
Roundup
Miles City
Yellowstone
Forsyth
Baker
.1076 Bowman
Linton
Oakes
Fergus Falls
Brainerd
381
Lake Mille
Ashland
Ironwood
Lemmon
Frederick
.500
James
Alexandria
Little Falls
ingston
Billings
Hardin
Ashland
Ekalaka
Sisseton
MINNESOTA
Pine City
Rhinelander
Granite Peak 3917
Broadus
Powder
Buffalo
Bison
Mobridge
Selby
Aberdeen
Ortonville
St. Cloud
Rice Lake
Ladysmith
Merrill
.840
Gettysburg
Willmar
Lake Oahe
.619
Montevideo
Minneapolis-St. Paul
River Falls
Chippewa Falls
Wausau
45°N
Canyon
Sheridan
Grey Bull
SOUTH DAKOTA
Watertown
Minnesota
Eau Claire
Marshfield
Cody
Cloud Peak 4016
Buffalo
Spearfish
Marshall
Red Wing
Snake River
Sundance
Pierre
Huron
Brookings
New Ulm
Faribault
Mississippi
WISCONSIN
Worland
Gillette
.2184
Rapid City
Black Hills
Des Moines
Mankato
Rochester
Tomah
WYOMING
Newcastle
Cheyenne
Chamberlain
Mitchell
Sioux Falls
Worthington
Fairmont
Albert Lea
Austin
La Crosse
Moran
Kaycee
White River
Big Sioux
Estherville
Decorah
Portage
Jackson
UNITED STATES
Hot Springs
.416
Spencer
Mason City
4202 Gannett Peak
Riverton
Missouri
Yankton
Storm Lake
Madison
Daniel
Lander
Casper 1561
Douglas
Lusk
Chadron
Valentine
Niobrara
Sioux City
Fort Dodge
.436
Janesville
Alcova
Bassett
O'Neill
Randolph
Webster City
Waterloo
Dubuque
ntpelier
Sweetwater
Muddy Gap
Alliance
Norfolk
IOWA
Eden
Scottsbluff
Thedford
700
Denison
Ames
Cedar Rapids
Rockford
Rawlins
NEBRASKA
Des Moines
Newton
Iowa City
Rochelle
Kemmerer
Chugwater
Columbus
Davenport
Moline
Evanston
Rock Springs
Laramie
Cheyenne 1848
Sidney
North Platte
Platte
Omaha
Council Bluffs
Creston
Knoxville
Oskaloosa
Princeton
Uinta Mts. 4123
Walden
South Platte
Grand Island
Lincoln
Ottumwa
Galesburg
r City
Vernal
Craig
Fort Collins
Greeley
Sterling
Kearney
Nebraska City
Shenandoah
Burlington
Peoria
Green River
Steamboat Springs
Front Range
Fort Morgan
Imperial
Hastings
.500
Missouri
Bethany
Keokuk
Bloomington
Boulder 1655
McCook
Republican
Beatrice
Kirksville
300
ILLINOIS
40°N
Denver 1608
Norton
Concordia
St. Joseph
Quincy
Jacksonville
Glenwood Springs
Atchison
Chillicothe
Macon
Springfield
Rifle
Limon
Burlington
Hannibal
Mack
Grand Junction
COLORADO
Oakley
Manhattan
Leavenworth
307
Litchfield
Mt. Elbert 4399
Colorado Springs 1833
Junction City
Topeka
Kansas City
Independence
Marshall
Columbia
Alton
Green River
Pikes Peak 4300
Kit Carson
Hays
Salina
Sedalia
St. Louis
Vandalia
.1500
Moab
Montrose
Canon City
.1000
KANSAS
Ottawa
Clinton
Jefferson City
San
Pueblo 1431
Sangre de Cristo Range
Emporia
MISSOURI
Festus
Mount Vernon
Colorado
Juan
La Junta
Lamar
Garden City
Hutchinson
Lake Ozark
Sullivan
nte
Monticello
Mountains
Walsenburg
Dodge City
El Dorado
Fort Scott
Nevada
Rolla
Perryville
Blanding
Durango
Bucklin
Pratt
Wichita
Verdigris
Lebanon
540
San Juan
Trinidad
Springfield
.512
Bolivar
Cape Girardeau
Mexican Water
Farmington
Liberal
Cimarron
Arkansas City
Independence
Joplin
Springfield
Cairo
Wheeler Peak 4009
Raton
Boise City
North Canadian
Alva
Bartlesville
Ponca City
Miami
Neosho
Aurora
Cabool
Poplar Bluff
Guymon
Branson
.411
Chinle
NEW MEXICO
Clayton
Perryton
Woodward
Enid
Fayetteville
Ozark Plateau
Hardy
oenkopi
Wagon Mound
Dalhart
Canadian
Tulsa
Oologah Lake
Dyersburg
Los Alamos
Guthrie
Marshall
Jonesboro
Santa Fe 2132
Las Vegas
TEXAS
OKLAHOMA
Muskogee
Gallup
Canadian
Clinton
Oklahoma City
Henryetta
Fort Smith
Clarksville
Newport
West Memphis
Memphis
hreys Peak
Houck
Grants
Albuquerque 1509
Tucumcari
.1516
Amarillo
Shamrock
Shawnee
Arkansas
ARKANSAS
Brinkley
Winslow
Santa Rosa
Hobart
Chickasha
McAlester
Little Rock
35°N
Holbrook
Belen
Rio Grande
Pecos
Hereford
Tulia
Altus
.500
Ada
Washita
.722
White
Mississippi
Vaughn
Clovis
Lawton
Show Low
Pine Bluff

n o p q r s t u v w x y z

16
19
20

0 100 200 300 miles
Average linear scale
0 100 200 300 400 500 Km

13
a b c d e f g h i j k l m
95°W 90°W 85°W 80°W 75°W
MANITOBA
Shamattawa
Fort Severn
Severn
Winisk
Hudson Bay
Belcher Islands
Great Whale
Kuujjuarapik
55°N
Cape Henrietta Maria
Gods Lake
Island Lake
Bearskin Lake
Big Trout Lake
Winisk
Lake River
Point Louis XIV
.168
James Bay
Kanaaupscow
Chisasibi
La Radisson
Sakami
La
Sandy Lake
.276
.88
Attawapiskat
Winisk Lake
Akimiski Island
.195
North Caribou Lake
Wunnummin Lake
.100
Eastmain
Eastmain
Pipangikum Lake
.396
Cat Lake
Attawapiskat Lake
Fort Albany
James Bay
Fort Hope
268.
Ogoki
Albany
Hannah Bay
Fort Rupert
Rupert
Red Lake
.359
Moosonee
Lake St. Joseph
Abitibi
Lac Seul
Missinaibi
Lake Evans .232
Q
U
Pinawa
.317
358.
Armstrong
Kesagami Lake
Chibougamau
50°N
Kenora
Dryden
Sioux Lookout
O N T A R I O
Nakina
Longlac
Geraldton
Lake Nipigon
Hearst
Fraserdale
.556
Chapais
Trans Canada Highway
Harricana
Matagami
Lake of the Woods
C A N A D A
Middleboro
English River
.500
Nipigon
Manitouwadge
Kapuskasing
Cochrane
Monts Deloge .533
Lake Abitibi
Schreiber
Marathon
Gouin Reservoir
Fort Frances
International Falls
Atikokan
390.
Timmins
Senneterre
Thief River Falls
Upper Red Lake
Rainy Lake
Thunder Bay
Isle Royale
White River
Noranda
Val-d'Or
358.
Lower Red Lake
Grand Marais
Wawa
Kirkland Lake
.609
.646
Chapleau
Lake Superior
.640
New Liskeard
Cabonga Reservoir
Hibbing
Virginia
Copper Harbor
.693
Bemidji
436.
Grand Rapids
Apostle Islands
Houghton
Témiscaming
471.
Duluth
Superior
Cloquet
.603
Marquette
Sault Ste. Marie
.665
Mont Laurier
Fergus Falls
381.
Brainerd
Mille Lacs
Ashland
Ironwood
.573
Seney
Blind River
Espanola
Sudbury
Sturgeon Falls
Maniwaki
Ottawa
960
Iron Mountain
.196
North Bay
Ste-Agathe-des-Monts
MINNESOTA
Little Falls
.454
322
Little Current
Manitoulin Island
Alexandria
Pine City
Rhinelander
Mackinaw City
Lake Huron
Georgian Bay
Pembroke
Buckingham
Hull
Montréal
St. Cloud
Rice Lake
Ladysmith
Escanaba
Parry Sound
Huntsville
Ottawa
Willmar
Merrill
Petoskey
Tobermory
Bancroft .419
Cornwall
Salaberry-de-Valleyfield
45°N
Minneapolis-St. Paul
River Falls
Chippewa Falls
Wausau
Menominee
Marinette
Alpena
Eau Claire
Stevens Point
Sturgeon Bay
Lake Michigan
Owen Sound
Perth
Smith's Falls
Brockville
15
Minnesota
Red Wing
Marshfield
Green Bay
Grayling
Port Elgin
Orillia
L. Simcoe
Marshall
Faribault
WISCONSIN
Appleton
Midland
Barrie
Peterborough
.510
New Ulm
Mankato
Mississippi
Winona
Manitowoc
Cadillac
.573
Kingston
Oshkosh
.223
Sheboygan
Belleville
Watertown
Adirondack Mountains
Rochester
Tomah
Fond du Lac
Ludington
MICHIGAN
Toronto
Oshawa
637.
Fairmont
La Crosse
Goderich
Lake Ontario
Worthington
Albert Lea
Austin
Bay City
Midland
Saginaw
Kitchener-Waterloo
Estherville
494.
Portage
.369
Hamilton
Spencer
Decorah
Mason City
.300
Madison
Muskegon
Flint
Sarnia
St. Catharines
Rochester
Syracuse
NEW
U N I T E D
Milwaukee
.385
London
Niagara Falls
Utica
Storm Lake
Fort Dodge
Webster City
Janesville
Racine
Grand Rapids
Lansing
St. Thomas
Buffalo
Seneca Lake
Schenectady
Albany
YORK
Dubuque
.436
Beloit
Kenosha
Ann Arbor
Detroit
Fredonia
Dansville
Ithaca
Waterloo
Freeport
Kalamazoo
358
Windsor
Jamestown
Catskill Mountains
Denison
Ames
.290
IOWA
Cedar Rapids
Rockford
Elgin
Chicago
Battle Creek
Jackson
Lake Erie
775.
Olean
Binghampton
Des Moines
Newton
Iowa City
Rochelle
De Kalb
Toledo
Ashtabula
Erie
Kane
Mansfield
Kingston
Omaha
Davenport
Moline
Gary
South Bend
Sandusky
Meadville
.424
Allegheny
Newburgh
Scranton
Council Bluffs
Knoxville
Oskaloosa
Ottumwa
Princeton
Joliet
Morris
Napoleon
Cleveland
Williamsport
Danbury
Bridgeport
Creston
Kankakee
Fort Wayne
Findlay
Clearfield
Susquehanna
Shenandoah
Burlington
Galesburg
Wabash
Youngstown
Paterson
Nebraska City
Missouri
236.
Peoria
PENNSYLVANIA
Newark
Elizabeth
Lafayette
Canton
Mansfield
Bethany
Keokuk
Bloomington
Rantoul
Kokomo
Lima
Kenton
Pittsburgh
Johnstown
Altoona
Allentown
Kirksville
S T A T E S
Reading
40°N
300
ILLINOIS
Champaign
Danville
Muncie
Columbus
Newark
Wheeling
Greensburg
Harrisburg
Trenton
St. Joseph
Quincy
Beardstown
Decatur
Springfield
Anderson
OHIO
.424
Cambridge
.706
Lancaster
Macon
Hannibal
Springfield
Indianapolis
Richmond
Dayton
Washington Court House
Wilmington
Philadelphia
Atchison
Chillicothe
Jacksonville
Terre Haute
INDIANA
Chillicothe
Fairmont
Marietta
Cumberland
Hagerstown
MARYLAND
NEW JERSEY
367.
Leavenworth
307.
Litchfield
Athens
Clarksburg
Vineland
Atlantic City
Kansas City
Independence
Marshall
Columbia
Alton
Bloomington
322
Cincinnati
Hillsboro
Parkersburg
1222. Bickle Knob
Baltimore
Washington DC
Topeka
Effingham
412.
Elkins
DELAWARE
Cape May
Sedalia
St Louis
Vandalia
Bedford
Ohio
Portsmouth
WEST
Arlington
Alexandria
Annapolis
Clinton
Jefferson City
Maysville
VIRGINIA
Culpeper
Cambridge
Fredericksburg
Ottawa
MISSOURI
Festus
Mt. Vernon
Jasper
Huntington
Charleston
1476. Spruce Knob
1234
Lexington
Park
Salisbury
Emporia
Sullivan
Louisville
Richwood
Allegheny Mountains
Evansville
Lexington
Morehead
Charlottesville
Lake Ozark
Rolla
Chesapeake Bay
Fort Scott
Nevada
Perryville
Mississippi
West Frankfort
Owensboro
Elizabethtown
Beckley
VIRGINIA
1241
Lebanon
.314
Williamson
Richmond
300.
540
Berea
Roanoke
James
Parsons
Bolivar
Springfield
Cape Girardeau
Hazard
Pikeville
Bluefield
Petersburg
Independence
.510
Cabool
Cairo
Central City
KENTUCKY
Lynchburg
Williamsburg
Hampton
Newport News
Joplin
Paducah
Somerset
28.
Miami
Aurora
.532
Wytheville
Portsmouth
Bartlesville
Neosho
Poplar Bluff
Bowling Green
Glasgow
Lake Cumberland
Marion
Martinsville
Norfolk
Branson
.1743
Danville
Vinita
307.
Clarksville
Kingsport
Mount Rogers
Mount Airy
Emporia
Ozark Plateau
Hardy
Cumberland
.784
Roanoke Rapids
Fayetteville
1077
Oak Ridge
High Knob
Winston Salem
Greensboro
Ahoskie
Elizabeth City
Tulsa
Tahlequah
Marshall
Blytheville
Tennessee
Nashville
Cumberland Plateau
Knoxville
1916
Durham
Jonesboro
Dyersburg
Murfreesboro
Newport
Roanoke
Muskogee
ARKANSAS
Rockwood
Hickory
Raleigh
Rocky Mount
Jackson
Columbia
2025
Clingmans Dome
Blue Ridge
Greenville
Clarksville
TENNESSEE
Great Smoky Mountains
Appalachian
Henryetta
Fort Smith
Newport
Asheville
Kannapolis
Goldsboro
Pamlico Sound
.100
Lawrenceburg
.879
1290
NORTH
CAROLINA
Poteau
West Memphis
Savannah
Cleveland
Hendersonville
Charlotte
50
Cape Hatteras
35°N
Conway
Little Rock
Memphis
Fayetteville
Chattanooga
Fayetteville
New Bern
McAlester
.100
Florence
Huntsville
Rockingham
Brinkley
Greenville
Spartanburg
Morehead City
95°W 90°W 85°W 80°W 75°W
n o p q r s t u v w x y z
19 20

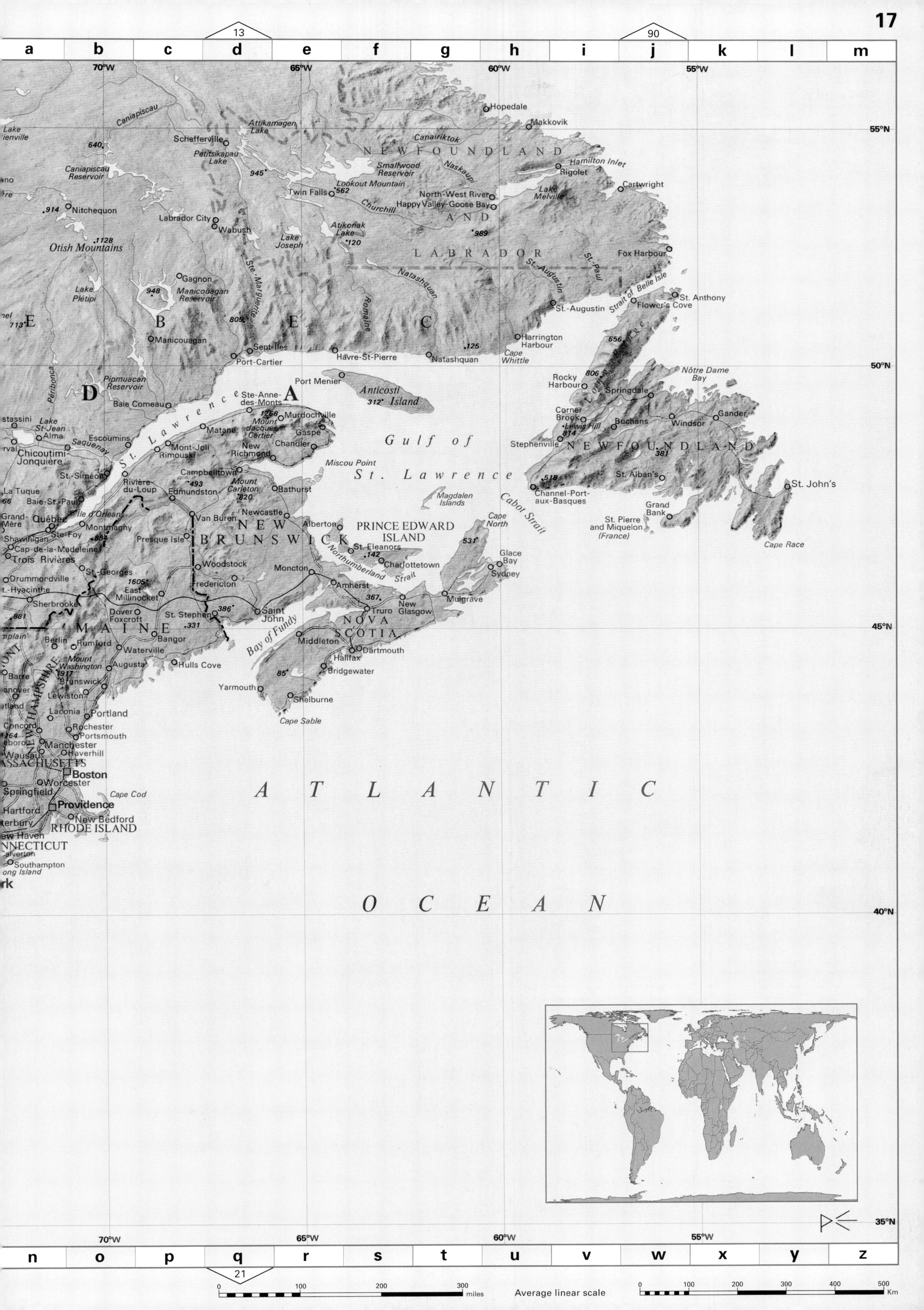

13
90
a b c d e f g h i j k l m
70°W 65°W 60°W 55°W
55°N 50°N 45°N 40°N 35°N
Hopedale
Makkovik
Schefferville
Attikamagen Lake
Caniapiscau
Canairiktok
NEWFOUNDLAND
AND
LABRADOR
Petitsikapau Lake
Smallwood Reservoir
Naskaupi
Hamilton Inlet
Rigolet
Caniapiscau Reservoir
Lookout Mountain
Twin Falls
562
Cartwright
Lake Melville
North-West River
Happy Valley-Goose Bay
Churchill
Nitchequon
Labrador City
Wabush
Atikonak Lake
Lake Joseph
Otish Mountains
Fox Harbour
St. Augustin
St. Paul
Natashquan
Gagnon
Manicouagan Reservoir
Lake Plétipi
Ste. Marguerite
Romaine
Strait of Belle Isle
St. Anthony
Flower's Cove
St.-Augustin
QUEBEC
Manicouagan
Sept-Îles
Port-Cartier
Havre-St-Pierre
Natashquan
Harrington Harbour
Cape Whittle
Rocky Harbour
Nôtre Dame Bay
Springdale
Pipmuacan Reservoir
Péribonca
Port Menier
Anticosti Island
Baie Comeau
Ste-Anne-des-Monts
Murdochville
Mount Jacques Cartier
Gaspé
St. Lawrence
Corner Brook
Lewis Hill
Buchans
Windsor
Gander
Lake St-Jean
Alma
Escoumins
Matane
Mont-Joli
Rimouski
New Richmond
Chandler
Gulf of St. Lawrence
Stephenville
NEWFOUNDLAND
Saguenay
Chicoutimi-Jonquière
Miscou Point
St.-Siméon
Campbellton
St. Alban's
Rivière du-Loup
Mount Carleton
Bathurst
Cabot Strait
Channel-Port-aux-Basques
St. John's
La Tuque
Edmundston
Magdalen Islands
Baie-St-Paul
Île d'Orléans
Québec
Montmagny
Ste-Foy
Van Buren
NEW BRUNSWICK
Newcastle
Alberton
PRINCE EDWARD ISLAND
Cape North
Grand Bank
St. Pierre and Miquelon (France)
Shawinigan
Cap-de-la-Madeleine
Trois Rivières
Presque Isle
St. Eleanors
Northumberland Strait
Charlottetown
Glace Bay
Sydney
Cape Race
Woodstock
Moncton
St.-Georges
Drummondville
St.-Hyacinthe
Fredericton
East Millinocket
Amherst
Sherbrooke
Dover Foxcroft
St. Stephen
Saint John
Mulgrave
New Glasgow
Truro
MAINE
NOVA SCOTIA
Bay of Fundy
Middleton
Dartmouth
Halifax
Bangor
Berlin
Rumford
Waterville
Augusta
Hulls Cove
Mount Washington
Bridgewater
Barre
NEW HAMPSHIRE
Brunswick
Lewiston
Yarmouth
Shelburne
Cape Sable
Laconia
Portland
Concord
Rochester
Portsmouth
Manchester
Haverhill
MASSACHUSETTS
Boston
Worcester
Springfield
Cape Cod
Hartford
Providence
New Bedford
RHODE ISLAND
CONNECTICUT
Southampton
ATLANTIC OCEAN
n o p q r s t u v w x y z
21
miles
Average linear scale
Km

14
15
a b c d e f g h i j k l m
n o p q r s t u v w x y z
35°N
30°N
25°N
20°N
15°N
120°W
115°W
110°W
105°W
160°W
CALIFORNIA
ARIZONA
NEW MEXICO
UNITED
MEX
Los Angeles
San Diego
Phoenix
Tucson
Albuquerque
El Paso
Ciudad Juárez
Chihuahua
Hermosillo
Tijuana
Mexicali
Ensenada
Culiacán
Mazatlán
Durango
Torreón
Gulf of California
Lower California
Western Sierra Madre
Channel Islands
Santa Barbara
Santa Cruz
Santa Rosa
San Nicolas
San Clemente
Santa Catalina
Long Beach
Santa Ana
Anaheim
San Bernardino
Palm Springs
Oceanside
Escondido
El Cajon
Calexico
Yuma
Rosarito
La Misión
San Vicente
Colnett
Barstow
Lancaster
Mojave
Ludlow
Blythe
Kingman
Lake Havasu City
Cottonwood
Prescott
Flagstaff
Mayer
Holbrook
Show Low
Mesa
Globe
Casa Grande
Safford
Silver City
Lordsburg
Las Cruces
Deming
Belen
Magdalena
Socorro
Santa Rosa
Vaughn
Clovis
Portales
Roswell
Alamogordo
Artesia
Hobbs
Carlsbad
Sierra Blanca Peak
Guadalupe Peak
Pecos
Van Horn
Fort Stockton
Alpine
Salton Sea
Gila
Colorado
San Luis Desert
Rio Colorado
Pinacate Desert
Cerro Pinacate
Sonoyta
Puerto Peñasco
El Chinero
San Felipe
Agua de Chale
El Socorro
Tajito
Caborca
Altar Desert
Estación Trincheras
Tubutama
Magdalena
Santa Ana
Sásabe
Nogales
Sierra Vista
Douglas
Agua Prieta
Cananea
Arizpe
Bavispe
Janos
Ascensión
El Porvenir
Villa Ahumada
Nuevo Casas Grandes
Buenaventura
San Lorenzo
El Sueco
Estación Babicora
Madera
Temosachic
Bachiniva
Sahuaripa
Novillo Reservoir
Moctezuma
Carbó
Yaqui
Soyapa
Yepachic
Yécora
Nuri
La Junta
Cuauhtémoc
El Sauz
Aldama
San Guillermo
Meoqui
Delicias
Saucillo
El Carrizalillo
Ojinaga
El Chilicote
Bravo del Norte
Rio Grande
Conchos
Llanos de los Caballos Mesteños
Ciudad Camargo
Boquilla Reservoir
Jiménez
Hidalgo del Parral
Villa Ocampo
Yermo
Rosario de Arriba
Misión San Fernando
Cape Lobos
Angel de la Guarda
Tiburón
Guadalupe
Punta Prieta
Los Angeles
Rosarito
Sebastián Vizcaíno Bay
Cedros
La. Ojo de Liebre
El Arco
Volcán Tres Vírgenes
Vizcaíno Desert
Santa Rosalía
Laguna San Ignacio
Ballenas Bay
Canipole
San Juanico
Loreto
Carmen
Ejido Insurgentes
Santa Catalina
Los Burros
San Carlos
San José
Cape San Lázaro
Rocas Alijos
Espíritu Santo
La Paz B.
La Paz
Las Cruces
Quiñones
El Triunfo
Todos Santos
San Lucas
Cape San Lucas
Tropic of Cancer
Avispas
Empalme
Guaymas
Guásimas
Esperanza
Tórim
Ciudad Obregón
Navojoa
Huatabampo
Las Bocas
Alamos
Macuarichic
San Blas
Ahome
Los Mochis
Guasave
M. Hidalgo Reservoir
El Fuerte
Guadalupe y Calvo
Guazapares
San Francisco del Oro
Santa Bárbara
Santa María
Topia
Santiago Papasquiaro
Pericos
Navolato
El Dorado
Cosalá
Dimas
Villa Unión
El Salto
La Pañilla
Mezquital
Canatlán
Guadalupe Victoria
Rodeo
Nazas
Lerdo
Gómez Palacio
Bermejillo
La Vibora
Cuatrociénegas
Escuinapa
Acaponeta
Tecuala
Mesa del Nayar
Tres Marias Islands
María Madre
María Magdalena
María Cleofas
Tuxpan
Santiago Ixcuintla
Tepic
Compostela
Ixtlán
Etzatlán
Ameca
Puerto Vallarta
Banderas Bay
Juchipila
Aguascalientes
Tepatitlán
Guadalajara
Cocula
Ocotlán
L. de Chapala
Sayula
Tomatlán
Autlán
Cd. Guzmán
Nev. de Colima
Colima
Tenacatita
Barra de Navidad
Tecomán
Manzanillo
Apatzingán
Playa Azul
Hawaiian Islands
Kauai
Haena
Lihue
Mana
Niihau
Kaula
Oahu
Kaneohe
Pearl City
Waipahu
Honolulu
Molokai
Halawa
Maunaloa
Wailuku
Kahului
Maui
Lanai
Kahoolawe
Hawi
Mauna Kea
Hilo
Kailua
Hawaii
Mauna Loa
Naalehu
These islands lie approximately 4000 kilometres to the west of here, in the Pacific Ocean.
San Benedicto
Revilla Gigedo Islands
Roca Partida
Socorro
Clarión
PACIFIC
OCEAN

16

a b c d e f g h i j k l m

100°W 95°W 90°W 85°W 35°N

Altus Lawton Ada McAlester 722 Little Rock Corinth Florence Huntsville Decatur
OKLAHOMA
Arkadelphia Pine Bluff Arkansas Tupelo 246 Gainesville Marietta
Wichita Falls Red River Paris ARKANSAS Clarksdale Cullman Gadsden Atlanta
Sherman Sulphur Springs Columbus College Park
STATES Texarkana Greenville Winona Birmingham Bessemer .734
.448 Fort Worth Irving Garland El Dorado Lake Providence MISSISSIPPI .152 Tuscaloosa Opelika GEORGIA .425
Abilene Arlington Dallas Marshall Minden 67 Monroe Meridian ALABAMA Macon
Sweetwater Tyler Longview Shreveport Vicksburg Jackson Selma Phenix City Columbus Montgomery Warner Robins
.233 .113 Natchitoches Mississippi 149 Tombigbee Alabama Eufaula
Brownwood Waco Nacogdoches Toledo Bend Reservoir Red River Laurel Greenville Dawson Albany
San Angelo Temple Lufkin Alexandria Brookhaven Jackson Andalusia Tifton
TEXAS Killeen Brazos McComb Hattiesburg .105 Dothan Valdosta
Edwards Plateau Huntsville Jasper LOUISIANA Mobile Crestview Marianna Thomasville
Bryan Baton Rouge Hammond Biloxi FLORIDA Chattahoochee
Austin Lake Charles Slidell Gulfport Pascagoula Pensacola Fort Walton Beach Tallahassee
Kerrville .567 Beaumont Lafayette New Iberia Kenner New Orleans Panama City 30°N
784 Amistad Reservoir New Braunfels Colorado Houston Baytown Houma Cape San Blas Apalachee Bay
Del Rio San Antonio Gonzales Pasadena Texas City Galveston Mississippi Delta
Uvalde Lake Jackson
Piedras Negras Pearsall Victoria
Allende Rio Grande Three Rivers .50
Nueva Rosita Sabinas Corpus Christi
277 Laredo Kingsville
Nuevo Laredo .302
Monclova Falcon Reservoir Laguna Madre
Sabinas Hidalgo Ciudad Mier McAllen Harlingen Brownsville
2643 Reynosa Matamoros
Monterrey China
2736
Montemorelos
Rayones Laguna Madre 25°N
MEXICO Linares 1794 La Carbonera
Aramberri Jiménez
4054
Ciudad Victoria
Jaumave Llera Tula
Charcas Ciudad Mante Manuel
Sierra Madre Oriental
Salinas Cerritos San Luis Potosí Ciudad Valles Ciudad Madero Tampico
Rio Verde Cárdenas Pánuco
San Felipe Moctezuma Laguna Tamiahua
San Luis de la Paz Tamazunchale
Guanajuato Tuxpan
Salamanca Zacualtipan
Querétaro Poza Rica Papantla
Celaya Pachuca 2426 Huauchinango
Salvatierra Tula Martínez de la Torre
Acámbaro Tepeji Tulancingo Teziutlán Tlapacoyan
Maravatío Perote
Morelia Mexico City Apan Calpulalpan Jalapa
Pátzcuaro Toluca Texcoco Apizaco 4110
Zamora Puebla Citlaltépetl 5700 Veracruz
Cuernavaca 5452 Atlixco Coscomatepec
Taxco 2780 Izúcar de Matamoros Córdoba Orizaba Alvarado Tlacotalpan San Andrés Tuxtla
Jojutla Iguala Tezonapa Tierra Blanca
Ciudad Altamirano Acatlán Miguel Alemán Reservoir Tuxtepec 1879 Coatzacoalcos
Balsas Huajuapan de León Acayucan Minatitlán
Chilapa Coixtlahuaca Isthmus of Tehuantepec
Chilpancingo Nochixtlán Jesús Carranza Morelos
Atoyac Tierra Colorada Tlaxiaco Oaxaca 1546 Mitla Paso Real
Tecpan Southern Sierra Madre Matías Romero
Acapulco San Marcos Ejutla Ciudad Ixtepec
Pinotepa Nacional Miahuatlán Juchatengo 3139 Juchitán Tehuantepec
Salina Cruz
Puerto Escondido Puerto Angel Gulf of Tehuantepec

Gulf of Mexico

Desterrada Pérez
Arenas
Nuevo
Triangulos
Arcas
Bay of Campeche

Dzilam de Bravo Río Lagartos Cape Catoche Chiquilá
Progreso Tizimín Puerto Juárez
Mérida Motul Espita Cancun
Izamal Valladolid Puerto Morelos
Maxcanú Ticul Tekax Tulum Cozumel
Calkiní .100 Peto
Bolonchén de Rejón
Campeche Yucatán Peninsula 20°N
Sihochac Chunhuhub Felipe Carrillo Puerto Ascensión Bay
Champotón Espíritu Santo Bay
Ciudad del Carmen Laguna de Términos Mamantel 310
Frontera Hondo Chetumal
Comalcalco
Villahermosa Balancán Altamira
Macuspana Tenosique BELIZE
Grijalva Palenque Belize Turneffe Islands
Chichón 2224 Flores Belmopan
Usumacinta .1122
2727 San Cristóbal las Casas
Tuxtla Gutiérrez Chiapa de Corzo
Comitán
Arriaga Tonalá Mar Muerto
Angostura Reservoir GUATEMALA Puerto Barrios
Pijijiapan Huehuetenango
Motozintla 4220 Motagua
Huixtla Quezaltenango
Tapachula Guatemala .2865

CUBA
Pinar del Río
Guanahacabibes Peninsula
Straits of Yucatán
Isla de Pinos
Gulf of Honduras
Swan Island (Hond.)
Roatán Guanaja Bay Islands
Utila Trujillo
Puerto Cortés Tela La Ceiba
San Pedro Sula Patuca
HONDURAS Puerto Lempira 15°N

20

100°W 95°W 90°W 86°W

n o p q r s t u v w x y z

22

0 100 200 300 miles
Average linear scale
0 100 200 300 400 500 Km

This map shows 1/60 of the earth's surface

17
a
b
c
d
e
f
g
h
i
j
k
l
m
75°W
70°W
65°W
35°N
30°N
25°N
20°N
15°N
Goldsboro
New Bern
Jacksonville
Morehead City
Wilmington
Myrtle Beach
Hamilton
Bermuda (U.K.)
A T L A N T I C
Sargasso Sea
O C E A N
Grand Bahama Island
Great Abaco Island
Nicholls Town
Nassau
Eleuthera
New Providence
Behring Point
Andros Islands
Cat
San Salvador
BAHAMAS
Rum Cay
Great Exuma Island
Long Island
Tropic of Cancer
Bahama Bank
Crooked Island
Acklins
Mayaguana Island
Grand Caicos
Turks and Caicos Islands (U.K.)
Morón
Ciego de Avila
Nuevitas
Great Inagua Island
Victoria de las Tunas
Banes
Holguín
Bayamo
Palma Soriano
Baracoa
Manzanillo
2005
Guantánamo
Niquero
Santiago de Cuba
Port-de-Paix
DOMINICAN REPUBLIC
Cap-Haïtien
Puerto Plata
Gonaïves
Mao
Santiago
HAITI
La Vega
3175
San Francisco de Macoris
St-Marc
San Juan
Montego Bay
JAMAICA
Anse d'Hainault
Port-au-Prince
Puerto Rico (U.S.A.)
Virgin Islands (U.K.)
San Juan
Carolina
Bayamón
Santo Domingo
La Romana
Mayagüez
1338
Caguas
Ponce
Virgin Islands (U.S.A.)
2680
Jacmel
Les Cayes
Barahona
Spanish Town
May Pen
Kingston
Anguilla (U.K.)
St. Martin
Philipsburg
St. Croix (U.S.A.)
Netherlands Antilles
Codrington
Barbuda
ANTIGUA AND BARBUDA
Basseterre
ST KITTS NEVIS
St. John's
Antigua
Montserrat (U.K.)
Plymouth
Leeward Islands
Guadeloupe (France)
Pointe-à-Pitre
Basse-Terre
DOMINICA
Roseau
Antilles
BEAN SEA
n
o
p
q
r
s
t
u
v
w
x
y
z
23
0 100 200 300 miles
Average linear scale
0 100 200 300 400 500 Km

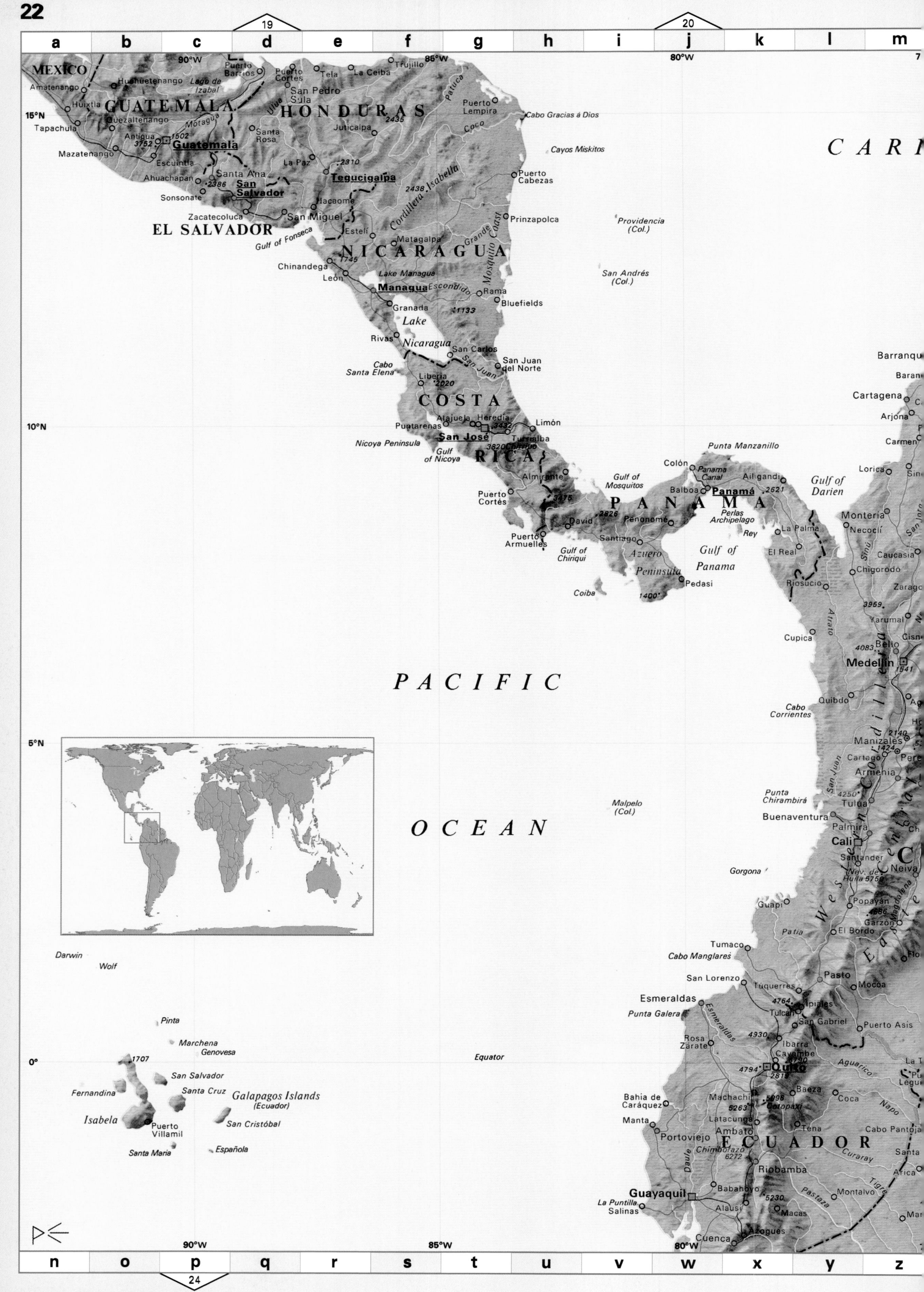
19
20
a b c d e f g h i j k l m
90°W
85°W
80°W
15°N
10°N
5°N
0°
MEXICO
GUATEMALA
HONDURAS
EL SALVADOR
NICARAGUA
COSTA RICA
PANAMA
ECUADOR
C A R I
PACIFIC
OCEAN
Guatemala
Tegucigalpa
San Salvador
Managua
San José
Panamá
Quito
Guayaquil
Medellín
Cali
Lake Nicaragua
Gulf of Fonseca
Gulf of Panama
Gulf of Darien
Galapagos Islands (Ecuador)
Isabela
Fernandina
San Salvador
Santa Cruz
San Cristóbal
Española
Santa Maria
Pinta
Marchena
Genovesa
Darwin
Wolf
Puerto Villamil
Equator
Malpelo (Col.)
Providencia (Col.)
San Andrés (Col.)
Cayos Miskitos
Cabo Gracias á Dios
n o p q r s t u v w x y z
24

21

a b c d e f g h i j k l m

70°W 65°W 60°W

DOMINICA
Roseau
Martinique-Passage
15°N
Martinique
(France)
Fort-de-France
St.-Lucia-Passage
Castries
SAINT LUCIA
St.-Vincent-Passage
SAINT
VINCENT
AND
GRENADINES
Kingstown
BARBADOS
Bridgetown
EAN SEA
Lesser Antilles
Cabo
Gallinas
Aruba
(Neth.)
Curaçao
Netherlands
Antilles
Bonaire
Willemstad
Guajira
Peninsula
820
Paraguaná
Peninsula
Saint George's
GRENADA
Punto Fijo
Islas
Los Roques
(Ven.)
Blanquilla
(Ven.)
ATLANTIC
Riohacha
Maicao
Gulf of
Venezuela
Puerto Cumarebo
Coro
Tobago
Scarborough
TRINIDAD
AND
TOBAGO
Margarita
La Asunción
Tortuga
San
Rafael
Churuguara
Tocuyo
Maiquetía
2765
Cabo Codera
Carúpano
Port
of Spain
940
Maracaibo
La
Concepción
Cabimas
Puerto
Cabello
Caracas
Cumaná
Güiria
Gulf of
Paria
Trinidad
San Fernando
1900
San
Felipe
Maracay
Puerto La Cruz
Caripito
Serpent's Mouth
Ciudad
Ojeda
Carora
Valencia
La Victoria
2660
Barcelona
10°N
3750
Machiques
Barquisimeto
San Juan
de los Morros
Píritu
Lake
Maracaibo
Acarigua
Unare
Maturín
3652
Anaco
Valera
El Sombrero
Valle de
la Pascua
Cantaura
Zaraza
Guanipa
Tigre
Manamo
OCEAN
Catatumbo
Boconó
El Banco
San Carlos
del Zulia
Guanare
El Baúl
Calabozo
El Tigre
Tucupita
Orinoco
Delta
Mérida
5007
Cojedes
Pariaguán
Barrancas
Boca Grande
Casigua
Barinas
Cordillera de Mérida
Amacuro
Delta
San José de Amacuro
Ocaña
Guárico
Suata
Orinoco
Ciudad Guayana
792
Boca del
Pao
Cúcuta
Bruzual
Apure
San Fernando
de Apure
Ciudad
Bolívar
Upata
Hossororo
San Cristóbal
Mantecal
Caicara de
Orinoco
Lago de Guri
Serranía de Imataca
Pamplona
4190
El Canton
Arauca
Maripa
Marlborough
El Callao
VENEZUELA
1863
1839
Barrancabermeja
Arauca
Capanaparo
La Urbana
Las
Trincheras
La Paragua
Suddie
Cuyuni
Llanos
El Dorado
Georgetown
Málaga
Santa Maria
Socorro
5493
Sierra de Maigualida
Cravo Norte
Casanare
Meta
La Venturosa
Puerto
Carreño
Sabana de
Cardona
Caura
Paragua
Caroní
Mayupa
Peters
Mine
Bartica
New Amsterdam
Angel Falls
Rockstone
Linden
Totness
2100
1890
La Escalera
Mazaruni
Sogamoso
100
2950
Nieuw
Nickerie
Tunja
Puerto
Nuevo
Puerto
Ayacucho
2285
Tomo
Puricama
Cavanayen
Pakaraima
2040
Tumatumari
Apoera
Yopal
Trinidad
Guiana Highlands
Gran
Sabana
2810
Roraima
Zipaquira
San Juan
GUYANA
Courantyne
5°N
Orocué
2030
Ventuari
El Oso
Meseta del
Cerro Jaua
Arabelo
Santa Elena
de Uairen
Essequibo
Kabalebo-
Reservoir
Bogotá
Sucuaro
1240
Meta
Orinoco
Pacaraima
Maturuca
SURINAME
Vichada
Villavicencio
San José
de Ocuné
San Fernando
de Atabapo
Catisimiña
Serra
Depósito
Apoteri
1026
Puerto
Lopez
Santa Barbara
Pakaraima Mountains
Juliana Top
1230
Eilerts de Haan Geb.
Puerto
Limon
Pavon
Santa
Rosa
Guaviare
Arrecifal
2579
Karanambo
Uraricoera
882
COLOMBIA
Inírida
Orinoco
2396
La Esmeralda
Serra Parima
Lethem
Uraricoera
Kanuku Mts.
Ariari
Victorino
Serra do Apiaú
Boa
Vista
Tacutu
Rupununi
Dadanawa
Oronoque
San Yanaro
San José
del Guaviare
San José
Boca
Mavaca
1047
Isherton
Guayabero
Casiquiare
Serra Curupira
Catrimani
New
Papai
Guainía
Morichal
El Mango
Calamar
Uainambi
San Carlos
Biloku
Serra Acarai
Buenos
Aires
Mesa de
Yambi
Vista Alegre
Caracaraí
Kamoa Mts.
Serra Tapirapecó
Branco
Demini
Miraflores
Mitú
Jibóia
Cucuí
734
Vaupés
São José
do Anauá
Anauá
Padauiri
Maloca
3014
Pico da
Neblina
Araçá
Cuñaré
Iuareté
Apaporis
Mapuera
Trombetas
Içana
Negro
Catrimani
Puerto
Huitoto
Macuje
Taracuá
Uaupés
Uaupés
0°
Paru de Oeste
Lérida
Calanaque
Jauaperi
Araracuara
São José
Tapurucuara
BRAZIL
Caquetá
La Chorrera
100
Barcelos
100
Cuiuni
La Pedrera
Tupanacca
Puerto
Miraña
Moura
Vila Bitencourtt
Unini
Santa
Maria
Nhamunda
Oriximiná
El Encanto
Marcelino
Maraã
Japurá
Jaú
Airão
Uatumã
Faro
U
San
Cristóbal
Arica
Amazon
Foz do Mamoriá
Fonte Boa
Santo
Antonio
Urucará
Parintins
Amazon
Santa Clotilde
Santa Clara
Tonantins
Solimões
Negro
70°W 65°W 60°W

n o p q r s t u v w x y z

26
25
26

0 100 200 300 miles
Average linear scale
0 100 200 300 400 500 Km

22

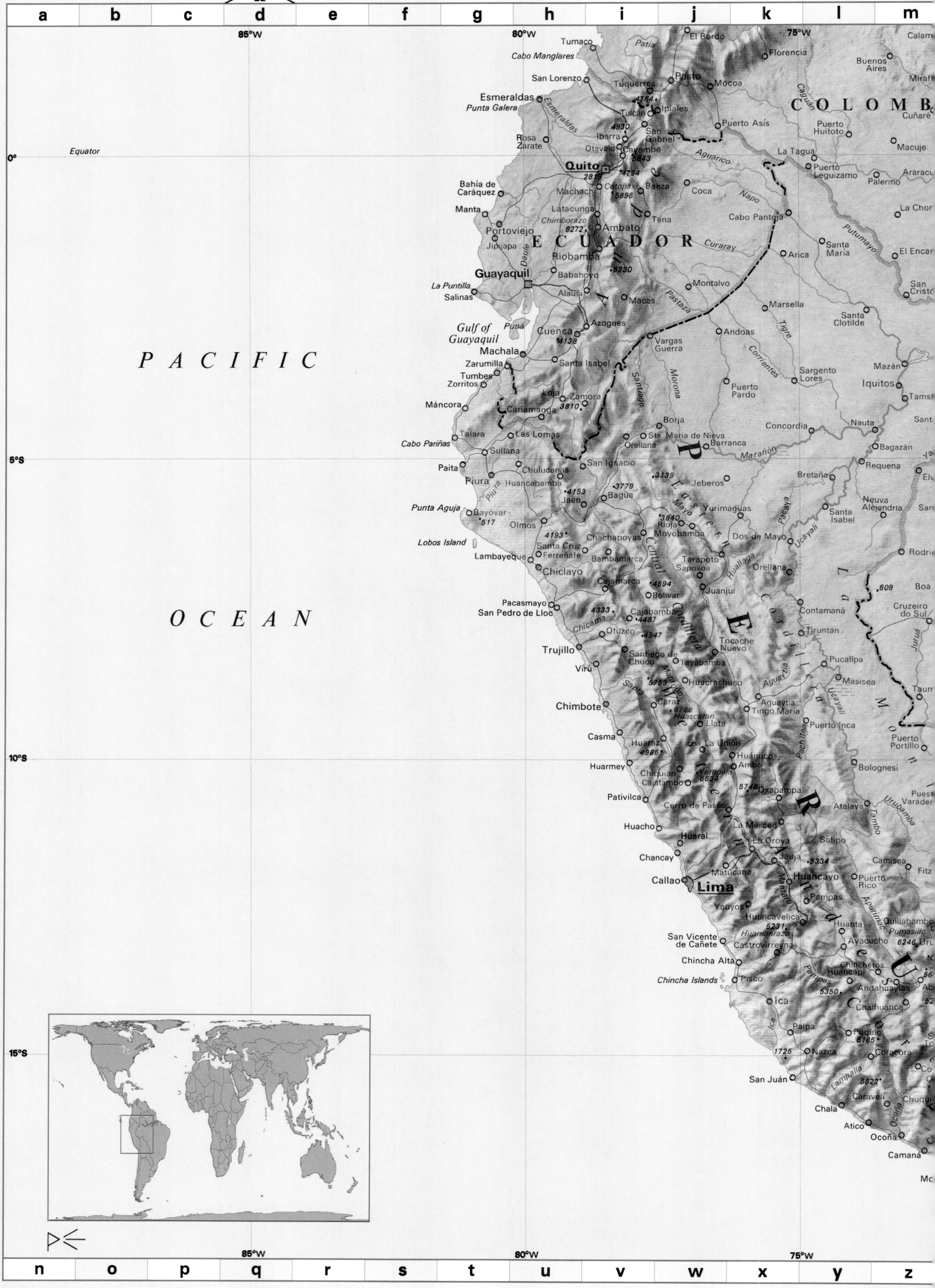

a b c d e f g h i j k l m

85°W 80°W 75°W

0° Equator

5°S

10°S

15°S

PACIFIC

OCEAN

ECUADOR

COLOMB

PERU

Tumaco
Cabo Manglares
Patia
El Bordo
Florencia
Calam
Buenos Aires
Miraf
San Lorenzo
Túquerres
Pasto
Mocoa
Caguán
Esmeraldas
Punta Galera
Esmeraldas
4764
Ipiales
Tulcán
Cuñare
4939
San Gabriel
Puerto Asís
Puerto Huitoto
Macuje
Rosa Zárate
Ibarra
Otavalo
Cayambe
5843
Aguarico
La Tagua
Puerto Leguizamo
Palermo
Araracu
Quito
2819
4794
Bahía de Caráquez
Machachi
Cotopaxi
5896
Baeza
Coca
Napo
Manta
Latacunga
Tena
Cabo Pantoja
La Chor
Portoviejo
Chimborazo
6272
Ambato
Putumayo
Jipijapa
Daule
Riobamba
Curaray
Santa María
Arica
El Encan
Guayaquil
Babahoyo
5230
Montalvo
La Puntilla
Salinas
Alausí
Macas
Pastaza
Marsella
San Cristo
Santa Clotilde
Gulf of Guayaquil
Puná
Cuenca
4138
Azogues
Vargas Guerra
Andoas
Tigre
Machala
Corrientes
Zarumilla
Santa Isabel
Mazán
Tumbes
Zorritos
Santiago
Morona
Sargento Lores
Iquitos
Loja
Zamora
Puerto Pardo
Tamsh
Máncora
Cariamanga
3810
Nauta
Sant
Borja
Concordia
Talara
Las Lomas
Sta. María de Nieva
Cabo Pariñas
Orellana
Barranca
Bagazán
Sullana
Marañón
Requena
Paita
San Ignacio
Ya
Piura
Chulucanas
3139
Jeberos
Bretaña
Elv
Piura
Huancabamba
3779
Eastern
4153
Bagua
Neuva Alejandría
Punta Aguja
Jaén
Yurimaguas
San
Bayóvar
517
Olmos
3840
Mayo
Pacaya
Santa Isabel
Rioja
Moyobamba
Ucayali
4193
Chachapoyas
Dos de Mayo
Lobos Island
Santa Cruz
Ferreñafe
Rodri
Lambayeque
Bambamarca
Tarapoto
Saposoa
Orellana
Chiclayo
Central
Huallaga
Cajamarca
4694
Juanjuí
609
Boa
Bolívar
Pacasmayo
San Pedro de Lloc
Cordillera
Contamana
Cruzeiro do Sul
4333
Cajabamba
4487
Chicama
Otuzco
4947
Tiruntán
Juruá
Trujillo
Tocache Nuevo
Santiago de Chuco
Tayabamba
Pucallpa
Virú
Masisea
5753
Huacrachuco
Santa
Aguaytía
Taum
Ucayali
Chimbote
Caraz
6768
Huascarán
Aguaytía
Tingo María
Llata
Puerto Inca
Casma
Huaraz
4986
La Unión
Pachitea
Puerto Portillo
Huánuco
Ambo
Huarmey
Bolognesi
Chiquián
Cajatambo
Yerupaja
6634
5748
Oxapampa
Puest
Varader
Pativilca
Cerro de Pasco
Atalaya
Urubamba
Tambo
Huacho
La Merced
Huaral
La Oroya
Satipo
Chancay
Jauja
5334
Camisea
Fitz
Matucana
Huancayo
Puerto Rico
Callao
Lima
Mantaro
Pampas
Yauyos
Apurímac
Quillabamba
Huancavelica
5231
Huamanrazo
Huanta
Pumasillo
San Vicente de Cañete
Castrovirreyna
Ayacucho
6246
Chincha Alta
Chinchero
Huancapi
Pampas
Chincha Islands
Pisco
5350
Andahuaylas
Ica
Chalhuanca
Ica
Palpa
Puquio
5185
1725
Nazca
Coracora
San Juán
Lampalla
5522
Caraveli
Chuqui
Chala
Ocoña
Atico
Ocoña
Camana
Mo

85°W 80°W 75°W

n o p q r s t u v w x y z

This map shows 1/60 of the earth's surface

23

a b c d e f g h i j k l m

70°W 65°W 60°W 55°W

Mesa de Yambi

Uainambi
San Carlos
El Mango
VENEZUELA
Serra Tapirapecó
Curupira
Pico de Neblina 3014
Caracaraí
Demini
Catrimani
GUYANA
Biloku
Kamoa Mts.
734
Serra Acaraí
Mitú
Jibóia
Vista Alegre
Cucuí
São José do Anauá
Anauá
Trombetas
Paru de Oeste
Vaupés
Iuaretê
Içana
Negro
Padauiri
Araçá
Catrimani
Taracuá
Uaupés
360
Apaporis
Lérida
São José
Negro
Tapurucuara
Calanaque
Branco
Jauaperi
Mapuera
Caquetá
La Pedrera
Puerto Miraña
Barcelos
Cuiuni
Tupanacca
Vila Bittencourt
Marcelino
Japurá
Maraã
Unini
Moura
Santa Maria
Nhamunda
Oriximiná
Óbidos
Airão
Uatumã
Faro
Foz do Mamoriá
Santo Antônio
Jaú
Santarém
Santa Clara
Fonte Boa
Urucará
Parintins
Belterra
Puerto San Agustín
Tontantins
Içá
Piorini
Negro
Tupinambarama
Santo Antônio de Içá
Alvarães
Badajós-See
Manaus
Itacoatiara
.100
Amazon (Solimões)
Manacapuru
Maués
Tefé
Piorini-See
Badajós
Anamã
Amazon
São Paulo de Olivença
Renascença
Codajás
Ilha
Brasília Legal
Nova Olinda do Norte
Mucajá
Caballococha
Leticia
Tefé
Coari
Itaituba
Benjamin Constant
Boca do Mutúm
Concórdia
Urucu
Purus
Diamantina
Madeira
Borba
Maués
Laranjal
San Luis de Tapajós
Caxias
.100
Jutaí
Carauari
Itaboca
Arumã
Canuma
Terra Preta
5°S
Santa Helena
Jutaí
Juruá
Coari
Prêto do Igapó-Açu
Novo Aripuanã
Jamanxim
Itui
Liberdade
Jaburu
Piranhas
S
Lajinha
Tapajós
Três Bocas
e
Tapauá
Manicoré
l
Canumã
v
a
Boca do Acará
Jacareacanga
.200
Posto Curuá
Soledade
Santos Dumont
Aliança
Ipixuna
Castanhal
Eirunepé
Tapauá
Aripuanã
Creporizinho
Canindé
Purus
Sucunduri
Sauré
B R A Z I L
Pirapetinga
Lábrea
Sucurundi
Tarauacá
Envira
Mamoriá
Prainha Nova
Barra do São Manuel
Manuelzinho
Boca do Moaco
Manjuriã
Humaitá
.100
Jatuarana
Samaumá
Tauacá
Pauini
Foz do Pauini
Mucuim
Calama
Curuá
Feijó
Boca do Curequetê
Ituxi
Teles Pires
Recreio
Envira
Manuel Urbano
Boca do Acre
Madeira
Pôrto Velho
Jamari
Jiparaná
Theodore Roosevelt
Jacaretinga
Juruena
Serra do Cachimbo
138
Purus
Pôrto Alegre
Tabajara
Arapari
Santa Rosa
Sena Madureira
Bom Jardim
Caratianas
Aripuanã
Gêlo
Jaciparaná
Cachimbo
404
Macapá
Serra dos Apiacás
Esperanza
Manoa
Abunã
Iracema
10°S
Purus
Iaco
Rio Branco
Taquaras
Ariquemes
Aarão
Aripuanã
Canamaria
Abunã
Acre
Xapuri
Villa Bella
Jarú
200
Serra do Norte
Arinos
Serra
Brasiléia
356
Puerto Rico
Antuérpia
Pôrto do Cajueiro
Iñapari
Cobija
Madre de Dios
Guajará Mirim
Serra dos Pacaas Novos
Rondônia
Fontanillas
Porvenir
Riberalta
800
Presidente Hermes
Pôrto dos Gauchos
Las Piedras
Iberia
Beni
Yata
Mamoré
Serra
Acampamento de Indios
dos
Providencia
Pimenta Bueno
Pôrto Atlântico
Fortaleza
Fortaleza
dos
Pouso Alegre
Manú
Madre de Dios
Puerto Maldonado
Puerto Heath
Guaporé
Costa Marques
José Bonifácio
Serra do Tombador
Parecis
Cavinas
Carmem
Serra Formosa
Santo Antônio
Vilhena
Juruena
Uriariti
Marape
Quince Mil
Madidi
100
Lago Rogaguado
San Joaquín
Baures
Itonamas
Mategua
Lucas
Astillero
Cord.
Auzangate 6394
Inambari
Ixiamas
Lago Rogagua
Magdalena
Pimenteiras
Santa Isabel
Ponta da Pedra
Urcos
Cerros de Bala
Lago de San Luis
El Carmen
San Martin
Puerto Alegre
Campo dos Parecis
Macusani
Santa Ana
Guaporé
Carabaya
Sandia
5852
Reyes
Blanco
Serra de Huachacaa
702
Mato
5443
5999 Palomani
Apolo
Rapulo
Apere
La Esperanza
Paraguá
Diamantino
Ayaviri
Azángaro
San Borja
San Ignacio
Trinidad
Perseverancia
Rosario Oeste
Grosso
Yauri
Huancané
Llanos de Mojos
La Noria
1095
Tapirapua
Rep. do Rio Manso
15°S
Plateau
5641
Puerto Acosta
6388
Loreto
Mato Grosso
Várzea Grande
Cuiabá
Juliaca
Lake Titicaca
Ancohuma
Santa Ana
B O L I V I A
Salinas
Puno
Beni
Puerto Marquez
Ascensión
1150
Pôrto Esperidião
Serra Aguapei
Cáceres
Poconé
Jaciara
5822
Achacachi
Coroico
Concepción
5486
Juli
La Paz
Chulumani
Ichilo
Rio Grande
San Javier
San Ignacio
283
San Matias
Arequipa
Mazo Cruz
3577
Illimani 6402
Cuiabá
2304
Guaqui
Viacha
Todos Santos
Puerto Villarroel
San Pablo
Descalvados
São Lorenço
Tambo
5781
Moquegua
5213
Corocoro
Eastern
Itiquira
5035
Cochabamba
Laguna Concepción
Laguna Uberaba
Calacoto
Sicasica
2558
Montero
Paraguai
Tarata
Umala
Quillacollo
3200
Portachuelo
.614
El Cerro
San José de Chiquitos
Pôrto Jofre
Charaña
Desaguadero
Cordillera
Totora
Santa Cruz
Santa Corazón
Pedro Gomes
5968 Vacora Putre
Sajama 6542
Oruro
Comarapa
Amolar
Tacna
5383
Aiquile
Llanos de Chiquitos
Roboré
Serra de Santiago
Arica
Uncia
Valle Grande
60°W
70°W 65°W 55°W

26

n o p q r s t u v w x y z

28

0 100 200 300 miles

Average linear scale

0 100 200 300 400 500 Km

23

a b c d e f g h i j k l m

25

n o p q r s t u v w x y z

28

This map shows 1/60 of the earth's surface

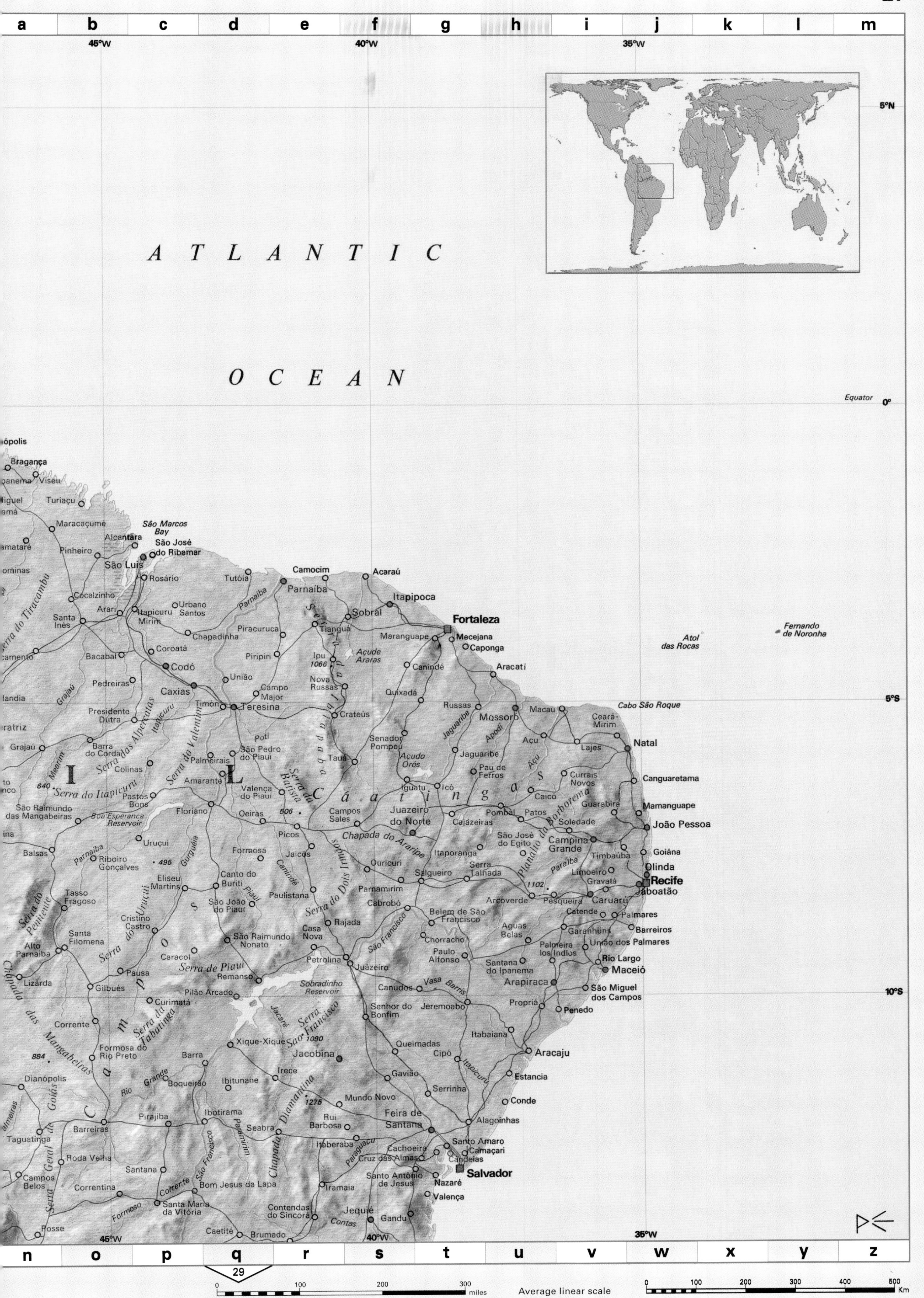
a b c d e f g h i j k l m
45°W
40°W
35°W
5°N
ATLANTIC
OCEAN
Equator
0°
5°S
10°S
Fernando de Noronha
Atol das Rocas
Cabo São Roque
São Marcos Bay
Bragança
Viseu
Turiaçu
Maracaçumé
Alcantara
São José do Ribamar
Pinheiro
São Luís
Rosário
Tutóia
Camocim
Acaraú
Parnaíba
Itapipoca
Sobral
Fortaleza
Mecejana
Caponga
Maranguape
Tianguá
Açude Araras
Ipu
1066
Canindé
Aracati
Nova Russas
Quixadá
Cocalzinho
Arari
Itapicuru Mirim
Urbano Santos
Santa Inês
Chapadinha
Piracuruca
Piripiri
Coroatá
Bacabal
Codó
União
Pedreiras
Caxias
Campo Major
Timon
Teresina
Presidente Dutra
Crateús
Russas
Mossoró
Macau
Ceará-Mirim
Natal
Açu
Lajes
Senador Pompeu
Jaguaribe
Apodi
Barra do Corda
São Pedro do Piauí
Palmeiras
Tauá
Açudo Orós
Pau de Ferros
Currais Novos
Canguaretama
Grajaú
Colinas
Amarante
Valença do Piauí
Serra do Itapicuru
Pastos Bons
640
Iguatu
Icó
Caicó
Guarabira
Mamanguape
Boa Esperança Reservoir
São Raimundo das Mangabeiras
Floriano
Oeiras
506
Campos Sales
Juazeiro do Norte
Cajàzeiras
Pombal
Patos
Soledade
João Pessoa
Picos
Chapada do Araripe
São José do Egito
Campina Grande
Goiána
Uruçui
Balsas
Ribeiro Gonçalves
495
Formosa
Jaicós
Itaporanga
Serra Talhada
Timbaúba
Olinda
Ouricuri
Salgueiro
Limoeiro
Recife
Eliseu Martins
Canto do Buriti
Gravatá
Jaboatão
Paulistana
Parnamirim
1102
Arcoverde
Pesqueira
Caruaru
Tasso Fragoso
São João do Piauí
Cabrobó
Belem de São Francisco
Catende
Palmares
Cristino Castro
Rajada
Aguas Belas
Garanhuns
Barreiros
Santa Filomena
São Raimundo Nonato
Casa Nova
Chorrocho
Palmeira los Indios
União dos Palmares
Alto Parnaíba
Caracol
Petrolina
Juàzeiro
Paulo Afonso
Santana do Ipanema
Rio Largo
Maceió
Pausa
Serra de Piauí
Remanso
Sobradinho Reservoir
Lizárda
Gilbués
Pilão Arcado
Canudos
Arapiraca
São Miguel dos Campos
Curimatá
Senhor do Bonfim
Jeremoabo
Propriá
Penedo
Corrente
Serra da Tabatinga
Itabaiana
Aracaju
884
Formosa do Rio Preto
Xique-Xique
1090
Queimadas
Barra
Jacobina
Cipó
Irece
Gavião
Estancia
Dianópolis
Boqueirão
Ibitunane
Serrinha
Mundo Novo
Conde
1275
Barreiras
Pirajiba
Ibotirama
Feira de Santana
Alagoinhas
Rui Barbosa
Taguatinga
Seabra
Itaberaba
Santo Amaro
Camaçari
Cachoeira
Roda Velha
Cruz das Almas
Candeias
Santana
Salvador
Campos Belos
Santo Antônio de Jesus
Nazaré
Correntina
Bom Jesus da Lapa
Iramaia
Valença
Santa Maria da Vitória
Contendas do Sincorá
Jequié
Gandu
Posse
Caetité
Brumado
Caatingas
Chapada Diamantina
Serra Geral de Goiás
Chapada das Mangabeiras
Serra do Penitente
Serra do Uruçuí
Serra das Alpercatas
Serra do Valentim
Serra da Ibiapaba
Serra da Baritsa
Serra do Dois Irmaos
Planalto da Borborema
Serra do Tiracambu
São Francisco
Parnaíba
Jaguaribe
Paraguaçu
Rio Grande
Contas
n o p q r s t u v w x y z
29
0 100 200 300 miles
Average linear scale
0 100 200 300 400 500 Km

25

a b c d e f g h i j k l m

PERU
BOLIVIA
PARAGUAY
ARGENTINA
CHILE
URU
PACIFIC OCEAN

Abancay
Urcos
Chalhuanca
Auzangate 6394
Sicuani
Macusani 5852
Sandia
5211
5443
Santo Tomás
Yauri
Ayaviri
Cord. de Carabaya
Palomani 5999
Apolo
Ixiamas
Cerros de Bala
Reyes
Lago Rogagua
Rapulo
Santa Ana
Apere
Lago de San Luis
El Carmen
Puerto Alegre
Ponta da Pedra
Campo dos Parecis
702
Diamantino
San Ignacio
Trinidad
San Borja
San Miguel
Blanco
La Esperanza
Perseverancia
Paraguá
Serra de Huanchaca
Guaporé
Tapirapua
Llanos de Mojos
Loreto
San Martín
La Noria
1095
Mato Grosso
15°S
Cotahuasi
Coropuna 6425
Chuquibamba
Chivay
Ampato 6310
Chachani 6075
5641
Juliaca
Huancané
Puerto Acosta
Illampu 6485
Ancohuma 6388
Santa Ana
Beni
Lake Titicaca
Puno
Achacachi
Coroico
Puerto Marquez
Mamoré
Río Grande
Ascensión
Salinas
1150
Pôrto Esperidião
Várzea Grande
Serra Aguapei
Cáceres
Poconé
Cuiabá
Ocoña
Camaná
5822
Arequipa
2304
5486
Juli
La Paz
3577
Chulumani
Viacha
Illimani 6402
Concepción
San Javier
San Ignacio
San Matías
283
Mollendo
Tambo
5593
5781
5213
Guaqui
Todos Santos
Ichilo
Puerto Villarroel
San Pablo
Descalvados
Moquegua
Calacoto
Umala
Sicasica
Quillacollo
5035
2558
Cochabamba
Eastern Cordillera
Montero
Itiquira
Paraguai
Tarata
Tacora 5988
Charana
Desaguadero
Totora
3200
Portachuelo
El Cerro
Laguna Concepción
San José de Chiquitos
Lagoa Uberaba
Pôrto Jofre
Ilo
Tacna
Sajama 6542
Oruro
Aiquile
Comarapa
Santa Cruz
Serra de Chiquitos
Santo Corazón
Grande
Amolar
Putre
5383
Corque
Uncia
Central
Río Grande
Valle Grande
Llanos de Chiquitos
Bañados del Izozog
Roboré
Serra de Santiago
Santa Ana
Pantanal
Arica
Guallatiri 6060
Altiplano
Lake Poopó
Challapata
5023
2790
Sucre
Tarabuco
Cordillera
Cabezas
Yuti
Fortín Ravelo
777
Fortín Max Paredes
Puerto Suárez
Corumbá
Cuya
Nama
Sabaya
5869
Lago de Coipasa
Río Mulatos
3976
Potosí
Lagunillas
Parapetí
Chaco
Pôrto Esperança
Aquidauana
Pisagua
Huara
Sillajhuay 5995
Azurduy
Camiri
Charagua
20°S
5218
Cord. de Tajsara
Guaicurus
Aquidabán
Iquique
5950
Vitichi
Salar de Uyuni
Uyuni
Cord. de Chichas
Pilcomayo
998
Timané
Bahia Negra
Pintados
5320
Fortín Gral. Eugenio Garay
Fortín Madrejón
Miranda
1590
Collahuasi
5739
Cotagaita
Machareti
Boreal
Fuerte Olimpo
Bonito
Lagunas
Ollagüe
Pilaya
Villa Montes
Fortín Garrapatal
Loa
6045
5865
Tupiza
Alota
San Pablo
Mojo
Tarija
Pôrto Murtinho
Quillagua
1825
Conchi
Cord. de Lipez
5695
Villazón
Yacuiba
Fortín Hernandarias
Mariscal Estigarribia
Puerto Sastre
Bella Vista
Tocopilla
5680
Chuquicamata
Tocorpuri 5833
La Quiaca
Tartagal
Dr. Pedro P. Peña
Filadelfia
María Elena
2293
Calama
Licancabur
5029
Bermejo
Verde
Puerto Piñasco
San Pedro de Atacama
5921
Abra Pampa
Rosario
San Ramón de la Nueva Orán
Mejillones
Carmen Alto
6050
Humahuaca
Juan Sola
Fortín Juan de Zalazar
Chaco
Fortín General Díaz
Salar de Atacama
Salinas Grandes
Embarcación
Los Blancos
Aquidabán
Concepción
Antofagasta
Augusta Victoria
5890
Catua
Libertador Gral. San Martín
Pozo Colorado Monte
Lima
Varillas
Pular 6225
5594
Chani 6200
Bermejo
Ingeniero Guillermo Nueva Juárez
Central
Lindo
San Pedro
Caleta el Cobre
Salar Punta Negra
Socompa 6031
Socompa
San Antonio de los Cobres
San Salvador de Jujuy
San Pedro
Los Vientos
Llullaillaco 6723
Salar de Arizaro
Salar de Pocitos
Cachi 6720
General Güemes
Las Lajitas
Las Lomitas
Teuco
Pozo del Tigre
Benjamín Aceval
25°S
Paposo
Salta
Joaquín V. González
5700
Cachi
Villa Hayes
Asunción
Taltal
Santa Catalina
Antofalla 6440
Palo Santo
Pirané
Clorinda
OCEAN
Salar de Antofalla
Cord. de Calalaste
Sierra de Aconquija
Metán
Rosario de la Frontera
Salado
Monte Quemado
Paraguarí
El Salvador
5070
Mojones 5990
Cafayate
Castelli
Laguna Vera
Diego de Almagro
Colorados
6049
Antofagasta de la Sierra
Nevada 6400
Pampa de los Guanacos
Tres Isletas
Formosa
Bermejo
Tebicuary
Chañaral
Santa María
San Miguel de Tucumán
Presidencia Roque Sáenz Peña
Pilar
San Juan Bautista
Inca de Oro
Monteros
Nev. de Aconquija
Tintina
Austral
San Ignacio
Caldera
Copiapo 6080
Ojos del Salado 6880
4920
Hondo Reservoir
Termes de Río Hondo
Charata
Presidencia de la Plaza
Paraná
Copiapó
Concepción
Belén
Andalgalá
Santiago del Estero
La Banda
Villa Angela
Resistencia
Corrientes
Castilla
Copiapó
Pissis 6858
Fiambalá
Bonete 6872
Suncho Corral
Empedrado
Apóstoles
Carrizal Bajo
Tinogasta
Salar de Pipanaco
Villa Guillermina
Huasco
5830
Sierra de Famatina
Catamarca
Villa San Martín
Añatuya
Bella Vista
Santo Tomé
Vallenar
Huasco
Jagüé
Frías
Cabo Bascuñan
Mejicana 6250
Chumbicha
Bandera
Domeyko
Toro 6380
Chilecito
Sierra de Velasco
Recreo
Pinto
Tostado
Reconquista
Goya
Mercedes
Villa Unión
La Rioja
Salinas Grandes
Villa Ojo de Agua
Vera
Corrientes
Yapeyú
La Higuera
Guandacol
Uruguaiana
La Serena
Rivadavia
Tórtolas 6332
Ceres
Salado
Chalchaquí
Curuzú Cuatiá
30°S
Coquimbo
Olivares 6282
Patquía
San Cristóbal
Esquina
San José de Jáchal
Chamical
Cruz del Eje
Deán Funes
Dulce
Bella Unión
5510
Ovalle
Precordillera
San Agustín de Valle Fértil
Laguna Chiquita
Morteros
La Paz
Artigas
Maitencillo
Bermejo
Zanjón
Chajarí
Combarbala
Arapey
5620
San Juan
Chepes
Córdoba
Rafaela
Concordia
Salto
Illapel
Salsacate
Alta Gracia
San Francisco
Esperanza
Santa Fe
Pampa de las Salinas
Tacuarembó
Salamanca
Mercedario 6770
Villa Media Agua
Villa Dolores
2880
Paraná
Villaguay
Quines
Oliva
Diamante
Chincolco
San Juan
Río Tercero
Villa María
Paysandú
San Felipe
Aconcagua 6959
Desaguadero
Santa Rosa del Conlara
Sierra de Córdoba
Cañada de Gómez
Victoria
Gualeguay
Concepción del Uruguay
Negro
La Calera
Mendoza
San Martín
San Luis
1599
Río Cuarto
La Carlota
Carcarañá
Rosario
Gualeguaychú
Paso de los Toros
Valparaíso
Santiago
Tupungato 6800
La Paz
Tunuyán
San Nicolás
Gualeguay
Uruguay
Mercedes
Durazno
San Antonio
San Bernardo
5830
Tunuyán
Mercedes
Vicuña Mackenna
Vanado Tuerto
San Pedro
Pergamino
Carmelo
Cardona
Rapel
Rancagua
5290
Justo Daract
Laboulaye
Zárate
Florida
Santa Cruz
San Rafael
Diamante
Salado
Buena Esperanza
Rufino
Junín
Martínez
C. del Sacramento
San Fernando
5160
4860
General Alvear
Huinca Renancó
Lincoln
Luján
Buenos Aires
35°S
Curicó
Atuel
General Villegas
Chacabuco
Chivilcoy
Lobos
La Plata
Montevideo
Constitución
70°W
Unión
65°W
60°W
Magdalena
River Plate
Pampas
Gran Chaco
Chaco Austral
Chaco Central
Chaco Boreal
Andes
Cord. de Olivares
Cord. de Ollita
Desert of Atacama

n o p q r s t u v w x y z

31

This map shows 1/60 of the earth's surface

B r a z i l i a n
P l a t e a u
C e n t r a l
P l a t e a u
M i n a s
G e r a i s
B R A Z I L
A T L A N T I C
O C E A N
Brasília
Belo Horizonte
Rio de Janeiro
São Paulo
Curitiba
Pôrto Alegre
Goiânia
Uberlandia
Uberaba
Montes Claros
Governador Valadares
Vitória
Vila Velha
Campos
Juiz de Fora
Niterói
Petrópolis
Santos
Campinas
Sorocaba
Ribeirão Preto
Londrina
Maringá
Ponta Grossa
Florianópolis
Joinville
Blumenau
Caxias do Sul
Pelotas
Rio Grande
Santa Maria
Passo Fundo
Ilhéus
Vitória da Conquista
Serra Geral do Paraná
Serra da Capivara
Serra das Divisoes
Serra da Mantiqueira
Serra Paranapiacaba
Serra das Araras
Serra do Mar
Serra do Espinhaço
Serra da Canastra
Lagoa dos Patos
Lagoa Mirim
Tres Marias Reservoir
Furnas Reservoir
Abrolhos Archipelago
Cabo Frio
Tropic of Capricorn
15°S
20°S
25°S
30°S
35°S
50°W
45°W
40°W
Average linear scale
miles
Km

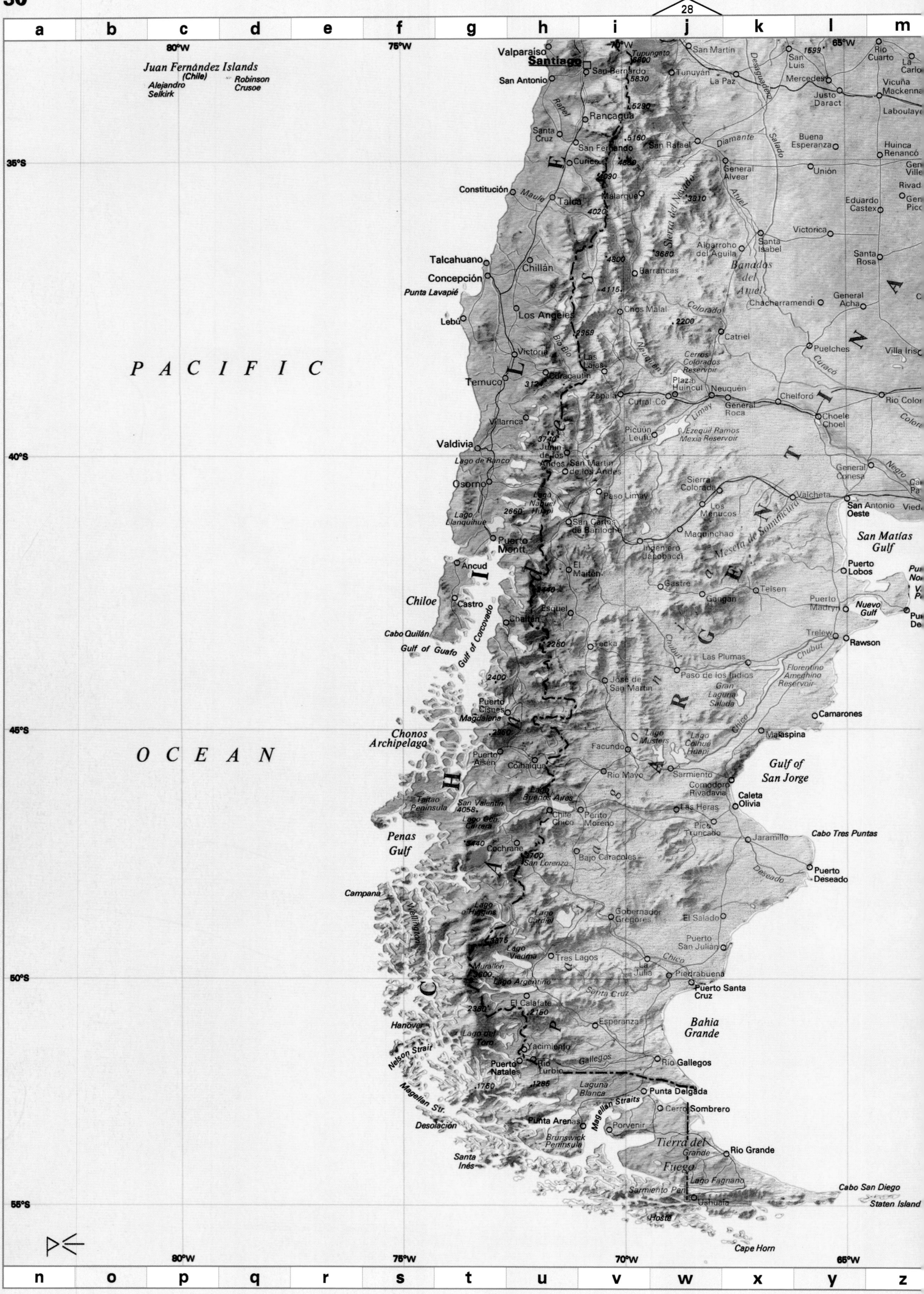

28

a b c d e f g h i j k l m

80°W 75°W 70°W 65°W

Juan Fernández Islands
(Chile)
Alejandro Selkirk
Robinson Crusoe

PACIFIC

OCEAN

35°S
40°S
45°S
50°S
55°S

Valparaíso
Santiago
San Antonio
San Bernardo
Tupungato 6800
5830
5290
Rancagua
Santa Cruz
Rapel
San Fernando
5160
San Rafael
Curicó
4680
4090
Constitución
Maule
Talca
Malargüe
4020
3810
San Martín
Tunuyán
La Paz
Desaguadero
San Luis
Mercedes
Justo Daract
1599
Río Cuarto
La Carlota
Vicuña Mackenna
Laboulaye
Diamante
Salado
Buena Esperanza
Huinca Renancó
General Alvear
Unión
Eduardo Castex
Atuel
Sierra del Nevado
Victorica
Santa Isabel
Algarroho del Aguila
Santa Rosa
Bañados del Atuel
3680
4800
Talcahuano
Concepción
Punta Lavapié
Chillán
Barrancas
4115
Chos Malal
Colorado
Chacharramendi
General Acha
Lebú
Los Angeles
2200
Catriel
2969
Neuquén
Biobío
Victoria
Las Lajas
Cerros Colorados Reservoir
Puelches
Curacó
Villa Iris
Temuco
3124
Curacautín
Plaza Huincul
Neuquén
Zapala
Cutral-Co
General Roca
Chelforó
Río Colorado
Choele Choel
Limay
Villarrica
Picuún Leufú
Ezequil Ramos Mexia Reservoir
Valdivia
3740
Junín de los Andes
Lago de Ranco
San Martín de los Andes
General Conesa
Negro
Osorno
Lago Nahuel Huapi
Paso Limay
Sierra Colorada
Los Menucos
Valcheta
San Antonio Oeste
Lago Llanquihue
2660
San Carlos de Bariloche
Maquinchao
Meseta de Somuncurá
San Matías Gulf
Puerto Montt
Ingeniero Jacobacci
Ancud
El Maitén
Puerto Lobos
Gastre
Telsen
Chiloé
Castro
Gangan
Puerto Madryn
Nuevo Gulf
2440
Esquel
Chaitén
Cabo Quilán
Gulf of Corcovado
Gulf of Guafo
2260
Tecka
Trelew
Rawson
Chubut
Las Plumas
Florentino Ameghino Reservoir
2400
José de San Martín
Paso de los Indios
Gran Laguna Salada
Puerto Cisnes
Magdalena
Camarones
2360
Chonos Archipelago
Lago Musters
Lago Colhué Huapí
Chico
Malaspina
Facundo
Puerto Aisén
Coihaique
Río Mayo
Sarmiento
Gulf of San Jorge
Comodoro Rivadavia
Caleta Olivia
Lago Buenos Aires
Taitao Peninsula
San Valentín 4058
Chile Chico
Perito Moreno
Las Heras
Lago Gral. Carrera
Pico Truncado
Cabo Tres Puntas
Penas Gulf
3440
Cochrane
Jaramillo
3700 San Lorenzo
Bajo Caracoles
Deseado
Puerto Deseado
Campana
Lago O'Higgins
Wellington
Lago Cardiel
Gobernador Gregores
El Salado
3375
Lago Viedma
Tres Lagos
Puerto San Julián
Chico
Murallón 3600
Lago Argentino
La Julia
Piedrabuena
Puerto Santa Cruz
Santa Cruz
El Calafate
2380
2150
Esperanza
Bahía Grande
Hanover
Lago del Toro
Yacimiento
Nelson Strait
Puerto Natales
Río Turbio
Gallegos
Río Gallegos
1750
1265
Laguna Blanca
Punta Delgada
Magellan Str.
Magellan Straits
Cerro Sombrero
Punta Arenas
Desolación
Brunswick Peninsula
Porvenir
Tierra del Fuego
Grande
Río Grande
Santa Inés
Lago Fagnano
Cabo San Diego
Sarmiento Peak
Ushuaia
Staten Island
Hoste
Cape Horn

CHILE

ARGENTINA

n o p q r s t u v w x y z

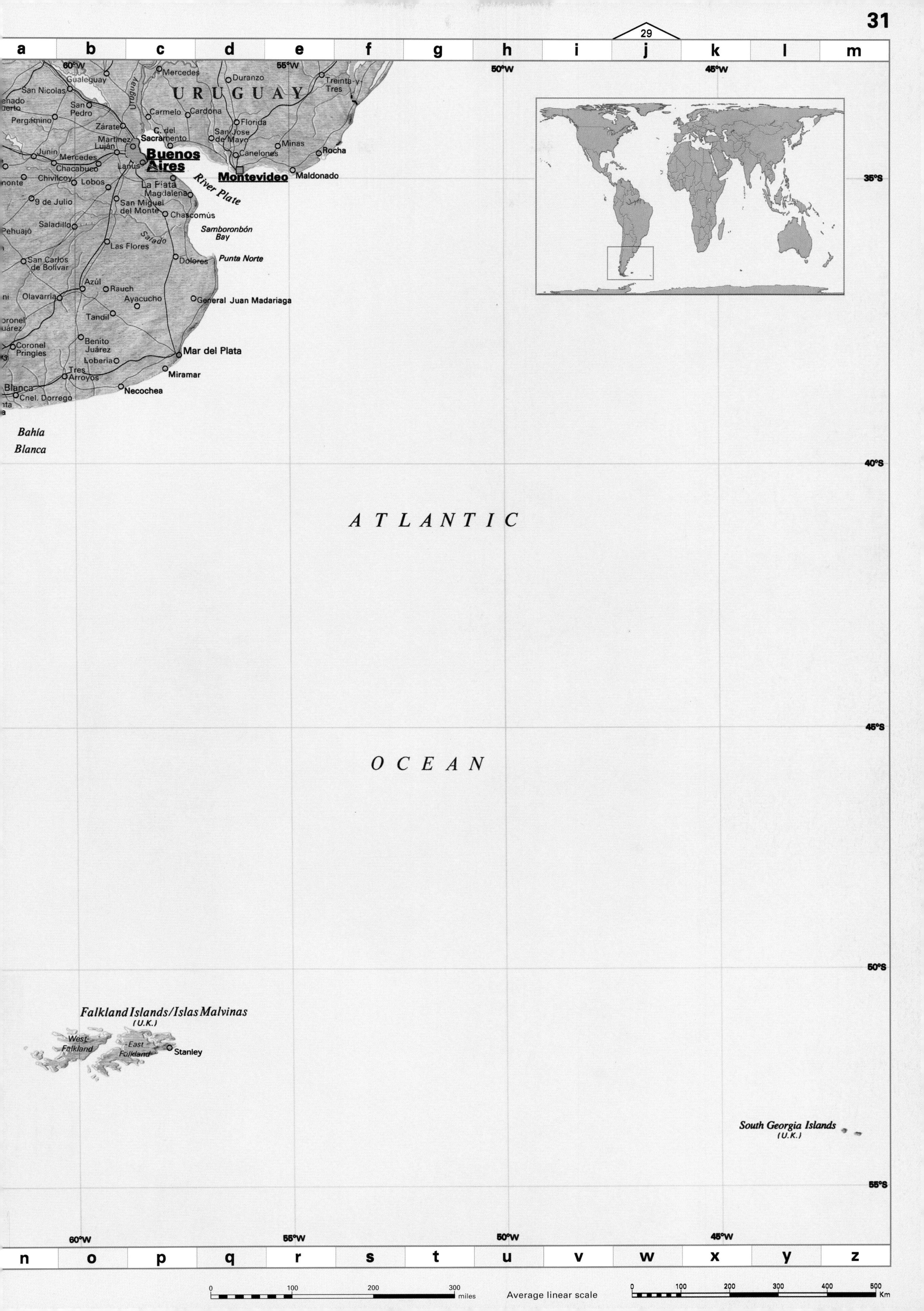
29
a
b
c
d
e
f
g
h
i
j
k
l
m
60°W
55°W
50°W
45°W
Gualeguay
Mercedes
Duranzo
Treinta-y-Tres
URUGUAY
Uruguay
San Nicolas
San Pedro
Pergamino
Carmelo
Cardona
Florida
Zárate
C. del Sacramento
San Jose de Mayo
Martinez
Luján
Minas
Rocha
Junin
Mercedes
Chacabuco
Lanús
Buenos Aires
Canelones
Montevideo
Maldonado
Chivilcoy
Lobos
La Plata
Magdalena
River Plate
9 de Julio
San Miguel del Monte
Chascomús
Saladillo
Samborombón Bay
Pehuajó
Salado
Las Flores
San Carlos de Bolivar
Dolores
Punta Norte
Azúl
Rauch
Olavarria
Ayacucho
General Juan Madariaga
Tandil
Benito Juárez
Coronel Pringles
Mar del Plata
Loberia
Tres Arroyos
Miramar
Blanca
Cnel. Dorrego
Necochea
Bahía Blanca
35°S
40°S
ATLANTIC
OCEAN
45°S
50°S
Falkland Islands/Islas Malvinas
(U.K.)
West Falkland
East Falkland
Stanley
South Georgia Islands
(U.K.)
55°S
60°W
55°W
50°W
45°W
n
o
p
q
r
s
t
u
v
w
x
y
z
0
100
200
300
miles
Average linear scale
0
100
200
300
400
500
Km

91

a b c d e f g h i j k l m

20°W 15°W 10°W 5°W 0°

70°N 65°N 60°N 55°N 50°N

Greenland (Denmark)
Scoresbysund
Scoresby Sound
Jan Mayen (Norway)
Denmark Strait
ARCTIC OCEAN
Arctic Circle
Cape Horn
Ísafjördur
Húna Bay
Akureyri
Fontur
Breidhi Fjord
ICELAND
1765
1400
Vatnajökull
2119
Djúpivogur
Akranes
Faxa Bay
Keflavík
Reykjavík
Reykjanes

Faeroe Islands (Denmark)
Shetland Islands
Lerwick
Orkney Islands
Cape Wrath
Pentland Firth
Thurso
ATLANTIC OCEAN
Lewis
Minch
Hebrides
Moray Firth
Skye
Inverness
Loch Ness
Elgin
Highlands
1309
Aberdeen
SCOTLAND
Fort William
1343
Mull
Perth
Dundee
Stirling
Edinburgh
Islay
Glasgow
Berwick upon Tweed
GREAT BRITAIN AND NORTHERN IRELAND
Ayr
North Channel
Newcastle upon Tyne
Sunderland
Londonderry
NORTHERN IRELAND
Carlisle
Lough Neagh
Belfast
Middlesbrough
Donegal Bay
Portadown
Sligo
Isle of Man
Douglas
Dundalk
Leeds
York
Westport
Blackpool
Kingston upon Hull
IRISH REPUBLIC
Irish Sea
Athlone
Liverpool
Manchester
Holyhead
Sheffield
Galway
Dublin
886
Anglesey
Shannon
Stoke-on-Trent
1085
Roscrea
Nottingham
ENGLAND
Derby
Norwich
Arklow
Limerick
Birmingham
Leicester
920
Wexford
WALES
Cambridge
Aberystwyth
Coventry
Ipswich
Killarney
Waterford
St. George's Channel
Fishguard
Oxford
Luton
Cork
Swansea
Thames
Southend-on-Sea
Cape Clear
Cardiff
Bristol
Reading
London
Ostend
Bristol Channel
Dover
Southampton
Strait of Dover
Calais
Bournemouth
Brighton
Exeter
Isle of Wight
Lille
Valenciennes
Plymouth
Land's End
Penzance
English Channel
Abbeville
Amiens
Cherbourg
Le Havre
Rouen
Guernsey
Channel Islands (U.K.)
Jersey
Caen
Évreux
Seine
Gulf of St. Malo
Granville
Paris
Brest
St. Brieuc
Chartres
Alençon
Rennes
FRANCE
Le Mans
Orléans
Lorient
Angers
Loire
St. Nazaire
Nantes
Tours
NORTH SEA

90

n o p q r s t u v w x y z

34

This map shows 1/60 of the earth's surface

92

a b c d e f g h i j k l m

NORWAY SWEDEN FINLAND RUSSIA ESTONIA LATVIA LITHUANIA BELARUS POLAND GERMANY DENMARK CZECH REPUBLIC SLOVAKIA AUSTRIA HUNGARY ROMANIA UKRAINE

Lappland

BALTIC SEA

Gulf of Bothnia

Gulf of Finland

Gulf of Riga

Skagerrak

Kattegat

50

n o p q r s t u v w x y z

35

0 100 200 300 miles Average linear scale 0 100 200 300 400 500 Km

32

a b c d e f g h i j k l m

5°W 0° 5°E 10°E

50°N 45°N 40°N 35°N

GREAT BRITAIN
Plymouth
Exeter
Bournemouth
Southampton
Isle of Wight
Brighton
Dover
Strait of Dover
Land's End
Penzance
English Channel
Channel Islands (U.K.)
Guernsey
Jersey
Gulf of St. Malo
Cherbourg
Le Havre
Rouen
Caen
Évreux
Granville
Calais
Lille
Abbeville
Amiens
St Quentin
Compiègne
Reims
Sedan
Ghent
Brussels (Bruxelles)
BELGIUM
Valenciennes
Charleroi
Liège
LUXEMBOURG
Luxembourg
Thionville
Metz
Aachen
Cologne (Köln)
Bonn
Koblenz
Wiesbaden
Trier
Saarbrücken
Bad Hersfeld
Erfurt
Plauen
Frankfurt
Würzburg
Main
Bamberg
GERMANY
Mannheim
Nuremberg (Nürnberg)
Karlsruhe
Stuttgart
Danube
Regensburg
Augsburg
Munich (München)
Memmingen
Lake Constance
Zugspitze 2963
Innsbruck
Rhine
Strasbourg
Nancy
Chaumont
Troyes
Paris
Seine
Chartres
Orléans
Alençon
Brest
St. Brieuc
Rennes
Lorient
Le Mans
Angers
Tours
Loire
St. Nazaire
Nantes
Auxerre
Dijon
Besançon
Mulhouse
Freiburg
1493
Basle
Zürich
Vaduz
LIECHTENSTEIN
Landeck
3774
Lucerne
Berne
SWITZERLAND
Chur
Lausanne
4185
Lake Geneva
Geneva (Genève)
Rhône
Mont Blanc 4807
Monte Rosa 4634
Annecy
Aosta
Como
Bernina 4050
Bolzano
Dolomites
3342
3554
Lake Garda
Bergamo
Brescia
Vicenza
Milan (Milano)
Novara
Turin (Torino)
Verona
Padua (Padova)
Po
Piacenza
Parma
Ferrara
Alessandria
Modena
Bologna
Genoa (Genova)
Gulf of Genoa
2120
La Spezia
SAN MARINO
Cuneo
3847
Maritime Alps
Digne
Nice
MONACO
Imperia
Pisa
Livorno (Leghorn)
Florence (Firenze)
Arezzo
Siena
Grosseto
Elba
Viterbo
Civitavecchia
Rome (Roma)
LIGURIAN SEA
Bastia
Corsica (France)
2710
Ajaccio
Bonifacio
Str. of Bonifacio
Olbia
Sassari
Nuoro
Sardinia (Italy)
1834
Oristano
Cágliari
Cape Teulada
TYRRHENIAN SEA
FRANCE
ATLANTIC OCEAN
Bay of Biscay
La Roche-sur-Yon
288
La Rochelle
Poitiers
Châteauroux
Chalons-sur-Saône
Moulins
Loire
Mâcon
Lyons
Saintes
Limoges
Angoulême
Gironde
Clermont-Ferrand
Allier
1885
St. Étienne
Massif Central
Le Puy
Grenoble
Pelvoux 4102
Valence
Brive
Aurillac
Bordeaux
Dordogne
Lot
Cévennes
Rhône
Landes
Agen
Garonne
Adour
Nîmes
Avignon
Montpellier
Arles
Aix-en-Provence
Marseilles
Toulon
Bayonne
Pau
Tarbes
Toulouse
Carcassonne
Narbonne
Gulf of Lion
Perpignan
ANDORRA
2504
3404 Pico de Aneto
2923
Pyrenees
Corunna
Gijón
Oviedo
2417
Cantabrian Mountains
2583
Santander
San Sebastián
Bilbao
Vitoria
Pamplona
Jaca
Cape Finisterre
Santiago de Compostela
Orense
Ponferrada
León
Burgos
Logroño
2142
Huesca
Vigo
Miño
Braga
Soria
Saragossa
Lérida
Ebro
Gerona
Costa Brava
Barcelona
Zamora
Valladolid
Duero
Oporto (Porto)
Douro
Salamanca
Calatayud
Tarragona
1382
Segovia
Aveiro
Guarda
Avila
2430
Guadalajara
Tortosa
Madrid
2592
Teruel
SPAIN
2020
Castellón
Coimbra
Talavera de la Reina
Toledo
Tagus
La Almarcha
Valencia
Leiria
Cáceres
Trujillo
PORTUGAL
Santarém
Lisbon (Lisboa)
Mérida
Ciudad Real
Guadiana
Júcar
Albacete
Gandia
Cabo de la Nao
Badajoz
Puertollano
Setúbal
Évora
Segura
Sierra Morena
Alicante
Costa Blanca
Sines
Beja
Córdoba
Ubeda
Murcia
2036
Odemira
Mértola
Aracena
Jaén
Algarve
Guadalquivir
Écija
Lorca
Cartagena
Sagres
Lagos
Huelva
Sevilla
Baza
Aguilas
Cape St. Vincent
Faro
Gulf of Cádiz
Antequera
Granada
3478
Sierra Nevada
Almería
Jerez de la Frontera
Málaga
Motril
Cádiz
Costa del Sol
Algeciras
Gibraltar (U.K.)
Str. of Gibraltar
Ceuta (Sp.)
Tangier
Tétouan
Asilah
Chechaouen
Al Hociema
Melilla (Sp.)
Ksar el Kebir
MOROCCO
Ouergha
Aknoul
Oujda
Beni Saf
Balearic Islands (Spain)
Menorca
Alcudia
Mallorca
Palma
Ibiza
MEDITERRANEAN SEA
Trapani
Egadi
Marsala
Bizerte
Gulf of Tunis
Cape Bon
Kélibia
Tunis
Tabarka
Annaba
Béja
Pantelleria (Italy)
Zaghouan
Teboursouk
Dellys
Bejaia
Skikda
Algiers (Alger)
Tizi Ouzou
586
Ténès
Khemis Miliana
Blida
Medea
Constantine
Setif
Souk-Ahras
Le Kef
Bordj Bou-Arreridj
Mostaganem
Cheliff
1983
Tell Atlas
Oran
Mohammadia
Relizane
Ain Oussera
Atlas Mountains
Batna
Ain Beida
1357
Sousse
Kairouan
Pelagie Islands (Italy)
Lampedusa
Mahdia
Sidi-bel-Abbes
Tiaret
Chellala
767
Bou Saada
2326
Khenchela
Tebessa
Kasserine
El Djem
Cape Kaboudia
Tlemcen
Marhoum
Djelfa
Monts des Ouled Nail
Biskra
Feriana
Sfax
Kerkenna Islands
Hauts Plateaux
Chott ech Chergui
Aflou
Saharan Atlas
Ouled Djellal
Chott Melrhir
Gafsa
Gulf of Gabès
Bougtob
1977
Messaad
Laghouat
El Meghaier
-40
Tozeur
Gabès
Djerba
El Bayadh
Chott Djerid
Kebili
Mechéria
Diamaa
El Oued
Médenine
Zarzis
ALGERIA
TUNISIA
Brézina
Tilrhemt
Touggourt
Ben Guerdane
2236
Ain Sefra
Ghardaia
Chebka du Mzab
Guerara
238
Metlili Chaamba
Bordj Bourguiba
Remada
Al Aziziyah
Safi
Ouargla
Hassi Messaoud
Nalut
688
Jadu
Great Western Erg
502
-145
306
Great Eastern Erg
Bir Zar
Essaouira
Amizmiz
Jebel Igdet 3615
Toubkal 4165
Rissani
Taouz
Abadla
834
El Goléa
Sinawan

n o p q r s t u v w x y z

36

This map shows 1/60 of the earth's surface

33

a b c d e f g h i j k l m

15°E 20°E 25°E 30°E 35°E

Dresden Wałbrzych Wrocław Częstochowa Kielce Vistula Zamość Lutsk Rovno Korosten Kiev Reservoir Nezhin Sumy
Sudety 1603 Hradec Králové Kolín 1490 Katowice POLAND Krakow Rzeszow Przemyśl L'viv Novograd-Volynskiy Kiev (Kyyiv) Dnieper Priluki Akhtyrka Khar'kov
CZECH REPUBLIC Olomouc Ostrava Bielsko-Biala Zakopane 1725 Gerlachovsky 2663 Stryy Ternopol Zhitomir UKRAINE Lubny Poltava Valki
Jihlava Vltava Znojmo Brno Žilina 1592 Prešov Košice 2043 1346 Ivano Frankovsk Khmel'nitskiy Vinnitsa Belaya Tserkov Cherkassy Kremenchugskoye Reservoir Kremenchug Pereshchepino
Linz Danube Vienna (Wien) 2075 Bratislava Zvolen SLOVAKIA Užgorod Mukachevo Carpathians Dniester Kolomyya Kamenets-Podol'skiy Mogilev Podol'skiy Uman' Znamenka Dneprodzerzhinsk Kirovograd Novomoskovsk Dnepropetrovsk
AUSTRIA Győr Danube Miskolc Nyíregyháza Chernovtsy Pervomaysk South Bug
Budapest Satu Mare Baia Mare Pietrosu 2305 Suceava Botoşani Bălţi MOLDOVA Orgejev Krivoy Rog Nikopol Zaporozh'ye
Leoben Graz Veszprém Balaton Lake HUNGARY Debrecen Oradea Dej 2103 Iaşi Prut Chişinău Tiraspol Nikolayev Kakhovskoye Reservoir
Klagenfurt Maribor Nagykanizsa Danube Békéscsaba Cluj Napoca Tîrgu Mureş Bacău Siret Dniester Odessa Kherson Novaya Kakhovka Melitopol
Ljubljana SLOVENIA Varaždin Pécs Szeged Arad ROMANIA 18.
Zagreb CROATIA Drava Subotica Tisza Mures Timişoara Deva Sibiu Braşov Belgorod Karkinitskiy Bay Dzhankoy Sea of Azov
Krk Karlovac Osijek Zrenjanin Negoiu 2548 Galaţi Bolgrad Mouths of the Danube Crimea
Cres Brod Sava Novi Sad Vršac 2509 Tulcea Simferopol Feodosiya 45°N
Pag Bihać Banja Luka Mitrovica Belgrade (Beograd) Danube Turnu-Severin Piteşti Ploieşti Danube 1259
BOSNIA-HERZEGOVINA Tuzla Smederevo Bucharest (Bucureşti) Sevastopol Jalda
Zadar Zenica SERBIA Svetozarevo Craiova Constanţa
Šibenik 2107 Sarajevo Užice Morava Roşiori
ADRIATIC SEA Split Jablanica Dinaric Alps Vidin Danube Ruse BLACK SEA
Brač Hvar Korčula Pelješac 2387 MONTE-NEGRO Niš Vraca Pleven Kolarovgrad Varna
Dubrovnik Leskovac KOSOVO Pirot Balkan Tŭrnovo
Podgorica 2693 Ivangrad Priština Sofia (Sofiya) 2376 Sliven BULGARIA Burgas
Corno Pescara Prizren Musala 2925 Stara Zagora
2793 Shkodër 2747 Kumanovo Skopje Blagoevgrad Plovdiv Cape Ince Sinop
Campobasso Veles MACEDONIA (F.Y.R.) Rhodope Kŭrdzali Edirne
Foggia Durrës ALBANIA Ohrid Lake Bitola 2191 Luleburgaz Bosporus
Benevento Tirana (Tiranë) Serrai Komotini Tekirdag Istanbul Üsküdar Ereğli Karabük Kastamonu Samsun
Bari Prespa Lake Edessa Kavalla Sea of Marmara Adapazari 2565
Salerno Potenza Brindisi Korça Thessaloniki Thasos Izmit Bolu Gerede 2068
Taranto Lecce Vlorë Kozáni 2633 2911 Chalkidike Bandirma Bursa Kizilirmak Çorum Turhal
Sapri Gulf of Taranto Pindus Gulf of Thermai Imbros Canakkale 2543 Bilecik Kirikkale 40°N
2248 Capo Santa Maria di Leuca Corfu Lemnos Troy Eskişehir Ankara Yozgat
Corigliano Corfu Janina Trikala Lárisa AEGEAN Balikesir Kütahya 2345
Cosenza Vólos Northern Sporades Ayvalik Anatolia
Catanzaro Levkas Lamia Lesbos Akhisar Afyon Lake Tuz Kayseri
Lipari Islands IONIAN SEA Cephalonia Agrinion Euboea Delphi SEA Manisa TURKEY 3916
Messina 1965 Reggio Chalkis Chíos Izmir Alaşehir 2446 Akşehir Aksaray
Etna 3340 Riposto Patras (Pátrai) Gulf of Corinth Athens (Athínai) Lake Eğridir Konya
Enna Catania 2224 Korinth Piraeus Ándros Sámos Aydin Menderes Denizli Lake Beyşehir Niğde
Zante Pyrgos Tripolis Cyclades Tínos Southern Sporades Ereğli Kozan
Gela Syracuse (Siracusa) Kalamai Náxos Muğla 3488 Karaman Adana
Noto Cape Akrítas Mílos Antalya Taurus Ceyhan
Cape Matapan Cape Maléa Fethiye 3086 Gulf of Antalya Alanya Mersin (İçel) İskenderun
Kíthira Rhodes Finike Anamur Silifke 1795
Valletta MALTA Sea of Crete 1215 Rhodes Cape Anamur Antakya
Kánea Kárpathos Cape Andreas
Melambes 2456 Iráklion CYPRUS Latakia
Crete Cape Arnauti Nicosia Famagusta 1385
Olympus 1951 Larnaca 35°N Tartus
Paphos Limassol Tripoli
MEDITERRANEAN SEA 3087 LEBANON
Beirut Zahlé
Damascus
Sur Qunaitra Golan Heights
Al Bayda Darnah Haifa Dar'a
Khoms Al Marj 882 Al Jabal al Akhdar Hadera Irbid
Misurata Benghazi Al Abyar Tobruk ISRAEL WEST BANK 1247 Zarqa
Beni Walid Gulf of Sirte Qaminis LIBYA 169 Al Adam Al Burdi Sīdi Barrānī Mersa Matruh Nile Delta Baltīm Dumyat Tel-Aviv-Jaffa Jerusalem Ammar
Buerát el Hsun Marmarica Sallūm Rosetta Al Mansura Port Said Suez Canal GAZA Gaza Ar Arish Dead Sea
Sirte 15°E 20°E 25°E Alexandria Damanhūr 30°E Beer Sheba 35°E Karak
AFRICA EGYPT

n o p q r s t u v w x y z

54

37 38

0 100 200 300 miles Average linear scale 0 100 200 300 400 500 Km

34

a b c d e f g h i j k l m

15°W 10°W 5°W 0°

PORTUGAL SPAIN

Sines Ourique Odemira Aljustrel Mértola Aracena Córdoba Jaén Huéscar Lorca Murcia Cartagena Águilas Portimão Sagres Lagos Faro Tavira Huelva Sevilla Marchena Lucena Granada Baza Mulhacén 3478 Almería Jerez de la Frontera Gulf of Cádiz Cádiz Ubrique 1654 Ronda Antequera Málaga Motril Algeciras Gibraltar (U.K.) Str. of Gibraltar Ceuta (Sp.) Tangier Tetuan Asilah

Madeira (Portugal) Funchal Desertas

ATLANTIC OCEAN

Canary Islands (Spain) Tenerife Santa Cruz Pico de Teide 3718 Gomera Lanzarote Arrecife Puerto del Rosario Fuerteventura Las Palmas Guia 1949

Mostaganem Oran Relizane Mohammadia Mascara Melilla (Sp.) Al Hoceima Nador Beni Saf Ghazaouet Sidi-bel-Abbes Chechaouen El Arisch Ksar el Kebir Rif Ouezzane Ouergha Midar Aknoul Oujda Tlemcen Marhoum Ras-al-Ma El Aricha Haut Plateaux Chott ech Chergui Bougtob El Bayadh Méchéria Mehdia Salé Rabat Kenitra Sidi Kacem Fès Taza Guercif Taourirt Ain Benimathar Sefrou Debdou Casablanca Khemisset Meknès Azrou 3190 Moulouya Outat-el-Hadj Tendrara Azemmour El Jadida Berrechid Rommani Khouribga Oued Zem Middle Atlas Missour Ain Sefra 2236 Saharan Atlas Oualidia Sidi Bennour Settat Oum er Rbia Benguerir Beni Mellal Midelt 3741 Ksabi Talsinnt Rich 2670 Bou Arfa Mengoub Figuig Beni Cunif Safi Tensift Demnate Ksar es Souk Essaouira Chichaoua Marrakech 4071 Ighil Goulmina Kenadsa Colomb Bechar Amizmiz High Atlas Tinerhir Erfoud Rissani Taouz Hamada du Guir Abadla Taghit 834 Jebel Igdet 3616 Toubkal 4165 Agdz Tazzarine Igli Agadir Inezgane Taroudannt Tazenakht Zagora Beni Abbès Anti Atlas Tagounite 30°N Tiznit 2359 Tata Bani Sidi Ifni 1250 Djebel Bani Hamada of Dra Tabelbala Erg er Raoui Kahal Tabelbala Great Western Erg Kerzaz 757 Bou Izakarene Tinfouchy 890 Ksabi Timimoun Tantan Foum el Hassane Bou Akba Dj. Ouarkziz Charouine Gourara MOROCCO

C. Yubi Tarfaya Tindouf Ouahila ALGERIA Adrar Tamentit El Mansour Titaf Sbaa Daora Hagunia El Aaiún Al Farcia 437 Iguidi Erg Mcherrah Bordj Flye Sainte Marie Aoulef Sali Reggane Lemsid Smara Aftout El Eglab 680 C. Bojador Bojador Tifariti Chenachane Amasin Ain ben Tili Yetti Erg Chech WESTERN SAHARA Bir Oum Greine 701 Chegga Guelta Zemmur Zemmour 25°N Rhallamane Aioun Abd el Malek Bir el Khzaim Hank .370 Kreb en Naga 305 Sahara Erg el Ahmar 315 Aoukar Tanezrouft Dakhla Aargub Bir Enzarah Zedness 500 Karet El Mreiti Rio de Oro Imilili 639 Hamada Safia G. de Cintra Zouerate Fdérik Aguelt el Melah Agueraktem Hamada el Haricha Taoudenit El Maia Cap Barbas Agailas 250 Hammami 273 361 Makteir 322 296 Bir Gandus Tichla Zug 647 330 Ouarane Er Mreyer El Djouf El Khenachich Bordj-Mokhtar Choûm Nouâdhibou Cape Nouâdhibou Güera Azefal Guelb er Richat Ouadane 321 Ksar Torchane Atar Chinguetti Jafene Douaouir 450 Tessalit 20°N Tidra Akchar MAURITANIA 284 Oujeft .501 Akjoujt Faye Mabrouk Timétrine Tilemsi C. Timiris Nouamrhar Adafer 750 Aguelhok MALI .88 Meraia Azaouad Tamassoumit Tidjikja Dabai Tichitt Tichitt Akreijit 23 Tagant 554 Nouakchott Dahar Oualata Trarza 334 Moudjeria Aouker Anéfis Boutilimit In Alay Oudeïka Anou Mellene Tamchaket Oualata Agamor 318 Aleg Mal Bamba Niger Bourem 15°W 10°W 5°W 0°

n o p q r s t u v w x y z

40 40

35

a b c d e f g h i j k l m

5°E 10°E 15°E 20°E

Algiers Dellys Bejaia Skikda Annaba Bizerte Mateur Tabarka Beja Tunis Kélibia Cap Bon Gulf of Tunis Hammam Lif Tebourbouk Zaghouan

Boufarik Blida Medea Tizi Ouzou Djidjelli Constantine Guelma Setif Souk-Ahras Le Kef Ksar el Boukhari Bordj-Bou-Arrerid Ain Beida Ain Oussera Batna Khenchela Tebessa Kasserine Feriana Sousse Msaken Mahdia El Djem Kairouan Sfax Cape Kaboudia Kerkenna Islands

Bou Saada Biskra Tolga Ouled Djellal Messaad Chott Melrhir Redeyef Gafsa Metlaoui Tozeur Nefta Kebili Chott Djerid Gabès Gulf of Gabès Djerba Zarzis Médenine Ben Guerdane

Djelfa Monts des Ouled Nail Atlas Laghouat El Meghaier Djamaa Guémar El Oued Touggourt Temacine Tilrhemt Berriane Guerara Dzioua Ghardaia Chebka du Mzab Metlili Chaamba Ouargla Hassi Messaoud

Egadi Marsala Sciacca Agrigento Corleone Licata Gela Ragusa Caltanissetta Enna Etna Riposto Catania Syracuse (Siracusa) Noto Pachino Bova Marina ITALY Sicilian Channel Pantelleria (It.)

Linosa Pelagie Islands (Italy) Lampedusa Gozo Valletta MALTA

MEDITERRANEAN SEA

Zuwarah Janzur Tripoli Tajura Khoms Zliten Misurata Al Aziziyah Tarhuna Gharyan Jadu Nalut Zintan Beni Walid Mizda Remada Bordj Bourguiba Bir Zar Sinawan Tripolitania Buerat el Hsun Sirte Gulf of Sirte Benghazi Qaminis Ajdabiya Al Aqaylah Bu Ndjem Sirte Desert Al Qariyat Ash Shwayrif Darj Ghadames Bordj Messouda Hamada el Homra Waddan Hun Suknah Maradah Zella Tlisen

Goléa Hassi Maroket Hassi Inifel Hassi Touareg Great Eastern Erg Ez Zemoul el Akbar Tinrhert Hamada Hassi bel Guebbour Ohanet Bordj Omar Driss In Amenas Tiguentourine Edjeleh Hamadat Tingharat Awaynat Wanin Hamada

ALGERIA Irharharene Idhan Awbari Brak Idri Umm el Abid Semnu Sabha El Fugha LIBYA Harudj el Asued Ubari Jarma Fezzan Zuwaylah Timsah Terbu Tined Bu Haschischa Murzuq Amguid Illizi Tarat Tassili-n-Ajjer Al Awaynat Ghat El Barcat Wau el Kebir Wau en Namus Qatrun Idhan Murzuq Arak Bordj Fort Gardel Djanet Djebel Tahat Telertheba Anai Tajarhi In Ekker Ideles Djebel Serkout In Amguel Hirafok Ahaggar In Afeleh Ramlat Rabyana Tropic of Cancer Atakor Abalessa Tit Tamanrasset Silet Amsel Manguéni Plateau Tumu

Madama In Ebeggi Ténéré du Tafassâsset Djado Plateau Auzu Jabal Nuqay Pic Toussidé Bardai Aozi Yebbi Bou Tibesti Tarso Ahon Zouar Sherda Emi Koussi Gouro CHAD Tassili Oua-n-Hoggar In Azaoua Djado Dao Timni Seguedine Cheffadene Yeggueba Aney Achénoumma Dirkou Bilma Grand Erg de Bilma Zoo Baba Yarda Borkou Faya Kichi Kichi Dibella

Bouressa Tin Zaouatene In Guezzam Assamakka In Azaraf In Abalene Talak Iférouane NIGER Timia Aïr Bagzane Aouderas Agadez Fachi Erg du Ténéré Azaouak

38

n o p q r s t u v w x y z

41

0 100 200 300 miles Average linear scale 0 100 200 300 400 500 Km

35

a b c d e f g h i j k l m

20°E 25°E 30°E 35°E

35°N 30°N 25°N 20°N

GREECE
Kithira
Cape Maléa
Sea of Crete
1215 Rhodes
Rhodes
Kárpathos
2540 Finike
TURKEY
Anamur
Cape Anamur
Silifke
Gulf of Antalya
CYPRUS
Cape Andreas
Cape Arnauti
Nicosia
Famagusta
Olympus 1951
Larnaca
Paphos
Limassol
1795 Antakya
Aleppo
Idlib
Latakia
Asi
1385
Hama
Tartus
Homs
Tripoli
Tall Kalakh
3087 Nabk
Baalbek
Zahle 2659
Beirut
LEBANON
Damascus
Syria
Sur
Qunaitra
Golan Heights
Haifa
Dera
1735
Hadera
Irbid
1247
Mafraq
WEST BANK
Tel-Aviv-Jaffa
Zarqa
Amman
Jerusalem
Gaza
GAZA
Dead Sea
Al Karak
Beer Sheba
ISRAEL
JORDAN
850
1641
Bayir
Al Isawiya
1615
Ma'an
Wadi Araba
Petra
Elat
Aqaba
Al Mudauwara
Al Mughayra

MEDITERRANEAN

SEA

Gulf of Sidra
Benghazi
Al Mekhily
Tobruk
Al Adam
Al Burdi
Sîdi Barrâni
Mersa Matruh
Nile Delta
Baltîm
Dumyat
Port Said
Qaminis
.169
Sallûm
Marmarica
Rosetta
Alexandria
Damanhûr
Cyrenaica
Fuka
Al Mansura
Suez Canal
Ar Arish
Ajdabiyah
Libyan Plateau
Al Alamein
Al Hammam
Tanta
Zagazig
Ismâiliya
Great Bitter Lake
Lower Egypt
Wadi al Arish
Quseima
30
Wadi al Hamim
Fort Qarain
Shubra al Kheima
Cairo
Giza
Suez
Nakhl
Qattara Depression
113
Al Jaghbub
Jaghbub Oasis
Pyramids
Memphis
Helwân
Qara
-123
Ain Sukhna
Sudr
Awjilah
Jalu
Jalu Oasis
Siwa
Siwa Oasis
Al Faiyûm
Beni Suef
Sinai
Nile
Nuweiba
1626
Peninsula
Katherina
2637 Dahab
Gulf of Aqaba
Al Bir
2580
Al Bad
Tabuk
Bahariya Oasis
Bawîti
Beni Mazâr
Ras Ghârib
Gulf of Suez
Wadi al Akhdar
Al Minyâ
Ofira
Ash Sharmah
Gemsa
Ras Muhammad
Dairût
1990
Duba
Hurghada
Farafra
Farafra Oasis
Asyût
Abu Tig
Arabian Desert
LIBYA
Libyan Desert
Akhmîm
Wadi Qena
Port Safaga
Sohâg
Qena
Al Wajh
EGYPT
Al Balyana
Qusair
Qus
Karnak
Thebes
Luxor
Wadi al Hamd
184
Tazirbu
Qasr
Dakhla Oasis
Mût
Al Khârga
Al Khârja Oasis
Isna
RED
Zighan
Upper Nile
Egypt
Idfu
Marsa Alam
Umm Lajj
Ras Abu Madd
Kufrah Oasis
Bâris
Kom Ombo
Al Jawf
Rabyanah
Wadi Garara
1977
Berenice
Ras Banas
Yanbu al Bahr
.625
1st. Cataract
Aswân
Tropic of Cancer
Gilf Kebir Plateau
Lake Toshka
Lake Nasser
Wadi Hadein
Bir al Hasa
Ras Abu Dara
Wadi Allaqi
Abu Simbel
Wadi Ibib
Admin. by Sudan
Halaib
1893
Jweinat
Ras Hadarba
2217
Admin. by Egypt
2nd. Cataract
Wadi Halfa
Nile
Wadi Gabgaba
Wadi Oko
Dungunab
Ras Abu Muhammad
2218 Erba
Kosha
Nubian Desert
Jebel Oda 2260
Delgo
Kerma
Abu Hamed
Port Sudan
Ounianga Kebir
Erdi
Dongola
3rd. Cataract
Umm Mirdi
Amur
Suakin
Sinkat
CHAD
SUDAN
Al Khandaq
Karima
4th. Cataract
Merowe
5th. Cataract
Tokar
Debba
Korti
Berber
Musmar
Haiya
Atbara
Mourdi Depression
Fada
545
Ennedi
Baiyuda
Derudeb
Archei
Adarama
Atbara
Howar
al Milk
White Nile
Shendi
Barka
Haouach
Wadi Seidna
6th. Cataract
.517
738
Mitatib
Omdurman
Khartoum North
ERITREA

n o p q r s t u v w x y z

37

42

This map shows 1/60 of the earth's surface

54

a b c d e f g h i j k l m

Al Hasakah 1463 Sinjar Tall Afar Mosul Arbil Zab Saqqez Banéh Qojur Takestan Qazvin Amol Damavand 5671 Ghaem Shahr Elburz Mts. Damghan Mayamey
Raqqah Suwar Sharqat Lesser Kirkuk Sulaimaniyah Bijar Razan Karaj Zarand Tehran Damavand Semnan Torud
Dayr az Zawr Wadi ath Tharthar Tigris 1097 Sanandaj Saveh Garmsar
Hayl Euphrates Anah Tikrit Tuz Khurmatu Ravansar Kangavar 1775 Hamadan 3572 Qareh Su Daryacheh-ye-Namak Dasht-e-Kavir 35°N
Abu Kamal Al Hadithah Jalaula Bakhtaran 3393 Malayer Qom
Lake Tharthar Diyala Karand 1322 Eslamabad-e-Gharb Arak Mahallat 3365 Kashan
Wadi Hawran Ba'qubah 2656 Ilam Borujerd 3075 Azna Khor Tabas
Ar Ramadi Baghdad Khorramabad Natanz
IRAQ Ar Rutbah Mehran 2041 Karkheh Dez 4328 Oshtoran Meymeh 3899 Ardestan Anarak Dasht-e-Lut
866 Wadi al Ghadaf Bahr al Milh Al Aziziyah Keshvar Daran Nain
Karbala Al Hillah Al Kut Dehloran Najafabad 1590 Isfahan Aliabad
Wadi Tibal An Nukhaib Dezful 4294 Zard 4548 Shahr-e-Kord IRAN Ardakan 3197
An Najaf Ad Diwaniyah Tigris Al Amarah Shush Karun Masjed-e-Soleiman 4298 Qomsheh Izadkhast Shir 4074 Yazd Mehriz Bafq Darband
Al Jalamid Wadi Ar'ar 321 Ar Rifa'i Ahwaz Ramhormoz 3746 Abadeh Abarqu Ravor
1070 Ar'ar As Samawah Euphrates Al Qurnah Ramshir Dinar 4432 3965 Zarand 3143
Ash Shubaiyai An Nasiriyah Hawr al Hammar Behbahan Dehbid Anar Kerman
Abu Qasr As Salman Basra Khorramshahr Bandar-e-Khomeini 3472 Rafsanjan Baghin
Ad Duwaid Al Busaiyah Umm Qasr Abadan Hendijan Nurabad Saadatabad Zagros Mountains
Sakakah Al Faw Bubiyan 3218 1539 Persepolis Daryacheh-ye-Tashk Hoseinabad 30°N
Al Jawf Rafha KUWAIT Jahra Kuwait Bandar-e-Rig Kazerun Shiraz Daryacheh-ye-Bakhtegan Sirjan Laleh Zar 4374 Baft
Ansab 299 Wadi al Batin Ahmadi Borazjan Neiriz
Linah Bushehr Ras Halileh Firuzabad 3188 Fasa Aliabad
908 Ad Dibdibah Al Wafra Mina Saud Khormuj 1960
An Nafud Al Qaisumah Jahrom Dowlatabad
Jubbah Safaniyah ARABIAN Mand Juyom Hajiabad
Al Maiyah Zeydan Kangan Qotbabad 3279
Ha'il 1500 Qaryat al 'Ulya Abu Hadriyah Lar 2804 Kul
Bir Shari Gavbandi Bastak
Tabah Ad Dahna Al Hasa Al Jubayl Bandar-e-Margam Dezhgan Bandar Abbas Minab
SAUDI Al Artawiyah Sarar Ras Tannurah Qeshm Qeshm Str. of Hormuz
Samirah Buraidah Az Zilfi Ash Shumlul Dammam Qeys GULF Bandar-e-Lengeh Musandam Pen.
Nafud Dhahran Manama Ar Ruwais Ash Sha'am 2081
Hulaifah Uqlat al Suqur Unaizah Al Majm'ah Abqaiq BAHRAIN Ras al Khaimah OMAN
Khayber Wadi ar Rimah Ushairah G. of Bahrain QATAR Dibba
Al Hufuf Dukhan Doha Sharjah Dubai Gulf of
Shaqra Khurais Al Udailiyah Karana Jebel Ali Fujairah Oman 25°N
Buwatah Al Hanakiyah Wadi al Jarir Khuff Umm Sa'id
Al Qurain Durma Riyadh Salwa Abu Dhabi Hajar al Gharbi Shinas
Medina Ad Dawadimi Sulaimaniyah Marawah Abu al Abyad Al Batinah Sohar
Afif Harad Al Jafurah Jabal adh Dhanna As Sila Tarif Al Khaznah Al Ain
Muhairiqah UNITED ARAB Habshan Al Khaburah
Hunain Sabkhat Matti Bu Hasa
As Sidr Mahd adh Dhahab Halaban Al Hillah EMIRATES Ibri 3018
Masturah Taraq An Nashash Bahla
Liwa Oasis
Zalim As Sawadah Jabal Tuwayq Ad Dahna
Wadi Aswad
Khulais ARABIA Layla 1012
Madrakah As Suq Umm as Samim
Mecca 1630 Taif Al Uruq al Mutaridah
Arafat 2386 Turabah Ar Rauda Wadi ad Dawasir Wadi Bishah Ar Rub' al Khali
Qunfudhah Al Lith Bani Sar Wadi Tathlith Al Khamasin As Sulaiyil 20°N
Qal'at Bishah
Al Ulaya Tathlith
Al Qunfudhah An Nimas RED SEA Asir
Sahil al Jazir Sauqira Bay
Khay Wadi bin Khawtar OMAN
3133 Abha Khamis Mushait Hima' Wadi Qitbit Sharbithat Ras Sharbithat
Kasar Ad Darb Wadi Shihan Dhofar
Zahran Sanaw Thamarit Kuria Muria Islands
Mersa Teklay Najran Thamud Jabal al Qamar Jabal al Qara
Farasan Is. Jizan Sa'dah Salalah Raisut Mirbat Ras Mirbat
YEMEN ARABIAN SEA
Midi Huth Al Ghaydah
Dahlak Islands Al Hazm Wadi al Jawf Wadi al Jiz Qamar Bay
3360 Haylan Haynan Sayun Wadi Masilah Ras Fartak
40°E 45°E 50°E 55°E

n o p q r s t u v w x y z

60

43

0 100 200 300 miles
Average linear scale
0 100 200 300 400 500 Km

36

a b c d e f g h i j k l m

15°W 10°W 5°W

MAURITANIA
Trarza
Aoukâr
Montagnes de l'Affolé
Moudjéria
Boûmdeïd
Tamchaket
Boutilimit
Mederdra
Aleg
Mâl
Oualâta
Rosso
Senegal
Bogué
Dagana
Kiffa
Ayoûn el Atroûs
Néma
Kaédi
Mbout
Timbedgha
Amourj
Kankossa
Kobenni
Bassikounou
Hamoud
Sélibabi
Maghama
Matam
St. Louis
Louga

SENEGAL
Linguère
Fourdou
15°N
Dakar
Cape Verde
Thiès
Mbaké
Diourbel
Ferlo
Mbour
Kaolack
Maleme-Hodar
Koumpentoum
Tambacounda
Bamba
Kidira
Banjul
Sere Kunda
Brikama
Karang
GAMBIA
Georgetown
Gambia
Basse Santa Su
Dialakoto
Niokolo Koba
Saraya
Kédougou
Bignona
Casamance
Kolda
Ziguinchor
Farim

GUINEA-BISSAU
Mansôa
Bafatá
Bissau
Corubal
Bissagos Islands
Catió
Koundara

GUINEA
Gaoual
Fouta Djalon
Tougué
Labé
Pita
Télimélé
Boké
Kogon
Fatala
Fria
Konkouré
Dalaba
Mamou
Kindia
Kavendou
Boffa
10°N
Conakry
Forécariah
Dinguiraye
Bafing
Siguiri
Dabola
Kouroussa
Kankan
Sanouyah
Faranah
Kabala
Bohodoyou
Kissidougou
Kérouané
Guéckédou
Macenta
Beyla
Nzérékoré
Tibé
Loma Mts.

SIERRA LEONE
Little Scarcies
Rokel
Kambia
Makeni
Koidu
Freetown
Pendembu
Bo
Kenema
Sewa
Moa
Sherbro Island
Pujehun
Mano River

LIBERIA
Bomi Hills
St. Paul
Lofa
Zorzor
Gbarnga
Ganta
Bong
Mani
Monrovia
Buchanan
Tapeta
Tchien
Cess
Juarzon
Greenville
Grain Coast
Harper
Plibo
Cavally

MALI
Oualâta
Iriqui
Sahel
Nioro du Sahel
Balle
Nara
Kayes
Koniakari
Diamou
Bafoulabé
Dialafara
Satadougou
Kita
Diéma
Didiéni
Toukoto
Kolokani
Banamba
Bamako
Kati
Koulikoro
Fana
Baguinéda
Ségou
Niger
Bani
San
Bla
Mpessoba
Koutiala
Ouéléssébougou
Bougouni
Garalo
Sikasso
Manankoro
Sokolo
Nampala
Macina
Mopti
Djenné
Tombouctou
Lake Faguibine
Goundam
Niafounké
Lac Débo
Bandiagara
Douentza
Hombori
Bamba
In Alay
Oudeika
Tuareg

BURKINA FASO
Djibo
Ouahigouya
Yako
Nouna
Dédougou
Koudougou
Ouagadougou
Koundougou
Houndé
Bobo-Dioulasso
Léo
Banfora
Gaoua
Tumu
Navrongo
Bolgatanga
Lawra
Wa
Yala
Po
Toéssé

CÔTE D'IVOIRE
Samatiguila
Odienné
Ouangolodougou
Pogo
Boundiali
Ferkéssédougou
Korhogo
Kong
Bako
Morondo
Kanawolo
Koro
Kani
Touba
Katiola
Séguéla
Biankouma
Man
Kossou Reservoir
Bouaké
Danané
Bouaflé
Yamoussoukro
Duékoué
Daloa
Guiglo
Toulépleu
Dimbokro
Abengourou
Toumodi
Akoupé
Agboville
Gagnoa
Lakota
Soubré
Tai
Sassandra
Niénokoué
Bandama
Dabou
Abidjan
Grand-Lahou
Grand-Bassam
Grabo
Tabou
San Pédro
Ivory Coast
Komoé
Bouna
Bondoukou
Goumeré
Agnibilékrou
Koutouba

GHANA
Black Volta
White Volta
Sawla
Bole
Maluwe
Bamboi
Kintampo
Techiman
Berekum
Sunyani
Kumasi
Konongo
Nkawkaw
Obuasi
Awaso
Tano
Dunkwa
Prestea
Tarkwa
Sekondi-Takoradi
Gold Coast
5°N

ATLANTIC OCEAN

0°
Equator

n o p q r s t u v w x y z

This map shows 1/60 of the earth's surface

37

a b c d e f g h i j k l m

5°E 10°E 15°E

.500
Anou Mellene
Tegguidda-n-Tessoum
Aouderas
Akrereb
Dibella
Vallée de L'Azaouak
Agadez
Agadem
Azaouak
.500
Ouyu Bezze Denga
Homodji
In Talak
Tillia
Mazalet
Termit N
Massif de Termit .710
Toro Doum
Ménaka
Tchin-Tabaradene
Ansongo
Tilemsès
Aderbissinat
Ngourti
.280
NIGER
Tanout
Moul
15°N
Abala
Tahoua
Task
Idaye
.255
Nokou
Illéla
Rig-Rig
Niger
.550
Tillabéri
Kanem
Ouallam
Filingué
Madaoua
.403
Nguigmi
Téra
Am Raya
.302
Matankari
Burni-Nkonni
Tessaoua
Zinder
Gouré
Goudoumaria
Mao
Mondo
Moussoro
Dogondoutchi
Illela
Bosso
Ngouri
Niamey
Maradi
Lake Chad
Torodi
Say
Dosso
Mainé-Soroa
Komadugu
Sokoto
Katsina
Dungas
Baga
Massakori
Talata Mafara
Gashua
Geidam
Kantchari
Sokoto
Argungu
Kaura Namoda
Nguru
Monguno
Massaguet
Ngoura
Birnin-Kebbi
Djermaya
Karmé
CHAD
Dapchi
Damakar
Koulou
Hadejia
Diapaga
Gaya
Anka
Gusau
Kano
Hadejia
Dikwa
296
N'Djamena
.442
Fort Foureau
Kamba
Faskari
Wudil
Damaturu
Maiduguri
Potiskum
Chari
Funtua
Massenya
Bama
Zuru
Foggo
Kari
Buni
Niger
.1141
Kandi
Zaria
Mokolo
Erguig
Yelwa
Gongola
Guélengdeng
Tanguiéta
.550
Birnin Gwari
Biu
Maroua
Sansanné-Mango
Natitingou
Béroubouay
Zalanga
Kainji Reservoir
Kontagora
Kaduna
.1594 Goura
Gombe
Bousso
Bongor
Boukombé
Bembéréké
Mubi
Bauchi
Bara
Wuyo
Tegina
Moutouroua
Ham
BENIN
Wawa
NIGERIA
Jos
10°N
Djougou
Lama-Kara
Yashikera
Kafanchan
Minna
1625 Kagora
Numan
Parakou
1518
Pankshin
Garoua
Bassari
Kaiama
Kaduna
Pala
Kelo
Lai
Abuja
Yola
Logone
Ouémé
Jebba
Bida
Zamko
772
Sokodé
Bassila
Akwanga
Wamba
Jalingo
Koumra
TOGO
Igbetti
Guidjiba
Moundou
Doba
Kilibo
Agoaré
Niger
Baro
Poli
Blitta
Ilorin
Lafia
Tchollire
Ogbomosho
Kpessi
Savé
Benue
Ibi
Touboro
Iseyin
Oshogbo
Lokoja
Mbé
Baibokoum
Gore
.845
Savalou
Oyo
Ede
Ilesha
Wukari
Beli
Ado Ekiti
Iwo
Ife
Kabba
Makurdi
Atakpamé
Ikerre
Okene
Ayangba
Adamaoua Highlands
Béka
Ngaoundéré
Béboura
Ogun
Ibadan
Akure
Abomey
Owo
Takum
Abeokuta
Oturkpo
Nuatja
Ondo
Doualayel
Bélel
Bocaranga
Ijebu Ode
Kpalimé
Banyo
Ilaro
Porto Novo
Lekki
Ogoja
Bossangoa
Tsévié
Lagos
Enugu
Nkambe
Meiganga
Bozoum
1890
Tibati
Ouidah
Cotonou
Benin City
Lomé
Ikom
CENTRAL
Onitsha
3008
Bamenda 2335
Garoua Boulaï
Slave Coast
Bouar
Sapele
Bombale
Mamfe
Foumban
Baboua
Afikpo
2740
Cross
Bafoussam
Bétaré-Oya
Niger
AFRICAN
Warri
Yoko
Dschang
Ughelli
CAMEROON
Goyoum
Aba
Bafang
Carnot
5°N
Port Harcourt
Nkongsamba
Ndjolé
2060
Bossembélé
Bafia
REPUBLIC
Calabar
Sanaga
Bertoua
Yabassi
Nanga Eboko
Batouri
Brass
Bonny
Buea
Berberati
Mt. Cameroon 4100
Kenzou
Limbe
Douala
Bania
Yaoundé
Abong Mbang
Malabo
Edéa
.2890
Boumba
Nola
Eséka
Bioko Island
Luba
Nyong
Mbalmayo
2662
Yokadouma
Gulf of Guinea
Dja
Ebolowa
Sangmélima
Bayanga
Lokomo
Ambam
EQUATORIAL GUINEA
Ebebiyin
Bitam
Ntam
Moloundou
Bomassa
Souanké
Bata
Niefang
937
Principe
Oyem .1200 Tembo
Ouesso
Mbini .1200
Mbini
Sembé
Evinayong
Likouala-aux-Herbes
SÃO TOMÉ AND PRINCIPE
Nkolabona
Liouesso
Sangha
Cocobeach
Mékambo
Mitzic
Makokou
Pikounda
Libreville
Kougouleu
.980
Lalara
CONGO
São Tomé
São Tomé
2024
Gabon
Ndjolé
Booué
Likouala
Makoua
0°
Kellé
GABON
.500
Owando
Ogooué
Kouyou
Port-Gentil
Lambaréné
Okondja
Ewo
.875
Lastoursville
Ogooué
Lake Onangue
Boundji
Congo
Koulamoutou
Mossaka
Annobón (Equa.Guinea)
Okoyo
Mimongo
Moanda
Omboué
Franceville
Mouila
5°E 10°E 15°E

n o p q r s t u v w x y z

42

44

0 100 200 300 miles

Average linear scale

0 100 200 300 400 500 Km

38

a b c d e f g h i j k l m

20°E 25°E 30°E

Fada
Ennedi
Bodélé
Archei
Haouach
Wadi Howar
Wadi al Milk
Nile
Shendi
Adara
6th Cataract
Wadi Seidna
Koro Toro
Ouagat
Karma
Kapka .1220
Tinê
Umm Saggat
Sindi
Magrur
.517
Omdurman
Khartoum North
Khartoum
Umm Inderaba
Haraza .1127
15°N
Salal
Mâba
Haddad
Hamrat al Shaikh
Kordofan
Blue Nile
White Nile
El Hasaheisa
Sodiri
Umm Saiyala
El Gezira
Wad Medani
Sahel
Biltine
Gurgei .2351
al Dueim
Rime
Abéché
Kebkabiya
El Fasher
Umm Keddada
al Ouaday
al Junayna
Adré
Wadi al Kû
Bara
Sennar
Oum Hadjer
Ati
Dam Gamad
CHAD
Zalingei .3071
Jebel Marra
Wad Banda
El Obeid
Tendelti
Kosti
Singa
Rabak
Batha
Menawashei
Kirim .640
Er Rahad
Umm Ruwaba
Azum
El Jebelein
Guedi .1506
Mangalmé
Mongo
Kass
al Nahûd
Abu Zabad
Abu Habl
Blue Nile
Goz Beïda
Gandi
Ghubeish
Dilling
1613
Bitkine
Azoum
Nyala
al Udaiya
Nuba Mountains
Turum .1122
Renk
Ed Damazin
Lake Roseires
Abou Deïa
'Idd al Ghanam
al Da'ain
al Fûla
.1325
Rahad al Berdi
Babanusa
al Muglad
.842
Kadugli
Mêlfi
Zakouma
Am Timan
Ibra
Talodi
White Nile
SUDAN
Kurmuk
Erguig
Kendégué
Birao
Dango 790.
Bahr al Arab
.1093
Tungaru
Paloich
10°N
Haraze
Oulou
Yata
Sudan
Sumaih
Salamat
Ouandja
Tiroungoulou
Bambesi .2185
Chari
Kéita ou Doka
Bentiu
Tonga
Malakal
Toussoro 1330
Sobat
Sarh
Aouk
Télé
Sudd
Koumra
Ouham
Massif des Bongos
Aweil
Nasir
Maro
Bangoran
Bora
Raga
Bahr al Ghazel
Dembi
Ndélé
Gambela
.850
Ouadda
1050
Gribingui
Kotto
Ndji
Sopo
Akobo
Kabo
Bamingui
Wau
Pongo
Toni
Batangafo
Ouandago
Haute Kotto
Kongor
Kaga Bandoro
Mbrès
Boungo
Busseri
Rumbek
Pibor Post
White Nile
CENTRAL AFRICAN REPUBLIC
Bossangoa
Bouca
Dékoa
Yalinga
Bria
Vovodo
Maridi
Mvolo
Bor
Kenamuke Swamp
Sibut
Bambari
Ouara
Tambura
Angeleri .838
Bogangolo
Tomi
Ouaka
Kotto
Chinko
Mbari
Sue
Ibba
Obo
Li Yuba
Mundri
Mongalla
Medi
5°N
Damara
Bossembélé
Kouango
Alindao
Dembia
Rafaï
Zémio
M'bomou
Kongbo
Gambo
Bangassou
Doruma
Yambio
Maridi
Juba
Ngangala
.1940
Kapoeta
Bangui
Oubangui
Mobaye
Garmabe
Bimbo
Zongo
Gbadolite
Bosobolo
Matundu
1067
1065.
Torit
Loki
Mbaïki
Mobayi-Mbongo
Monga
Bili
Ango
Ese
Yei
Lalyo
Kinyeti .3187
Zinga
Lua Dekere
Yakoma
Uele
Api
Bambili
Niangara
Aba
Kajo-Kaji
Lobaye
Boyabo
Libenge
Abumonbazi
Bondo
Api
Faradje
Laropi
Nimule
Kak
Ebola
Angu
Dungu
Busingo
Gemena
Dua
Likati
Bambesa
Baranga
Rungu
Kaabong .2381
Enyélé
Bodala
Titule
Poko
Watsa
Arua .1310
Atiak
Kitgum
Loyoro
CONGO
Aketi
Dulia
Buta
Isiro
Aru
Albert Nile
Pajule
Kotid
Kungu
Tele
Gombari
Gulu
Budjala
Mongola
Modjamboli
Ibembo
Anaka
Moro
Dongou
Bumba
Zambeke
Medje
Nepoko
Lisala
Izimbiri
Kole
Wamba
2446
Mahagi
Lira
Am
Mobeka
DEMOCRATIC
Ituri
Fataki
Lake Albert
UGANDA
Makanza
Busu-Djanoa
Masindi
Soroti
Impfondo
Giri
Congo
Lopori
Bomili
Lake Kyoga
Bongandanga
Banalia
Nia Nia
Komanda
Bunia
Hoima
Mbale
Likouala-aux-Herbes
Oubangui
Lulonga
Basoko
Aruwimi
Lindi
Mambasa
Nakasongola
Basankusu
Yahuma
Yambuya
Kamuli
REPUBLIC
Waka
Isangi
Yangambi
Bengamisa
Batama
Bafwabalinga
Hoyo .1450
Ntoroko
Kaliro
Iganga
Tororo
Lulonga
Bolomba
Lingomo
Yekana
Fort Portal
Kyanjojo
Kayunga
Befale
Maringa
Djolu
Kisangani
Madula
Boyoma Falls
Beni
Margherita Peak .5109
Mubende
Kampala
Jinja
Samba
Befori
Yatolema
Maiko
Opienge
Butembo
Kasese
Kaka
0°
Mbandaka
Yali
OF
Entebbe
Kisum
Ruki
Busira
Ingende
Boende
Watsi
Wema
Ekoli
Ubundu
Pene-Tungu
Lake Edward
.2197
Masaka
Kalamba
Lubero
Bushenyi
Sese Islands
Lomela
Tshuapa
Opala
.956
Lubutu
2341
Bikoro
Momboyo
Watsi-Kengo
Kirundu
Mitumba Mountains
Ishasha River
Mbarara
Kis
Lake Tumba
Busanga
Congo
Lomani
Lake
Kilko
CONGO
Ikela
Likoto
Rutshuru
Kabale
Kikagati
Victoria
Lowa
Walikale
Masisi
Karisimbi .4507
Ruhengeri
Kyaka
Bukoba
Tarime
Yandja
Bolia
Yalifafu
Yolombo
Goma
Gisenyi
Musoma
Inongo
Kiri
Monkoto
Punia
Kabunga
Lake Kivu
Kigali
Kayonza
TANZANIA
Lake Mai-Ndombe
RWANDA
Ukerewe Island
Ntadembele
Kavumu
Nansio
Banagi
20°E 25°E 30°E

41

n o p q r s t u v w x y z

44

This map shows 1/60 of the earth's surface

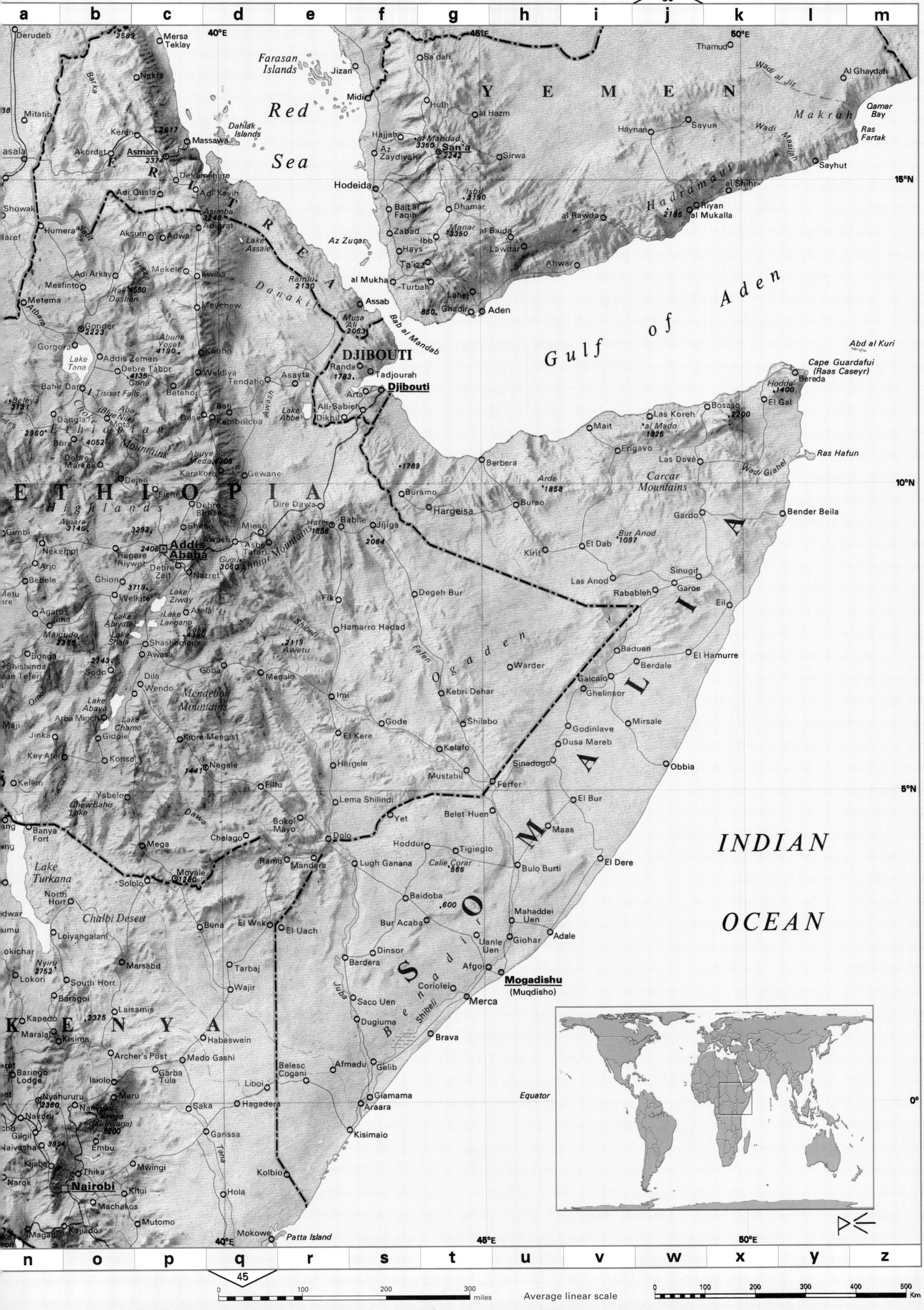

39

a b c d e f g h i j k l m

40°E 45°E 50°E

15°N 10°N 5°N 0°

Derudeb
Mersa Teklay
Nakfa
Barka
Mitatib
Keren
2617
Massawa
Akordat
Asmara 2374
ERITREA
Dekemhare
Adi Quala
Adi Keyih
Showak
Humera
Setit
Aksum
Adwa
Adigrat
Asimba 3248
Lake Assale
Mekele
Kwiha
Adi Arkay
Tekeze
Mesfinto
Ras Dashen 4550
Metema
Atbara
Gonder 2223
Gorgora
Lake Tana
Addis Zemen
Debre Tabor
4135 Guna
Abune Yosef 4190
Kobbo
Weldiya
Bahir Dar
Tisisat Falls
Beleya 3131
Abay (Blue Nile)
Dangla
Chok
Mota
Bure
2960
4052
Debre Markos
Ethiopian Mountains
Betehor
Bati
Dese
Kembolcha
Abuye Meda 4205
Karakore
Gewane
Dejen
Fiche
ETHIOPIA
Highlands
Debre Birhan
Amara 3146
3292
Gimbi
Sheno
Nekemt
Addis Ababa
2408
Hagere Hiywot
Debre Zeit
Nazret
Awash
Mieso
Asbe Tafari
Gugu 3060
Arjo
Bedele
Ghion
3719
Welkite
Lake Ziway
Asela
Agaro
Jima
Lake Abiyata
Lake Langano
4180
Lake Shala
Maigudo 2386
Shashemene
Awasa
Bonga
Shishinda
2743
Sodo
Dila
Wendo
Omo
Lake Abaya
Arba Minch
Lake Chamo
Gidole
Jinka
Key Afer
Konso
Mendebo Mountains
Kibre Mengist
1441
Negele
Kelem
Yabelo
Chew Bahir Lake
Dawa
Banya Fort
Mega
Chelago
Lake Turkana
Sololo
Moyale 1280
North Horr
Chalbi Desert
Loiyangalani
Nyiru 2752
Lokori
Marsabit
South Horr
Baragoi
Laisamis
2375
Kapedo
KENYA
Maralal
Kisima
Archer's Post
Baringo Lodge
Garba Tula
Isiolo
Nyahururu 2360
Meru
Nakuru
Nanyuki
Kenya (Kirinyaga) 5200
Gilgil
3994
Naivasha
Embu
Kijabe
Thika
Mwingi
Narok
Nairobi
Kitui
Machakos
Mutomo
Magadi
Kajiado

Farasan Islands
Jizan
Midi
Red Sea
Dahlak Islands
Sa'dah
Huth
al Hazm
YEMEN
Hajjah
al Mahdad 3360
San'a 2242
Az Zaydiyah
Sirwa
Hodeida
Ishil 3190
Bait al Faqih
Dhamar
Zabad
Manar 3350
al Baida
Ibb
Hays
Ta'izz
Lawdar
Az Zuqar
Ramlu 2130
Danakil
al Mukha
Turbah
Lahej
Assab
850
Ghadir
Aden
Musa Ali 2063
Bab al Mandab
DJIBOUTI
Randa
1783
Tadjourah
Djibouti
Tendaho
Asayta
Awash
Arta
Lake Abbe
Ali-Sabieh
Dikhil
Ahwar
al Rawda
Hadramaut
2185
Riyan
al Mukalla
al Shihr
Sayhut
Thamud
Wadi al Jiz
Al Ghaydah
Qamar Bay
Makrah
Haynan
Sayun
Wadi Masilah
Ras Fartak

Gulf of Aden
Abd al Kuri
Cape Guardafui (Raas Caseyr)
Hodda 1400
Bereda
El Gal
Bosaso
2200
Las Koreh
Mait
al Mado 1826
Erigavo
Las Davo
Ras Hafun
Wadi Ghahel
Berbera
1789
Arde 1858
Carcar Mountains
Buramo
Hargeisa
Burao
Bender Beila
Dire Dawa
Harer 1856
Babile
Jijiga
2064
Ahmar Mountains
Gardo
Bur Anod 1097
El Dab
Kirit
Sinugif
Las Anod
Garoe
Rabableh
Eil
Degeh Bur
Fik
Hamarro Hadad
Shibeli
2119 Awetu
Fafen
Ogaden
Baduen
El Hamurre
Warder
Berdale
Goba
Megalo
Galcaio
Ghelinsor
Imi
Kebri Dehar
SOMALIA
Gode
Shilabo
Mirsale
Godinlave
El Kere
Dusa Mareb
Kelafo
Hargele
Sinadogo
Obbia
Mustahil
Ferfer
Filtu
Lema Shilindi
El Bur
Yet
Belet Huen
Bokol Mayo
Maas
Dolo
Hoddur
Tigieglo
Ramu
Mandera
Lugh Ganana
Calie Corar 586
Bulo Burti
El Dere
Baidoba
600
Mahaddei Uen
Bur Acaba
Buna
El Wak
El Uach
Tarbaj
Dinsor
Uanle Uen
Giohar
Adale
Bardera
Benadir
Afgoi
Wajir
Mogadishu (Muqdisho)
Coriolei
Juba
Saco Uen
Merca
Shibeli
Dugiuma
Habaswein
Brava
Mado Gashi
Afmadu
Gelib
Belesc Cogani
Liboi
Giamama
Araara
Saka
Hagadera
Garissa
Tana
Kisimaio
Kolbio
Hola
Mokowe
Patta Island

INDIAN OCEAN

Equator

n o p q r s t u v w x y z

45

0 100 200 300 miles

Average linear scale

0 100 200 300 400 500 Km

41 42

a b c d e f g h i j k l m

SÃO TOMÉ AND PRINCIPE
São Tome
Libreville
GABON
CONGO
DEMOCRATIC REPUBLIC OF CONGO
CABINDA (Angola)
Brazzaville
Kinshasa
Luanda
ANGOLA
NAMIBIA
BOTSWANA
ATLANTIC OCEAN

Cocobeach, Mitzic, Mékambo, Makokou, Kougouleu, Lalara, Ndjolé, Booué, Port-Gentil, Lambaréné, Lake Onangué, Ogooué, Okondja, Bonda (Lastoursville), Koulamoutou, Omboué, Mimongo, Moanda, Franceville, Mouila, Boumango, Ndendé, Mayoko, Bambama, Tchibanga, Mossendjo, Kibangou, Mayumba, Mapati, Sibiti, Sounda, Kouilou, Loubomo, Madingou, Bas-Kouilou, Pointe-Noire, Kinkala, Boko, Luozi, Tshela, Lândana, Cabinda, Lukula, Seke Banza, Isangila Falls, Boma, Matadi, Muanda, Soyo, Kimpese, Mbanza-Ngungu, Nsdinga, Kimvula, Madimba, Inkisi-Kisantu, Mayamba

Pikounda, Sangha, Likouala, Kéllé, Makoua, Kouyou, Owando, Ewo, Boundji, Okoyo, Mossaka, Gamboma, Bouanga, Nsah, Ngo, Inoni, Congo, Masia-Mbio, Cuango, Fatunda, Kenge, Masi-Manimba, Kikwit, Kwilu, Bulungu, Popokabaka, Feshi, Gungu, Kasongo-Lunda, Bandundu, Bagata, Kasai, Lukenie, Oshwe, Dekese, Kapia, Idiofa, Mpata, Banda, Kilembe, Ilebo, Mweka, Kakenge, Bena Dibele, Bena-Tshadi, Sankuru, Lusa, Luebo, Demba, Dimbelenge, Kananga, Mbuji-, Kazumba, Tshikapa, Kamiji, Gandajika, Mwen, Luiza, Kaniam

Lulonga, Waka, Bolomba, Mbandaka, Ruki, Busira, Ingende, Befale, Lingomo, Yekana, Djolu, Befori, Samba, Maringa, Yali, Boende, Watsi, Lomela, Wema, Tshuapa, Momboyo, Irebu, Kalamba, Bikoro, Lake Tumba, Lukolela, Watsi Kengo, Busanga, Ikela, Yalifafu, Yandja, Bolia, Kiri, Inongo, Lake Mai-Ndombe, Monkoto, Yolombo, Ntadembele, Nioki, Kutu, Lomela, Lodja, Oubangui

Maquela do Zombo, M'Pala, M'Banza-Congo, Quimbele, Tombôco, Damba, Bembe, Sanza Pombo, N'Zeto, Mussera, Uige, Negage, Ambriz, Quitexe, Nambuangongo, Camabatela, Quibaxe, Caxito, Samba Caju, Catete, Lucala, Kalandula, N'Dalatando, Malange, Muxima, Dondo, Cuanza, Calulo, Porto Amboim, Quibala, Gabela, Mussende, Sumbe, Waco-Kungo, Lobito, Benguela, Alto Hama, Bailundo, Balombo, Caala, Catengue, Ganda, Lucira, Caconda, Negola, Cacula, Gambos, Lubango, Chibia, Chianje, Mulundo, Quiteve, Oncócua, Roçadas, Cunene, Naulila, Ruacana, N'Giva, Oshakati, Ondangwa, Obombo, Opuwa, Purros, Cape Frio, Etosha Pan, Namutoni, Keibeb

Kahemba, Warnba, Luachimo, Forte Carumbo, Verissimo Sarmento, Lucapa, Luremo, Caungula, Camaxilo, Cuango, Xa-Muteba, Xinge, Chicapa, Saurimo, Nova Gaia, Quitapa, Cacolo, Mona-Quimbundo, Muriege, Kasai, Sandoa, Tshimbala, Kapanga, Lubilanji, Muconda, Luau, Dilolo, Dala, Andulo, Buçaco, Cassai, Luena, Moxico, Chicala, Camacupa, Cuemba, Kuito, Cachingues, Chitembo, Mumbuè, Lucusse, Cassamba, Luzi, Luvuei, Lumbala, Zambezi, Lutembo, Sessa, Lumbala N'Guimbo, Lungue-Bungu, Lukulu, Kabom, Cazombo, Ikeleng, Mwin, Menongue, Longa, Capelongo, Cuchi, Cassinga, Cuito Cuanavale, Chiume, Mongu, Cuvelai, Caiundo, Cubango, Mavinga, Savate, Rito, Rivungo, Senang, Luengué, Cuito, Chibaranda, Cuando, Luiana, Cuangar, Mucusso, Rundu, Shakamku, Shakawe, Kongola, Caprivi Strip, Sepopa, Numkaub, Sesheke

Equator, 0°, 10°E, 15°E, 20°E, 5°S, 10°S, 15°S
980, 500, 875, 820, 975, 834, 798, 1150, 1265, 1190, 1160, 1096, 1784, 2024

n o p q r s t u v w x y z

48

43

a b c d e f g h i j k l m

UGANDA
KENYA
SOMALIA
RWANDA
BURUNDI
TANZANIA
MALAWI
ZAMBIA
MOZAMBIQUE
ZIMBABWE

INDIAN OCEAN

Lake Victoria
Lake Tanganyika
Lake Malawi (Lake Nyasa)
Lake Rukwa
Lake Mweru
Lake Bangweulu
Lake Kivu
Lake Edward
Lake George
Lake Natron
Lake Chilwa
Kariba Reservoir
Cabora Bassa Reservoir
Sese Islands
Ukerewe Islands
Pemba Island
Zanzibar Island
Mafia Island
Patta Island
Cape Delgado

Mitumba Mountains
Marungu Mountains
Muchinga Mountains
Livingstone Mountains
Serengeti Plain
Rift Valley
Masai Steppe
Nyiri Desert
Tsavo National Park

Kisangani, Madula, Opienge, Pene-Tungu, Kirundu, Lubutu, Punia, Walikale, Kabunga, Masisi, Goma, Bukavu, Kalima, Kindu, Pangi, Kasambule, Kampene, Kingombe, Kipaka, Kasongo, Samba, Kitutu, Mwenga, Kalole, Kwanga, Tshofa, Lubao, Kabalo, Nyunzu, Niemba, Katompi, Kaloko, Manono, Kiambi, Sange, Mulongo, Pidi, Kikondja, Malemba Nkulu, Mitwaba, Kabondo Dianda, Mukana, Luena, Busanga, Bunkeya, Luambo, Kambove, Likasi, Minga, Lubumbashi, Kipushi, Solwezi, Mokambo, Chililabombwe, Chingola, Mufulira, Kitwe, Ndola, Luanshya, Kasempa

Beni, Margherita Peak, Fort Portal, Kyanjojo, Mubende, Kasese, Butembo, Kasindi, Lubero, Bushenyi, Mbarara, Masaka, Ishasha River, Kikagati, Rutshuru, Karisimbi, Kabale, Ruhengeri, Gisenyi, Kigali, Kibuye, Gitarama, Kayonza, Cibitoke, Kayanza, Bujumbura, Kibondo, Kasulu, Kigoma, Uvinza, Kalemie, Moba, Kapona, Pweto, Chiengi, Nchelenge, Kasembe, Kawambwa, Mporokoso, Kasenga, Mununga, Mansa, Samfya, Chembe, Mukuku, Kapalala, Serenje, Chifwefwe, Kanona

Kampala, Entebbe, Kayunga, Kaliro, Iganga, Jinja, Tororo, Eldoret, Kakamega, Kisumu, Kisii, Kericho, Kilkoris, Narok, Bukoba, Kyaka, Musoma, Tarime, Banagi, Nansio, Mwanza, Geita, Ngudu, Nyakanazi, Shinyanga, Kahama, Nzega, Ibologero, Tabora, Sikonge, Itigi, Manyoni, Mpanda, Kitunda, Rungwa, Namanyere, Sumbawanga, Kipembawe, Kasanga, Sumbu, Mpulungu, Mbala, Nakonde, Tunduma, Isoka, Kasama, Kapatu, Luwingu, Mbesuma, Chinsali, Chama, Chisoso, Mpika, Chilonga, Chikwa, Lundazi, Chibembe, Makongolosi, Chunya, Mbeya, Chimala, Uyole, Rungwe, Makambako, Njombe, Itungi, Karonga, Chilumba, Livingstonia, Rumphi, Mzuzu, Mzimba, Nkhata Bay, Jenda, Dwangwa, Nkhotakota, Kasungu, Chipata, Mchinji, Lilongwe, Salima, Dedza, Mangochi, Balaka, Zomba, Limbe, Blantyre, Mulanje

Loruk, Baringo Lodge, Kapsabet, Nyahururu, Nakuru, Gilgil, Naivasha, Kijabe, Thika, Nairobi, Kajiado, Magadi, Machakos, Kitui, Mutomo, Namanga, Oloitokitok, Kilimanjaro, Meru, Moshi, Arusha, Makuyuni, Oldeani, Mbulu, Ndareda, Babati, Katesh, Singida, Kondoa Irangi, Dodoma, Kongwa, Mpwapwa, Gairo, Mvomero, Kilosa, Mikumi, Morogoro, Mbuyuni, Iringa, Ifakara, Sao Hill, Mahenge, Luhombero, Lukumburu, Gumbiro, Songea, Tunduru, Chamba

Archer's Post, Isiolo, Meru, Nanyuki, Kenya (Kirinyaga), Embu, Mwingi, Garba Tula, Mado Gashi, Saka, Garissa, Hagadera, Liboi, Hola, Mokowe, Tsavo, Mtito Andei, Manyami, Malindi, Kilifi, Mwatate, Mombasa, Kwale, Same, Mkomazi, Lushoto, Tanga, Korogwe, Segera, Handeni, Msata, Chalinze, Bagamoyo, Zanzibar, Dar-es-Salaam, Kisarawe, Kibiti, Kilindoni, Mohoro, Nangurukuru, Lindi, Mingoyo, Mtwara, Nachingwea, Masasi, Nangomba, Newala, Masuguru, Mueda, Diaca, Mocimboa da Praia, Afmadu, Belesc Cogani, Kisimaio, Kolbio, Equator

Nantulo, Macomia, Montepuez, Metoro, Pemba, Marrupa, Litunde, Nungo, Lichinga, Malanga, Maniamba, Massangulo, Maúa, Namapa, Mandimba, Cuamba, Nacaroa, Nacala, Ribáuè, Mutuali, Namialo, Nampula, Monapo, Lumbo, Moçambique, Gurué, Liupo, Nametil, Alto Molócuè, Erregó, Angoche, Mocuba, Mucubela, Moma, Pebane, Namacurra, Quelimane, Chinde, Mopeia, Caia, Inhaminga, Gorongosa, Sena, Nsanje

Kabwe, Kapiri Mposhi, Petauke, Katete, Nyimba, Kachalola, Chitunde, Bene, Fingoe, Landless Corner, Lubungu, Mumbwa, Rufunsa, Lusaka, Zumbo, Chiuta, Songo, Namwala, Mazabuka, Tete, Mkumbura, Changara, Tambara, Kariba, Choma, Kalomo, Karoi, Mhangura, Mount Darwin, Mvurwi, Nyamapanda, Bindura, Guro, Mutoko, Banket, Harare, Binga, Livingstone, Victoria Falls, Hwange, Gokwe, Chegutu, Kadoma, Inyanga, Catandica, Rusape, Dete, Gwai River

Lindi, Maiko, Lowa, Ulindi, Congo, Lukuga, Luvua, Malagarasi, Ugalla, Manonga, Igombe, Rungwa, Kisigu, Chambeshi, Luangwa, Luapula, Lunga, Kafue, Zambezi, Sanyati, Angwa, Luangua, Rufiji, Ruvuma, Lugenda, Messalo, Lúrio, Mara, Tana, Galana, Tsavo

5109, 2197, 2341, 4507, 3044, 956, 1040, 1047, 1019, 2073, 2670, 2373, 2460, 1052, 1139, 2418, 2775, 2277, 3100, 1662, 5895, 4556, 2942, 3188, 3420, 2124, 2193, 2287, 2646, 2576, 2072, 2961, 2606, 1475, 1836, 2035, 2133, 3000, 2054, 2419, 200, 560, 1204, 1350, 1261, 1279, 220, 1472, 2596, 1868, 105

30°E, 35°E, 40°E, 0°, 5°S, 10°S, 15°S

46

n o p q r s t u v w x y z

49

0 100 200 300 miles
Average linear scale
0 100 200 300 400 500 Km

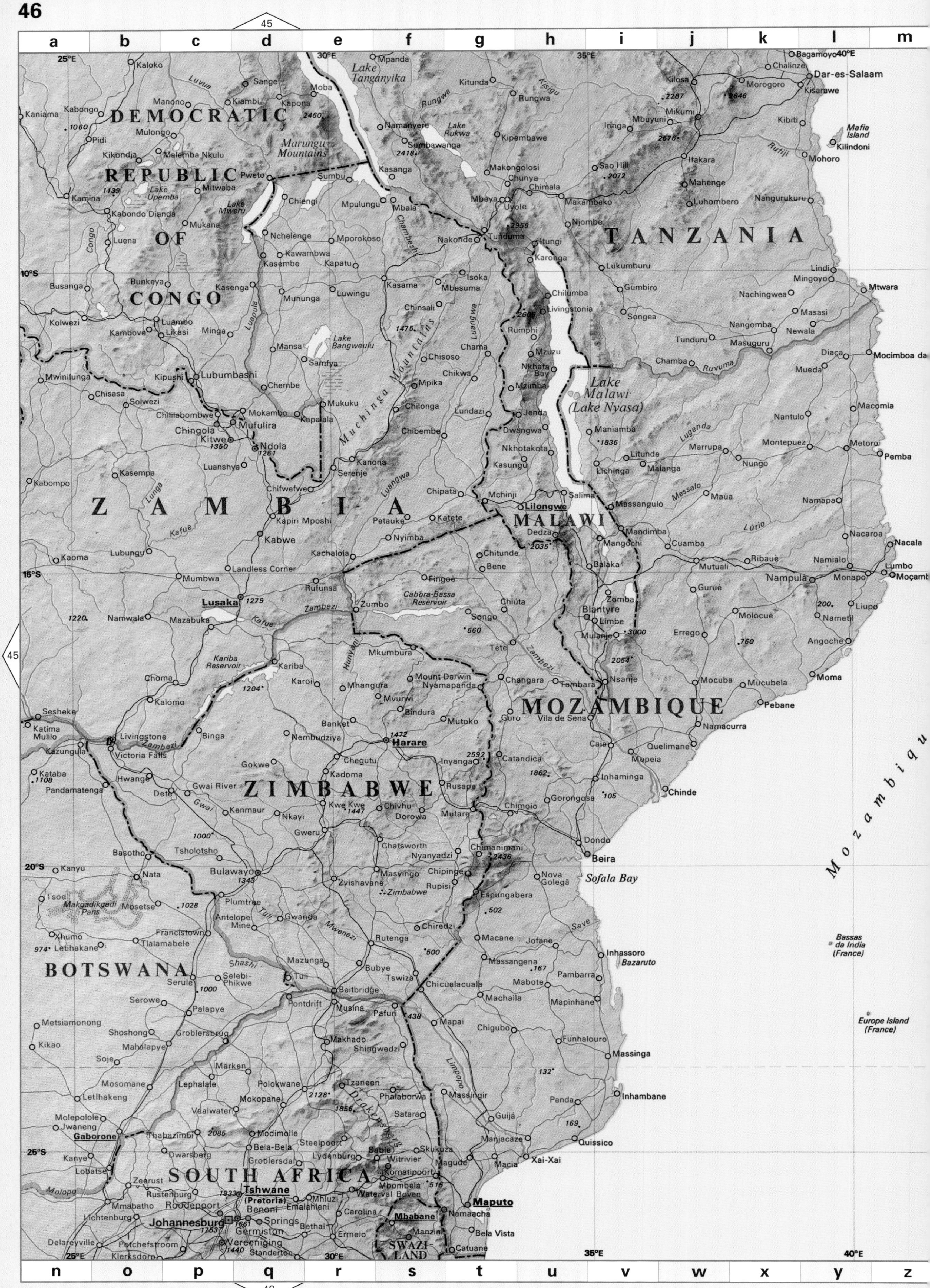

45

a b c d e f g h i j k l m

25°E 30°E 35°E 40°E

Kaloko Mpanda Lake Tanganyika Bagamoyo Chalinze Dar-es-Salaam

Luvua Sange Moba Kitunda Kilosa Morogoro Kisarawe

Manono Kiambi Kapona Rungwa Kisigu .2287 2646

Kaniama Kabongo DEMOCRATIC 2460 Rungwa Mikumi Kibiti

.1060 Pidi Namanyere Lake Rukwa Iringa Mbuyuni Mafia Island Kilindoni

Mulongo Marungu Mountains Sumbawanga Kipembawe 2576 Rufiji

Kikondja Malemba Nkulu 2418 Ifakara Mohoro

REPUBLIC Pweto Sumbu Kasanga Makongolosi Sao Hill

Lake Upemba 1139 Mitwaba Chunya .2072 Mahenge

Kamina Lake Mweru Chiengi Mpulungu Mbala Mbeya Chimala Makambako Luhombero Nangurukuru

Kabondo Dianda Uyole

Mukana OF Chambeshi 2959 Njombe

Congo Luena Nchelenge Mporokoso Nakonde Tunduma Itungi TANZANIA

Kawambwa Kasembe Kapatu Karonga Lukumburu Lindi

10°S Busanga Bunkeya Kasenga Isoka Mingoyo

CONGO Munanga Luwingu Kasama Mbesuma Chilumba Gumbiro Nachingwea Mtwara

Kolwezi Luapula Chinsali Luangwa 2606 Livingstonia Songea Masasi

Luambo Kambove Likasi Minga 1475 Rumphi Nangomba Newala

Mansa Lake Bangweulu Chama Mzuzu Tunduru Masuguru

Samfya Chisoso Nkhata Bay Chamba Ruvuma Diaca Mocimboa da

Mwinilunga Kipushi Lubumbashi Chikwa Lake Malawi (Lake Nyasa) Mueda

Chisasa Chembe Muchinga Mountains Mpika Mzimba

Solwezi Mukuku Chilonga Lundazi Macomia

Chililabombwe Mokambo Kapalala Jenda Nantulo

Chingola Mufulira Chibembe Dwangwa Maniamba Lugenda

Kitwe Ndola .1836 Montepuez Metoro

1350 1261 Nkhotakota Litunde Marrupa Pemba

Kasempa Luanshya Kanona Kasungu Lichinga Malanga Nungo

Kabompo Serenje Chipata Messalo Maúa

Lunga Chifwefwe Mchinji Lilongwe Salima Massangulo Namapa

ZAMBIA Kapiri Mposhi Petauke Katete MALAWI Mandimba Lúrio

Kafue Kabwe Nyimba Dedza Mangochi Cuamba Nacaroa Nacala

Kaoma Lubungu Kachalola 2035 Chitunde Ribauè Namialo

Landless Corner Bene Balaka Mutuali Lumbo

15°S Mumbwa Rufunsa Fingoe Nampula Monapo Moçamb

Lusaka 1279 Zumbo Cabora-Bassa Reservoir Chiúta Gurué

Zambezi Songo Zomba 200. Liupo

1220. Namwala Mazabuka Kafue Blantyre Limbe Molócue Nametil

.560 Mulanje .3000 Erego .760 Angoche

Kariba Reservoir Mkumbura Tete Zambezi 2054

Kariba Hunyani Mount Darwin Changara Tambara Nsanje Mocuba Mucubela Moma

Choma Karoi Mhangura Nyamapanda

1204 Mvurwi MOZAMBIQUE Pebane

Kalomo Bindura Guro Vila de Sena

Sesheke Banket Mutoko Namacurra

Katima Mulilo Livingstone Binga Nembudziya 1472 Harare Caia Quelimane

Kazungula Zambezi Victoria Falls Chegutu Inyanga 2592 Catandica Mopeia

Kataba .1108 Hwange Gokwe Kadoma 1862 Inhaminga

Pandamatenga Dete Gwai River ZIMBABWE Rusape Chinde

Gwai Kenmaur Nkayi Kwe Kwe 1447 Chivhu Dorowa Mutare Chimoio Gorongosa .105

1000 Gweru Dondo

Basotho Tsholotsho Chatsworth Nyanyadzi Chimanimani Beira

20°S Kanyu Nata Bulawayo 1343 Masvingo Chipinge 2436 Nova Golegã Sofala Bay

Zvishavane Rupisi Zimbabwe Espungabera Mozambique

Tsoe Makgadikgadi Pans Mosetse .1028 Plumtree Tuli Gwanda Mwenezi .502 Save

Antelope Mine Chiredzi

Xhumo Francistown Rutenga Macane Jofane Inhassoro Bassas da India (France)

974 Letlhakane Tlalamabele .500 Bazaruto

BOTSWANA Shashi Mazunga Bubye Massangena .167

Serule Selebi-Phikwe Tuli Tswiza Mabote Pambarra

.1000 Beitbridge Chicualacuala

Serowe Pontdrift Musina Pafuri 438 Machaila Mapinhane

Palapye

Metsiamonong Shoshong Groblersbrug Mapai Chigubo Europe Island (France)

Kikao Mahalapye Makhado Shingwedzi Limpopo Funhalouro

Soje Massinga

Marken 132

Mosomane Lephalale Polokwane Tzaneen

Letlhakeng Mokopane 2128 Drakensberg Phalaborwa Massingir Panda Inhambane

Vaalwater 1866

Molepolole Jwaneng Satara Guijá 169

Gaborone Thabazimbi 2085 Modimolle Manjacaze Quissico

Bela-Bela Steelpoort Skukuza

25°S Kanye Dwarsberg Groblersdal Lydenburg Sabie Witrivier Magude Xai-Xai

Lobatse Zeerust SOUTH AFRICA Komatipoort Macia

Molopo Rustenburg 1833 Tshwane (Pretoria) Mbombela Waterval Boven 515

Mmabatho Roodepoort Benoni Emalahleni Mhluzi Carolina Maputo

Lichtenburg Johannesburg 1661 Springs Bethal Mbabane Namaacha

1753 Germiston Ermelo SWAZILAND Manzini Bela Vista

Delareyville Potchefstroom Vereeniging 1440 Standerton

25°E Klerksdorp 30°E Catuane 35°E 40°E

n o p q r s t u v w x y z

49

This map shows 1/60 of the earth's surface

a b c d e f g h i j k l m
45°E 50°E 55°E 60°E
INDIAN OCEAN
Aldabra Island (Seychelles)
10°S
15°S
20°S
25°S
Moroni
COMOROS
Moheli
Anjouan
Comoros Islands
Dzaoudzi
Mayotte (France)
Channel
Juan de Nova (France)
MADAGASCAR
Antsiranana
Ambilobe
Nosy-Bé
Hell-Ville
Iharaña
Tsaratanana Mountains
2876
Sambava
Andapa
Antalaha
Antsohihy
Befandriana Av.
1218
Ambohitralanana
Maroantsetra
Mahalevona
Mahajanga
Port-Bergé-Vaovao
Mandritsara
Marovoay
Mampikony
Mananara
1301
Betsiboka
Mahavavy
Miarinarivo
Nosy Boraha
1325
Maevatanana
Andriamena
1545
Morafenobe
Vohidiala
Toamasina
Ankazobe
Antsalova
Tsiroanomandidy
Antananarivo
1381
2643
Manambolo
Mandoto
Betafo
Antsirabe
Mahanoro
Tsimafana
Tsiribihina
2140
Fandriana
Morondava
Mahabo
Ambositra
Mandabe
Mananjary
Fianarantsoa
Irondro
Morombe
Mangoky
Isalo Mountains
Ambalavao
Manakara
2658
Ankazoabo
1348
Ihosy
Ivohibe
Manombo
Mananara
Farafangana
Toliara
Andranovory
Betroka
1824
Vangaindrano
Tropic of Capricorn
Betioky
Ampanihy
1957
Antanimora
Taolañaro
Tsihombe
Ambovombe
Port Louis
MAURITIUS
Saint-Denis
3069
Réunion (France)
n o p q r s t u v w x y z
0 100 200 300 miles
Average linear scale
0 100 200 300 400 500 Km

44

a b c d e f g h i j k l m

10°E 15°E 20°E

Namibe Chibia Cassinga Chiume Mongu
Cuito Cuanavale
Tômbua (Porto Alexandre) Chianje Mulundo Cuvelai Caiundo
900 Tambor Quiteve 1265 ANGOLA Senanga
Rivungo
1190 Cubango Savate Rito Luengué
Oncócua Roçadas
Iona Cuando
Foz do Cunene N'Giva 1160 Chibaranda
2195 Cunene Naulila
Ruacana Cuangar Luiana
Xamavera
Oshakati Ondangwa Mucusso Kong
Orupembe Obombo Ovamboland Rundu
Opuwa 1096
Cape Frio Shakamku Shakawe
1784 Okavango
Etosha Pan Namutoni Keibeb Numkaub Sepopa
Purros 1093 950
Kowares Okaukuejo Tsumeb
Otavi 2149 Grootfontein Gumare Okavango Delta
869 Kamanjab Tsumkwe Mount Aha 1070
20°S Terrace Bay Goreis Outjo Tsau
Torra Bay Khorixas Otjiwarongo To
1932 Okakarara
Kalkfeld
Brandberg 2579 NAMIBIA
Uis Mine Hochfeld Dekar
Omaruru Ghanzi
2350
Cape Kruis Steinhausen
Usakos Okahandja
Henties Bay Namib 1537 Buitepos Kalkfontein BOT
Windhoek 1654 Witvlei Gobabis
Swakopmund Anschluss Takatshwaane
160 Dordabis Kule
Walvis Bay Kalaha
Tropic of Capricorn Rehoboth 1000
2334 Leonardville Ukwi Kang
Derm
Abbabis Kalkrand Aranos Tshane
ATLANTIC
Stampriet
Mariental
Sesriem Desert Khakh
25°S Maltahöhe Gochas
Naribis Zaris 1046 Mpaathutlwa Pan Makopong
Asab Asanib Twee Rivier Nossob Ter Firr
Helmeringhausen Koës
Great Tiraz 1867 1185 Tshabong
Bethanie Molopo
Lüderitz Aus Aroab Twee Rivieren Frylincks
Goageb Keetmanshoop Gr. Karasberge 2202
Narubis Gemsbok Hot
Pomona
1000 Kuruma
Witputz 1107 Sishen 1832
Grünau Karasburg Daniëls
Orange Ariamsvlei 903 Upington
1341 Augrabies Falls Keimoes Postmasbur
Alexander Bay Onseepkans Kakamas Orange
Vioolsdrif Groblershoop Griekwast
Port Nolloth Steinkopf Pofadder Kenhardt S
OCEAN
Nababeep Namies Marydale
Springbok
30°S Pri Copperton
Platbakkies Van Wyksvlei
Garies Brandvlei Vosburg Brits
Loeriesfontein Sak Carnarvon Vict We
Bitterfontein Nieuwoudtville Williston Loxton
Calvinia
Vanrhynsdorp Fish Fraserburg A
Clanwilliam
Sutherland Komsberg 1721 Beaufort West
Citrusdal
Slippers Bay Kiewie
Vredenburg 1040 Gr. Winterhoek 2078 Prince Albert Road Great Karoo
Saldanha Laingsburg Little Swartberge Willown
2325
Malmesbury Touws River Oudtshoorn
Wellington Haarlen
Little Karoo
Cape Town Worcester Swellendam George
Strand Kn
Cape of Good Hope Caledon Mossel Bay
Witsand Stil Bay
Cape Agulhas Agulhas
35°S

10°E 15°E 20°E

n o p q r s t u v w x y z

This map shows 1/60 of the earth's surface

45

a b c d e f g h i j k l m

25°E 30°E 35°E 40°E

Mumbwa Rufunsa Fingoè Zomba Guruè Nampula Moçambique

Lusaka 1279 Zambezi Zumbo Cabora Bassa - Reservoir Chiúta MALAWI

1220 Namwala Lake Kafue Kafue Mazabuka Songo 560 Blantyre Limbe 3000 Mulanje Liupo 200 Molócuè Nametil Errego Angoche

AMBIA Kariba Hunyani Mkumbura Tete Zambezi 2054 760

Kariba Reservoir Karoi Mhangura Mount Darwin Changara Tambara Nsanje Mocuba Mucubela Moma

Kalomo Choma 1204 Mvurwi Nyamapanda Guro Vila de Sena Pebane

Sesheke Bindura Namacurra

Katima Mulilo Livingstone Zambezi Binga Banket Mutoko Caia Quelimane

Kazungula Victoria Falls 1472 Harare Mopeia

Kataba 1108 Hwange Chegutu 2592 Catandica Inhaminga

Pandamatenga Gokwe Kadoma Inyanga 1862 Chinde

Dete Gwai River ZIMBABWE Rusape 105

Gwai Kenmaur Nkayi Kwe Kwe 1447 Chivhu Mutare Gorongosa

Dorowa Chimoio

1000 Gweru Dondo

Tsholotsho Chatsworth Nyanyadzi Chimanimani Beira

Basotho Chipinge Sofala Bay Mozambique Channel

Kanyu Bulawayo 1343 Masvingo Rupisi Nova Golegã 20°S

Nata Plumtree Zvishavane Espungabera

Tsoe Mosetse 1028 Tuli MOZAMBIQUE

Gwanda Mwenezi Chiredzi 502

Xhumo 974 Tlalamabele Francistown Antelope Mine Rutenga Macane Save

Letlhakane Shashi 500 Jofane Inhassoro Bassas da India (France)

Mazunga Bubye Massangena 167 Bazaruto

Serule Selebi-Phikwe Tuli Tswiza Mabote Pambarra

WANA Serowe 1000 Pontdrif Beitbridge Chicualacuala Machaila

Metsiamonong Palapye Musina Pafuri 438 Mapinhane Europe Island (France)

Kalamare Groblersbrug Mapai Chigubo

Kikao Mahalapye Makhado Shingwedzi Limpopo Funhalouro

Soje Marken Massinga

Lephalale Tzaneen Phalaborwa 132

tlwe Mosomane Polokwane (Pietersburg) 2128 Drakensberg Massingir Inhambane

Letlhakeng Molepolole Vaalwater Mokopane (Potgietersrus) 1856 Satara Panda

ma Gaborone 2085 Thabazimbi Modimolle (Nylstroom) Guijá 169

waneng Kanye Bela-Bela (Warmbaths) Steelpoort Skukuza Manjacaze

1479 Lobatse Dwarsberg Sabie Magude Quissico

Grobiersdal Lydenburg Witrivier Macia Xai-Xai 25°S

46

Molopo Zeerust Tshwane (Pretoria) 1333 Middelburg Nelspruit 515 Komatipoort

Mmabatho Rustenburg Waterval Boven Maputo

rokweng Lichtenburg Roodepoort Benoni Witbank Carolina Namaacha

Johannesburg 1753 Germiston Springs Mbabane Bela Vista

Delareyville Potchefstroom Soweto 1661 Vereeniging Bethal Ermelo Manzini

Vryburg Klerksdorp 1440 Standerton SWAZILAND Catuane

Parys Vaal Piet Retief

Schweizer-Reneke Wolmaransstad Vaal Reservoir Frankfort Lavumisa

Heilbron Volksrust 2277 Pongola

Bloemhof Reservoir Kroonstad Utrecht Mkuze

eivilo Bloemhof Reitz Newcastle Vryheid Lake St. Lucia

Christiana Welkom Bethlehem 1532

arrenton Bultfontein Harrismith Dundee Ulundi

Barkly West Winburg Senekal Mont aux Sources Ladysmith Mtubatuba

Kimberley Ficksburg 3285 Drakensberg Eshowe INDIAN

Clocolan Estcourt Tugela Richards Bay

TH 1426 Maseru Greytown

Mangaung (Bloemfontein) 3096 3482

town Luckhoff Fauresmith Wepener LESOTHO Himeville Pietermaritzburg

Trompsburg Mafeteng Durban

P. K. le Roux Reservoir Smithfield Orange Drakensberg Ixopo 30°S

Orange Zastron Moyeni 1000 Umzinto

Colesberg Gariep Reservoir Kokstad Harding

Aar Aliwal North Lady Grey Mount Fletcher Port Shepstone

Hanover Burgersdorp Maclear Mount Frere Port Edward

2052 Barkly East Elliot

iddelburg Steynsburg 1677

ICA Lady Frere Umtata Port St. Johns

Queenstown Coffee Bay

Graaff-Reinet Cradock Idutywa

Great Fish Stutterheim

rdeen Fort Beaufort King William's Town

Somerset East 500 Grahamstown East London

Kirkwood Bell

Steytlerville 527 Port Alfred

Uitenhage Port Elizabeth OCEAN

ansdorp Jeffreys Bay

35°S

25°E 30°E 35°E 40°E

n o p q r s t u v w x y z

0 100 200 300 miles

Average linear scale

0 100 200 300 400 500 Km

92

a b c d e f g h i j k l m

35°E 40°E 45°E 50°E 55°E 60°E 65°E

85°N 80°N 75°N 70°N 65°N 60°N

ARCTIC

Alexandra Land
George Land
Salisbury I.
Jackson I.
Rudolf I.
Karla-Aleksandra
La Rons'yer
Yeva-Liv
Luidzhi
Hooker I.
Hall I.
McClintock I.
Sal'm
Wilczek Land
Graham Island
Franz Josef Land
Zem

Russkaya Gavan
Smidovich
Novaya Zemlya
Sedova 1115
Stolbovoy
Litke
260
Krasino
Proliv Karskiye Vorota
162 Vajgač
Cape Ue
Amderma
Baide...

BARENTS SEA

KARA SEA

Pechora Sea

Kolgujev 166
Cape Kanin Nos
242
Kanin Peninsula
Česa Bay
Mezen' Gulf
Volonga
Malozemel'skaya Tundra
Chernaya
Dresvyanka
201
Bol'shezemel'skaya Tundra
Nar'yan Mar
106
Kolva
Koreyver
Velikovisochnoye
Pay-Khoy
467
Ust'-Kara
Yangarey
Khal'mer-Yu
Vorkuta
Yeletskiy
Adzva
Makarikha
Trosh
Usa
Abez'
Pay-yer 1499
Labytn...
Inta
Kosyu
Narodnaya 1894
Northern Ural
Pechora
Kadzherom
Saranpaul'
Vanzev...
Berezovo
North Sosva
Patrasuy
Kyrta
1617
Muligort
Sergino
Nyaksimvol
Murmansk
Mončegorsk 1191
Kirovsk
397
Kola
Kandalakša
Arctic Circle
Kandalakša Gulf
White Sea
Dvina Bay
Onega Bay
Belomorsk
Severodvinsk
Archangel
Pinega
Dvina
Onega
Segeža
Mezen'
Slafonovo
Nonburg
Ust'Tsil'ma
155
Azopol'ye
Izhma
Timan Ridge
463
164
Voyvozh
Kedva
Politovo
Mezen'
Pinega
Vendenga
Shomvukva
Vym'
Ukhta
Pechora
Loptyuga
324
Vey Vozh
Troitsko-Pechorsk
1108
Zheleznodorozhnyy
Mikun
RUSSIA
Puzla
Porog
Vyčegda
Irta
259
North Dvina
Verkhnyaya Toyma
417
Medvežjegors
Lake Onega
Petrozavodsk
Kargopol'
Vaga
Ustya
Syktyvkar
Ust'Kulom
Kur'ya 303
Suyevatpaul
Sovetskiy
Komsomol'ski
Pionerskiy
Khango...
Trapsul
Kizema
Kotlas
239
Konoša
Vel
Vel'sk
Vizinga
Kolva
1027
Polunochnoye
Ivdel'
Podporoze
Sukhona
Velikiy Ustyug
213
Kazhim
Noshul'
Uvaly
Kama
Pyatigory
Cherdyn
1493
Denezhkin Kamen
Totma
Nizhniy Yenangsk
Northern Uvaly
Krasnoturinsk
162
Tichvin
Nikol'sk
Kirs
Solikamsk
Berezniki
Central Ural
Lobva
Sos'va
Tavda
Čerepovec
Vologda
Murashi
Kudymkar
Kamskoje Reservoir
292
Vetluga
Kirov
Novo-Vyatsk
Gubakha
883
Rybinsker Reservoir
Bui
Vetluzhskiy
Kotel'nich
Dobryanka
Tura
Verkhniy Tura
321
Rybinsk
Glazov
Krasnokamsk
Perm'
Turinsk
Pizhma
Kez
Nizhniy Tagil
Kostroma
Unža
Vyšni Voloček
Jaroslavl
Volga
Kinešma
Uren
Nolinsk
Igra
343
Yaransk
Artemovskiy
Kalyazin
Kungur
Ostashkov
Votkinsk
Ivanovo
Krasnyye-Baki
Yekaterinburg
Torzhok
Tver
Dubna
Kil'mez
Izhevsk
Pervoural'sk
Talitsa
Bogdanovich
Klin
Dmitrov
Malmyzh
217
Krasnoufimsk
Yalutoro...
Nelidovo
Staritsa
Rzev
Yoshkar Ola
115
Kovrov
Nizhniy Novgorod
Agryz
Sarapul
Degtyarsk
Sysert'
Kamensk-Ural'skiy
Sergiyev Posad
Vladimir
Volokolamsk
Mytišci
Balashikha
Noginsk
Dzerzhinsk
Volga
Arsk
Neftekamsk
Nyazepetrovsk
Shadrinsk
Cheboksary
Ufa
Moscow (Moskva)
Orechovo Zujevo
Elektrostal
Odintsovo
Oka
235
Yadrin
Murom
Kazan'
Mamadysh
Naberezhnyje Čelny
Kasli

33

n o p q r s t u v w x y z

54

55

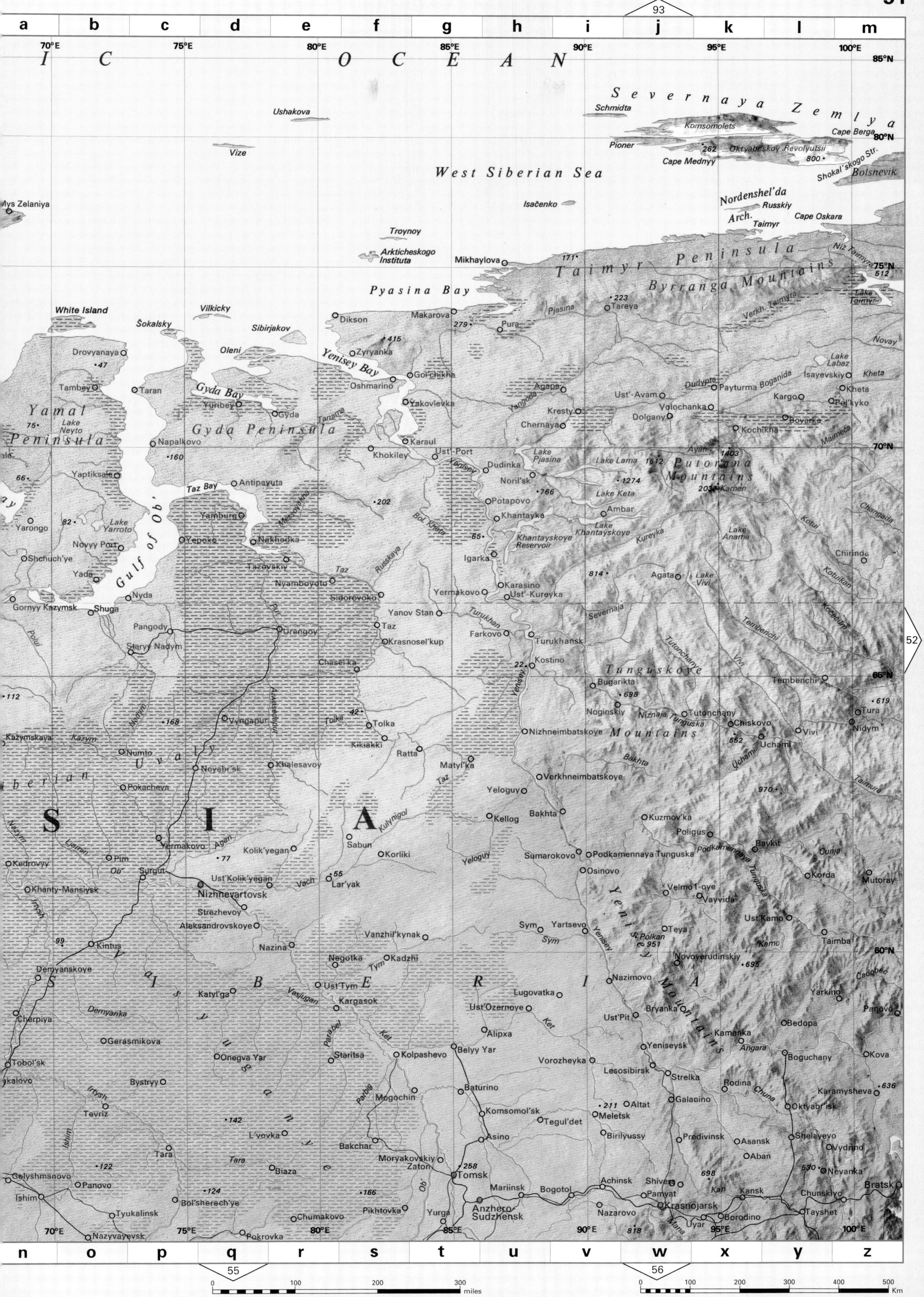

93

a b c d e f g h i j k l m

70°E 75°E 80°E 85°E 90°E 95°E 100°E 85°N 80°N 75°N 70°N 65°N 60°N

I C O C E A N

Severnaya Zemlya
Schmidta
Komsomolets
Cape Berga
Pioner
262
Oktyabr'skoy Revolyutsii
800
Cape Mednyy
Shokal'skogo Str.
Bolsnevik
Ushakova
Vize
West Siberian Sea
Isačenko
Nordenshel'da Arch.
Russkiy
Taimyr
Cape Oskara
Mys Zelaniya
Troynoy
Arktičeskogo Instituta
Mikhaylova
171
Taimyr Peninsula
Byrranga Mountains
Niz Taimyra
512
Lake Taimyr
Verkh. Taimyra
223
Tareya
Pjasina
Pyasina Bay
Makarova
279
Pura
Novay
White Island
Šokalsky
Vilkicky
Sibirjakov
Dikson
415
Zyryanka
Drovyanaya
47
Oleni
Yenisey Bay
Gol'chikha
Oshmarino
Lake Labaz
Isayevskiy
Kheta
Tambey
Taran
Gyda Bay
Yuribey
Gyda
Yakovlevka
Tanama
Yangoda
Agapa
Ust'-Avam
Dudypta
Payturma
Boganida
Kargo
Kheta
Pol'kyko
Yamal Peninsula
75
Lake Neyto
Gyda Peninsula
Kresty
Chernaya
Volochanka
Dolgany
Kochikha
Boyarka
Maimeča
Napalkovo
160
Karaul
Khokiley
Ust'-Port
Yenisey
Dudinka
Noril'sk
Lake Pjasina
Lake Lama
1612
Ayan
1403
Putorana Mountains
1274
2037
Kamen
Yaptiksale
66
Antipayuta
Taz Bay
Messoyakha
202
766
Potapovo
Lake Keta
Ambar
Changada
Yarongo
82
Lake Yarroto
Yamburg
Bol. Kheta
Khantayka
Lake Khantayskoye
Khantayskoye Reservoir
Kureyka
Kotui
Lake Anama
Novyy Port
Yepoko
Nakhodka
65
Igarka
Chirinda
Shchuch'ye
Gulf of Ob'
Tazovskiy
Russkaya
814
Agata
Lake Vivi
Kotuikan
Yada
Taz
Nyamboyoto
Karasino
Gornyy Kazymsk
Nyda
Sidorovoko
Yermakovo
Ust'-Kureyka
Shuga
Yanov Stan
Turukhan
Severnaja
Kočečumo
Poilui
Pangody
Pur
Taz
Urengoy
Farkovo
Turukhansk
Tembenchi
Staryy Nadym
Krasnosel'kup
Tutonchany
Vivi
Chasel'ka
22
Kostino
Tunguskoye
Tembenchi
112
Bugarikta
698
619
Tura
Nadym
168
Vyngapur
Aivasedapur
Tolka
42
Tolka
Noginskiy
Nizhnaja Tunguska
Tutonchany
Chiskovo
Nidym
Kazymskaya
Kazym
Kikiakki
Nizhneimbatskoye
Mountains
552
Uchami
Vivi
Numto
Ratta
Matyl'ka
Bakhta
Uchami
Uvaly
Noyabr'sk
Khalesavoy
Taz
Verkhneimbatskoye
970
Taimura
Siberian
Pokacheva
Yeloguy
S I A
Kulyngol
Kellog
Bakhta
Kuzmov'ka
Nazym
Yermakovo
Agan
Sabun
Poligus
Lyamin
Kolik'yegan
Podkamennaya Tunguska
Baykit
Pim
77
Korliki
Yeloguy
Sumarokovo
Podkamennaya Tunguska
Čunja
Kedrovyy
Ob'
Surgut
Ust'Kolik'yegan
55
Osinovo
Korda
Mutoray
Vach
Lar'yak
Yenisey
Khanty-Mansiysk
Nizhnevartovsk
Velmo 1-oye
Vayvida
Irtysh
Strezhevoy
Ust'Kamo
Aleksandrovskoye
Sym
Yartsevo
Teya
Vanzhil'kynak
Sym
Polkan
951
Kamo
Taimba
99
Kintus
Nazina
Novoyerudinskiy
695
Negotka
Kadzhi
Tym
Demyanskoye
Nazimovo
Cadobec
Ust'Tym
E R I
Vasyugan
Katyl'ga
Lugovatka
Yarkino
Kargasok
Vasyuganye
Demyanka
Ust'Ozernoye
Bryanka
Yenisey Mountains
Panovo
Ust'Pit
Cherpiya
Ket
Bedopa
Alipxa
Gerasmikova
Parabel
Ket
Kamenka
Belyy Yar
Yeniseysk
Angara
Tobol'sk
Onegva Yar
Staritsa
Kolpashevo
Vorozheyka
Boguchany
Kova
Lesosibirsk
Strelka
Bystryy
Baturino
Rodina
Chuna
Karamysheva
636
Irtysh
Parbig
Mogochin
211
Altat
Galanino
Oktyabr'sk
Tevriz
Ishim
Komsomol'sk
142
Tegul'det
Meletsk
L'vovka
Asino
Birilyussy
Predivinsk
Asansk
Shelaevo
Vydrino
Tara
Bakchar
Aban
Tara
Moryakovskiy Zaton
258
122
Biaza
Tomsk
698
530
Nevanka
Golyshmanovo
Panovo
Achinsk
Shiveo
Mariinsk
Bogotol
Pamyat
Kan
Kansk
Chunskiyo
Bratsk
Ishim
124
166
Ob'
Anzhero-Sudzhensk
Krasnoyarsk
Tayshet
Bol'sherech'ye
Pikhtovka
Yurga
Nazarovo
Uyar
Borodino
Tyukalinsk
Chumakovo
Mana
Nazyvayevsk
Pokrovka
818

52

n o p q r s t u v w x y z

55

56

0 100 200 300 miles

0 100 200 300 400 500 Km

94

a b c d e f g h i j k l m

105°E 110°E 115°E 120°E 125°E 130°E

75°N 70°N 65°N 60°N 55°N

Byrranga Mountains
Vezdekhodnaya
Laptev Sea
Lake Taimyr
Bol. Balakhnya
Begichev
Nordvik
Cape Nordvik
Korennoye
Khorgo
Kozhevnikovo
Novoryonye
Novyy
Uryung-Khaya
Uele
Suolama
Popigay
Sagyr
Khatanga
Lukunskiy
Novay
Bychez
Khatanga
Fomich
Popigay
Saskylakh
Anabar
268
Star. Kayakhnyy
Amakinskiy
Kotuykan
Popigay
536
Dzhelinde
Central
Ulgumun
Ust'-Olenëk
Stannakh-Khocho
Turkannakh
Dunay
Sagastyr
Antipinskiy
Trofimovsk
Orto-Aya
Ary
211. Taymylyr
52
Ot-Siyen
Pur
Sklad
Tit-Ary
Tiksi
128
Bor-Yuryakh
405
921
Chekurovka
Kyusyur
Khasalakh
Govorovo
Kuoyka
Siktyakh
Sakhandzha
Kel'
Lena
1291
Tukalan
Kirbey
Mongolo
Ukukit
Molodo
Sukhana
Motorchuna
Dzhardzhan
Yessey
Kotui
Dzhara
Ylas-Yuryakh
Kirbey
Olenëk
Kyuekh-Bulung
Moyero
Arga-Sala
Siligir
Olenëk
Menkere
Sencha
Dzhelon
Murukta
Siberian
Olenëk
Muna
Kystatyam
Menkere
2389
Zhigansk
Tirekh
Vilyuy Mountains
Eyakit-Tërdë
Udachnyy
Aykhal
Onkuchakh
R U S
Kharalakh
Khoronnokh
Verkhoyanskiy
Ekonda
Uplands
Eyik
Tyung
Bakhynay
Vilyuy
Markha
Andyngda
Amysakh
Borolgustakh
Endybal
823
Kochechum
Markoka
Tyukyan
Bagadzha
Mastakh
Linde
Tungus-Khaya
Kyrgyday
Niznaya Tunguska
Yeyka
Engerdyakh
Dalgoye
Kananda
Amo
501
Ygyatta
Malykay
Ulgumdzha
Vilyuy
Khampa
Vilyuysk
Lena
Batamay
Kobya
Ust'-Ilimpeya
Yukta
Ankacho
Kysyl-Yllyk
Nyurba
Verkhnevilyuysk
Ebe
Ilbenge
Taimura
Chuyengo
Chernyshevskiy
Novyy
Khordogoy
Sheya
Khochot
Tyugene
Kangalassay
Ilimpeya
Simenga
Mirnyy
Almaznyy
Viljujskoje Reservoir
Kiriyestyakh
Tunor
Suntar
Oleng-Sala
Tongulakh
Yakutsk
Bugorkan
Tenke
Kerekyano
Čunja
Ayan
Dzhunkun
Tas-Yuryakh
Pokrovsk
Strelka-Čunja
Botuobuya
Sinyaya
Yerbogachen
Ergedzhey
Atakh-Yuryakh
Yet-Kyuyel'
Kytyl-Zhura
Kachikattsy
Niznaya Tunguska
Chamcha
Lena
Sosna
Ulakhan Botuobuya
Lensk
Nyuya
Olekminsk
Uritskoye
Sangyyakhtakh
Taloye
Vanavara
Kulinda
Khomokashevo
Yerema
Dulga-Kyuyel'
Khamra
Nyuya
Khabalakh
Bol. Patom
Patom
Cherendey
Tuolba
Tegyulte-Terde
Tetere
Ust'-Chayka
Tokko
Kudu-Kyuyel'
Chemdal'sk
Ayan
Nepa
Tolon
Vitim
1639
Andreyevskiy
Berezovskaya
Kamanga
Ika
Nepa
Chuya
Patomskoye Plateau
Khoppuruo
Amga
Ugoyan
Verkhnyaya Amga
Panovo
Kureyskaya
Vorontsovka
Polovinka
Chara
Olekma
Dikimdya
Tommot
Angara
Kata
Bur
Cherkashina
Chara
Torgo
Aldan
Ugun
Ust'-Timpton
Ichera
Kropotkin
1771
Severomuysk
Usmun
1612
Volokon
Lena
Vitimskiy
Tokko
Yenyuka
Suon-Tit
Timpton
Ust'Ilimsk
Chuya
Mama
Bodaybo
Vitim
Berezovka
Bol. Khatymy
Kirensk
Sinyuga
Lake Nichatka
Gynym
Vorob'yeva
Chaya
Gorno-Chuyskiy
North
Karalon
Oran
Ul'Kan
Baykal
Chara
Khani
Garmenka
Romanova
Kirenga
Plateau
Udokan
Neryungri
Gonam
Ust'-Kut
Yermaki
Ilimsk
Riga
Injaptuk
2579
Ust'-Muya
2467
Taluma
Ust'Nyukzha
Berkakit
Bratsk
Vidim
Suvorka
Orlinga
Uoyan
Tonnel'nyy
1870
Kazachinskoye
Kalar
Nagornyy
Sutam
Bratskoye Reservoir
Lena
Yukhta
Nizhneangarsk
Vitim
Bambuyka
Sredniy Kalar
Lopcha
Chapa
Baykal'skoje
Baunt
Kadali
1592
Vetekhtina
Stanovoy Mount
Baykal'skiy Mountains
Plateau
Kalakan
Larba
Tynda
Atalanka
Žigalovo
Sugdža
Oron
Olokma
Gulya
Belen'kaya
Ugagli
Zeyski
763
Ust'-Kada
Mogoito
Bagdarin
Ust'-Karenga
Koltovkinda
Zeya
Angara
Sosnovka
2523
Jeleninskij
Vitim
Tupik
Bam
Solov'yevsk
Ogoron
Balagansk
2068
Zel'onoje Ozero
Amazar
Urusha
Skovorodino
Zeya
Zima
Bol. Onguren
Barguzin
Bugunda
Chulugli
1911
Mogoča
Zalari
Manzurka
Lake Baykal
Ust'-Džilinda
Silka
Dzhalinda
Magdagachi
Maksimicha
Romanovka
Nerča
Mošegda
Luoguhe
557
Petropavlovka
Čeremchovo
774
Yablonovyy Mountains
Bukačača
1249
Usolje Sibirskoje
Isinga
Ust'-Karsk
Gulian
Ershiyizhan
Oktyabrskiy
Ust'-Ordynskij
2049
Telemba
Yimuhe
Walagan
Ushumun
Angarsk
Chaim
1322
Černyševsk
Amur
Irkutsk
Selichov
Uda
Veršino-Darasunskij
Sretensk
Borshchovochnyy Mountains
Qiqian
Novorossiyka
Tataurovo
Chonnsk
Chita
Kurleja
Mangui
Shimanovsk
3266
Kyren
Kamensk
C H I N A
Listv'anka
Ulan-Ude
Silka
Ingoda
Baley
Sl'ud'anka
Chilok
Darasun
Karymskoje
Nerchinskiy Zavod
Argun He
Linhai
Svobodnyy
Novokiyev
Tanchoj
2304
Selenga
Petrovsk Zabajkal'skij
Ulety
Mordaga
Jinhe
Huma
Zeya
Gusinoozersk
Tanga
Il'a
Olov'annaja
Klin
Shisanzhan
827
Amur
Belogorsk
Džida
Chilok
Argun Youqi
Kalagi
Oroqen Zizhiqi
Džida
1248
Gol
Zakamensk
Jamarovka
Priargunsk
Tuline

n o p q r s t u v w x y z

51

57

58

This map shows 1/60 of the earth's surface

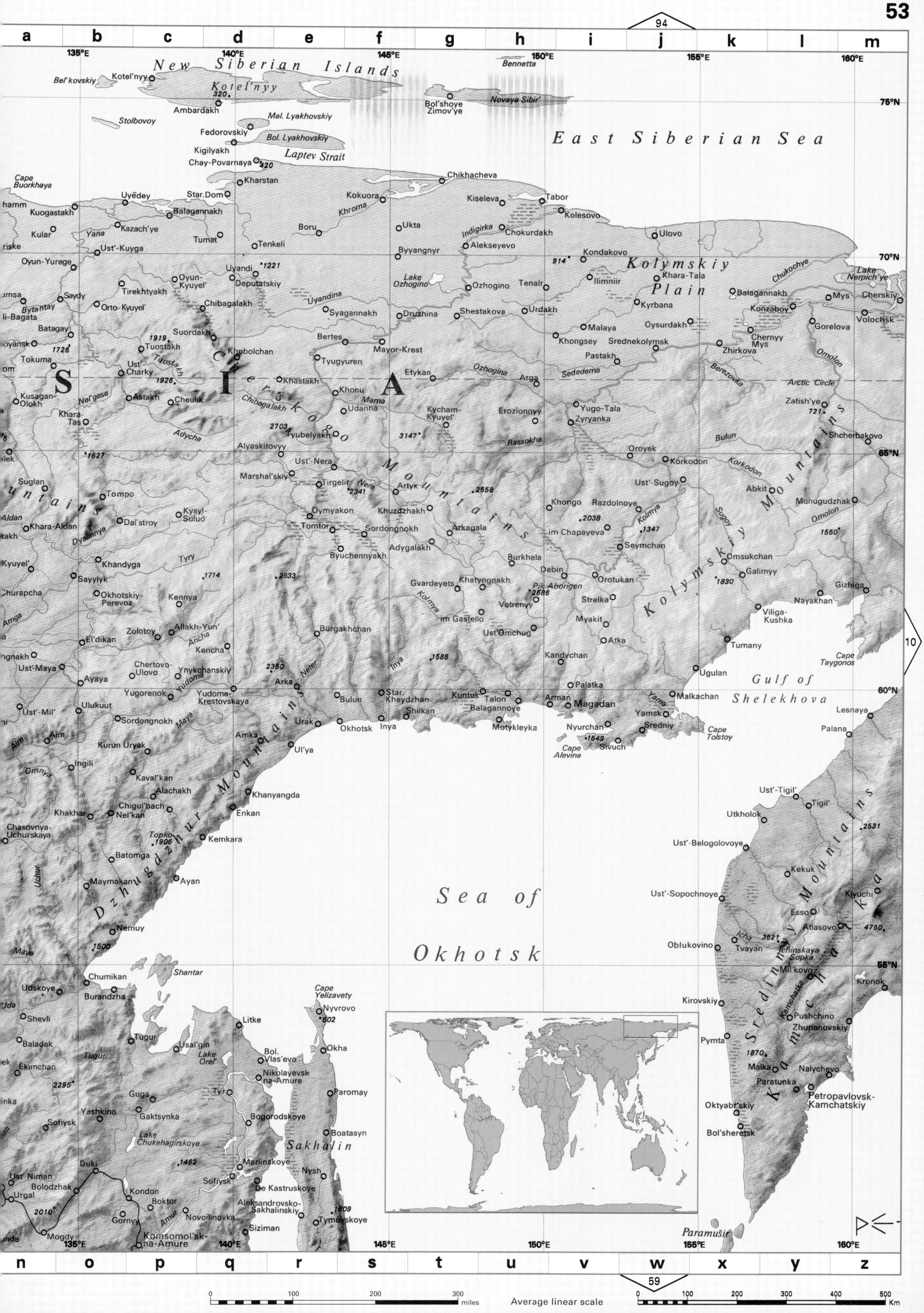

94

a b c d e f g h i j k l m

135°E 140°E 145°E 150°E 155°E 160°E

New Siberian Islands

Bel'kovskiy
Kotel'nyy
Kotel'nyy
320
Bennetta
Bol'shoye Zimov'ye
Novaya Sibir'
75°N
Ambardakh
Stolbovoy
Mal. Lyakhovskiy
Fedorovskiy
Bol. Lyakhovskiy
Kigilyakh
Laptev Strait
Chay-Povarnaya
420

East Siberian Sea

Cape Buorkhaya
Kharstan
Chikhacheva
Uyëdey
Star.Dom
Kokuora
Kiseleva
Tabor
Kuogastakh
Balagannakh
Khroma
Kolesovo
Kular
Yana
Kazach'ye
Boru
Ukta
Indigirka
Chokurdakh
Tumat
Tenkeli
Alekseyevo
Ulovo
Ust'-Kuyga
Byyangnyr
Kondakovo
Oyun-Yurege
Uyandi
1221
914
70°N
Kolymskiy Plain
Khara-Tala
Chukochye
Lake Nerpich'ye
Oyun-Kyuyel'
Deputatskiy
Lake Ozhogino
Ozhogino
Tenalr
Ilimniir
Tirekhtyakh
Balagannakh
Mys
Cherskiy
Saydy
Uyandina
Bytantay
Orto-Kyuyel'
Chibagalakh
Kyrbana
Konzabov
Volochsk
Batagay
Syagannakh
Druzhina
Shestakova
Urdakh
Oysurdakh
Gorelova
Suordakh
Malaya
Bertes
Khongsey
Srednekolymsk
Chernyy Mys
1919
Tuostakh
Mayor-Krest
Zhirkova
1726
Khobolchan
Tokuma
Tyugyuren
Pastakh
Omolon
Tuostakh
Ust'-Charky
Etykan
Ozhogina
Sededema
1926
Arga
Berezovka

S I A

Khastakh
Arctic Circle
Kusagan-Olokh
Nel'gese
Astakh
Khonu
Cheulik
Mama
Zatish'ye
Khara-Tas
Chibagalakh
Udanna
Yugo-Tala
721
Kycham-Kyuyel'
Eroziоnnyy
Zyryanka
2703
Tyubelyakh
3147
Adycha
Rassokha
Bulun
Shcherbakovo
Alyaskitovyy
Oroyek
1627
Ust'-Nera
Korkodon
65°N
Korkodon
Marshal'skiy
Suglan
Tirgelir
Nera
Artyk
Ust'-Sugoy
2341
Abkit
2558
Tompo
Khongo
Razdolnoye
Munugudzhak
Kysyl-Suluo
Oymyakon
Khuzdzhakh
Sugoy
2038
Omolon
Aldan
Khara-Aldan
Dal'stroy
Tomtor
Sordongnokh
Arkagala
im Chapayeva
Kolmya
1347
1550
Dyalinnya
Adygalakh
Seymchan
Byuchennyakh
Omsukchan
Khandyga
Tyry
Burkhala
Galimyy
Debin
1830
Kyuyel'
Sayylyk
1714
2933
Khatyngnakh
Orotukan
Gizhiga
Gvardeyets
Pik Aborigen
Churapcha
Okhotskiy-Perevoz
Kennya
2586
Nayakhan
Strelka
Kolyma
Vetrenyy
Viliga-Kushka
im Gastello
Myakit
Amga
Allakh-Yun'
Zolotoy
Ust'Omchug
El'dikan
Burgakhchan
Ancha
Atka
Tumany
Kencha
Kandychan

10

Cape Taygonos
Chertovo-Ulovo
Ynykchanskiy
Inya
1585
Ust'-Maya
2350
Ugulan
Ayaya
Yudoma
Arka
Neter
Gulf of Shelekhova
Palatka
Yugorenok
Yudoma-Krestovskaya
Bulun
Star. Kheydzhan
Kuntuk
Talon
Arman
Malkachan
60°N
Ust'-Mil'
Ulukuut
Shilkan
Balagannoye
Magadan
Yamsk
Lesnaya
Sordongnokh
Maya
Urak
Okhotsk
Inya
Motykleyka
Nyurchan
Sredniy
Palana
Aim
Amka
Cape Tolstoy
Kurun Uryak
Ul'ya
1549
Sivuch
Cape Alevina
Ingili
Omnya
Kaval'kan
Alachakh
Dzhugdzhur Mountains
Khanyangda
Ust'-Tigil'
Tigil'
Khakhar
Chigul'bach
Nel'kan
Enkan
Utkholok
Chasovnya-Uchurskaya
Topko
1906
2531
Kemkara
Ust'-Belogolovoye
Batomga
Uchur
Kekuk
Maymakan
Ayan
Ust'-Sopochnoye

Sea of Okhotsk

Klyuchi
Esso
Nemuy
Atlasovo
4750
Icha
3621
Oblukovino
Tvayan
Ichinskaya Sopka
1500
Maya
Mil'kovo
55°N
Chumikan
Shantar
Kronok
Udskoye
Cape Yelizavety
Burandzha
Nyvrovo
602
Kirovskiy
Shevli
Litke
Pushchino
Zhupanovskiy
Baladek
Tugur
Usal'gin
Lake Orel'
Okha
Pymta
Sredinnyy Mountains
Tugur
Bol. Vlas'evo
1870
Ekimchan
Nikolayevsk-na-Amure
Malka
Nalychevo
2295
Paromay
Paratunka
Kamchatka
Petropavlovsk-Kamchatskiy
Tyr
Guga
Yashkino
Gaktsynka
Bogorodskoye
Oktyabr'skiy
Sofiysk
Lake Chukchagirskoye
Boatasyn
Bol'sheretsk
Sakhalin
1462
Duki
Marlinskoye
Nysh
Ust' Niman
Sofiysk
Bolodzhak
Kondon
De Kastruskoye
Urgal
Boktor
Aleksandrovsk-Sakhalinskiy
1609
2010
Amur
Novoilinovka
Tymovskoye
Gornyy
Siziman
Mogdy
Komsomol'sk-na-Amure
Paramušir

n o p q r s t u v w x y z

59

0 100 200 300 miles
Average linear scale
0 100 200 300 400 500 Km

50

a b c d e f g h i j k l m

35°E 40°E 45°E 50°E

Ostashkov
Jaroslavl'
Kineshma
Krasnyye-Baki
Uren
Yaransu
Kilmez
Votkinsk
Torzhok
Tver
Ivanovo
Yoshkar Ola
Izhevsk
Malmyzh
.217
Agryz
Sarapul
Dubna
Dmitrov
Kovrov
Nizhniy Novgorod
.115
Nelidovo
Rzev
Staritsa
Klin
Sergiyev Posad
Balashikha
Noginsk
Dzerzhinsky
Volga
Cheboksary
Arsk
Volokolamsk
Mytišči
Vladimir
Kazan'
Nefte
Gagarin
Moscow (Moskva)
Orechovo-Zujevo
.235
Yadrin
Mamadysh
Naberezhnyje Celny
.320
Vyaz'ma
Mozhaysk
Odintsovo
Elektrostal'
Oka
Murom
Sergach
Kanash
Chistopol'
Obninsk
Podolsk
Kolomna
Arzamas
R U S
55°N
Dnieper
Smolensk
242.
Serpuchovo
Temnikov
Tetyushi
Kuybyshevskoje Reservoir
Al'met'yevsk
Oktyabr's
Kaluga
Tula
Rjazan
Alatyr
Ul'yanovsk
Bugul'ma
Davleka
Sukhinichi
Chekalio
Novomoskovsk
Nazarovka
Zubova-Polyana
Saransk
Severnoye
Roslavl
Zhizdra
Plavsk
Skopin
Shatsk
Inza
Yelkhovka
Aksakovo
Belev
Troyekurovo
Morshansk
Barysh
Togliatti
Abdulino
Bryansk
Kadnoye
Michurinsk
Kamenka
Kuznetsk
Krotovka
Buguruslan
Karchev
Navlya
Orel
Zmiyevka
Yelets
Lipetsk
Gryazi
Tambov
Kirsanov
Penza
Syzran
Samara
Sharly
Klintsy
Desna
Livny
Chapayevsk
Buzuluk
Rtishchevo
Maryevka
Sorochinsk
Mordovo
Tugolukovo
Vyzakova
Vol'sk
Balakovo
Bol'shaya Glushitsa
Andreyevka
Obščij Syrt
Shostka
Fatzeh
Khomutovka
Seym
Kursk
Gorshechnoye
Voronezh
Atkarsk
Saratov
Engel's
Orenburg
Borisoglebsk
Kalininsk
Starryy Oskol
Rogachevka
Balashov
.122
Dergachi
Ural'sk
Ilek
Aksay
Sol'-Iletsk
Konotop
Oboyan
Pushkino
Priluki
Sumy
Belgorod
Alekseyevka
Kamenka
Buturlinovka
Uryupinsk
Rudnya
Rovnoye
Novo Uzensk
Pavlovsk
Akhtyrka
Valuyki
Kamyshin
Chapayevo
Lubny
Valki
Don
Mikhaylovka
Nikolayevsk
Mergenevo
50°N
Khar'kov
Poltava
Kantemirovka
Veshenskaya
Kaysatskoye
Furmanovo
UKRAINE
Kremenchugskoye Reservoir
Izyum
Volga
Log
Kremenchug
Pereshchepino
Primorsk
Elton
.67
Lake Aralsor
Antonovo
Znamenka
Novomoskovsk
Kramatorsk
Donets
Masteksay
Dneprodzerzhinsk
Pavlogard
Millerovo
Kalach-na-Donu
Volgograd
Krasnoslobodsk
Inderborskiy
Kirovograd
Dnepropetrovsk
Gorlovka
Stakhanov
Lugansk
Kamensk Shakhtinskiy
Morozovsk
Kapustin Yar
Yenakiyevo
Solodniki
Kulagino
Chaplino
Makeyevka
Tsimlyanskoye Reservoir
Chernyy Yar
Krivoy Rog
Zaporozh'ye
Donetsk
Novoshakhtinsk
Nikopol
Kakhovskoye Reservoir
Shakhty
Novocherkassk
Volgodonsk
Kotel'nikovo
Makat
Dnieper
Taganrog
Rostov
Dubovskoye
Mikhaylova
Caspian Depression
Sarychik
Kharabali
Volga
.-12
Isking
Bataysk
Atyrau
.27
Kul's
Melitopol'
Berdyansk
Mariupol
Zhagaly
Novaya Kakhovka
Yeysk
Sal'sk
Kherson
Lake Manych Gudilo
Elista
Utta
Krasnyy Yar
Astrakhan
Canyushikino
Karaton
18.
Sea of Azov
Sosyko
Yashkul'
Zelenga
Opornyy
Karkinitskiy Bay
Dzhankoy
Primorsko-Akhtarsk
Tikhoretsk
Divnoye
Mumra
Poqva
Crimea
Kerch'
Kavkaz
Kropotkin
Kugulta
Kuban'
Ulan Khol
Kultay
45°N
Simferopol'
Armavir
Stavropol'
.1259
Feodosiya
Krasnodar
Kuma
Velichayevskoye
U
Sevastopol'
Jalda
Novorossiysk
Kulaly
Maykop
Kochubey
.44
Fort-Ševčenko
Čerkessk
C A S P I A N S E A
Tuapse
Pjatigorsk
Say Ute
2867
.3238
Shetpe
P
Kislovodsk
Terek
Prochladnyj
Kiz'lar
Soči
Elbrus 5642
Nal'čik
-.25
Aktau
.-132
Groznyj
Suchumi
Machačkala
Vladikavkaz
Kyzyk
BLACK SEA
5047
Kazabeg
.4151
Fetisovo
Cape Ince
Sinop
Kutaisi
Poti
GEORGIA
Derbent
Kara-Bogaz-Gol
Batumi
Chasuri
Tbilisi
Pontic Mountains
Bekdaš
Kastamonu
Karabuk
Samsun
Hopa
.3438
Mamneuli
Rustavi
Kuba
Kara-Bogaz-Gol
.2565
Fatsa
Trabzon
Ardahan
Kazach
Mingecaurskoje Reservoir
.2205
Gerede
2062
Giresun
3065
Kars
1337
Gjumri
Wanadsor
Gandsya
Sumgait
Karshi
Kizilirmak
Corum
Turhal
Gümüshane
Lake Sevan
Jevlach
Achsu
Omchal
ARMENIA
AZERBAIJAN
Baku
Žioj
40°N
Ankara
Kirikkale
Sivas
Erzurum
Kagizman
Yerevan
Agdam
Turkmenbashi
Yozgat
2740
Kara Dağ
Erzincan
Askale
Aras
Ararat 5156
Agri
Saljany
2345
Divriği
AZER.
Araks
Čeleken
Anatolia
Patnos
Maku
Nachičevan
Ožalilabad
TURKEY
Bingöl
.4434
Keban Lake
Ogurčinskij
Lake Tuz
Gürün
Elaziğ
Murat Nehri
Lake Van
Van
Ahar
Kayseri
Tatvan
Khoy
Marand
Astara
Aksaray
.3916
Ardebil
Malatya
.4811
Tabriz
1362
Konya
.2500
Ar Dağ
Diyarbakir
Lake Urmia
Niğde
Kurtalan
Siverek
Hakkâri
Oroumieh
Maragheh
Mianeh
Bandar Anzeli
Maras
Kozan
.4168
Rasht
Çekm
Ereğli
3488
Karaman
Ceyhan
Gaziantep
Urfa
Kiziltepe
Nusaybin
Tigris
.3080
Sefid
Adana
Mersin
Ramsar
Taurus
Iskenderun
Elburz Mountains
.1798
Al Hasakah
.1463
Mahabad
Zanjan
.3719
Amol
Ghaem Shahr
SYRIA
Mosul
Silifke
Anamur
Antakya
Aleppo
Maskana
Sinjar
Tall 'Afar
Arbil
Saqqez
Qazvin
Takestan
Damavand
Cape Anamur
MEDITERRANEAN SEA
Raqqah
IRAQ
Zab
Baneh
Qojur
I R
5671
Damavand
CYPRUS
Cape Andreas
Idlib
Euphrates
Suwar
Lesser Zab
Bijar
Lattakia
Asi
Sharqat
Kirkuk
Sulaimaniyah
Razan
Karaj
Tehran
Semnan
35°E 40°E 45°E 50°E

35

39

n o p q r s t u v w x y z

This map shows 1/60 of the earth's surface

51

a b c d e f g h i j k l m

60°E 65°E 70°E 75°E 55°N 50°N 45°N 40°N

Kungur, Pervoural'sk, Yekaterinburg, Talitsa, Bogdanovich, Kamensk-Ural'skiy, Krasnoufimsk, Degtyarsk, Sysert', Nyazepetrovsk, Kasli, Shadrinsk, Yalutorovsk, Golyshmanovo, Ishim, Panovo, Tevriz, Tara, Bol'sherech'ye, Tyukalinsk, Biaza, Pokrovka, Nazyvayevsk, Tatarsk, Barabinsk

Asha, Min'yar, Suleya, Zlatoust, Chelyabinsk, Shumikha, Kurgan, Makushino, Petukhovo, Petropavlovsk, Lyubinskiy, Omsk, Isil'kul, Kalachinsk, Chistoozernoye, Lake Chany, Kupino

Chernikovsk, Ust' Kata, Kurtamysh, Chudinovo, Plast, Troitsk, Ust'-Uyskoye, Presnovka, Presnogor'kovka, Petrovka, Lake Ul'kenkaroy, Cherlak, Beloretsk, Verkhneural'sk, Komsomolets, Borovskoye, Dem'yanovka, Mar'yevka, Krasnoarmeysk, Kzyltu, Zhelezinka, Karasuk, Magnitogorsk, Varna, Kaga, Kartaly, Kustanay, Uritskiy, Peski, Volodarskoye, Kokchetav, Lake Selety-Tengiz, Lake Azhbulat, Kachiry

Stavropolka, Ruzayevka, Stepnyak, Aydabul, Makinsk, Aksu, Bestobe, Shuga, Tobol, Baymak, Bredy, Kushmurun, Pavlodar, Jamyševo, Troitskoye, Saraktash, Krasnoyarskiy, Dznetygara, Yesil, Dzhaksy, Atbasar, Alekseyevka, Tortkuduk, Yermak, Ekibastuz, Orsk, Terensay, Akkarga, Dzhambul, Naurzum, Zhaltyr, Zholymbet, Yermentau, Maykain, Mednogorsk, Novoishimskiy, Astana, Novodolinka, Dombarovskiy, Derzhavinsk, Sabyndy, Karashoky

S t e p p e, Tolybay, Aksuat, Arkalyk, Lake Tengiz, Kurgal'dzhino, Aktau, Ajryk, Khrom-Tau, Karabutak, Temirtau, Ul'yanovskoye, Alga, Akkabak, Turgay, Shenber, Sonaly, Saran, Karaganda, Abay, Korobovskiy, Kiikkaškan, Karagayly, Kajnar

Temir, Emba, Irgiz, Saga, Brali, Ulu Tau, Zahksykan, Dar'inskiy, Kazak Uplands, Atasu, Uspenskiy, Nura, Myylybulak, Aulkeldy, Shakhty, Ulutau, Kyzyl-Dzhar, Agadyr, Kyzyluy, Nikol'skiy, Baykonyr, Dzhezkazgan, Zharkamys, Chelkar, Ayshirak, Zhamshi

KAZAKHSTAN

Togyz, Kiik, Mointy, Chushakyl', Beleuty, Sarysu, Sajak, Sokyrbulak, Akespe, Aral'sk, Balkhash, Karazhingil, Lake Balkhash, Tomar, Karabas, Kokaral, Betpak-Dala Steppe, Bet-Pak-Dala, Kulanoy, Bugun, Lake Arys, Kashkanteniz, Karoy, Barsa-Kel'mes, Kazalinsk, Leninsk, Dzhusaly, Aral Sea, Vozrozhdeniya, Zhanay, Mynaral, Kuyygan, Burylbaytal, Uštobe, Kamkaly, Taldy-Kurgan, Kyzyl-Orda, Erimbet, Uzynkair, Algatart, Aksumbe, Furmanovka, Chilli, Syrdarja, Khantau, Aktogaj, Saryozek, Yany-Kurgan, Uyuk, Kapčagajskoje Reservoir, Kapčagaj, Šatlyk, Muinak, Urga, Lake Sudočje, Kazakdarya, Kentau, Čilik, Čemolgan, Almaty, Turkestan, Kara Tau, Tatty, Kaskalen, Karaozek, Chimbay, Kungrad, Kyzyl Kum, Džambul, Lugovoi, Bishkek, Kara-Balta, Ala Tau, Chodzeili, Lake Sarykamyšskoje, Bolševik, Kun'a Urgenč, Mynbulak, Uchkuduk, Arys, Čimkent, Issyk-Kul', Lake Issyk-Kul', Prževal'sk, Ottuk, Tašauz, Turtkul', Urgenč, Zarafshan, Toktogul, Toktogul Res., Čajek, Tien Shan, Naryn, Karagay, Karasaj, Čirčik, Tashkent, Taš-Kumyr, KYRGYZSTAN, UZBEKISTAN, Lebap, Cardara, Jangijul', Angren, Namangan, Kok-Jangak, Pik Dankowa, Toxkan, Čatyrtaš, Cardarinskoje Reservoir, Gulistan, Andižan, Kokand, Margilan, Oš, Lake Catyrk'ol, Sari Bulak, Akqi

Darvaza, Gorel'de, Amudarja, Gizhduvan, Navoi, Džizak, Khujand, Kajrakkumskoje Reservoir, Bekabad, Fergana, Gul'ča, Bucharа, Kagan, Kattakurgan, Samarkand, Ura Jube, Daraut Kurgan, Irkeštam, Sugun, Sanchakou, TURKMENISTAN, Jerbent, Kara Kum, Kabakly, Alat, Mubarek, Ajni, Alai, Lenina, Kashgar, Yopurga, Čardžou, Šachrisabz, TAJIKISTAN, Lake Karakul, Opal, Novabad, Mt. Communism, Dzirgatal, Kizyl Arvat, Bachardok, Karši, Markit, Repetek, Dushanbe, Vichary, Arkbajtal, Murgab, Bulunkol, Kungur, Shache, Arčman, Denau, Pamir, CHINA, Ashgabat, Tezejet, Mary, Karakumskiy Canal, Bajram Ali, Kerići, Kul'ab, Mamazair, Hasalbag, Yecheng, Tedžen, Kurgan-T'ube, Chorog, Muji, Artyk, Bojnurd, Dušak, Iolotan, Murgab, Termez, Dusti, Quchan, Andkhoy, Aqcha, Khulm, Kunduz, Faizabad, Qala Panja, Misgar, Mazar, Sheberghan, Mazar-i-Sharif, Taliqan, Zebak, Kush, Karakoram Range, Sarakhs, Afghanistan, Sabzevar, Mashhad, Takhta Bazar, Sar-i-Pul, Aibak, Hindu Kush, Tirich Mir, Mastuj, Rakaposhi, Neishabur, Maimana, Baghlan, PAKISTAN, Gilgit, Chitral, K2, Qaisar, Bala Murghab, Doab-i Mikhe Zarin, Dosh, Balkhab, Drosh, Indus, Chilas, Ronda

4876, 4503, 5382, 4929, 4641, 5509, 7134, 7495, 4643, 7719, 2227, 6083, 7487, 2243, 3117, 3416, 6525, 7690, 7788, 7228, 8611, 5715, 293, 2165, 2176, 1506, 3817, 592, 335, 473, 146, 81, 224, 343, 59, 391, 447, 633, 621, 603, 887, 122, 124, 142, 316

56

n o p q r s t u v w x y z

60

0 100 200 300 miles — Average linear scale — 0 100 200 300 400 500 Km

51

a	b	c	d	e	f	g	h	i	j	k	l	m

Onegva Yar
Vasyuganye
80°E
Staritsa
Kolpashevo
Ket
85°E
Belyy Yar
Mogochin
Baturino
Komsomol'sk
Vorozheyka
90°E
Yeniseysk
Lesosibirsk
Strelka
Galanino
95°E
Angara
Boguchany
Rodina
Čuna
Oktyabr'skiy
Parabel
Parbig
•142
L'vovka
Čulym
Tegul'det
Meletsk
•211
Altat
Birilyussy
Predivinsk
Asansk
Shelayevo
Bakchar
Asino
Moryakovskiy Zaton
•258
Tara
Biaza
•166
Tomsk
Yurga
Mariinsk
Bogotol
Achinsk
Pamyat
Shivera
•898
Aban
Kan
Kansk
Neva
Pikhtovka
Anzhero-Sudzhensk
Nazarovo
Krasnojarsk
Pokrovka
Chumakovo
Ob
Kemerovo
Uzhur
•818
Mana
Uyar
Borodino
Zamzor
Barabinsk
Chulym
Ob'
Tsentral'nyy
Čulym
Krasnoyarskoye Reservoir
Bujedžul'
Mina
Sum
Nižneudinsk
55°N
Novosibirsk
Tom'
R
Sira
U
Bellyk
•1778
S
Gutara
Lake Chany
Chistoozernoye
Ordynskoye
Krasnobrodskij
Kuragino
Art'omsk
Kupino
Cherepanovo
T'agun
Kisel'ovsk
Sorsk
Kazyr
Pokrovsk
Kamen-na-Obi
Suzun
Tal'menka
Prokopjevsk
Novokuzneck
•2178
Černogorsk
Minusinsk
Burgon
Grandioznyj •2922
Alygdžer
Karasuk
Ob'
Meždurečensk
Birikčul
Abakan
Lake Azhbulat
Khabary
Barnaul
Pavlovsk
Troickoje
Mundybaš
Abakan Mts
Šušenskoje
Vostochnyy
Len'ki
•286
Rebricha
•621
Taštagol
Taštyp
Sajanogorsk
Sayan Mountains
Kachiry
Lake Kulundinskoje
Blagoveščenka
Bijsk
Bija
•2456
Idžim
Bujba
Seyi
Kulunda
Rodino
Alejsk
Turočak
Abakan
Sayan
Jenisej
•2682
Toora-Chem
Bol'Yenisey
Pavlodar
Jamyševo
Pospelicha
Alej
Altejskij
Altai
Kyzyl
Saryg-Sep
Ust'-Bel'dir
Kyzyl-Chem
Yermak
Kulundinskaya
Ščerbakty
Steppe
Rubcovsk
Čaryšskoje
Lake Teleckoje
Čel'us
•2930
Sagonar
Ak-Dovurak
Cadan •2972
Balgazyn
Irtysh
•1206
Gorn'ak
Tuekta
Čodro
•3487
Tannu Mountains
•2668
Ajryk
Molgary
Dolon
Bel'agaš
Semonaicha
•2820
Inja
Kyzyl-Chaja
Orog Nuur
Uvs Nuur
Samagaltaj
•2584
Šagan
•606
Semipalatinsk
•2776
Katun
Kuraj
Koš-Agač
Turgen
Naryn
50°N
Ust'-Kamenogorsk
•4506
Argut
•4029
Malčin
Cagaan-Uul
Kiikkaškan
Serebr'ansk
Čarsk
Georgiievka
•1608
Cagaan Nuur
Chovd Gol
Chirgis Nuur
Baruun Turuun
Bajan-Uul
Kajnar
Bestamak
Bol'šenarymskoje
Cagaan Gol
Ulgij
Ölgij
•2928
Ojgon Nuur
Karaaul
Kokpekty
Lake Markakol'
•2648
Erdene Büren
Telmen Nuur
Dagandel
Žarma
Kurčum
Chungui Gol
K A Z A K H S T A N
•1305
Korti Linchang
Altaj
Tolbo Nuur
Char Nuur
•2896
Jaruu
Hangay
Madenijet
Ajaguz
Lake Zajsan
Buran
Altay
Kobdo
Char Us Nuur
Aldar
Uliastaj
•3906
Tarbagatay Mts
Belaja Škola
Ertix He
Burqin
•3743
Dörgön Nuur
Dzavchan Gol
Bujant
Taskesken
Zajsan
Muz Tau
Beitun
Fuyun
Manchan
Dzereg
M
O
Sajak
Aktogaj
Urdžar
•2992
•3816
Ulungur Hu
Jili Hu
Sarbulak
•4362
Mönch Chajrchan
•3578
Cagaan-Olom
Lake Sasykkol
Tacheng
Utubulak
Ovoot
Lake Balkhash
Ajaguz
Lepsy
•756
Žarsuat
Lake Alakol'
Ulungur
Ertai
Tamč
Delger
Karabas
Mataj
Urho
Bulgan
Altaj
55
Karatal
•603
Toli
Karamay
Manas Hu
Türgen
Sarkand
•2923
Tachakou
Junggar Pendi
Bugat
Beger
Uštobe
•4442
Ebinur Hu
•3479
Dzachuj
Taldy-Kurgan
45°N
Wenquan
Bole
Altaj
Bajan-Ondor
Sayram Hu
Jinhe
Jiangjumiao
Gov'Chonin
•3802
Aktogaj
Alatau
Sarydzek
Usu
Shihezi
Jungarian Gobi
Santanghu
Panfilov
Borohoro Shan
Manas
Ganhezi
Qitai
Kapčagaj
Kapčagayskoye Reservoir
Yining
Nilka
Changji
Bogda Feng •5445
Barkol Hu
Nom
Ili He
Qapqal
•5500
Urumchi
Qijiaojing
Barkol Kazak
Čilik
Cundža
Tekes
Xinyuan
Narat
•3951
Cemolgan
Almaty
•3638
Zhaosu
Houxia
Qiquanhu
Liaodun
Yiwu
Kaskelen
•4876
Kegen
Tien Shan
Baiyanghe
Ewirgol
Shanshan
Liushuquan
Hami
•4925
Karlik Shan
Rubačje
Ananjevo
Narynkol
Kaidu
Bulguntay
Turpan
Metgol
Lake Issyk-Kul'
Prževal'sk
•4553
•154
Yandun
Mingshui
G
Ottuk
Pik Pobedy
Keyi
Yengisar
Yanqi Huizu Zizhixian
Gongpoquan
KYRGYZSTAN
•7439
Qarqi
Bosten Hu
Weiya
Xingxingxia
Naryn Taragay
Yakrik
Kuqa
Dalaoba
Korla
•1524
•2584
Jiangjuntai
Karasaj
Karayulgun
Xinho
Yuli
Daquan
Pik Dankova
Toxkan
Aksu
Tarim
Kongi
•1238
Bei Shan
Hongliuyuan
Jiangquanzi
Čatyrtaš
•5982
Akqi
Tarim
Tarim Liuchang
•1762
Anxi
Qiaowan
•4929
Sari Bulak
Awat
Aral
Lop Nur
Zhangjiaquan
Shule
Yengisu
Kumkuduk
Dunhuang
Choushi
40°N
Sanchakou
Bachu Liuchang
Dongbatu
Sugun
•1066
Ikanbujmal
Shazaoyuan
Jiayuguan
Tarim Basin
•1082
•1099
C
H
Changma
Jiuquan
Yarkant
Hotan
Luobuzhuang
Dongluk
Aksay
Qilian Shan
•5547
Yopurga
Takla Makan Desert
Miran
•5798
Tai
Markit
Qargan
Ruoqiang
Altun Shan
Lo Liang
Suhai Hu
Shule
Aktaz
Waxxari
Xorkol
Niubiziliang
•2774
Lenghu
Huahaizi
East Turkestan
•1570
Tongguzbasti
•5827
Shacha
•5810
Youshashan
Tsaidam Basin
Iqe
Har Hu
Koxlax
Qiemo
Gas Hu
Hasalbag
Yecheng
Hadilik
Ayakkum Hu
Mangnai
Shaliangzi
Da Qaidam
Muji
Andirlangar
•6140
Suli Hu
•3099
•5030
Zangguy
Zawa
Hotan
Qira
Minfeng
Tura
Aqqikkol Hu
Hoit Tana
Delingha
Xitieshan
Karakax
Bostan
•6748
Aktag
Nur Turu
Dabsan Hu
Qarhan
Ulan
Mazar
Tekiliktag •5466
Yutian
Karasay
Kunlun Shan
Muztag •7723
De Juh
Nan Hulsan Hu
Yarkant
Golmud
Bolontay
•7228
Kangxiwar
80°E
Pulu
85°E
90°E
•7720
95°E
Nomhon
•5026
Du
Xiang

n	o	p	q	r	s	t	u	v	w	x	y	z

61

This map shows 1/60 of the earth's surface

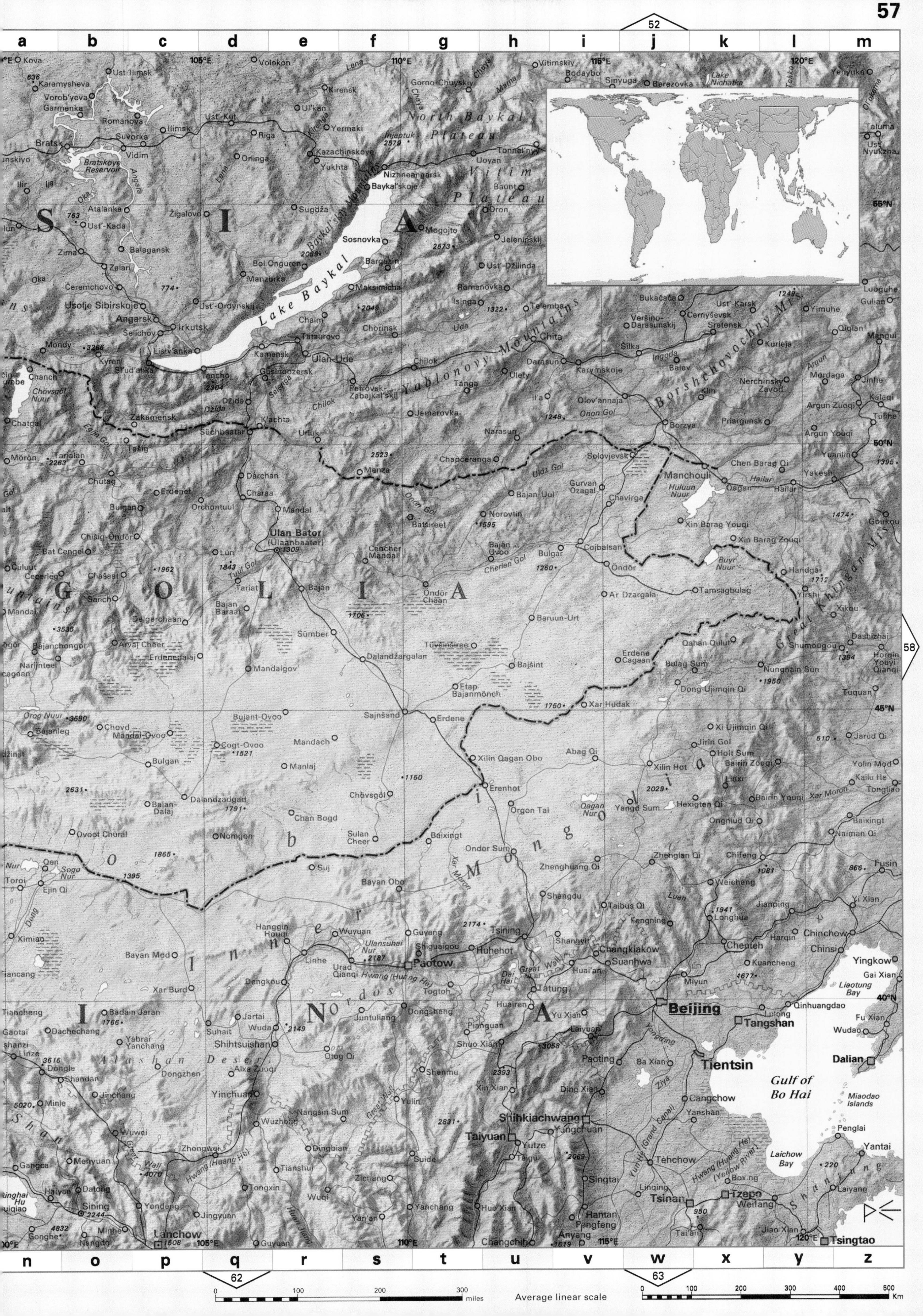

52
a b c d e f g h i j k l m
105°E
110°E
115°E
120°E
Kova
Ust'-Ilimsk
636
Karamysheva
Vorob'yeva
Garmenka
Romanova
Suvorka
Ilimsk
Ust'-Kut
Riga
Volokon
Lena
Kirensk
Ul'kan
Yermaki
Kirenga
Gorno-Chuyskiy
Chuya
Chaya
Vitimskiy
Bodaybo
Sinyuga
Berezovka
Lake Nichatka
Tokko
Yenyuka
Olekma
Bratsk
Vidim
unskiyo
Bratskoye Reservoir
Angara
Orlinga
Kazachinskoye
Yukhta
Injaptuk
2579
North Baykal Plateau
Tonnel'nyy
Uoyan
Taluma
Ust'-Nyukzha
Nizhneangarsk
Baykal'skoje
Vitim Plateau
Baunt
Ilir
Ija
Oka
Atalanka
763
Ust'-Kada
Žigalovo
Sugdža
Oron
55°N
S I A
Baykal'skiy Mountains
Mogojto
Sosnovka
Jeleninskij
2573
Zima
Balagansk
Zalari
Bol Onguren
2089
Barguzin
Ust'-Džilinda
Manzurka
774
Lake Baykal
Oka
Čeremchovo
Maksimicha
Romanovka
Usolje Sibirskoje
Ust'-Ordynskij
Isinga
1322
Talemba
2049
Angarsk
Šelichov
Irkutsk
Chaim
Chorinsk
Uda
Chita
Mondy
3286
Listv'anka
Tataurovo
Kyran
Kamensk
Ulan-Ude
Sl'ud'anka
Tencho
Gusinoozersk
Chilok
Yablonovy Mountains
Uleto
Darasun
Karymskoje
2304
Selenga
Petrovsk-Zabajkal'skij
Tanga
Chanch
Chövsgöl Nuur
Džida
Chilok
Il'a
Olov'annaja
Zakamensk
K'achta
Jamarovka
1248
Onon Gol
Chatgal
Egiin Gol
Süchbaatar
Urluk
Narasun
Bukačača
Ust'-Karsk
1249
Yimuhe
Veršino-Darasunskij
Černyševsk
Srotensk
Qiqian
Šilka
Kurleja
Mangui
Ingoda
Balej
Borshchovochny Mts.
Nerchinsky Zavod
Klin
Argun
Mordaga
Jinhe
Kalagi
Argun Zuoqi
Borzya
Priargunsk
Tulihe
Argun Youqi
50°N
Tesig
Mörön
Tarjalan
2263
2523
Chapčeranga
Solovjevsk
Manchouli
Chen Barag Qi
Yuanlin
1395
Manza
Uldz Gol
Hailar
Qagan
Yakeshi
Chutag
Darchan
Gurvan Ozagal
Huluun Nuur
Erdenet
Charaa
Bajan-Uul
Chavirga
Bulgan
Orchontuul
Mandal
Onon Gol
Noroytin
1474
Goukou
Batšireet
1595
Xin Barag Youqi
Chišig-Öndör
Ulan Bator (Ulaanbaatar)
1309
Xin Barag Zuoqi
Bat Cengel
Lün
Cencher Mandal
Bajan Ovoo
Bulgar
Cojbalsan
Buyr Nuur
Great Khingan Mts.
Čuluut
1962
1843
Tuul Gol
Cherlen Gol
1260
Öndör
Handgai
1712
Cecerleg
Čašaat
Tariat
Yirshi
Sanch
G O L I A
Bajan
Öndör Chaan
Ar Dzargala
Tamsagbulag
Mandal
Bajan Baraat
1706
Xikou
Delgerchaan
Baruun-Urt
3535
Dashizhai
Bajanchongor
Arvaj Cheer
Sümber
Tüvšinširee
Qahan Qulut
Shumougou
58
1394
Horqin Youyi Qianqi
Narijnteel
Erdenedalaj
Dalandžargalan
Erdene Cagaan
Bulag Sum
Nungnain Sum
cagaan
Mandalgov'
Bajšint
1950
Etap Bajanmönch
Dong Ujimqin Qi
Tuquan
Orog Nuur
3590
Bujant-Ovoo
Sajnšand
1750
Xar Hudak
45°N
Bajanleg
Chovd
Mandal-Ovoo
Erdene
Xi Ujimqin Qi
Mandach
Jarud Qi
Cogt-Ovoo
1521
Jirin Gol
510
Bulgan
Manlaj
Xilin Qagan Obo
Abag Qi
Holt Sum
Xilin Hot
Bairin Zuoqi
Yolin Mod
1150
Kailu He
2631
Dalandzadgad
Chövsgöl
Erenhot
2029
Linxi
Bajan-Dalaj
1791
Orgon Tal
Qagan Nur
Yangd Sum
Hexigten Qi
Bairin Youqi
Xar Moron
Tongliao
Chan Bogd
Ongniud Qi
Baixingt
Ovoot Chural
Nomgon
Sulan Cheer
Baixingt
Naiman Qi
1865
Ondor Sum
Nur
Oen
Sogo Nur
1395
Suj
Zhenglan Qi
Chifeng
Fusin
Torol
Ejin Qi
Zhenghuang Qi
1081
866
Bayan Obo
Xar Moron
Weichang
Yi Xian
Inner Mongolia
Dong
Shangdu
Taibus Qi
Luan
Jianping
Fengning
1941
Longhua
Ximiao
Hanggin Houqi
Wuyuan
Guyang
2174
Tsining
Shangyi
Chinchow
Harqin
Ulansuhai Nur
Shiguaigou
Changkiakow
Chengteh
Chinsi
Bayan Mod
Linhe
2187
Huhehot
Suanhwa
Kuancheng
Yingkow
Urad Qianqi
Paotow
Dai Hai
Great Wall
Huai'an
Gai Xian
Tiancang
Hwang (Huang He)
4677
Liaotung Bay
Dengkou
Togtoh
Tatung
Miyun
Xar Burd
Ordos
40°N
Huairen
Beijing
Qinhuangdao
Tiancheng
Badain Jaran
1766
Jartai
Juntuliang
Dongsheng
Yu Xian
Tangshan
Fu Xian
Gaotai
Dachechang
Suhait
Wuda
2149
Pianguan
Laiyuan
Wudao
Yabrai Yanchang
Shihtsuishan
Shuo Xian
3058
Yongding
shanzi
Linze
Otog Qi
Dalian
3616
Dongle
Alashan Desert
Paoting
Ba Xian
Tientsin
Shandan
Dongzhen
Alxa Zuoqi
Shenmu
2393
Gulf of Bo Hai
Jinchang
Yinchuan
Ziya
5020
Minle
Xin Xian
Ding Xian
Miaodao Islands
Yulin
Cangchow
Nangsin Sum
2831
Shihkiachwang
Yanshan
Wuwei
Wuzhong
Yangchuan
Penglai
Great Wall
Dingbian
Taiyuan
Yutze
Yantai
Menyuan
Zhongwei
Hwang (Huang He)
Suide
2069
Tehchow
Laichow Bay
Gangca
4070
Tianshui
Taigu
Yun He (Grand Canal)
Hwang (Huang He) (Yellow River)
220
Haiyan
Datong
Zichang
Singtai
Box ng
Tongxin
Linqing
Laiyang
Qinghai Hu
Sining
Wuqi
Tsinan
Tzepo
Weifang
Shantung
2244
Yondeng
Huan Jiang
Yanchang
Huo Xian
Hantan
950
Jingyuan
Yan'an
Fengfeng
4832
Minhe
Lanchow
Tai'an
Jiao Xian
Gonghe
Nangdoi
1508
Guyuan
Changchih
Anyang
Tsingtao
1619
n o p q r s t u v w x y z
62
63
0 100 200 300 miles
Average linear scale
0 100 200 300 400 500 Km

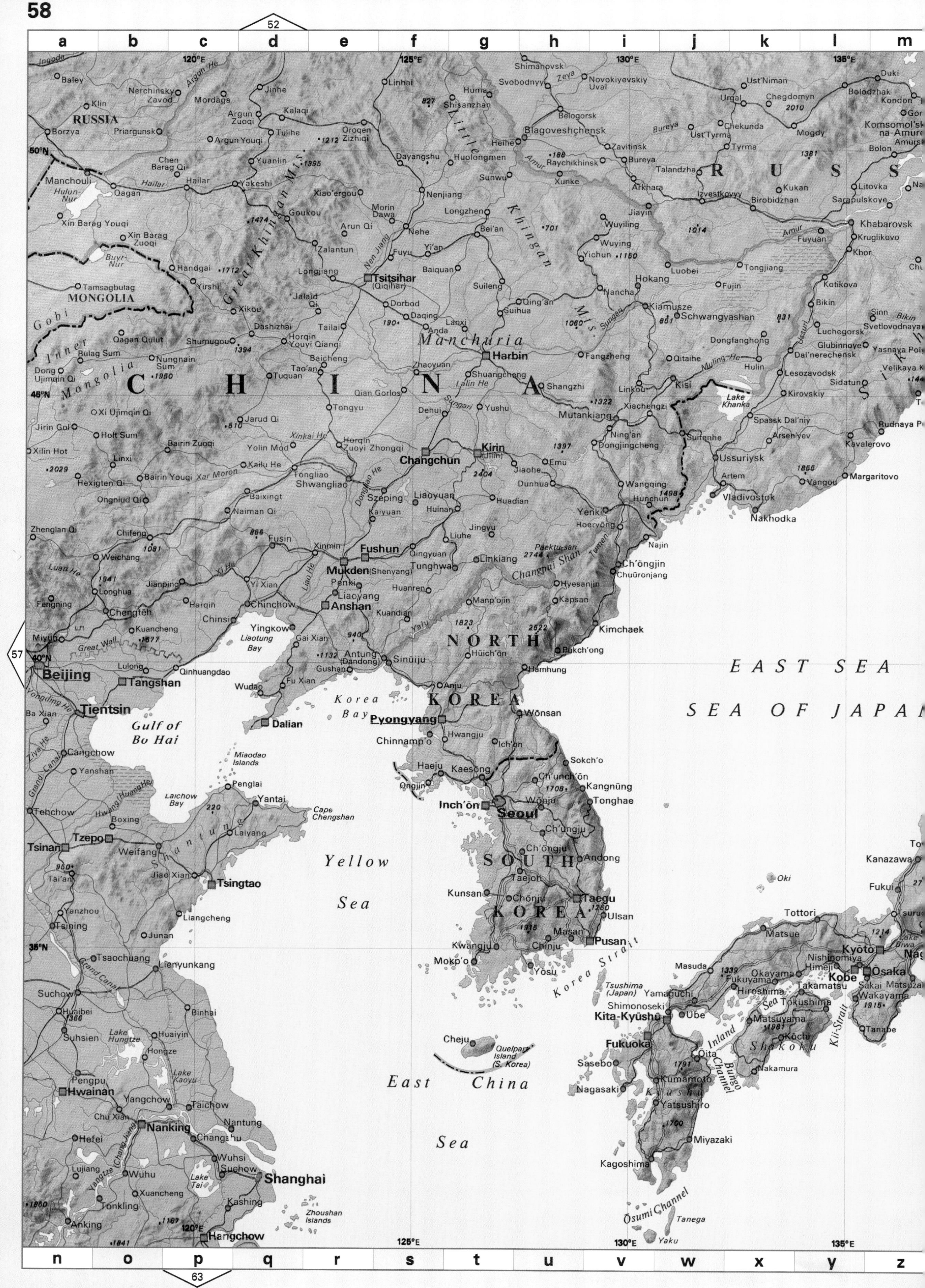
52
a b c d e f g h i j k l m
120°E
125°E
130°E
135°E
RUSSIA
MONGOLIA
CHINA
Manchuria
Inner Mongolia
Gobi
Great Khingan Mts.
Little Khingan
Khingan Mts.
NORTH KOREA
SOUTH KOREA
RUSS
EAST SEA
SEA OF JAPAN
Yellow Sea
East China Sea
Gulf of Bo Hai
Korea Bay
Liaotung Bay
Laichow Bay
Korea Strait
Inland Sea
Bungo Channel
Kii Strait
Ōsumi Channel
Shikoku
Kyushu
Shantung
Changpai Shan
Paektu-san 2744
Great Wall
Amur
Sungari
Nen Jiang
Hailar
Argun He
Zeya
Bureya
Ussuri
Muling He
Tumen
Yalu
Liao He
Xi He
Luan He
Xar Moron
Xinkai He
Dongliao He
Lalin He
Yongding He
Ziya He
Grand Canal
Hwang (Huang) He
Yangtze (Chang Jiang)
Ingoda
Hulun Nur
Buyr Nur
Lake Khanka
Lake Hungtze
Lake Kaoyu
Lake Tai
Miaodao Islands
Zhoushan Islands
Cape Chengshan
Quelpart Island (S. Korea)
Tsushima (Japan)
Oki
Tanega
Yaku
50°N
45°N
40°N
35°N
Baley
Klin
Borzya
Priargunsk
Nerchinsky Zavod
Mordaga
Argun Zuoqi
Argun Youqi
Jinhe
Kalaqi
Tulihe
Oroqen Zizhiqi
Linhai
Huma
Shisanzhan
Shimanovsk
Svobodnyy
Novokiyevskiy Uval
Belogorsk
Blagoveshchensk
Heihe
Zavitinsk
Raychikhinsk
Bureya
Xunke
Arkhara
Ust'Niman
Urgal
Chegdomyn
Chekunda
Ust'Tyrma
Tyrma
Talandzha
Duki
Bolodzhak
Kondon
Komsomol'sk na-Amure
Amursk
Mogdy
Bolon
Gor
Manchouli
Qagan
Chen Barag Qi
Hailar
Yakeshi
Yuanlin
Dayangshu
Huolongmen
Sunwu
Nenjiang
Xiao'ergou
Morin Dawa
Longzhen
Goukou
Arun Qi
Nehe
Bei'an
Yi'an
Fuyu
Zalantun
Xin Barag Youqi
Xin Barag Zuoqi
Handgai
Yirshi
Tamsagbulag
Longjiang
Tsitsihar (Qiqihar)
Baiquan
Suileng
Jalaid Qi
Dorbod
Xikou
Daqing
Anda
Lanxi
Suihua
Qing'an
Dashizhai
Tailai
Qagan Qulut
Shumugou
Horqin Youyi Qiangqi
Bulag Sum
Nungnain Sum
Dong Ujimqin Qi
Baicheng
Tao'an
Tuquan
Zhaoyuan
Harbin
Shuangcheng
Qian Gorlos
Shangzhi
Tongyu
Dehui
Yushu
Xi Ujimqin Qi
Jarud Qi
Jirin Gol
Holt Sum
Xilin Hot
Bairin Zuoqi
Yolin Mod
Horqin Zuoyi Zhongqi
Linxi
Kailu He
Changchun
Kirin (Jilin)
Bairin Youqi
Hexigten Qi
Tongliao
Shwangliao
Szeping
Liaoyuan
Ongniud Qi
Baixingt
Naiman Qi
Kaiyuan
Huinan
Huadian
Jiaohe
Emu
Dunhua
Zhenglan Qi
Chifeng
Weichang
Fusin
Xinmin
Fushun
Qingyuan
Tunghwa
Liuhe
Jingyu
Linkiang
Mukden (Shenyang)
Penki
Liaoyang
Anshan
Jianping
Yi Xian
Chinchow
Harqin
Longhua
Fengning
Chengteh
Chinsi
Kuancheng
Huanren
Kuandian
Manp'ojin
Miyun
Yingkow
Gai Xian
Antung (Dandong)
Sinŭiju
Gushan
Fu Xian
Wudao
Dalian
Beijing
Lulong
Tangshan
Qinhuangdao
Tientsin
Ba Xian
Cangchow
Yanshan
Tehchow
Tzepo
Tsinan
Boxing
Weifang
Penglai
Yantai
Laiyang
Jiao Xian
Tsingtao
Tai'an
Yanzhou
Tsining
Liangcheng
Junan
Tsaochuang
Lienyunkang
Suchow
Huaibei
Binhai
Suhsien
Huaiyin
Hongze
Pengpu
Hwainan
Yangchow
Taichow
Chu Xian
Nanking
Nantung
Hefei
Changshu
Wuhsi
Suchow
Lujiang
Wuhu
Shanghai
Xuancheng
Tonkling
Kashing
Anking
Hangchow
Yichun
Wuyiling
Wuying
Jiayin
Hokang
Luobei
Tongjiang
Fujin
Nancha
Kiamusze
Schwangyashan
Fangzheng
Qitaihe
Dongfanghong
Linkou
Kisi
Hulin
Mutankiang
Xiachengzi
Ning'an
Dongjingcheng
Suifenhe
Wangqing
Hunchun
Yenki
Hoeryŏng
Najin
Ch'ŏngjin
Chuŭronjang
Hyesanjin
Kapsan
Kimchaek
Pukch'ong
Hŭich'ŏn
Hamhung
Anju
Wŏnsan
Pyongyang
Hwangju
Chinnamp'o
Ich'ŏn
Haeju
Kaesong
Ongjin
Sokch'o
Ch'unch'ŏn
Kangnŭng
Tonghae
Wŏnju
Inch'ŏn
Seoul
Ch'ungju
Ch'ŏngju
Andong
Taejon
Kunsan
Chŏnju
Taegu
Ulsan
Masan
Pusan
Chinju
Kwangju
Mokp'o
Yŏsu
Cheju
Birobidzhan
Kukan
Izvestkovyy
Khabarovsk
Kruglikovo
Khor
Litovka
Sarapulskoye
Fuyuan
Kotikova
Bikin
Luchegorsk
Svetlovodnaya
Glubinnoye
Dal'nerechensk
Yasnaya Pol
Lesozavodsk
Sidatun
Kirovskiy
Velikaya K
Spassk Dal'niy
Arsen'yev
Rudnaya P
Ussuriysk
Kavalerovo
Artem
Vangou
Margaritovo
Vladivostok
Nakhodka
Kanazawa
Fukui
Tottori
Matsue
Kyōto
Nishinomiya
Himeji
Kobe
Ōsaka
Masuda
Okayama
Fukuyama
Hiroshima
Takamatsu
Sakai
Matsuzaka
Wakayama
Tanabe
Tokushima
Matsuyama
Kōchi
Nakamura
Yamaguchi
Shimonoseki
Kita-Kyūshū
Ube
Fukuoka
Ōita
Sasebo
Kumamoto
Nagasaki
Yatsushiro
Miyazaki
Kagoshima
57
63
n o p q r s t u v w x y z

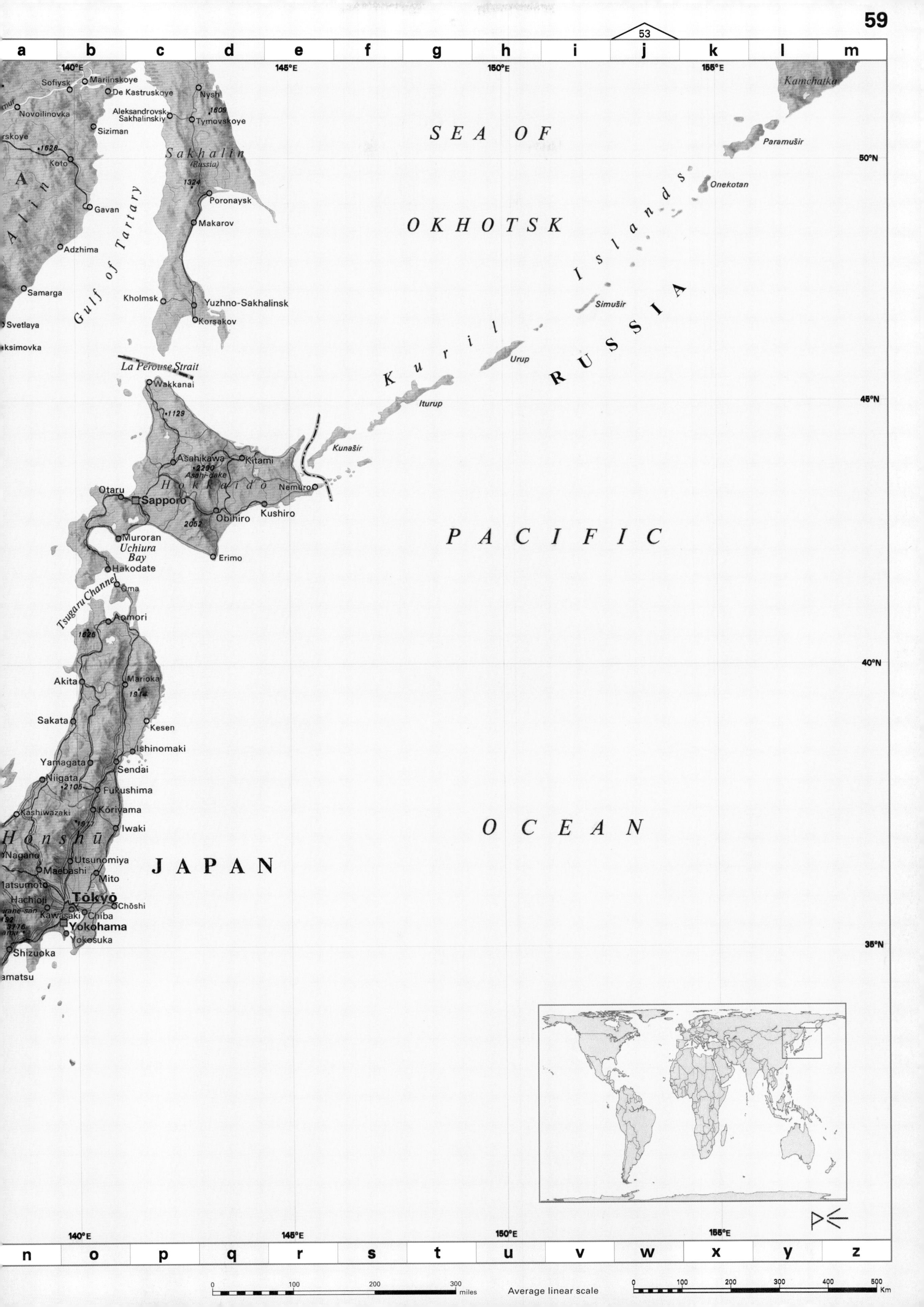

SEA OF
OKHOTSK
PACIFIC
OCEAN
JAPAN
RUSSIA
Kuril Islands
Sakhalin
(Russia)
Gulf of Tartary
La Perouse Strait
Hokkaido
Honshū
Tsugaru Channel
Uchiura Bay
Kamchatka
Paramušir
Onekotan
Simušir
Urup
Iturup
Kunašir
Mariinskoye
De Kastruskoye
Sofiysk
Novoilinovka
Aleksandrovsk-Sakhalinskiy
Tymovskoye
Nysh
Siziman
Koto
Gavan
Poronaysk
Makarov
Adzhima
Samarga
Kholmsk
Yuzhno-Sakhalinsk
Korsakov
Svetlaya
Wakkanai
Asahikawa
Asahi-dake
Kitami
Nemuro
Otaru
Sapporo
Obihiro
Kushiro
Muroran
Erimo
Hakodate
Ōma
Aomori
Akita
Morioka
Sakata
Kesen
Ishinomaki
Yamagata
Sendai
Niigata
Fukushima
Kashiwazaki
Kōriyama
Iwaki
Nagano
Utsunomiya
Maebashi
Mito
Tōkyō
Chōshi
Hachioji
Kawasaki
Chiba
Yokohama
Yokosuka
Shizuoka
140°E
145°E
150°E
155°E
50°N
45°N
40°N
35°N
a b c d e f g h i j k l m
n o p q r s t u v w x y z
53
0 100 200 300 miles
Average linear scale
0 100 200 300 400 500 Km

55
a b c d e f g h i j k l m
TURKMENISTAN
Quchan
Dusak
Mayamey
Sabzevar
Neishabur
Mashhad
Sarakhs
Andkhoy
Aqcha
Mazar-i Sharif
Khulm
Kunduz
Taliqan
Faizabad
Qala Panja
Sheberghan
Sar-i-Pul
Aibak
Baghlan
Zebak
Mastuj
Tirich Mir
Chitral
Takhta Bazar
Maimana
Qaisar
Bala Murghab
Kushka
Doshi
Doab-i Mikhe Zarin
Charikar
Drosh
Bardeskan
Kashmar
Torbat-e-Heidariye
Torbat-e-Jam
Quala-i Nau
Jawand
Bamian
Asadabad
Besham Qila
Dashi-e-Kavir
Tayebad
Chaghcharan
Daulat Yar
Kuh-e-Baba
Sorabai
Kabul
Jalalabad
Mardan
Bidokht
Ferdows
Herat
Hari Rud
Qarah Tarai
Behsud
Kota-i-Ashro
Khyber Pass
Peshawar
Abbottabad
Islamabad
Tabas
Qaen
Ghorian
Farsi
AFGHANISTAN
Sangan
Khurd
Gardez
Ghazni
Rawalpindi
Yazdan
Shindand
Deihuk
Matun
Kalabagh
Bannu
Aliabad
Khusf
Birjand
Qaisar
Farahrod
Uruzgan
Zarghunshar
Tarin Kot
Shahjui
Naiband
Sarbisheh
Farah
Nauzad
Dilaram
Qalat-i-Chilzai
Razmak
Lakki
Mianwali
Darband
Bafq
Ravor
Nehbandan
Lasho Joayin
Girishk
Kandahar
Khash Rud
Lashkargah
Arghandab
Tarnak
Sargodha
Dera Ismail Khan
Sulaiman
Faisalabad
Jhang Maghiana
Toba & Kakar Ranges
Sakir
Fort Sandeman
Zarand
IRAN
Namakzar-e Shadad
Zabol
Zaranj
Safar
Registan
Chaman
Muslimbagh
Rafsanjan
Kerman
Siraj
Mirabad
Helmand
Rudbar
Zargun
Qila Saifullah
Quetta
Loralai
Kingri
Leiah
Multan
Baghin
Nosratabad
Mach
Dera Ghazi Khan
Hoseinabad
Sirjan
Zahedan
Ribat
Chagai Hills
Chagai
Nushki
Sibi
Kahan
Sutlej
Bahawalpur
Tahrud
Laleh Zar
Baft
Darzin
Bam
Shurgaz
Mirjaveh
Sultan
Ras Koh
Dalbandin
PAKISTAN
Dera Bugti
Uch
Rajanpur
Chenab
Aliabad
Sabzevaran
Dehak
Taftan
Nok Kundi
Kharan
Kalat
Indus
Pugal
Rajasthan Canal
Dowlatabad
Hajiabad
Kahnuj
Bazman
Khash
Qila Ladgasht
Besima
Surab
Jacobabad
Rahimyar-Khan
Qotbabad
Baluchistan
Shikarpur
Khuzdar
Sukkur
Tanot
Hamun-e Jaz Murian
Saravan
Larkana
Khairpur
Bandar Abbas
Jaghin
Bampur
Iranshahr
Kuhak
Patandar
Jebri
Wad
Kirthar Range
Sri Mohangarh
Shahgarh
Bap
Qeshm
Minab
Panjgur
Makran Range
Pokaran
Jaisalmer
Remeshk
Sarbaz
Central
Moro
39
Straits of Hormuz
Ras Musandam
Al Sha'am
OMAN
Ras al Khaimah
Dibba
Awaran
Sehwan
Sanghar
Myajlar
Phalsund
Nikshahr
Pishin
Turbat
Hoshab
Bela
Balotra
Sharjah
Dubai
Fujairah
Jask
Bahu Kalat
Dasht
Kikki
Hab
Mirpur Khas
Barmer
Ras Kuh Lab
Chabahar
Jiwani
Ras Nuh
Pasni
Ormara
Kotri
Hyderabad
Gurha
Aravalli Range
Shinas
Gulf of Oman
Hab Chauki
Karachi
Thatta
Luni
Sirohi
Guru Sikhar
Al Batinah
Sohar
Al Ain
Badin
Virawah
Tharad
Indus Delta
Jati
Rann of Kutch
Palanpur
Al Khaburah
As Suwaiq
Radhanpur
Muscat
As Sib
Tropic of Cancer
Lakhpat
Mahesana
Sumail
Quraiyat
Ibri
Al Hajar al Sharqi
Rampur
Bhuj
New Kandla
Ahmedabad
Nazwa
Izki
ARABIAN
Morvi
Mandvi
Sur
Ras al Hadd
Gulf of Kutch
Jamnagar
Dhandhuka
Wadi Aswad
Adam
Dwarka
Rajkot
Khambhat
OMAN
Al Kamil
Kathiawar
Ramlat al Wahiba
Al Ashkhirah
Bharuch
Umm as Samim
SEA
Porbandar
Bhavnagar
Junagadh
Surat
Gulf of Cambay (Gulf of Khambhat)
Veraval
Diu
Valsad
Daman
Masirah
Duqm
Jawhar
Thane
Ras Madrakah
Mumbai (Bombay)
Sahil al Jazir
Sharbithat
Ras Sharbithat
Janjira
Kuria Muria Islands
Chiplun
Ratnagiri
35°N
30°N
25°N
20°N
60°E
65°E
70°E
n o p q r s t u v w x y z
64

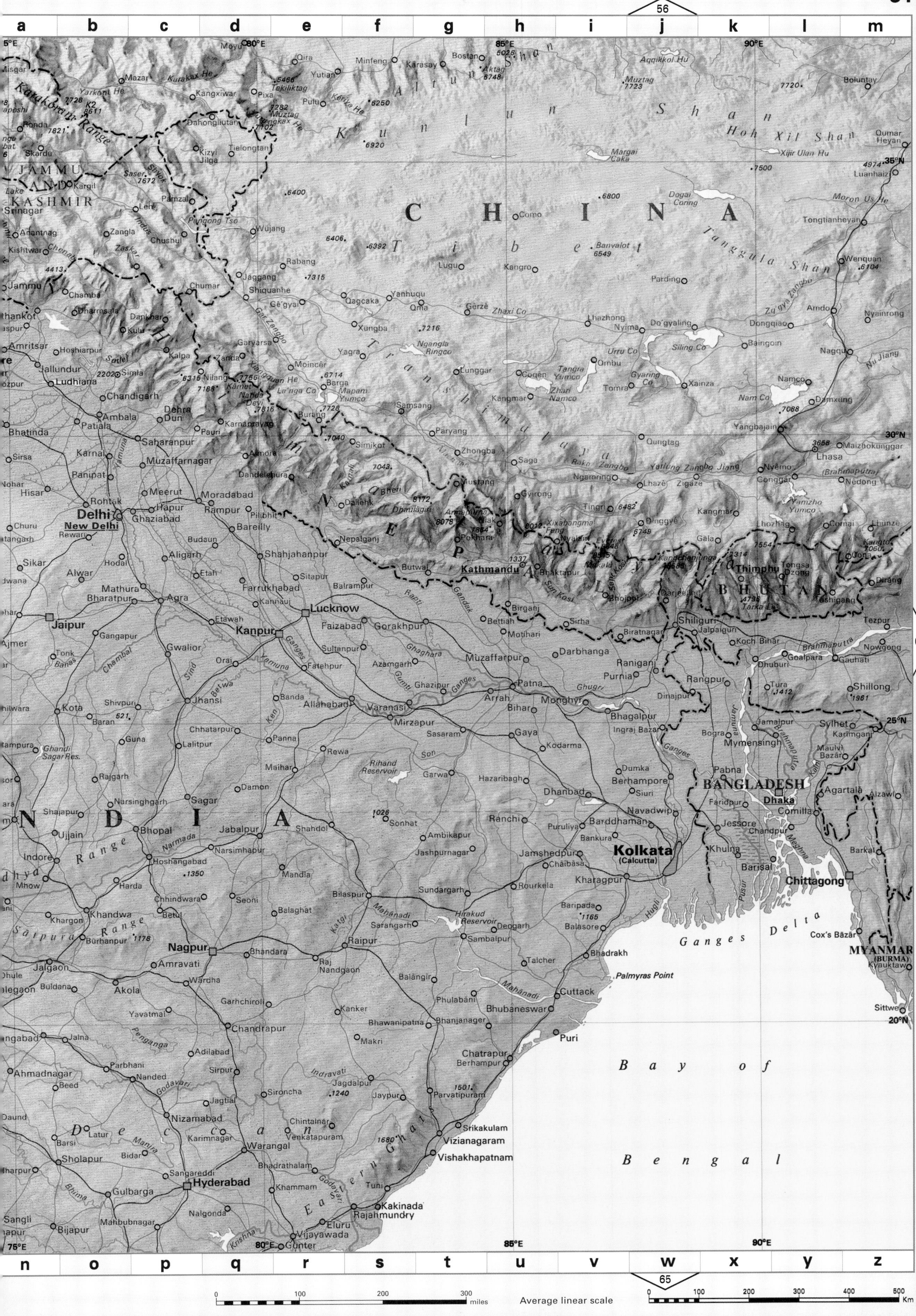

56

a b c d e f g h i j k l m

CHINA

Tibet

Kunlun Shan

Hoh Xil Shan

Tanggula Shan

Transhimalaya

JAMMU AND KASHMIR

NEPAL

BHUTAN

BANGLADESH

INDIA

MYANMAR (BURMA)

Kathmandu

Thimphu

Dhaka

New Delhi

Delhi

Lucknow

Kanpur

Jaipur

Kolkata (Calcutta)

Chittagong

Nagpur

Hyderabad

Ganges Delta

Bay of Bengal

Palmyras Point

Deccan

Eastern Ghats

62

n o p q r s t u v w x y z

65

0 100 200 300 miles

Average linear scale

0 100 200 300 400 500 Km

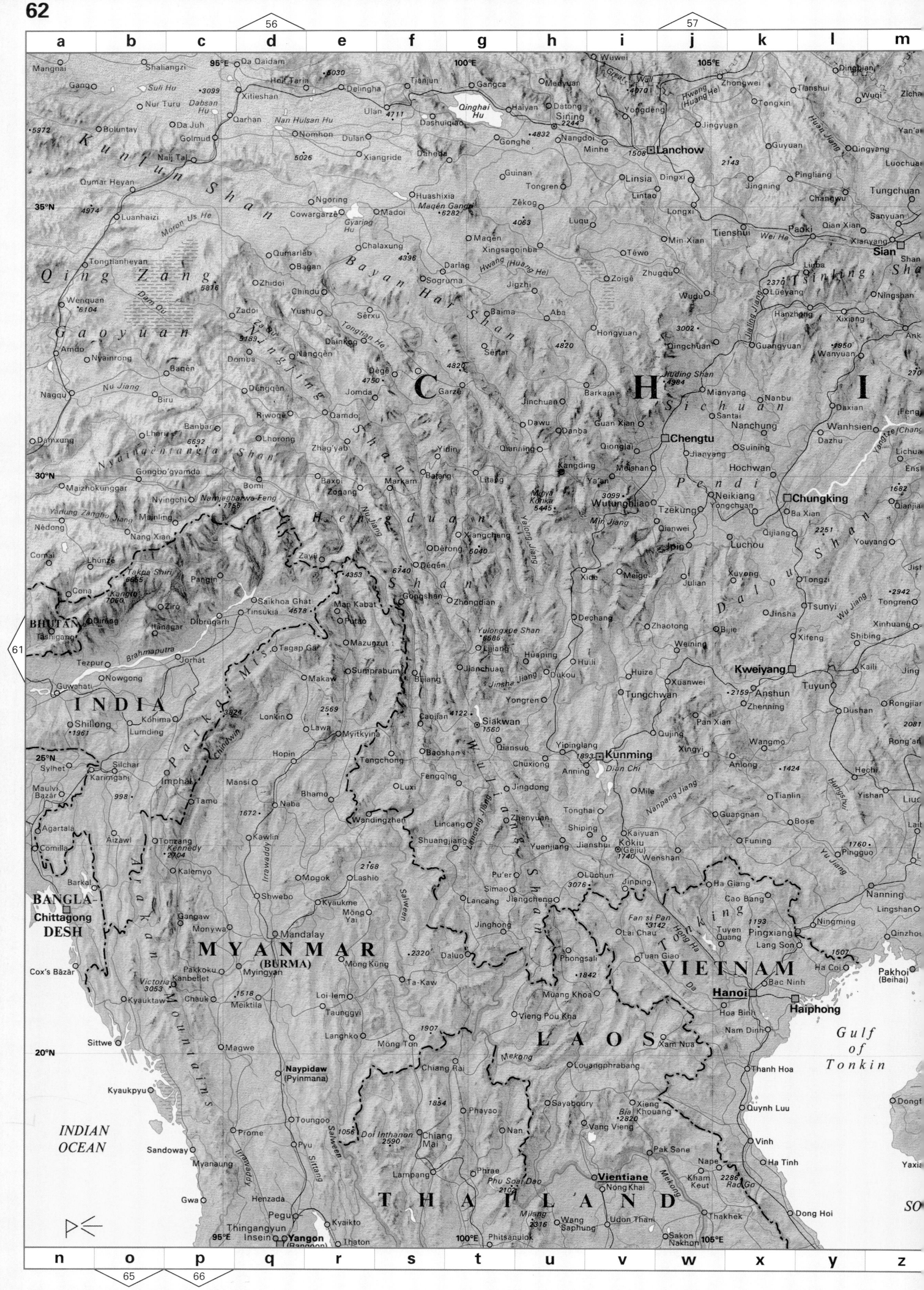

56
57
a b c d e f g h i j k l m
61
n o p q r s t u v w x y z
65
66
95°E
100°E
105°E
35°N
30°N
25°N
20°N
CHINA
INDIA
BHUTAN
BANGLADESH
MYANMAR (BURMA)
LAOS
THAILAND
VIETNAM
INDIAN OCEAN
Gulf of Tonkin
Kunlun Shan
Qing Zang Gaoyuan
Bayan Har Shan
Nyainqentanglha Shan
Hengduan Shan
Ningjing Shan
Sichuan Pendi
Dalou Shan
Tsinling Shan
Wuliang Shan
Arakan Mountains
Patkai Mts
Qinghai Hu
Brahmaputra
Irrawaddy
Salween
Mekong
Yangtze (Chang)
Hwang (Huang He)
Nu Jiang
Jinsha Jiang
Yalong Jiang
Min Jiang
Tonkin
Lanchow
Sining
Chengtu
Chungking
Kweiyang
Kunming
Sian
Nanning
Hanoi
Haiphong
Lhasa
Mandalay
Naypidaw (Pyinmana)
Yangon (Rangoon)
Vientiane
Chiang Mai
Chittagong
Shillong
Imphal
Kohima
Guwahati
Myitkyina
Lashio
Xam Nua
Louangphrabang
Pakhoi (Beihai)
Kokiu (Gejiu)
Siakwan
Lijiang
Garzê
Qamdo
Golmud
Yushu
Yibin
Luchou
Wanhsien
Nanchung
Mianyang

57
58

a b c d e f g h i j k l m

Taiyuan
Yutze
Taigu
Yangchuan
2069
115°E
Tehchow
120°E
Penglai
Yantai
Cape Chengshan
(S.Korea)
125°E
Ongjin
Kangnung
170
Inch'on
Seoul
Wonju
Ch'ungju
Ch'ongju
SOUTH
Andong
Taejon
Kunsan
Chonju
Taegu
KOREA
Masan
1915
Chinju
Kwangju
Mokp'o
Yosu
35°N
Singtai
Hantan
Huo Xian
Fengfeng
Changchih
1619
Anyang
Hohpi
Houma
2322
Jiaozuo
Sinsiang
Linqing
Boxing
Tsinan
Tzepo
950
Weifang
Shantung
220
Laiyang
Jiao Xian
Tsingtao
Tai'an
Yellow
Sea
Hwang (Huang He)
Heze
Yanzhou
Tsining
Liangcheng
Junan
Sanmenhsia
Loyang
1440
Chengchow
Kaifeng
Qi Xian
Shangkiu
Tsaochuang
Yun He
Lienyunkang
Suchow
Grand Canal
Binhai
Pingtingshan
Hsuchang
Zhecheng
Bo Xian
Huaibei
366
Shangnan
Nanzhao
Luohe
Shangshui
Suhsien
Huaiyin
Great
Lake Hungtze
Hongze
Zhenping
Nanyang
Tanghe
Fuyang
Xincai
Pengpu
Lake Kaoyu
Hwainan
Yangchow
Taichow
Huai He
1140
1612
Luoshan
Xinyang
Huangchuan
Chu Xian
Nanking
Nantung
Siangfan
Sui Xian
Hefei
Changshu
Nanzhang
N
A
Han Shui
Wuhsi
Macheng
Lujiang
Wuhu
Lake Tai
Suchow
Shanghai
Yunmeng
Plain
Yangtze (Chang Jiang)
Xuancheng
EAST CHINA
SEA
1860
Ichang
Wuhan
Anking
Tonkling
Kashing
Yidu
Shasi
Mianyang
Hwangshih
1187
Hangchow
Zhoushan Islands
1841
Li Xian
Tongshan
Shaohing
Ningpo
Guoju
30°N
Tunxi
Kiukiang
Xingzi
Kingtehchen
Xin'anjiang
Changteh
Yueyang
Lake Tungting
1596
Lake Poyang
Kinhwa
Xiushui
Linhai
Quzhou
Yiyang
Nanchang
Shangjiao
Gao'an
Lishui
Changsha
Cuixi
Xinyu
Fuzhou
Siangtan
Chuchow
2158
Yunhe
Wenchow
Pucheng
Liahyuan
Pingsiang
Gongxi
Zhenghe
Ryukyu Islands
1290
Nanfeng
Shaowu
Fuding
Shaoyang
Gan Jiang
Wuyi Shan
Hengyang
Ningde
Okinawa
Ji'an
Nanping
Naha
Ningdu
1871
(Japan)
Xiang Jiang
Leiyang
1199
Min Jiang
Sanming
1494
Minqing
Foochow
Kanchow
Ruijin
Quanzhou
Yong'an
Chen Xian
Ningyuan
Putian
Nanxiong
Longyan
Chilung
Taoyüan
Taipei
Hsinchu
25°N
Lian Xian
1902
Shaokwan
Miyako
Ilan
Pingle
Qiuling
1560
Changchow
3884
Xuewang
Taiwan Strait
Mei Xian
Amoy (Xiamen)
Taichung
Iriomote
Yingde
Changhua
Huaiji
Zhangpu
Hualien
Longchuan
Dong Jiang
Zhao'an
Jieyang
Chiai
TAIWAN
Tropic of Cancer
Wuchow
1282
Swatow
Chaoyang
3997
Xi (Xi Jiang)
Canton (Guangzhou)
Luoding
Foshan
Huizhou
Lufeng
Tainan
Shun-te
Kaohsiung
Pingtung
1704
Kongmoon (Jiangmen)
Fangshan
Chuhoi
Hong Kong
Macao
Hengchun
Yangjiang
Mowming
PACIFIC
Zhanjiang
Bashi Channel
Luzon Strait
Batan Islands
Strait
Haikow
20°N
OCEAN
Hainan
Wanning
Babuyan Islands
Cape Bojeador
Cape Engaño
Aparri
Laoag
Cordillera Central
Sierra Madre
Luzon
Bangued
Tuguegarao
Vigan
PHILIPPINES
Ilagan
Pulog
2934
115°E
120°E
125°E

n o p q r s t u v w x y z

67

0 100 200 300 miles
Average linear scale
0 100 200 300 400 500 Km

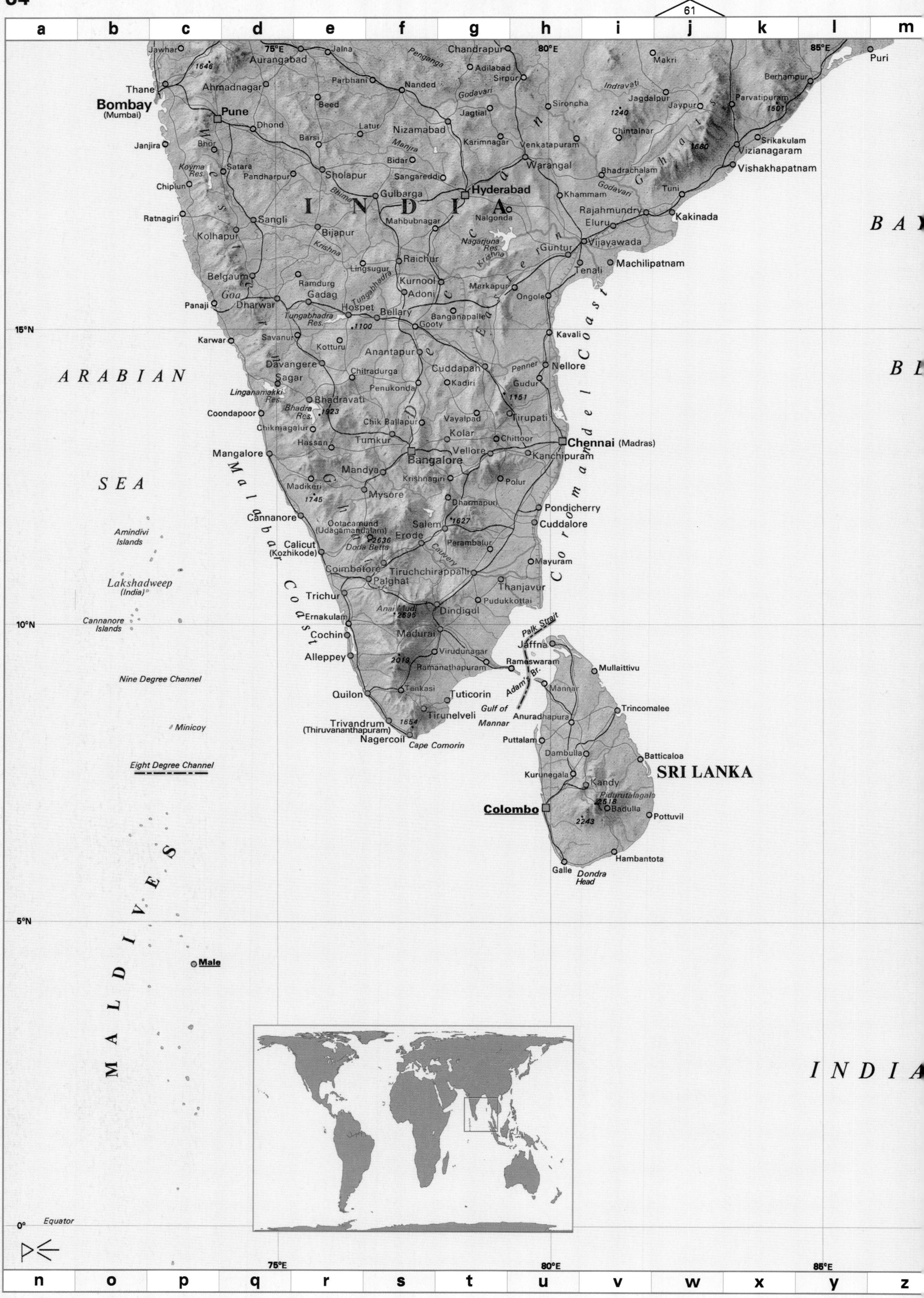

61

a b c d e f g h i j k l m

75°E 80°E 85°E

15°N 10°N 5°N 0° Equator

INDIA

ARABIAN SEA

BAY OF BENGAL

INDIAN OCEAN

SRI LANKA

MALDIVES

Jawhar, Aurangabad, Jalna, Chandrapur, Adilabad, Sirpur, Makri, Puri, Berhampur, Thane, Ahmadnagar, Parbhani, Nanded, Jagdalpur, Jaypur, Parvatipuram, Bombay (Mumbai), Pune, Beed, Sironcha, Jagtial, Dhond, Latur, Nizamabad, Karimnagar, Venkatapuram, Chintalnar, Srikakulam, Vizianagaram, Vishakhapatnam, Janjira, Bhor, Barsi, Bidar, Warangal, Satara, Pandharpur, Sholapur, Sangareddi, Bhadrachalam, Koyna Res., Chiplun, Gulbarga, Hyderabad, Khammam, Tuni, Ratnagiri, Sangli, Mahbubnagar, Nalgonda, Rajahmundry, Kakinada, Eluru, Kolhapur, Bijapur, Vijayawada, Guntur, Raichur, Nagarjuna Res., Tenali, Machilipatnam, Belgaum, Lingsugur, Ramdurg, Gadag, Kurnool, Adoni, Markapur, Ongole, Goa, Panaji, Dharwar, Hospet, Bellary, Tungabhadra Res., Banganapalle, Gooty, Kavali, Karwar, Savanur, Kotturu, Anantapur, Davangere, Chitradurga, Cuddapah, Nellore, Sagar, Penukonda, Kadiri, Gudur, Linganamakki Res., Bhadravati, Bhadra Res., Coondapoor, Chik Ballapur, Vayalpad, Tirupati, Chikmagalur, Kolar, Chittoor, Chennai (Madras), Hassan, Tumkur, Bangalore, Vellore, Kanchipuram, Mangalore, Mandya, Krishnagiri, Polur, Madikeri, Mysore, Dharmapuri, Pondicherry, Cannanore, Ootacamund (Udagamandalam), Salem, Cuddalore, Calicut (Kozhikode), Doda Betta, Erode, Perambalur, Mayuram, Coimbatore, Tiruchchirappalli, Palghat, Thanjavur, Trichur, Pudukkottai, Anai Mudi, Dindigul, Ernakulam, Cochin, Madurai, Alleppey, Virudunagar, Ramanathapuram, Rameswaram, Jaffna, Mullaittivu, Quilon, Tenkasi, Tuticorin, Mannar, Trincomalee, Tirunelveli, Anuradhapura, Trivandrum (Thiruvananthapuram), Nagercoil, Cape Comorin, Puttalam, Dambulla, Batticaloa, Kurunegala, Kandy, Pidurutalagala, Colombo, Badulla, Pottuvil, Hambantota, Galle, Dondra Head, Male

Amindivi Islands, Lakshadweep (India), Cannanore Islands, Nine Degree Channel, Minicoy, Eight Degree Channel

Western Ghats, Eastern Ghats, Deccan, Malabar Coast, Coromandel Coast, Palk Strait, Adam's Br., Gulf of Mannar

Penganga, Godavari, Indravati, Manjra, Bhima, Krishna, Tungabhadra, Penner, Cauvery

1646, 1501, 1240, 1680, 1100, 1151, 1923, 1745, 1627, 2636, 2695, 2019, 1654, 2518, 2243

n o p q r s t u v w x y z

62

a b c d e f g h i j k l m

90°E 95°E 100°E

Loikaw
Naypidaw (Pyinmana)
Muang Chiang Rai
1854
Mekong
Luang Prabang
Ban Ban
Ramree
Thayetmyo
Salween
Phayao
Sayaboury
Xieng Khouang
Vang Vieng
2820
Bia
Cheduba
Toungoo
1056
Inthanon 2590
Chiang Mai
Nan
LAOS
Prome
Pyu
Sittang
Pak Sane
Kham Keut
Myanaung
Irrawaddy
Lampang
Phrae
Vientiane
Soai Dao 2102
Mekong
MYANMAR (BURMA)
Nong Khai
Henzada
Thakhek
Pegu
Kyaikto
Wang Saphung
Miang 2316
Udon Thani
Sakon Nakhon
Insein
Thingangyun
Yangon (Rangoon)
Thaton
Mae Sot
Tak
Phitsanulok
OF
Bassein
Kanbe
Gulf of Martaban
Chum Phae
Kalasin
Moulmein
Khon Kaen
Pyapon
Phetchabun
Roi Et
Maha Sarakham
Mouths of the Irrawaddy
Chaiyaphum
Yasothon
Nakhon Sawan
Ubon Ratchathani
THAILAND
Tenasserim
Ye
Nakhon Ratchasima
Si Sa Ket
15°N
Preparis
Chao Phraya
Sing Buri
Lop Buri
Buriram
Surin
Suphan Buri
Khiaw 1282
AL
Chai Si
Kanchanaburi
Prachin Buri
849
Samrong
Cocos Islands (Burma)
Tavoy
Nakhon Pathom
Ban Pong
Bangkok (Krung Thep)
Sisophon
Thon Buri
Angkor
North Andaman
Chon Buri
Phetchaburi
Siracha
Battambang
Tonle Sap
Andaman Islands (India)
Tenasserim
1633
CAMBODIA
Middle Andaman
Mergui Archipelago
Kadan
Klaeng
Rayong
Mergui
Hua Hin
Chantaburi
Pursat
Kompong Chhnang
Laem Ngop
Chang
1813
South Andaman
Hat Lek
Kut
Andaman
Letsok-Aw
1251
Prachuap Khiri Khan
Phnom Penh
Gulf
Little Andaman
768
Lanbi
of
Chumphon
Kompong Som
Isthmus of Kra
Ranong
Thailand
Phu Quoc
Ten Degree Channel
St Matthew's
10°N
Sea
Phangan
Samui
66
Car Nicobar
Surat Thani
Ban Takua Pa
Ban Na San
Nakhon Si Thammarat
Khao Luang 1835
Cape Mau
Thap Put
Nicobar Islands (India)
Krabi
Malay Peninsula
Phuket
Katchall
Phatthalung
Thale Luang
Little Nicobar
Trang
Songkhla
Great Nicobar
Hat Yai
Pattani
Terutao
Satun
Sai Buri
Yala
Narathiwat
Langkawi
Alor Setar
Sungai Ko-lok
Kota Baharu
Sungai Petani
Banda Aceh
Pinang (George Town)
Butterworth
Pinang
2171 Chamah
Kelantan
Kuala Terengganu
Sigli
Lhokseumawe
Bireuen
Lhoksukon
2855 Geureudong
Idi
Peureulak
5°N
Taiping
Perak
Sungai Siput Utara
Dungun
Calang
Ipoh
MALAYA
Langsa
Kampar
2131
Kuala Lipis
Tapis 1512
Meulaboh
Pangkalanbrandan
Tanjungpura
Raub
Leuser 3404
Sumatra
Medan
Strait of Malacca
Kuala Kubu Baharu
Bentong
Kuantan
Kutacane
Tebingtinggi
MALAYSIA (WESTERN)
Tapaktuan
Tanjungbalai
Petaling Jaya
Kuala Lumpur
Kabanjahe
Pematangsiantar
Kelang
Putrajaya
Seremban
Segamat
Tioman
OCEAN
Lake Toba
Simeulue
Singkilbaru
Malacca
Muar
Blumut 1010
2300 Sihabuhabu
Rantauprapat
Johor
Tuangku
Tarutung
Kelusng
Sibolga
Rupat
Barumun
Dumai
Kulai
Johor Baharu
Duri
Padangsidimpuan
SINGAPORE
Balaipungut
Nias
Panyabungan
Riau Islands
INDONESIA
Hutanopan
Pakanbaru
Pini
Ophir 2912
Kampar
Lubuksikaping
0°
Lingga Islands
Bukittinggi
Payakumbuh
Rengat
Padangpanjang
Tanahbala
Indragiri
Singkep

90°E 95°E 100°E

n o p q r s t u v w x y z

68

0 100 200 300 miles
Average linear scale
0 100 200 300 400 500 Km

62

a b c d e f g h i j k l m

MYANMAR (BURMA)
THAILAND
LAOS
VIETNAM
CAMBODIA
MALAYA
MALAYSIA (WESTERN)

Toungoo, Prome, Pyu, Sittang, Myanaung, Henzada, Irrawaddy, Pegu, Kyaikto, Insein, Thingangyun, Yangon (Rangoon), Kanbe, Thaton, Gulf of Martaban, Pyapon, Moulmein, Salween, 1056, Inthanon 2590, 1854, Chiang Mai, Lampang, 100°E, Phayao, Nan, Phrae, Sayaboury, Vang Vieng, Bia 2820, Xieng Khouang, Soai Dao 2102, Vientiane, Nong Khai, Pak Sane, 105°E, Quynh Luu, Vinh, Ha Tinh, Kham Keut, Nape, 2286, Rao Go, Mekong, Gulf of Tongkin, Dongfang, Hainan (China), 110°E, 1879, Yaxian, Dong Hoi, Thakhek, Wang Saphung, Miang 2316, Udon Thani, Sakon Nakhon, Phitsanulok, Tak, Mae Sot, Chum Phae, Khon Kaen, Kalasin, Savannakhet, Sepone, Hue, Phetchabun, Roi Et, Maha Sarakham, Khemarat, 2500 Atouat, Da Nang, Nakhon Sawan, Chaiyaphum, Yasothon, Ubon Ratchathani, B.Thateng, Ye, 15°N, Chao Phraya, Nakhon Ratchasima, Si Sa Ket, Warin Chamrap, Pakse, 2009, Sing Buri, Lop Buri, Buriram, Surin, Det Udom, Phiafay, Attopeu, Suphan Buri, Khiaw 1282, Kontum, Tavoy, Kanchanaburi, Nakhon Pathom, Prachin Buri, 849, Samrong, Khong, Ban Pong, Bangkok (Krung Thep), Thon Buri, Sisophon, Angkor, Pleiku, 1570, An Tuc, Qui Nhon, Stung Treng, Chon Buri, Phetchaburi, Battambang, Siracha, Tonle Sap, Mekong, Tenasserim, Mergui Archipelago, Kadan, Klaeng, 1633, Ban Me Thuot, Mdrak, Chantaburi, Ban Pu Kroy, Rayong, Pursat, Kompong Chhnang, Kratie, Nha Trang, Mergui, Hua Hin, Laem Ngop, 1544, Chang, 1813, Kompong Cham, Da Lat, Cam Ranh, Kut, Hat Lek, Phnom Penh, 1532, Letsok-Aw, 1251, Khiri Khan Prachuap, Bao Loc, Di Linh, Basac, Phu Chong, Bien Hoa, Andaman, Gulf of Thailand, Kompong Som, Saigon (Ho Chi Minh), Lanbi, 758, Chau Phu, My Tho, Vung Tau, Chumphon, Long Xuyen, Phu Quoc, Rach Gia, Can-Tho, 10°N, Isthmus of Kra, St Matthew's, Ranong, Phangan, Mekong Delta, Khanh Hung, Samui, Surat Thani, Sea, Ban Takua Pa, Ban Na San, Nam Can, Nakhon Si Thammarat, Cape Mau, Spratly Islands, Luang 1835, Thap Put, Krabi, Phuket, Phatthalung, Thale Luang, Trang, Malay Peninsula, Songkhla, Hat Yai, Pattani, Terutao, Sai Buri, Satun, Yala, Narathiwat, Langkawi, Alor Setar, Sungai Ko-lok, Kota Baharu, Sungai Petani, Kelantan, Pinang (George Town), Butterworth, Banda Aceh, Sigli, Lhokseumawe, 2171 Chamah, Kuala Terengganu, Bireuen, Cheksukon, Idi, Pinang, 5°N, Geureudong 2855, Peureulak, Perak, Dungun, Taiping, Sungai Siput Utara, Calang, Ipoh, Langsa, Kampar, Pangkalanbrandan, 2131, Kuala Lipis, Tapis 1512, Meulaboh, Tanjungpura, Raub, Kuala Kubu Baharu, Kuantan, Leuser 3404, Medan, Kutacane, Tebingtinggi, Bentong, Tapaktuan, Kabanjahe, Pematangsiantar, Kuala Lumpur, Petaling Jaya, Kelang, Putrajaya, Tanjungbalai, Lake Toba, Tioman, Anambas Islands (Indonesia), Simeulue, Seremban, Singkilbaru, Sihabuhabu 2300, Malacca, Segamat, Blumut 1010, Tuangku, Tarutung, Rantauprapat, Muar, Keluang, Sibolga, Barumun, Dumai, Rupat, Kulai, Johor, Johor Baharu, Padangsidimpuan, Duri, SINGAPORE, Nias, Panyabungan, Balaipungut, Hutanopan, Riau Islands, Tambelan Islands, INDIAN, Pakanbaru, Kampar, 2912, Lubuksikaping, Pini, 0° Equator, Bukittinggi, Payakumbuh, Lingga Islands, OCEAN, Tanahbala, Padangpanjang, Rengat, Indragiri, Singkep, Berhala Strait, Solok, Siberut, Padang, 100°E, Cape Jabung, 105°E, Strait of Malacca, Sumatra, Indonesia, SOUTH, North Natuna, Natuna, Natuna Islands (Indonesia), South Natuna Islands, Binatang, Sarikei, Cape Datu, Datuk Bay, Kuching, Lupar, Sri Aman, Sambas, Pamangkat, Singkawang, Pinang, Ngabang, Sanggau, Kapuas, Pontianak, Bengkolan Bay, Maya, 110°E, Nanga Sokan

65

n o p q r s t u v w x y z

68

This map shows 1/60 of the earth's surface

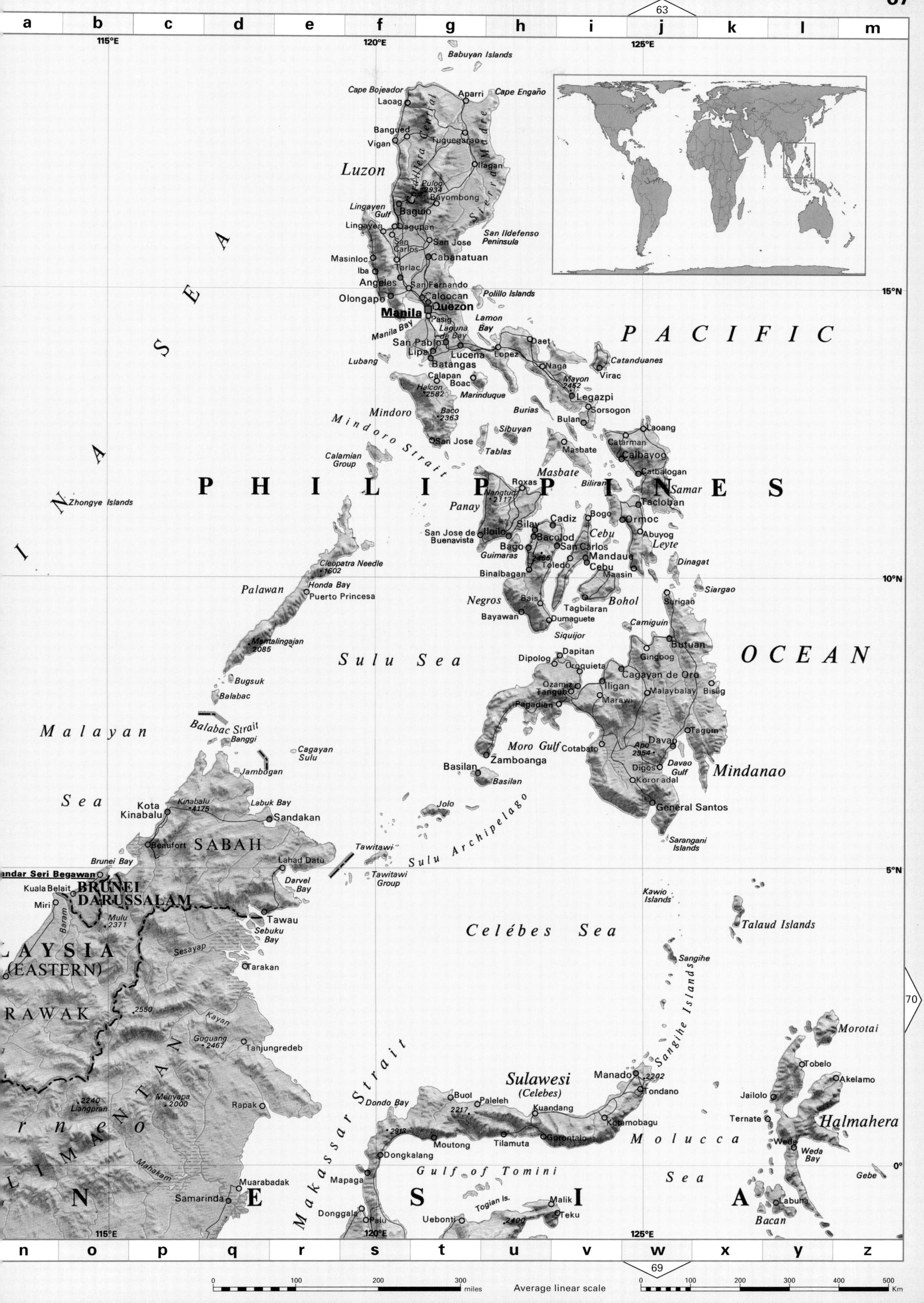

63

a b c d e f g h i j k l m

115°E 120°E 125°E

Babuyan Islands

Cape Bojeador
Aparri
Cape Engaño
Laoag
Cordillera Central
Sierra Madre
Bangued
Vigan
Tuguegarao
Ilagan
Luzon
Pulog 2934
Bayombong
Lingayen Gulf
Baguio
Lingayen
Dagupan
San Carlos
San Jose
San Ildefonso Peninsula
Masinloc
Tarlac
Cabanatuan
Iba
Angeles
San Fernando
Olongapo
Caloocan
Polillo Islands
Manila
Quezon
Pasig
Lamon Bay
Manila Bay
Laguna de Bay
San Pablo
Daet
Lipa
Lucena
Lopez
Batangas
Lubang
Naga
Catanduanes
Virac
Calapan
Boac
Halcon 2582
Marinduque
Mayon 2462
Legazpi
Mindoro
Baco 2363
Burias
Sorsogon
Bulan
Mindoro Strait
Sibuyan
Laoang
San Jose
Tablas
Masbate
Catarman
Calamian Group
Calbayog
Masbate
Catbalogan

P H I L I P P I N E S

Roxas
Biliran
Samar
Nangtud 2117
Panay
Tacloban
Zhongye Islands
Cadiz
Bogo
Ormoc
Silay
Cebu
San Jose de Buenavista
Iloilo
Bacolod
Abuyog
Leyte
Bago
San Carlos
Guimaras
2465
Mandaue
Toledo
Cebu
Dinagat
Cleopatra Needle 1602
Binalbagan
Maasin
Honda Bay
Siargao
Palawan
Puerto Princesa
Surigao
Negros
Bais
Bohol
Tagbilaran
Bayawan
Dumaguete
Camiguin
Siquijor
Butuan
Mantalingajan 2085
Dapitan
Sulu Sea
Dipolog
Gingoog
Oroquieta
Cagayan de Oro
Ozamiz
Iligan
Bislig
Tangub
Malaybalay
Marawi
Bugsuk
Pagadian
Balabac
Balabac Strait
Tagum
Banggi
Moro Gulf
Cotabato
Davao
Cagayan Sulu
Apo 2954
Zamboanga
Jambongan
Davao Gulf
Basilan
Digos
Mindanao
Koronadal
Basilan
Kota Kinabalu
Kinabalu 4175
Labuk Bay
Jolo
Sandakan
General Santos
Sulu Archipelago
Beaufort
SABAH
Sarangani Islands
Tawitawi
Brunei Bay
Lahad Datu
Bandar Seri Begawan
Tawitawi Group
Darvel Bay
Kuala Belait
BRUNEI DARUSSALAM
Kawio Islands
Miri
Baram
Mulu 2371
Tawau
Sebuku Bay
Talaud Islands
Celébes Sea
MALAYSIA (EASTERN)
Sesayap
Sangihe
Tarakan
Sangihe Islands
SARAWAK
2550
Kayan
Morotai
Guguang 2467
Tanjungredeb
Tobelo
Makassar Strait
Manado
2202
Akelamo
Sulawesi (Celebes)
Tondano
Jailolo
2240 Liangpran
Menyapa 2000
Buol
Paleleh
Rapak
Dondo Bay
2217
Kuandang
Ternate
Halmahera
Kotamobagu
KALIMANTAN
2913
Moutong
Tilamuta
Gorontalo
Molucca Sea
Weda
Weda Bay
Dongkalang
Gebe
Gulf of Tomini
Mahakam
Muarabadak
Mapaga
Togian Is.
Malik
INDONESIA
Samarinda
Labuha
Donggala
Palu
Uebonti
2400
Teku
Bacan

PACIFIC OCEAN

SOUTH CHINA SEA

Malayan Sea

Borneo

15°N 10°N 5°N 0°

70

n o p q r s t u v w x y z

69

0 100 200 300 miles

Average linear scale

0 100 200 300 400 500 Km

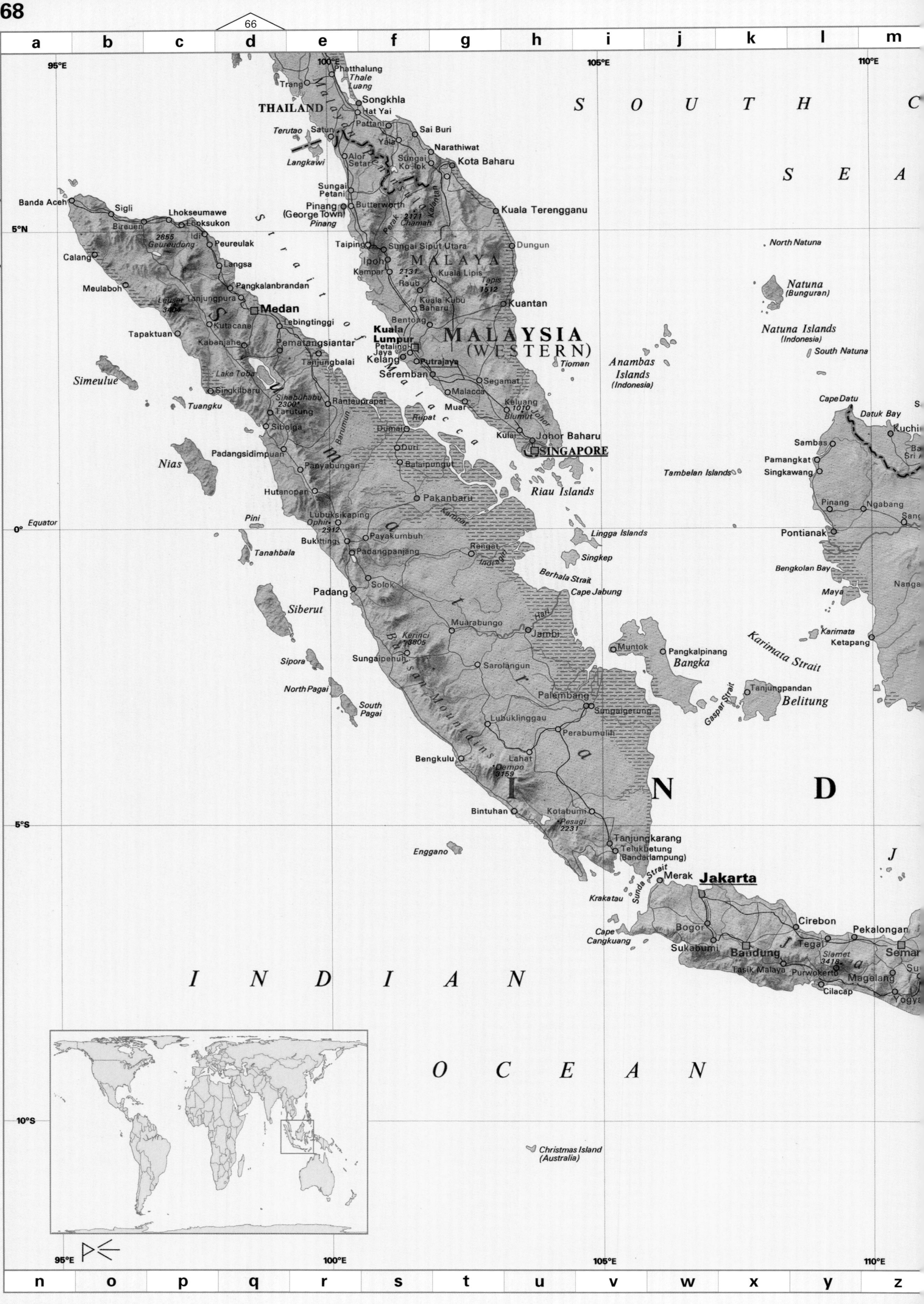

66

a b c d e f g h i j k l m

95°E 100°E 105°E 110°E

SOUTH CHINA SEA

THAILAND

Phatthalung
Thale Luang
Trang
Songkhla
Hat Yai
Pattani
Terutao
Satun
Sai Buri
Yala
Narathiwat
Langkawi
Alor Setar
Sungai Ko-lok
Kota Baharu
Sungai Petani
Pinang (George Town)
Pinang
Butterworth
Kuala Terengganu
2171 Chamah
Taiping
Sungai Siput Utara
Dungun
Ipoh
Kampar
MALAYA
2131
Kuala Lipis
Raub
Tapis 1512
Kuala Kubu Baharu
Kuantan
Bentong
Kuala Lumpur
Petaling Jaya
Kelang
Putrajaya
MALAYSIA (WESTERN)
Tioman
Seremban
Segamat
Malacca
Muar
Keluang
1010 Blumut
Kulai
Johor Baharu
SINGAPORE
Strait of Malacca

65

5°N
Banda Aceh
Sigli
Lhokseumawe
Lhoksukon
Bireuen
2855 Geureudong
Idi
Peureulak
Calang
Langsa
Meulaboh
Pangkalanbrandan
Leuser 3404
Tanjungpura
Medan
Tebingtinggi
Tapaktuan
Kutacane
Kabanjahe
Pematangsiantar
Tanjungbalai
Simeulue
Lake Toba
Singkilbaru
Sihabuhabu 2300
Rantauprapat
Tuangku
Tarutung
Sibolga
Rupat
Dumai
Duri
Nias
Padangsidimpuan
Panyabungan
Bagaipunggut
Hutanopan
Pakanbaru
Riau Islands
Pini
Lubuksikaping
Ophir 2912
Equator
0°
Bukittinggi
Payakumbuh
Rengat
Lingga Islands
Tanahbala
Padangpanjang
Singkep
Berhala Strait
Solok
Cape Jabung
Padang
Siberut
Muarabungo
Jambi
Kerinci 3805
Muntok
Pangkalpinang
Bangka
Sungaipenuh
Sipora
Sarolangun
North Pagai
South Pagai
Palembang
Sungaigerung
Lubuklinggau
Perabumulih
Bengkulu
Barisan Mountains
Dempo 3159
Lahat
SUMATRA
INDONESIA
Bintuhan
Kotabumi
Pesagi 2231
5°S
Tanjungkarang
Telukbetung (Bandarlampung)
Enggano
Sunda Strait
Merak
Jakarta
Krakatau
Cape Cangkuang
Bogor
Sukabumi
Bandung
Cirebon
Tegal
Pekalongan
Slamet 3418
Tasik Malaya
Purwokerto
Magelang
Cilacap
Yogya
Semar
JAVA

Anambas Islands (Indonesia)
North Natuna
Natuna (Bunguran)
Natuna Islands (Indonesia)
South Natuna
Cape Datu
Datuk Bay
Kuchi
Sambas
Pamangkat
Singkawang
Tambelan Islands
Pinang
Ngabang
Pontianak
Bengkolan Bay
Maya
Nanga
Karimata
Ketapang
Karimata Strait
Tanjungpandan
Belitung
Gaspar Strait

INDIAN OCEAN

10°S
Christmas Island (Australia)

95°E 100°E 105°E 110°E

n o p q r s t u v w x y z

This map shows 1/60 of the earth's surface

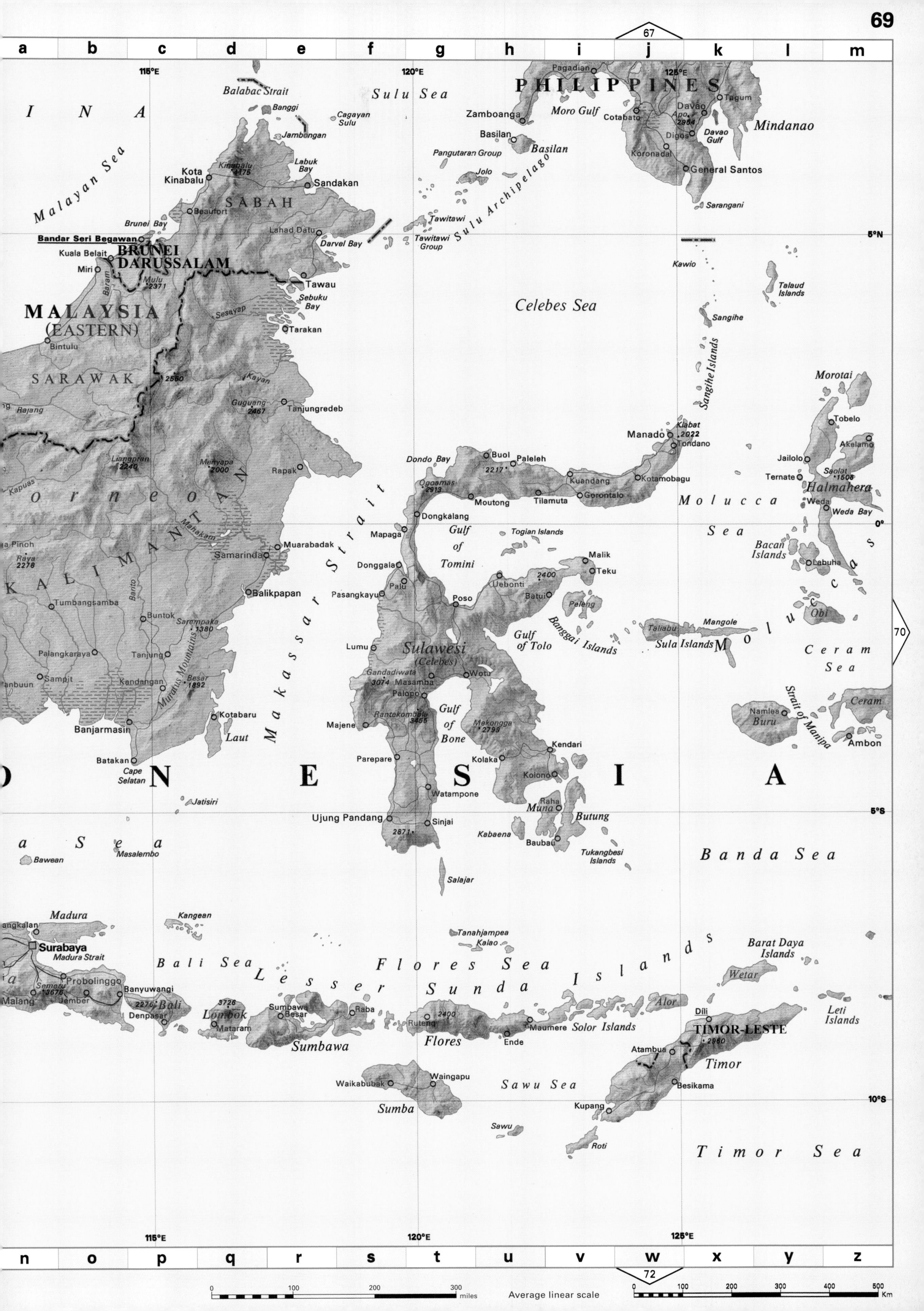

PHILIPPINES
Sulu Sea
Balabac Strait
Banggi
Cagayan Sulu
Jambongan
Zamboanga
Moro Gulf
Cotabato
Davao
Tagum
Apo 2954
Mindanao
Digos
Davao Gulf
Koronadal
General Santos
Sarangani
Basilan
Pangutaran Group
Jolo
Sulu Archipelago
Tawitawi
Tawitawi Group
Malayan Sea
Kota Kinabalu
Kinabalu 4175
Labuk Bay
Sandakan
SABAH
Beaufort
Brunei Bay
Bandar Seri Begawan
BRUNEI DARUSSALAM
Kuala Belait
Miri
Baram
Mulu 2371
Lahad Datu
Darvel Bay
Tawau
Sebuku Bay
Sesayap
Tarakan
MALAYSIA (EASTERN)
Bintulu
SARAWAK
2580
Kayan
Rajang
Guguang 2467
Tanjungredeb
Kawio
Talaud Islands
Celebes Sea
Sangihe
Sangihe Islands
Morotai
Tobelo
Manado
Klabat 2022
Tondano
Akelamo
Jailolo
Ternate
Saolat 1508
Halmahera
Weda
Weda Bay
Liangpran 2240
Menyapa 2000
Rapak
Kapuas
Borneo
KALIMANTAN
Dondo Bay
Buol
Paleleh
2217
Ogoamas 2913
Moutong
Tilamuta
Kuandang
Gorontalo
Kotamobagu
Molucca Sea
Dongkalang
Mahakam
Mapaga
Gulf of Tomini
Togian Islands
Bacan Islands
Labuha
Pinoh
Raya 2278
Muarabadak
Samarinda
Makassar Strait
Donggala
Malik
Teku
2400
Uebonti
Palu
Pasangkayu
Balikpapan
Poso
Batui
Peleng
Obi
Tumbangsamba
Barito
Buntok
Sarempaka 1380
Banggai Islands
Mangole
Taliabu
Sula Islands
Gulf of Tolo
Molucca
Ceram Sea
Palangkaraya
Tanjung
Meratus Mountains
Lumu
Sulawesi (Celebes)
Gandadiwata 3074
Masamba
Wotu
Sampit
Kandangan
Besar 1892
Palopo
Rantekombola 3455
Gulf of Bone
Mekongga 2798
Strait of Manipa
Namlea
Buru
Ceram
Ambon
Kotabaru
Laut
Majene
Banjarmasin
Kendari
Batakan
Cape Selatan
Parepare
Kolaka
Kolono
INDONESIA
Watampone
Raha
Muna
Butung
Jatisiri
Ujung Pandang
Sinjai
2871
Kabaena
Baubau
Tukangbesi Islands
Banda Sea
Java Sea
Bawean
Masalembo
Salajar
Madura
Kangean
Bangkalan
Surabaya
Madura Strait
Tanahjampea
Kalao
Barat Daya Islands
Bali Sea
Flores Sea
Lesser Sunda Islands
Wetar
Probolinggo
Semeru 3676
Malang
Jember
Banyuwangi
2276
Bali
Denpasar
3726
Lombok
Mataram
Sumbawa Besar
Raba
Ruteng
2400
Alor
Dili
Leti Islands
TIMOR-LESTE
2960
Maumere
Solor Islands
Sumbawa
Flores
Ende
Atambua
Timor
Waikabubak
Waingapu
Sawu Sea
Besikama
Sumba
Kupang
Sawu
Roti
Timor Sea
5°N
0°
5°S
10°S
115°E
120°E
125°E
67
70
72
a b c d e f g h i j k l m
n o p q r s t u v w x y z
0 100 200 300 miles
Average linear scale
0 100 200 300 400 500 Km

a b c d e f g h i j k l m
130°E
135°E
140°E
Yap Islands
Faraulep Atoll
Ngulu Atoll
Sorol Atoll
M I C
PALAU
Babel Thuap
Koror
Woleai Atoll
Ifalik Atoll
Eauripik Atoll
C a r o l i n e
Sonsorol
5°N
Pulo Anna
Merir
P A C I F
Tobi
Helen Reef
Morotai
Akelamo
Halmahera
Mapia Islands
Ayu Islands
O C E A
Waigeo
0°
69
Dampier Strait
Sorong
Kwoko 3000
Manokwari
Peg Ariak 2939
Biak
Cenderawasih
Yapen
Misool
990
Sarmi
Steenkool
C e r a m S e a
I N D O N E S I A
Babo
Gulf of Cenderawasih
Van Rees Mountains
Jayapura
Vanimo
Ceram
Bula
Fakfak
Bomberai
3019
WEST PAPUA
Mamberamo
Aitape
Tobo
Kaimana
Lumi
Dreikikir
Wewak
Ambon
Maoke Mountains
Wamena
Jaya 5029
Sepik
Ramu
Kokonau
New Guinea
Mangala 4702
5°S
Telefomin
Bismarck
Kai Islands
Banda Sea
Aru Islands
Kopiago
Wabag
Mount Hagen
Tanahmerah
Strickland
Mendi
Kubor 4369
2895
Digul
Lake Murray
Mappi
NEW
Damar
Tanimbar Islands
Kikori
Dolak Island
Fly
Babar
Sermata
Selaru
Cape Vals
Merauke
Gulf of Pap
Daru
Arafura Sea
Torres Strait
10°S
Badu
Moa
Prince of Wales Island
Cape York
n o p q r s t u v w x y z
73

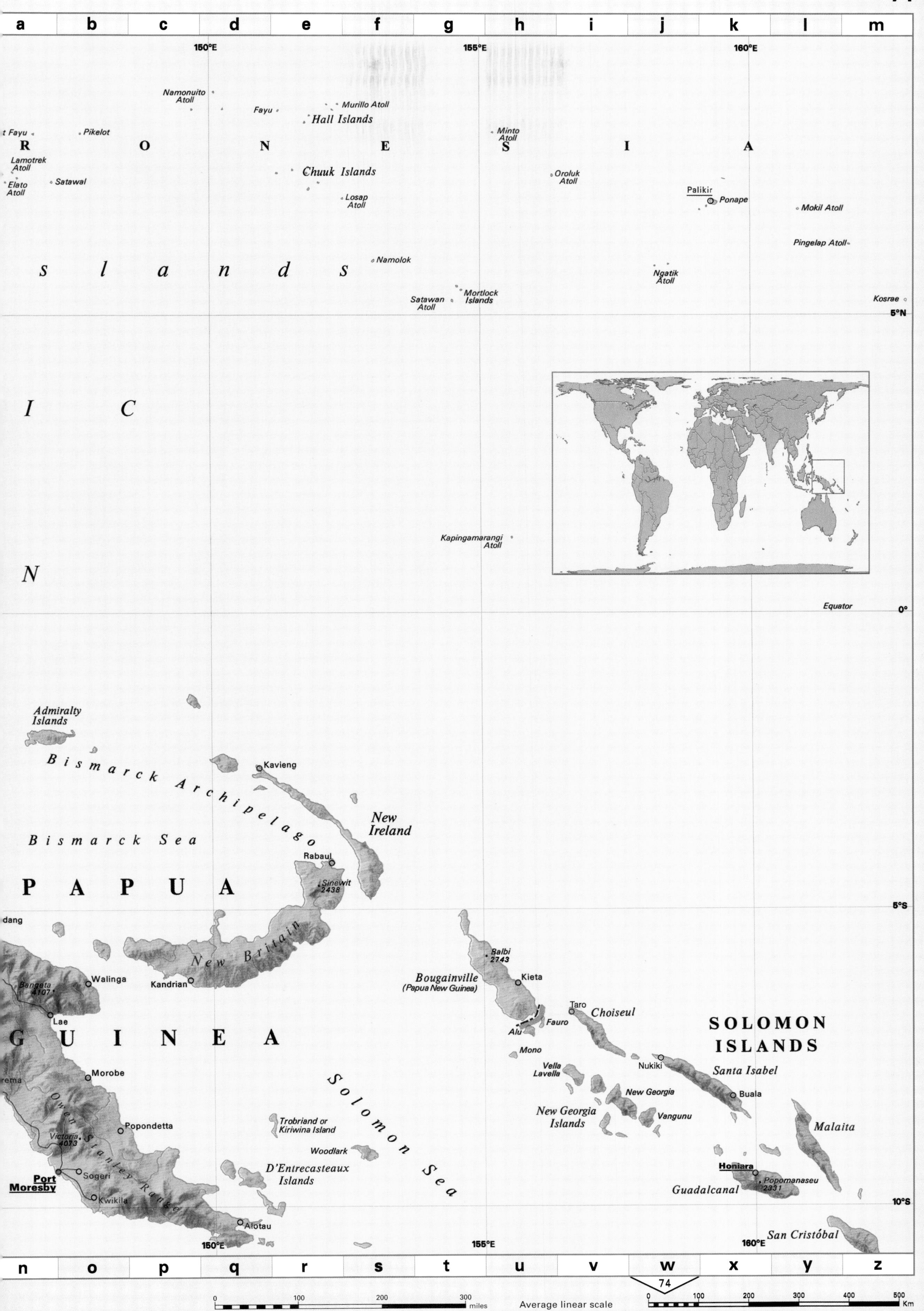
a b c d e f g h i j k l m
150°E
155°E
160°E
Namonuito Atoll
Fayu
Murillo Atoll
Hall Islands
Fayu
Pikelot
Minto Atoll
R O N E S I A
Lamotrek Atoll
Elato Atoll
Satawal
Chuuk Islands
Oroluk Atoll
Losap Atoll
Palikir
Ponape
Mokil Atoll
Pingelap Atoll
s l a n d s
Namolok
Ngatik Atoll
Satawan Atoll
Mortlock Islands
Kosrae
5°N
I C
N
Kapingamarangi Atoll
Equator
0°
Admiralty Islands
Bismarck Archipelago
Kavieng
New Ireland
Bismarck Sea
Rabaul
PAPUA
Sinewit 2438
5°S
dang
New Britain
Balbi 2743
Bougainville (Papua New Guinea)
Kieta
Walinga
Kandrian
Bangeta 4107
Lae
Taro
Choiseul
Fauro
Alu
SOLOMON ISLANDS
GUINEA
Mono
Vella Lavella
Nukiki
Santa Isabel
Morobe
rema
Solomon Sea
New Georgia
Buala
New Georgia Islands
Vangunu
Malaita
Popondetta
Owen Stanley Range
Victoria 4073
Trobriand or Kiriwina Island
Woodlark
Port Moresby
Sogeri
D'Entrecasteaux Islands
Honiara
Popomanaseu 2331
Guadalcanal
Kwikila
10°S
Alotau
150°E
155°E
160°E
San Cristóbal
n o p q r s t u v w x y z
74
0 100 200 300 miles
Average linear scale
0 100 200 300 400 500 Km

69
76
INDIAN OCEAN
INDONESIA
Java
Bali
Lombok
Sumbawa
Flores
Sumba
Timor
TIMOR LESTE
Sawu Sea
Denpasar
Mataram
Kupang
Broome
Derby
Port Hedland
Dampier
Onslow
Exmouth
Carnarvon
Newman
Meekatharra
Wiluna
Kalbarri
Denham
Great Sandy Desert
Gibson Desert
Tropic of Capricorn
WESTERN AUSTRALIA
Eighty Mile Beach
Dampier Land
King Leopold Ranges
Hamersley Range
Shark Bay
Exmouth Gulf
Barrow Island
North West Cape
Cape Cuvier
Cape Inscription
Cartier
Bonaparte Archipelago
Lake Disappointment
Lake Carnegie
110°E
115°E
120°E
125°E
10°S
15°S
20°S
25°S

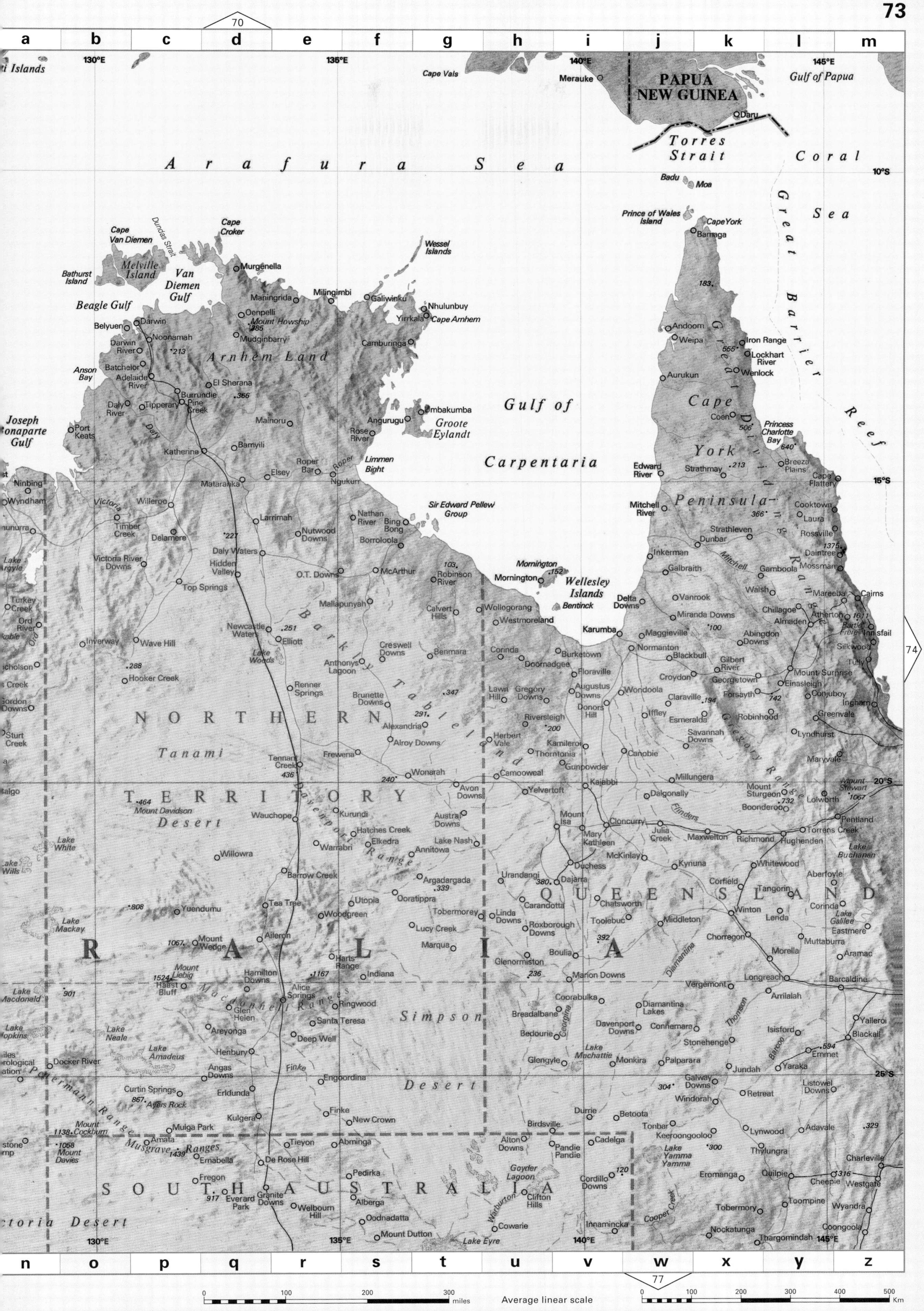

70

a b c d e f g h i j k l m

130°E 135°E 140°E 145°E 10°S 15°S 20°S 25°S

PAPUA NEW GUINEA

Arafura Sea

Torres Strait

Coral Sea

Great Barrier Reef

Gulf of Papua

Cape Vals

Merauke

Daru

Badu

Moa

Prince of Wales Island

Cape York

Bamaga

Gulf of Carpentaria

Cape York Peninsula

Great Dividing Range

Arnhem Land

Melville Island

Bathurst Island

Cape Van Diemen

Dundas Strait

Cape Croker

Van Diemen Gulf

Beagle Gulf

Anson Bay

Joseph Bonaparte Gulf

Wessel Islands

Groote Eylandt

Limmen Bight

Sir Edward Pellew Group

Wellesley Islands

Mornington

Bentinck

Princess Charlotte Bay

Darwin, Belyuen, Noonamah, Darwin River, Batchelor, Adelaide River, Daly River, Port Keats, Tipperary, Burrundie, Pine Creek, El Sherana, Murgenella, Maningrida, Milingimbi, Galiwinku, Nhulunbuy, Yirrkala, Cape Arnhem, Oenpelli, Mount Howship, Mudginbarry, Camburinga, Umbakumba, Angurugu, Rose River, Mainoru, Bamyili, Katherine, Mataranka, Elsey, Roper Bar, Ngukurr, Roper

Ninbing, Wyndham, Kununurra, Willeroo, Victoria, Timber Creek, Delamere, Victoria River Downs, Lake Argyle, Larrimah, Nutwood Downs, Nathan River, Bing Bong, Borroloola, Daly Waters, Hidden Valley, Top Springs, O.T. Downs, McArthur, Robinson River, Calvert Hills, Mallapunyah, Newcastle Waters, Elliott, Lake Woods, Inverway, Wave Hill, Turkey Creek, Ord River, Hooker Creek, Creswell Downs, Anthony's Lagoon, Benmara, Renner Springs, Brunette Downs, Alexandria, Alroy Downs, Frewena, Tennant Creek, Wonarah, Avon Downs, Barkly Tableland

NORTHERN TERRITORY

Tanami Desert

Mount Davidson

Wauchope, Kurundi, Hatches Creek, Elkedra, Warrabri, Davenport Range, Annitowa, Lake Nash, Austral Downs, Willowra, Barrow Creek, Argadargada, Ooratippra, Utopia, Tea Tree, Woodgreen, Tobermorey, Lucy Creek, Marqua, Yuendumu, Lake Mackay, Mount Wedge, Aileron, Harts Range, Indiana, Mount Liebig, Haast Bluff, Hamilton Downs, Alice Springs, Ringwood, Macdonnell Ranges, Glen Helen, Areyonga, Santa Teresa, Deep Well, Henbury, Lake Neale, Lake Amadeus, Docker River, Angas Downs, Finke, Engoordina, Erldunda, Curtin Springs, Ayers Rock, Kulgera, Finke, New Crown, Petermann Ranges, Mount Cockburn, Muiga Park, Amata, Mount Davies, Musgrave Ranges, Ernabella, Tieyon, Abminga, De Rose Hill, Fregon, Everard Park, Granite Downs, Welbourn Hill, Pedirka, Alberga, Oodnadatta, Mount Dutton, Lake Eyre, Simpson Desert, Lake White, Lake Wills, Lake Macdonald, Lake Hopkins, Sturt Creek, Balgo

RALIA

SOUTH AUSTRALIA

Victoria Desert

Andoom, Weipa, Iron Range, Lockhart River, Wenlock, Aurukun, Coen, Edward River, Strathmay, Breeza Plains, Cape Flattery, Mitchell River, Cooktown, Laura, Strathleven, Dunbar, Rossville, Daintree, Mossman, Inkerman, Galbraith, Gamboola, Walsh, Mitchell, Vanrook, Delta Downs, Miranda Downs, Chillagoe, Mareeba, Cairns, Atherton, Almaden, Bartle Frere, Innisfail, Karumba, Maggieville, Normanton, Abingdon Downs, Blackbull, Silkwood, Tully, Gilbert River, Croydon, Georgetown, Mount Surprise, Einasleigh, Forsayth, Conjuboy, Ingham, Claraville, Esmeralda, Robinhood, Greenvale, Iffley, Savannah Downs, Lyndhurst, Maryvale, Gregory Range, Wollogorang, Westmoreland, Corinda, Burketown, Doomadgee, Floraville, Augustus Downs, Wondoola, Donors Hill, Lawn Hill, Gregory Downs, Riversleigh, Herbert Vale, Kamileroi, Thorntonia, Gunpowder, Canobie, Camooweal, Kajabbi, Millungera, Dalgonally, Mount Sturgeon, Boonderoo, Mount Stewart, Lolworth, Pentland, Yelvertoft, Mount Isa, Mary Kathleen, Cloncurry, Julia Creek, Maxwelton, Richmond, Hughenden, Torrens Creek, Flinders, Lake Buchanan, McKinlay, Kynuna, Whitewood, Aberfoyle, Duchess, Dajarra, Urandangi, Corfield, Tangorin, Chatsworth, Carandotta, Winton, Corinda, Lake Galilee, Eastmere, Linda Downs, Roxborough Downs, Toolebuc, Middleton, Lerida, Muttaburra, Chorregon, Boulia, Morella, Aramac, Glenormiston, Marion Downs, Diamantina, Longreach, Barcaldine, Vergemont, Arrilalah, Coorabulka, Diamantina Lakes, Thomson, Breadalbane, Georgina, Connemara, Yalleroi, Bedourie, Davenport Downs, Isisford, Blackall, Stonehenge, Lake Machattie, Monkira, Palparara, Barcoo, Emmet, Glengyle, Jundah, Yaraka, Galway Downs, Retreat, Listowel Downs, Windorah, Durrie, Betoota, Birdsville, Tonbar, Keeroongooloo, Lynwood, Adavale, Alton Downs, Pandie Pandie, Cadelga, Lake Yamma Yamma, Thylungra, Charleville, Goyder Lagoon, Cordillo Downs, Eromanga, Quilpie, Cheepie, Westgate, Clifton Hills, Warburton, Cooper Creek, Toompine, Wyandra, Tobermory, Cowarie, Innamincka, Nockatunga, Thargomindah, Coongoola

QUEENSLAND

183, 555, 506, 640, 213, 366, 1375, 1611, 100, 194, 742, 200, 732, 1067, 380, 392, 236, 304, 300, 120, 594, 316, 329, 213, 385, 366, 227, 251, 288, 347, 291, 240, 436, 464, 339, 808, 1067, 1167, 1524, 901, 867, 1138, 1058, 1439, 917, 103, 152

n o p q r s t u v w x y z

74

77

0 100 200 300 miles

Average linear scale

0 100 200 300 400 500 Km

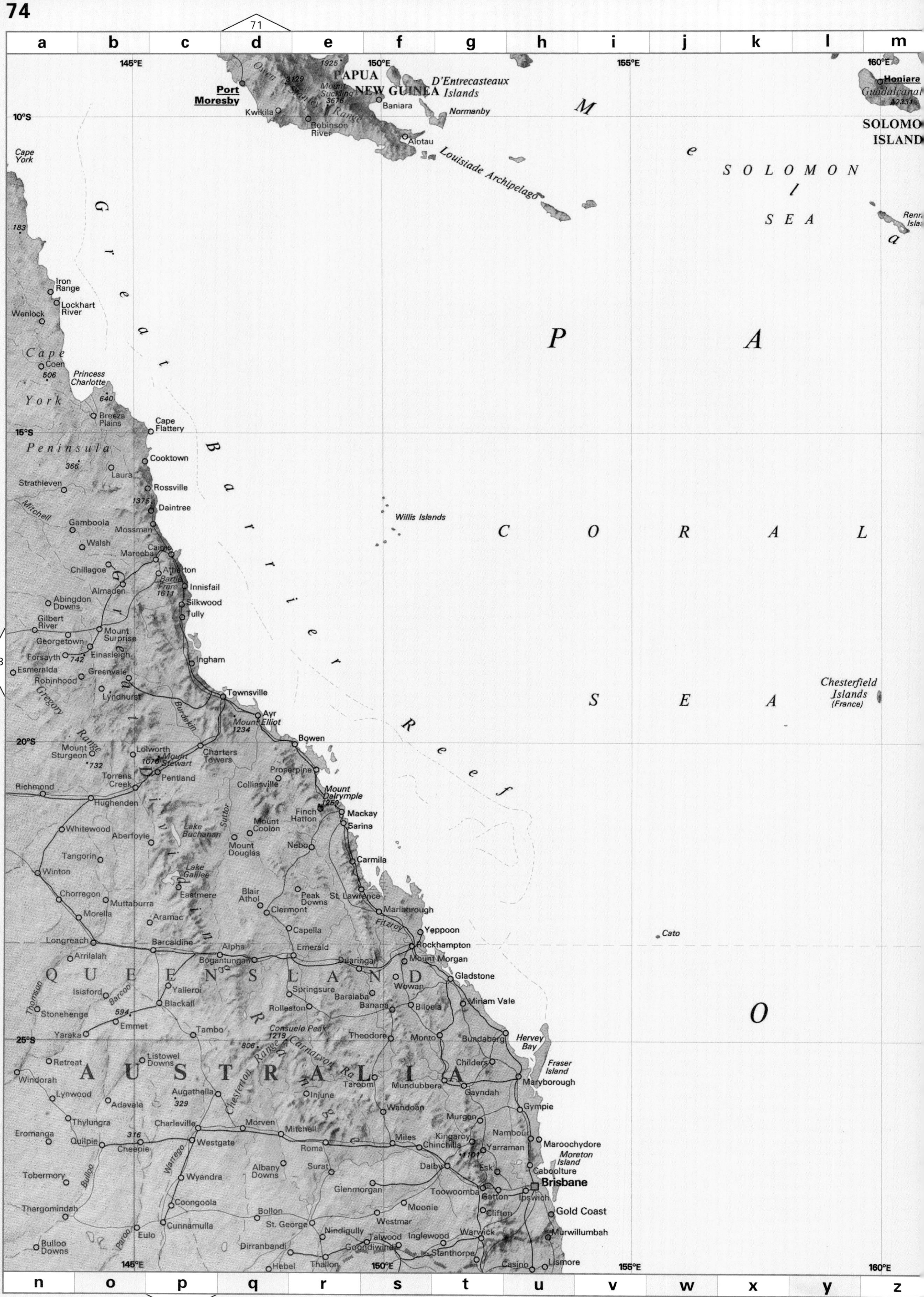

71

a b c d e f g h i j k l m

145°E 150°E 155°E 160°E

10°S 15°S 20°S 25°S

PAPUA NEW GUINEA

Port Moresby

Owen Stanley Range

3129

Mount Suckling 3676

1925

Kwikila

Robinson River

Baniara

D'Entrecasteaux Islands

Normanby

Alotau

Louisiade Archipelago

Honiara

Guadalcanal

2331

SOLOMON ISLANDS

Rennell Island

SOLOMON SEA

Melanesia

Cape York

183

Iron Range

Lockhart River

Wenlock

Cape York Peninsula

Coen

506

Princess Charlotte Bay

640

Breeza Plains

Cape Flattery

366

Laura

Cooktown

Strathleven

Rossville

1375

Daintree

Mitchell

Gamboola

Mossman

Walsh

Cairns

Mareeba

Chillagoe

Atherton

Bartle Frere 1611

Innisfail

Almaden

Silkwood

Abingdon Downs

Tully

Gilbert River

Mount Surprise

Georgetown

Einasleigh

Forsayth

742

Ingham

Esmeralda

Greenvale

Robinhood

Lyndhurst

Gregory Range

Townsville

Burdekin

Ayr

Mount Elliot 1234

Bowen

Mount Sturgeon

Lolworth

1076

Mount Stewart

Charters Towers

732

Torrens Creek

Pentland

Proserpine

Richmond

Hughenden

Collinsville

Mount Dalrymple 1259

Finch Hatton

Mackay

Sarina

Whitewood

Aberfoyle

Lake Buchanan

Suttor

Mount Coolon

Mount Douglas

Nebo

Tangorin

Carmila

Winton

Lake Galilee

Eastmere

Chorregon

Muttaburra

Blair Athol

Peak Downs

St. Lawrence

Clermont

Marlborough

Morella

Aramac

Capella

Fitzroy

Yeppoon

Longreach

Barcaldine

Alpha

Emerald

Rockhampton

Arrilalah

Bogantungan

Duaringa

Mount Morgan

QUEENSLAND

Great Dividing Range

Gladstone

Thomson

Isisford

Barcoo

Yalleroi

Springsure

Baralaba

Wowan

Blackall

Rolleston

Banana

Biloela

Miriam Vale

Stonehenge

594

Emmet

Consuelo Peak 1219

Yaraka

Tambo

Carnarvon Range

Theodore

Monto

Bundaberg

Hervey Bay

806

Retreat

Listowel Downs

Childers

Fraser Island

AUSTRALIA

Windorah

Taroom

Mundubbera

Gayndah

Maryborough

Augathella

Lynwood

Adavale

329

Injune

Chesterton Range

Wandoan

Gympie

Thylungra

Charleville

Morven

Murgon

Nambour

Mitchell

Eromanga

316

Kingaroy

Maroochydore

Quilpie

Cheepie

Westgate

Roma

Miles

Chinchilla

Yarraman

1101

Moreton Island

Warrego

Albany Downs

Surat

Dalby

Esk

Caboolture

Tobermory

Bulloo

Wyandra

Glenmorgan

Toowoomba

Brisbane

Gatton

Ipswich

Coongoola

Moonie

Gold Coast

Thargomindah

Bollon

Westmar

Clifton

Cunnamulla

St. George

Nindigully

Inglewood

Warwick

Murwillumbah

Paroo

Eulo

Talwood

Goondiwindi

Bulloo Downs

Dirranbandi

Stanthorpe

Hebel

Thallon

Casino

Lismore

Great Barrier Reef

Willis Islands

Chesterfield Islands (France)

Cato

PACIFIC

CORAL SEA

OCEAN

73

n o p q r s t u v w x y z

78

a b c d e f g h i j k l m
165°E
170°E
175°E
180°
San Cristóbal
Santa Cruz Island
P A C I F I C
Banks Islands
Espíritu Santo
1879
Malekula
New Hebrides
VANUATU
Efate
Vila
Erromango
New Caledonia
(France)
1650
Houailu
Bourail
Nouméa
Loyalty Islands
(France)
Vanua Levu
Lambasa (Labasa)
1032
FIJI
Koro Sea
Tavua
Nandi (Nadi)
Mount Victoria 1324
Viti Levu
Singatoka (Sigatoka)
Suva
15°S
20°S
25°S
Tropic of Capricorn
O C E A N
n o p q r s t u v w x y z
79
0 100 200 300 miles
Average linear scale
0 100 200 300 400 500 Km

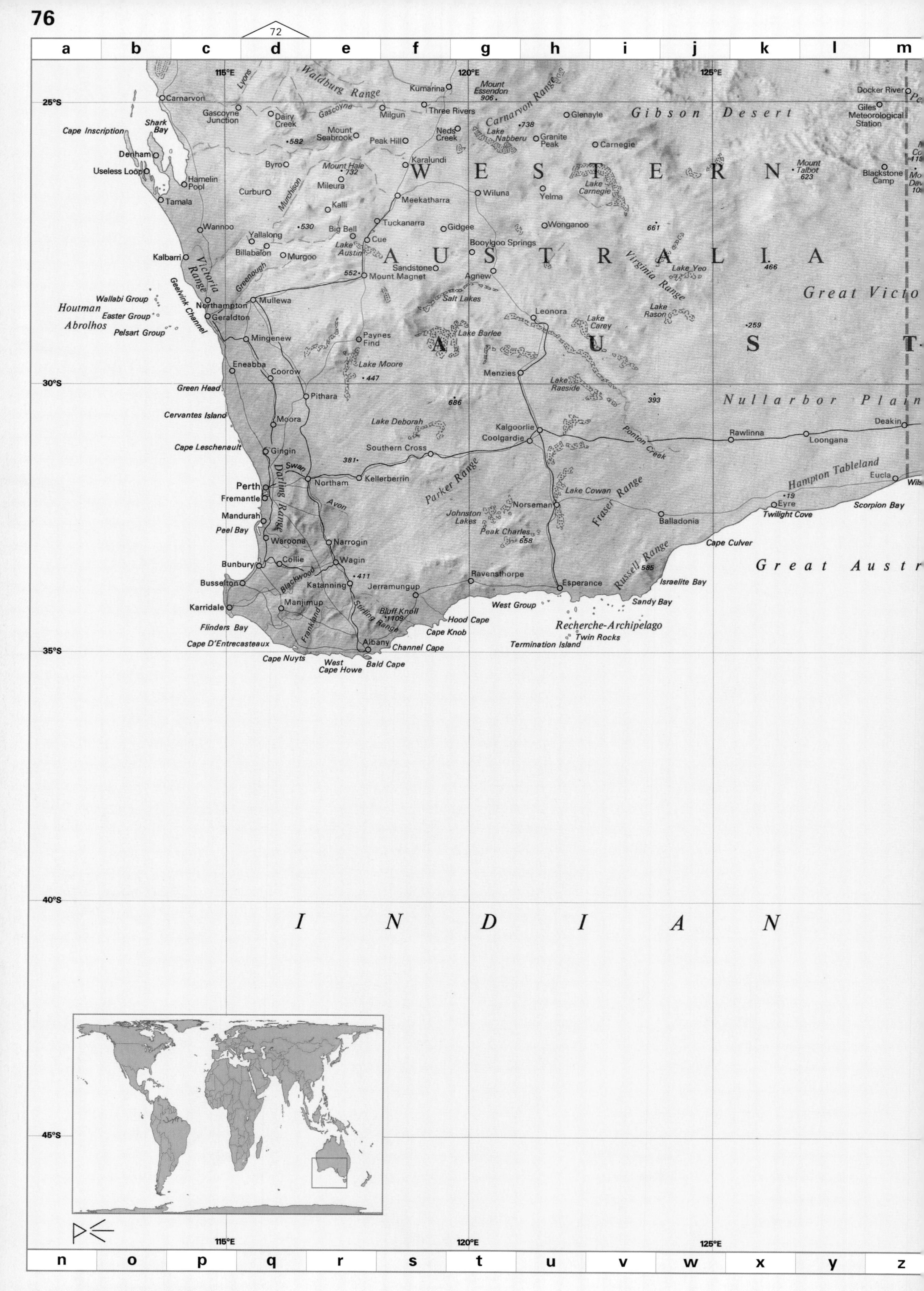
72
WESTERN AUSTRALIA
Gibson Desert
Great Victoria Desert
Nullarbor Plain
INDIAN
Carnarvon
Shark Bay
Cape Inscription
Denham
Useless Loop
Hamelin Pool
Tamala
Wannoo
Kalbarri
Victoria Range
Houtman Abrolhos
Wallabi Group
Easter Group
Pelsart Group
Geelvink Channel
Northampton
Geraldton
Mullewa
Mingenew
Eneabba
Coorow
Green Head
Cervantes Island
Moora
Cape Leschenault
Gingin
Perth
Fremantle
Mandurah
Peel Bay
Darling Range
Waroona
Bunbury
Collie
Busselton
Karridale
Flinders Bay
Cape D'Entrecasteaux
Cape Nuyts
West Cape Howe
Bald Cape
Albany
Channel Cape
Cape Knob
Hood Cape
Bluff Knoll 1109
Stirling Range
Frankland
Manjimup
Blackwood
Katanning
Wagin
Narrogin
Northam
Swan
Avon
Kellerberrin
Southern Cross
Jerramungup
Ravensthorpe
West Group
Recherche-Archipelago
Twin Rocks
Termination Island
Esperance
Russell Range
Israelite Bay
Sandy Bay
Cape Culver
Balladonia
Fraser Range
Norseman
Lake Cowan
Peak Charles 658
Johnston Lakes
Parker Range
Kalgoorlie
Coolgardie
Lake Deborah
Menzies
Lake Raeside
Leonora
Lake Carey
Lake Rason
Lake Barlee
Salt Lakes
Paynes Find
Lake Moore
Pithara
Mount Magnet
Sandstone
Agnew
Cue
Lake Austin
Big Bell
Murgoo
Billabalon
Yallalong
Greenough
Kalli
Mileura
Mount Hale 732
Curbur
Byro
Murchison
Tuckanarra
Meekatharra
Karalundi
Peak Hill
Mount Seabrook
Dairy Creek
Gascoyne Junction
Gascoyne
Lyons
Waldburg Range
Milgun
Kumarina
Three Rivers
Neds Creek
Mount Essendon 906
Carnarvon Range
Lake Nabberu
Granite Peak
Glenayle
Carnegie
Lake Carnegie
Yelma
Wiluna
Gidgee
Booylgoo Springs
Wongonoo
Virginia Range
Lake Yeo
Mount Talbot 623
Docker River
Giles Meteorological Station
Blackstone Camp
Ponton Creek
Rawlinna
Loongana
Deakin
Hampton Tableland
Eucla
Eyre
Twilight Cove
Scorpion Bay
115°E
120°E
125°E
25°S
30°S
35°S
40°S
45°S

73
78
a b c d e f g h i j k l m
n o p q r s t u v w x y z
135°E
140°E
145°E
25°S
30°S
35°S
40°S
45°S
NORTHERN TERRITORY
QUEENSLAND
SOUTH AUSTRALIA
NEW SOUTH WALES
VICTORIA
AUST. CAPITAL TERRITORY
TASMANIA
OCEAN
Bight
Lake Amadeus
Henbury
Angas Downs
Ayers Rock (Uluru)
867
Erldunda
Finke
Engoordina
Kulgera
Mulga Park
Musgrave Ranges
Ernabella
Fregon
917
Everard Park
Granite Downs
Tieyon
Abminga
New Crown
Simpson Desert
Pedirka
Alberga
Welbourn Hill
Oodnadatta
Mount Dutton
181
Coober Pedy
216
Lake Maurice
205
Lake Eyre
-16
11
Cowarie
Cooper
Alton Downs
Warburton
Goyder's Lagoon
Clifton Hills
Birdsville
Pandie Pandie
Cadelga
Durrie
Diamantina
Betoota
Glengyle
Lake Machattie
Monkira
120
Palparara
Stonehenge
Thomson
Jundah
Yaraka
Emmet
Blackall
725
Tambo
Chesterton Range
Galway Downs
304
Windorah
Retreat
Listowel Downs
Tanbar
Keeroongooloo
300
Adavale
Thylungra
Lake Yamma Yamma
Grey Range
Cordillo Downs
Cooper Creek
Innamincka
142
Gidgelpa
Sturt Desert
Nockatunga
Eromanga
Quilpie
316
Cheepie
Charleville
Moven
Mitchell
Westgate
Maranoa
914
Toompine
Bulloo
Thargomindah
Paroo
Wyandra
Coongoola
Cunnamulla
251
St. George
Bollon
Dirranbandi
Hebel
Thollon
Warrego
Bourke
Brewarrina
Walgett
Wanaaring
427
Milparinka
Marree
91
951
Leigh Creek
Lake Frome
Lake Torrens
Woomera
Flinders Ranges
1189
Wilpena
Passmore
Eurinilla
Barrier Ra.
Mount Robe
474
White Cliffs
221
Wilcannia
Cobar
419
Nyngan
Coonamble
1372
Crowl
Dubbo
Narromine
Ooldea
Tarcoola
Kingoonya
1186
Lake Everard
Lake Gairdner
Island Lagoon
Yalata
Head of Bight
Penong
168
Ceduna
Fowler's Bay
Nuyts Arch.
Streaky Bay
Gawler Ranges
472
307
Iron Knob
Quorn
Port Augusta
Cockburn
Broken Hill
Benda Ra.
Darling
Ivanhoe
Condobolin
Parkes
Forbes
Wudinna
250
Whyalla
Port Pirie
Peterborough
Anxious Bay
Eyre Peninsula
Cowell
Spencer Gulf
934
Burra
Morgan
Renmark
Murray
Wentworth
Lachlan
West Wyalong
732
Investigator Group
Pearson Island
Wallaroo
Kadina
Mildura
Balranald
Hay
Griffith
Leeton
Murrumbidgee
Cocoparra Range
Cootamundra
Whidby Point
Port Lincoln
Sir Joseph Banks Group
Thistle I.
Yorke Peninsula
St. Vincent Gulf
Gawler
Adelaide
Murray Bridge
Pinnaroo
101
Ouyen
Narrandera
Junee
Wagga Wagga
Cape Spencer
Investigator Strait
Kangaroo Island
Victor Harbor
Encounter Bay
Lake Alexandra
Swan Hill
Deniliquin
Billabong
Lake Hindmarsh
The Coorong
Nhill
Loddon
Donald
Corowa
Albury
Hume Res.
Lake Eucumbene
Mt. Kosciuszko
2228
Shepparton
Wangaratta
Lacepede Bay
Naracoorte
Horsham
Bendigo
Seymour
Buller
1804
1922
Great Dividing Range
1742
Omeo
Snowy
Mitchell
1320
Glenelg
Maryborough
Ararat
1011
Healesville
Millicent
220
Hamilton
Ballarat
Emu
Melbourne
Orbost
Mount Gambier
Geelong
Ninety Mile Beach
Discovery Bay
Portland
Warrnambool
Port Phillip Bay
Moe
Traralgon
Sale
Wonthaggi
Port Albert
Cape Otway
Apollo Bay
Waratah Bay
Wilson's Promontory
Kent Group
Flinders I.
Currie
King Island
Bass Strait
Furneaux Group
Hunter Island
Three Hummock Island
Banks Strait
Smithton
Burnie
Arthur
Herrick
Sandy Cape
Devonport
1573
St. Marys
Mt. Ossa
1617
Launceston
Coles Bay
Queenstown
Macquarie Harbour
Strahan
Oatlands
1444
Derwent
Swansea
Point Hibbs
Gordon
Hobart
New Norfolk
1327
Frankland Ra.
Elliot Bay
Port Arthur
Storm Bay
Port Davey
Maatsuyker Islands
South East Cape
Average linear scale
0
100
200
300
miles
400
500
Km

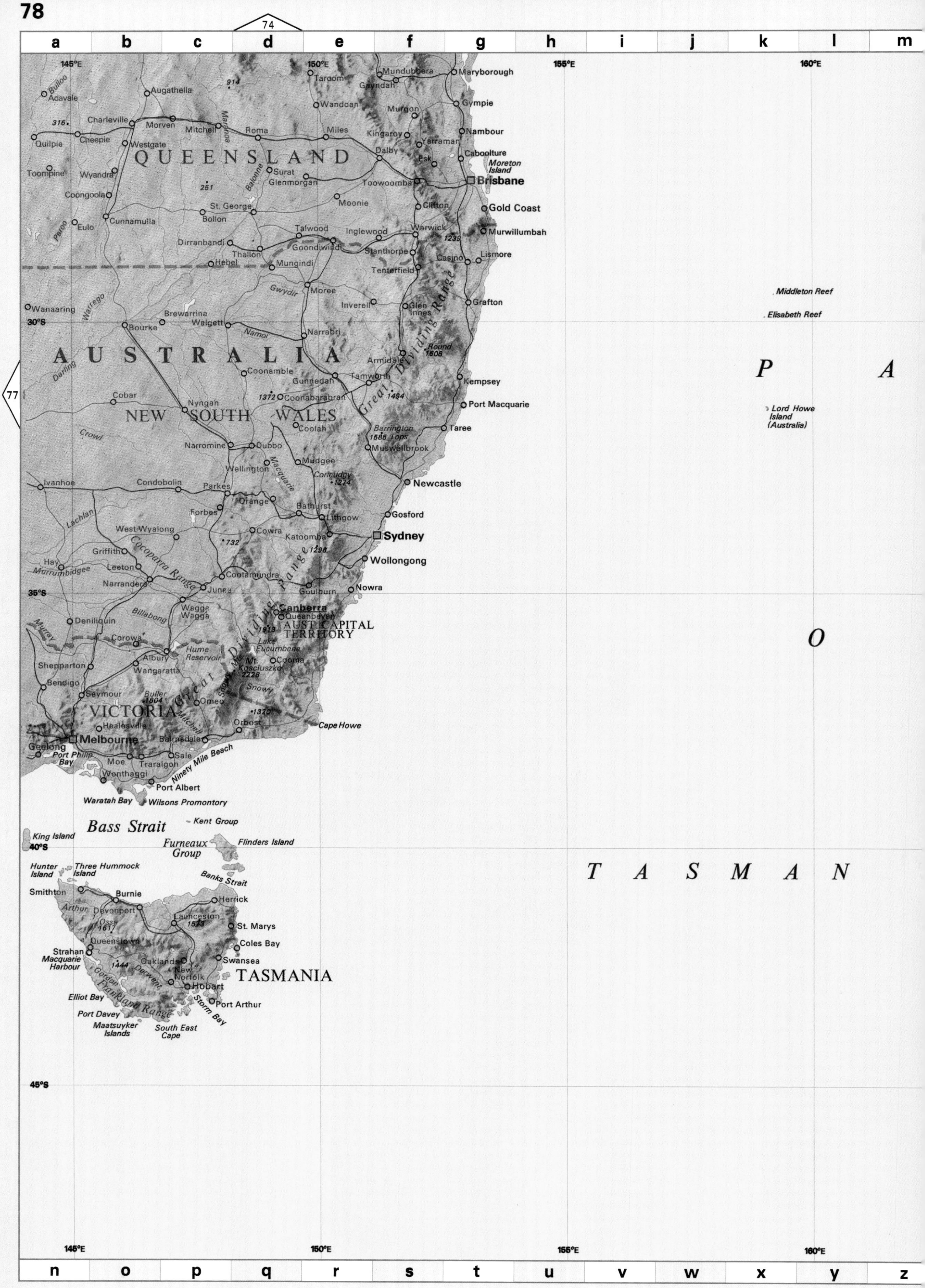

74

a b c d e f g h i j k l m

145°E 150°E 155°E 160°E

Bulloo
Adavale
Augathella
914
Taroom
Mundubbera
Maryborough
Gayndah
Wandoan
Gympie
Murgon
316
Charleville
Morven
Mitchell
Maranoa
Roma
Miles
Kingaroy
Nambour
Quilpie
Cheepie
Westgate
Yarraman
Dalby
Caboolture
Esk
Moreton Island
QUEENSLAND
Toompine
Wyandra
Surat
Balonne
Glenmorgan
Toowoomba
Brisbane
251
Cungoola
Moonie
Clifton
Gold Coast
St. George
Bollon
Paroo
Eulo
Cunnamulla
Talwood
Inglewood
Warwick
Murwillumbah
1235
Dirranbandi
Goondiwindi
Thallon
Stanthorpe
Lismore
Casino
Hebel
Mungindi
Tenterfield
Gwydir
Moree
Inverell
Glen Innes
Grafton
Wanaaring
Warrego
Brewarrina
Walgett
Namoi
Narrabri
Great Dividing Range
Round 1608
30°S
Bourke
Middleton Reef
Elisabeth Reef

77

AUSTRALIA
Darling
Coonamble
Gunnedah
Armidale
Tamworth
Kempsey
Cobar
1372
Coonabarabran
1494
Nyngan
Port Macquarie
NEW SOUTH WALES
Coolah
Barrington Tops 1586
Taree
Crowl
Narromine
Dubbo
Muswellbrook
Lord Howe Island (Australia)
Macquarie
Mudgee
Wellington
Coricudgy 1274
Newcastle
Ivanhoe
Condobolin
Parkes
Orange
Bathurst
Lachlan
Forbes
Lithgow
Gosford
West Wyalong
Cowra
732
Katoomba
Sydney
Cocoparra Range
1298
Griffith
Hay
Wollongong
Murrumbidgee
Leeton
Narrandera
Cootamundra
Junee
Goulburn
Nowra
35°S
Wagga Wagga
Canberra
Queanbeyan
Billabong
Deniliquin
AUST. CAPITAL TERRITORY
1913
Murray
Corowa
Lake Eucumbene
Albury
Hume Reservoir
Cooma
Mt. Kosciuszko 2228
Shepparton
Wangaratta
Snowy Mts.
Bendigo
Snowy
Seymour
Buller 1804
Omeo
Mitchell
1320
VICTORIA
Healesville
Orbost
Cape Howe
Melbourne
Bairnsdale
Geelong
Port Phillip Bay
Moe
Traralgon
Sale
Ninety Mile Beach
Wonthaggi
Port Albert
Waratah Bay
Wilsons Promontory
Kent Group
Bass Strait
King Island
Furneaux Group
Flinders Island
40°S
Hunter Island
Three Hummock Island
Banks Strait
Smithton
Burnie
Herrick
Arthur
Devonport
Launceston
Ossa 1617
1573
St. Marys
Queenstown
Coles Bay
Strahan
Macquarie Harbour
1444
Oaklands
Swansea
New Norfolk
Derwent
Gordon
Hobart
TASMANIA
Elliot Bay
Franklin Range
Storm Bay
Port Arthur
Port Davey
Maatsuyker Islands
South East Cape
45°S

P A
O
T A S M A N

145°E 150°E 155°E 160°E

n o p q r s t u v w x y z

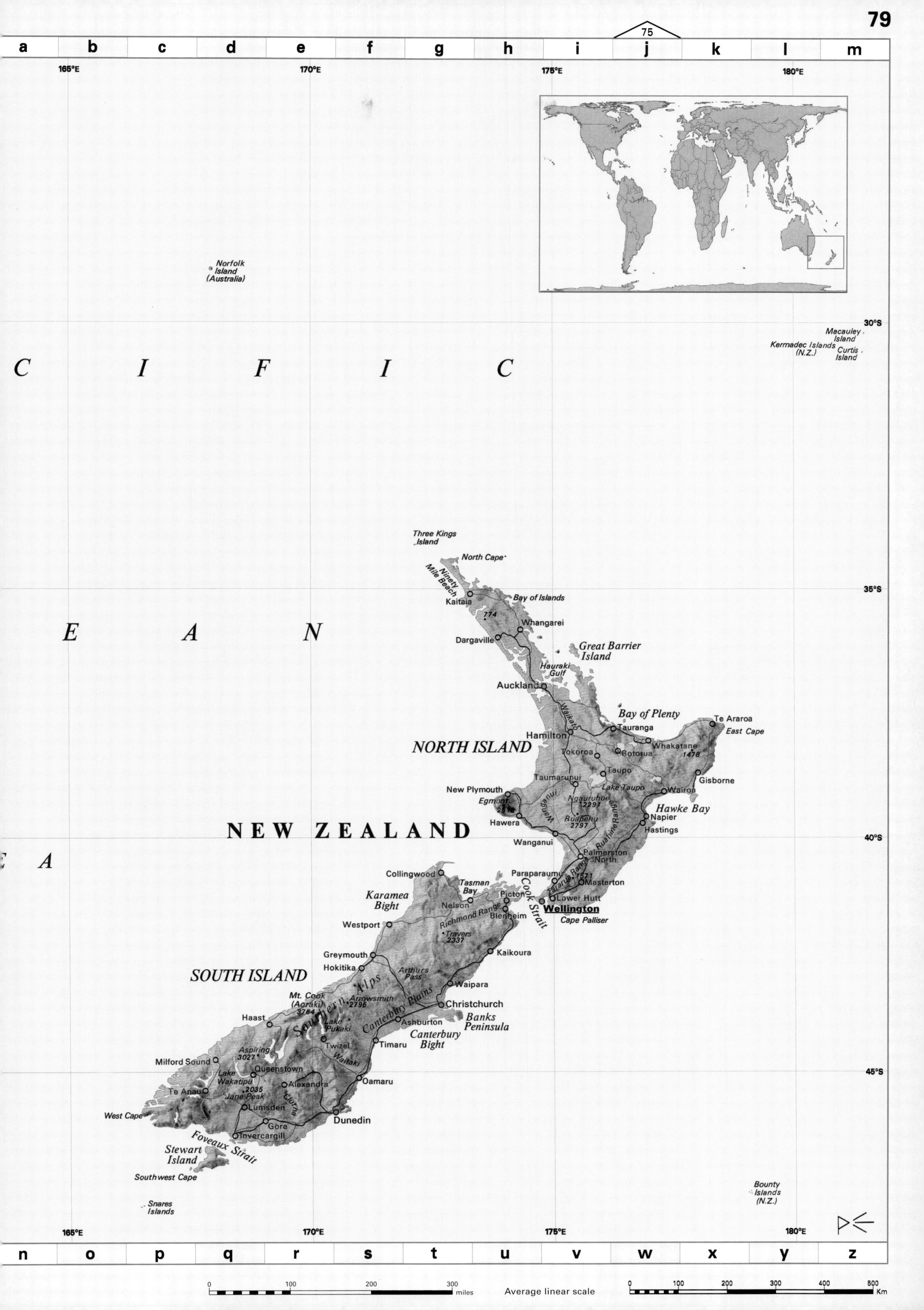
75
a b c d e f g h i j k l m
165°E
170°E
175°E
180°E
Norfolk Island (Australia)
30°S
Macauley Island
Kermadec Islands (N.Z.)
Curtis Island
C I F I C
E A N
Three Kings Island
North Cape
Ninety Mile Beach
Kaitaia
35°S
Bay of Islands
774
Whangarei
Dargaville
Great Barrier Island
Hauraki Gulf
Auckland
Waikato
Bay of Plenty
Te Araroa
East Cape
Tauranga
Hamilton
Whakatane
1478
NORTH ISLAND
Tokoroa
Rotorua
Taupo
Taumarunui
Lake Taupo
Gisborne
New Plymouth
Wairoa
Egmont
2518
Ngauruhoe 2291
Hawke Bay
Whanganui
Ruapehu 2797
Ruahine Range
Napier
Hawera
Hastings
NEW ZEALAND
Wanganui
40°S
A
Palmerston North
Paraparaumu
1571
Tararua Range
Masterton
Collingwood
Tasman Bay
Picton
Cook Strait
Lower Hutt
Karamea Bight
Nelson
Wellington
Blenheim
Richmond Range
Cape Palliser
Westport
Travers 2337
Greymouth
Kaikoura
Hokitika
Arthurs Pass
SOUTH ISLAND
Southern Alps
Waipara
Mt. Cook (Aoraki) 3764
Arrowsmith 2795
Christchurch
Haast
Canterbury Plains
Lake Pukaki
Ashburton
Banks Peninsula
Canterbury Bight
Aspiring 3027
Twizel
Timaru
Milford Sound
Waitaki
Queenstown
Lake Wakatipu
Alexandra
Oamaru
45°S
Te Anau
2035
Jane Peak
Clutha
Lumsden
West Cape
Dunedin
Gore
Invercargill
Foveaux Strait
Stewart Island
Southwest Cape
Snares Islands
Bounty Islands (N.Z.)
165°E
170°E
175°E
180°E
n o p q r s t u v w x y z
0 100 200 300 miles
Average linear scale
0 100 200 300 400 500 Km

82

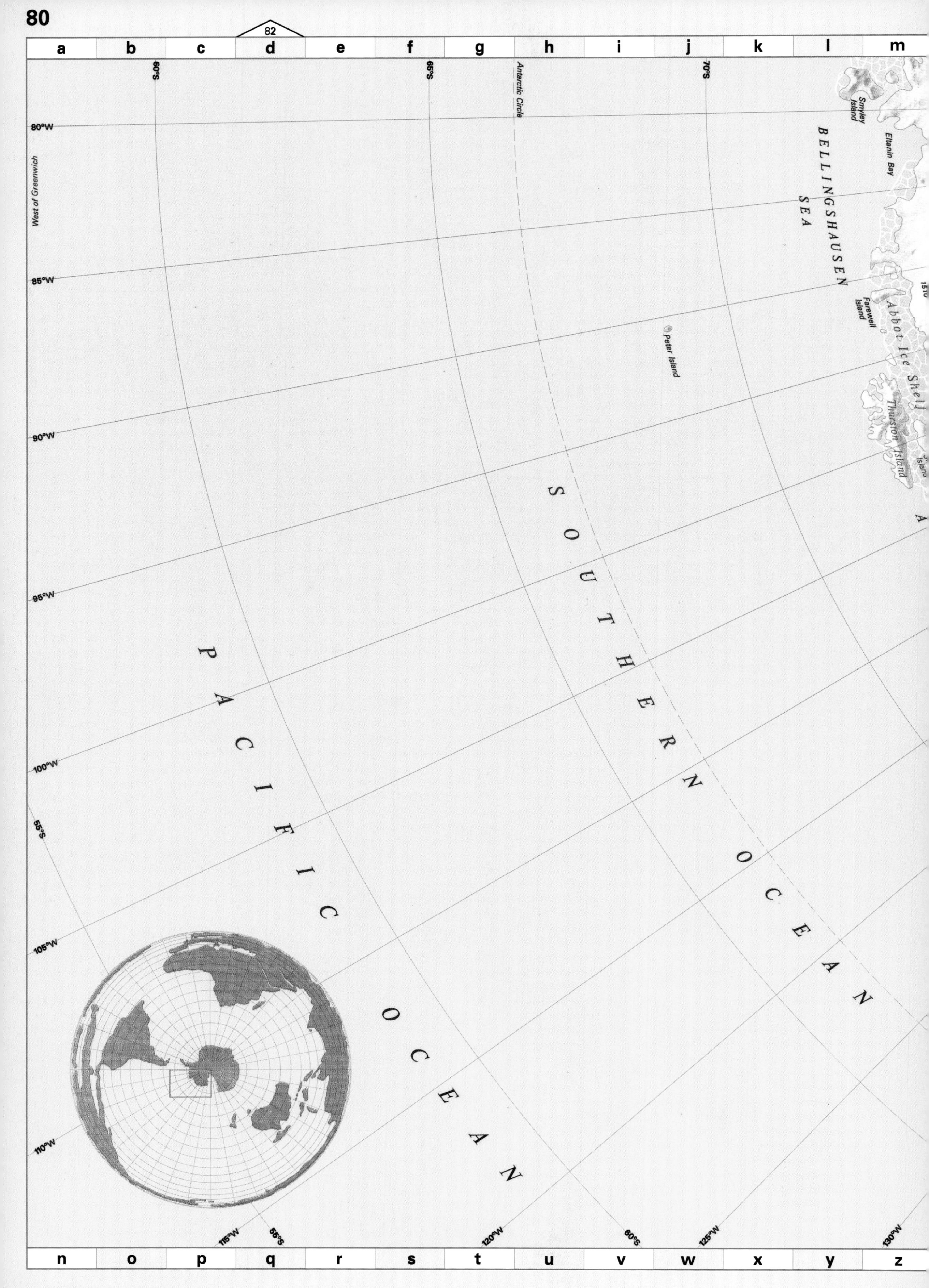

This map shows 1/60 of the earth's surface

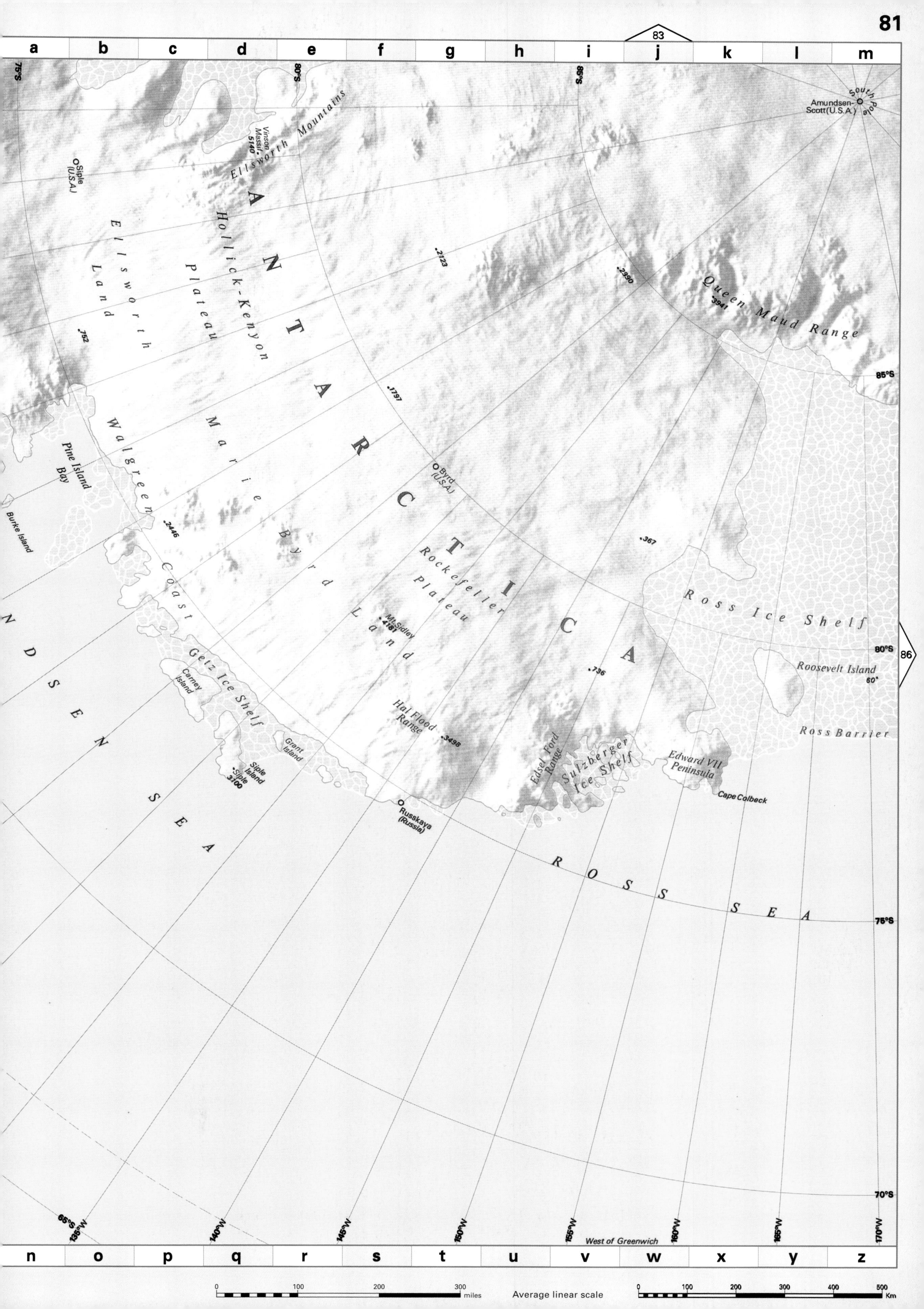
83
86
a
b
c
d
e
f
g
h
i
j
k
l
m
n
o
p
q
r
s
t
u
v
w
x
y
z
ANTARCTICA
Ellsworth Mountains
Vinson Massif 5140
Siple (U.S.A.)
Ellsworth Land
Hollick-Kenyon Plateau
Marie Byrd Land
Walgreen Coast
Pine Island Bay
Burke Island
Amundsen-Scott (U.S.A.)
South Pole
Queen Maud Range
3941
2990
2123
1797
752
2446
Byrd (U.S.A.)
Rockefeller Plateau
Mt. Sidley 4181
367
Ross Ice Shelf
Roosevelt Island
Ross Barrier
736
Hal Flood Range
3498
Edsel Ford Range
Sulzberger Ice Shelf
Edward VII Peninsula
Cape Colbeck
Russkaya (Russia)
Getz Ice Shelf
Carney Island
Siple Island
Siple 3100
Grant Island
AMUNDSEN SEA
ROSS SEA
75°S
80°S
85°S
70°S
65°S
135°W
140°W
145°W
150°W
155°W
160°W
165°W
170°W
West of Greenwich
0
100
200
300
miles
Average linear scale
0
100
200
300
400
500
Km

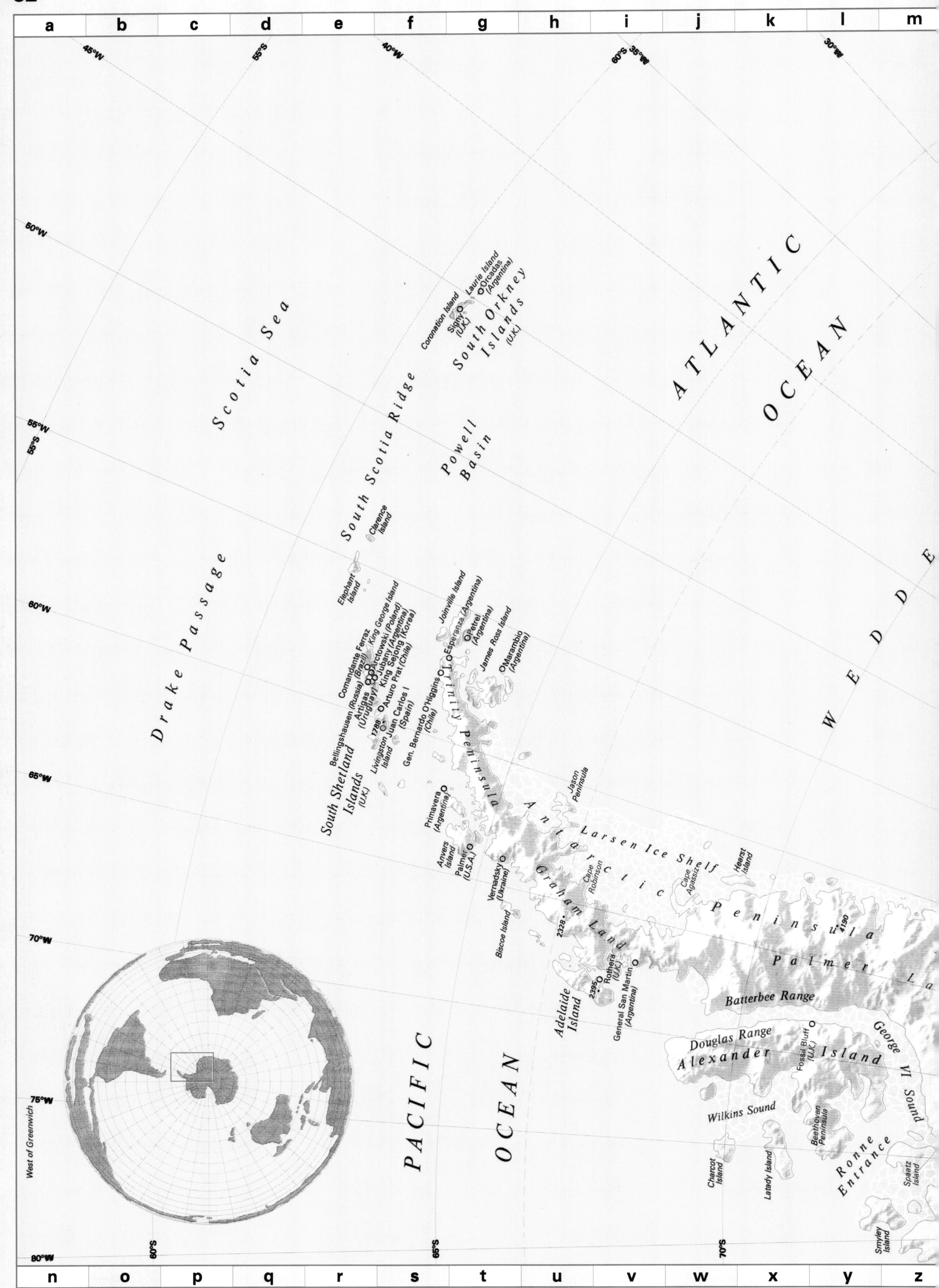

Scotia Sea
Drake Passage
South Scotia Ridge
Powell Basin
South Orkney Islands (U.K.)
Coronation Island
Laurie Island
Orcadas (Argentina)
Signy (U.K.)
ATLANTIC OCEAN
Clarence Island
Elephant Island
King George Island
South Shetland Islands (U.K.)
Livingston Island
Joinville Island
James Ross Island
Marambio (Argentina)
Esperanza (Argentina)
Petrel (Argentina)
Gen. Bernardo O'Higgins (Chile)
Juan Carlos I (Spain)
Arturo Prat (Chile)
King Sejong (Korea)
Jubany (Argentina)
Arctowski (Poland)
Comandante Ferraz (Brazil)
Artigas (Uruguay)
Bellingshausen (Russia)
1786
Trinity Peninsula
Primavera (Argentina)
Anvers Island
Palmer (U.S.A.)
Vernadsky (Ukraine)
Antarctic Peninsula
Jason Peninsula
Larsen Ice Shelf
Hearst Island
Cape Robinson
Cape Agassiz
Graham Land
Biscoe Island
2328
2395
Rothera (U.K.)
General San Martin (Argentina)
Adelaide Island
Palmer Land
4190
Batterbee Range
Douglas Range
Alexander Island
Fossil Bluff (U.K.)
George VI Sound
Wilkins Sound
Beethoven Peninsula
Charcot Island
Latady Island
Ronne Entrance
Spaatz Island
Smyley Island
PACIFIC OCEAN
WEDDE
West of Greenwich

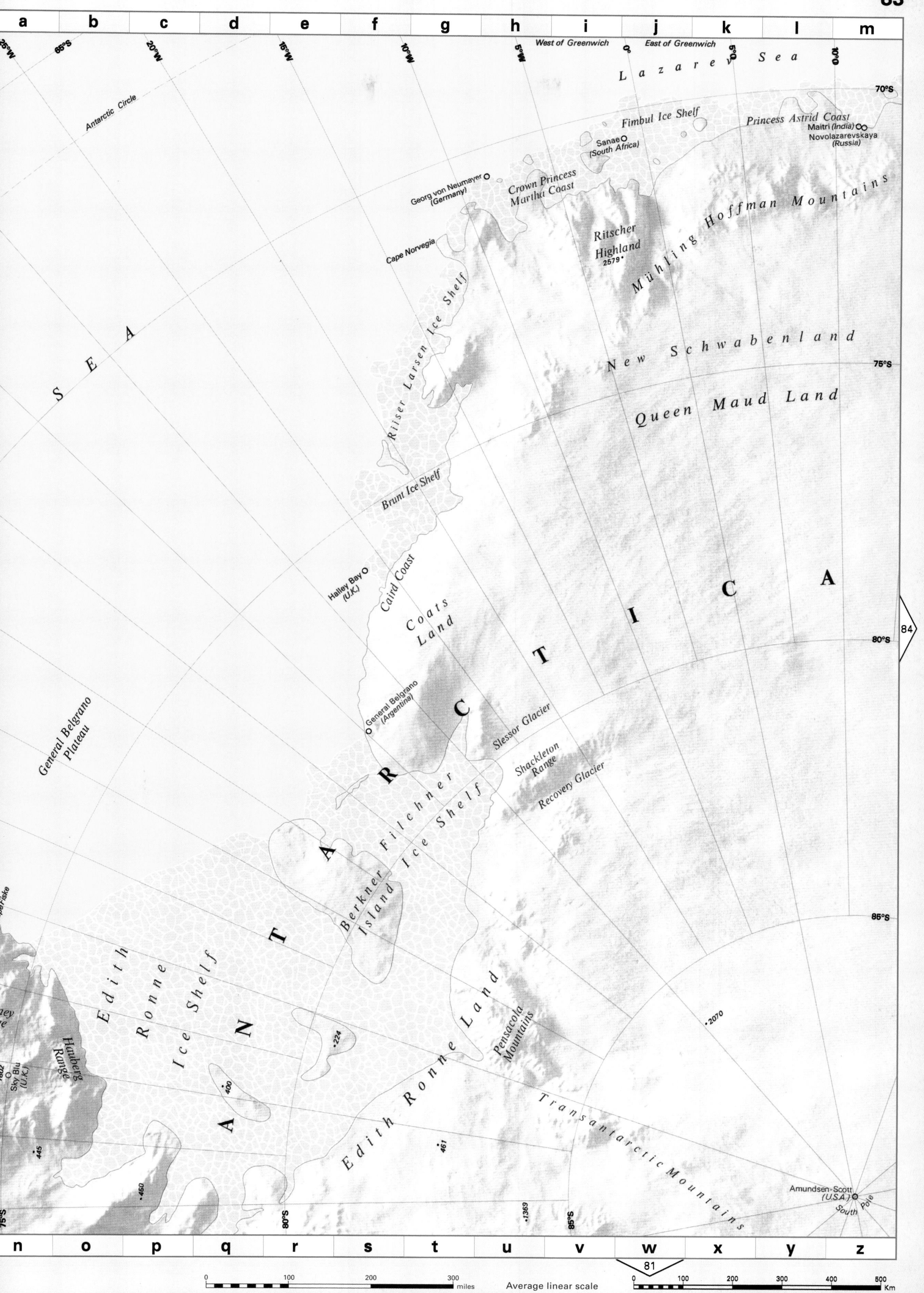
a
b
c
d
e
f
g
h
i
j
k
l
m
West of Greenwich
East of Greenwich
Lazarev Sea
Antarctic Circle
Fimbul Ice Shelf
Princess Astrid Coast
Maitri (India)
Novolazarevskaya (Russia)
Sanae (South Africa)
Georg von Neumayer (Germany)
Crown Princess Martha Coast
Ritscher Highland
2579
Mühling Hoffman Mountains
Cape Norvegia
Riiser Larsen Ice Shelf
SEA
New Schwabenland
Queen Maud Land
Brunt Ice Shelf
Halley Bay (U.K.)
Caird Coast
Coats Land
ANTARCTICA
General Belgrano (Argentina)
Slessor Glacier
Shackleton Range
Recovery Glacier
General Belgrano Plateau
Filchner Ice Shelf
Berkner Island
Edith Ronne Ice Shelf
Edith Ronne Land
Hauberg Range
Sky Blu (U.K.)
Pensacola Mountains
Transantarctic Mountains
Amundsen-Scott (U.S.A.)
South Pole
2070
224
400
445
460
461
1369
1802
70°S
75°S
80°S
85°S
84
n
o
p
q
r
s
t
u
v
w
x
y
z
81
Average linear scale
miles
Km

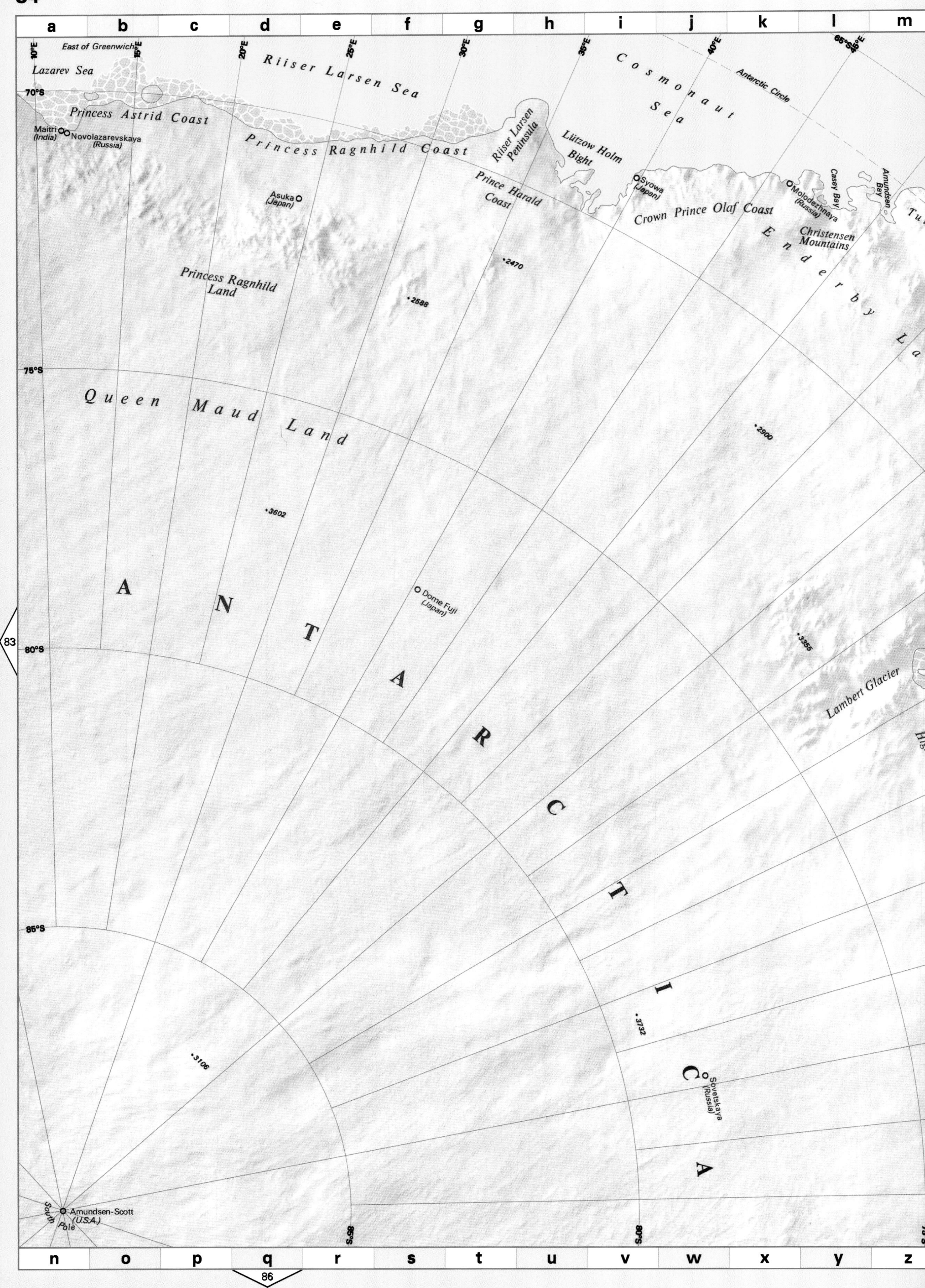

This map shows 1/60 of the earth's surface

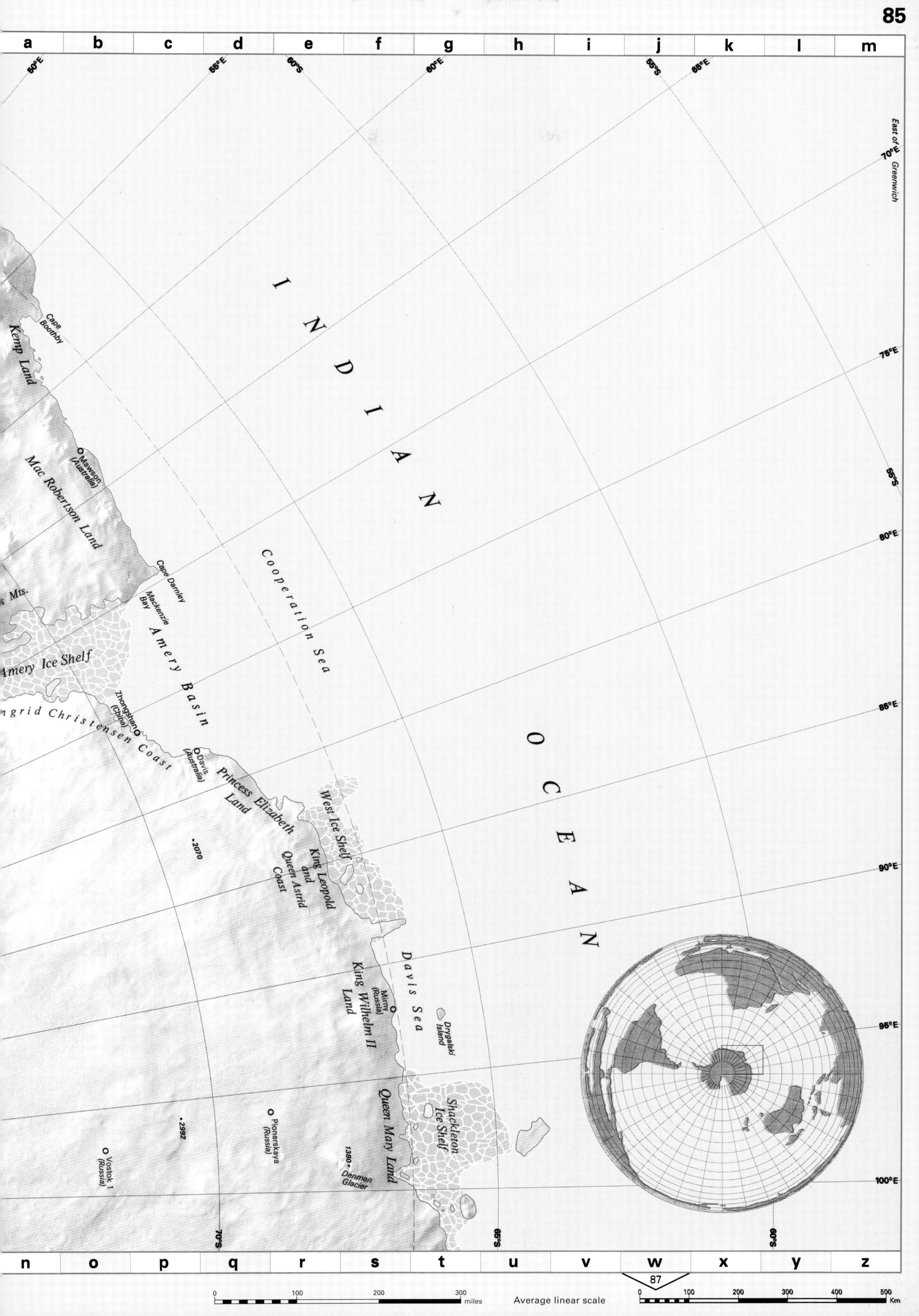
a b c d e f g h i j k l m
n o p q r s t u v w x y z
INDIAN OCEAN
Kemp Land
Cape Boothby
Mawson (Australia)
Mac Robertson Land
Cape Darnley
Mackenzie Bay
Amery Ice Shelf
Amery Basin
Cooperation Sea
Ingrid Christensen Coast
Zhongshan (China)
Davis (Australia)
Princess Elizabeth Land
West Ice Shelf
King Leopold and Queen Astrid Coast
King Wilhelm II Land
Mirny (Russia)
Davis Sea
Drygalski Island
Queen Mary Land
Shackleton Ice Shelf
Denman Glacier
Pionerskaya (Russia)
Vostok 1 (Russia)
2070
2992
1380
East of Greenwich
50°E
55°E
60°E
65°E
70°E
75°E
80°E
85°E
90°E
95°E
100°E
55°S
60°S
65°S
70°S
87
Average linear scale
0 100 200 300 miles
0 100 200 300 400 500 Km

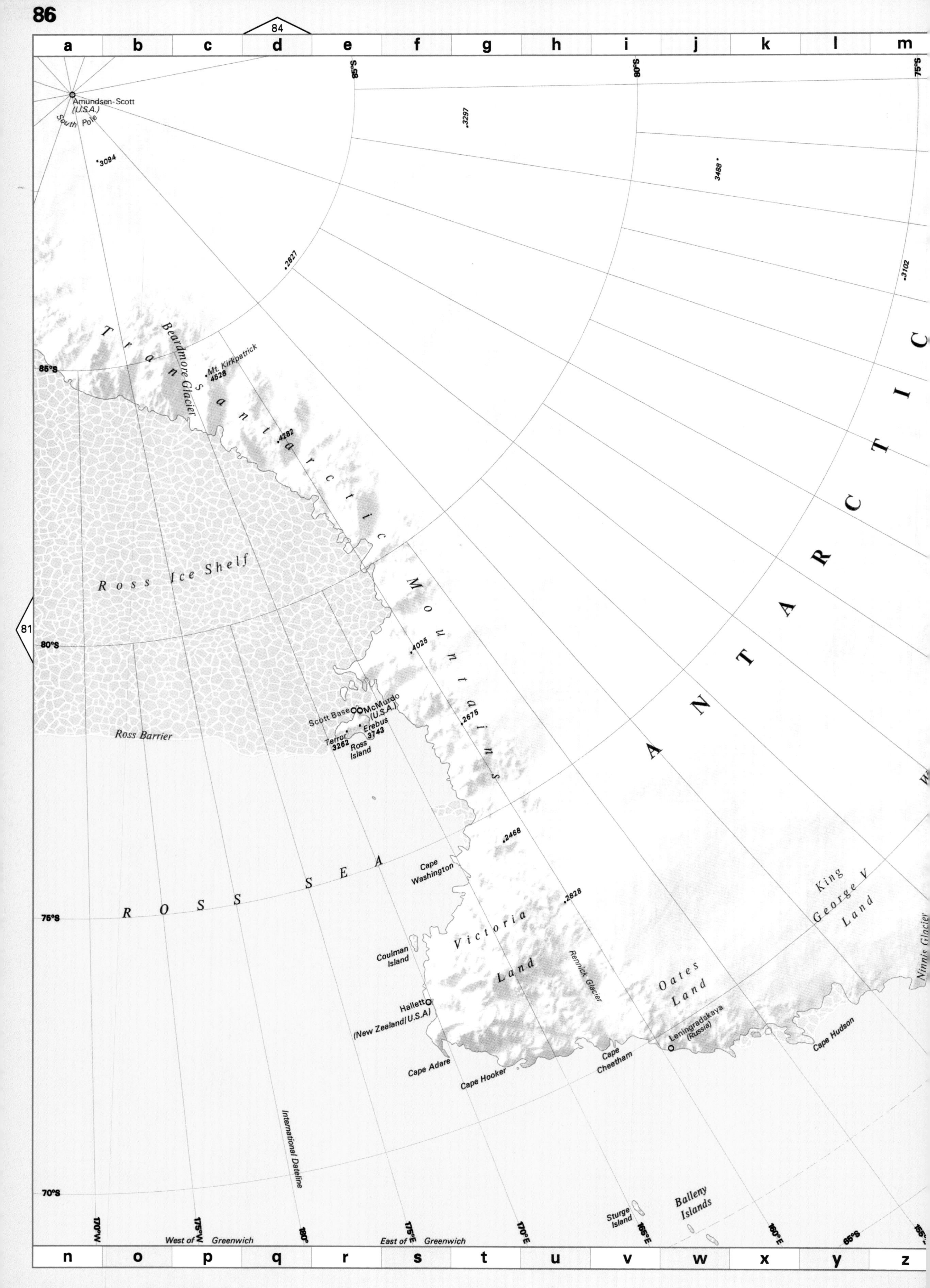

This map shows 1/60 of the earth's surface

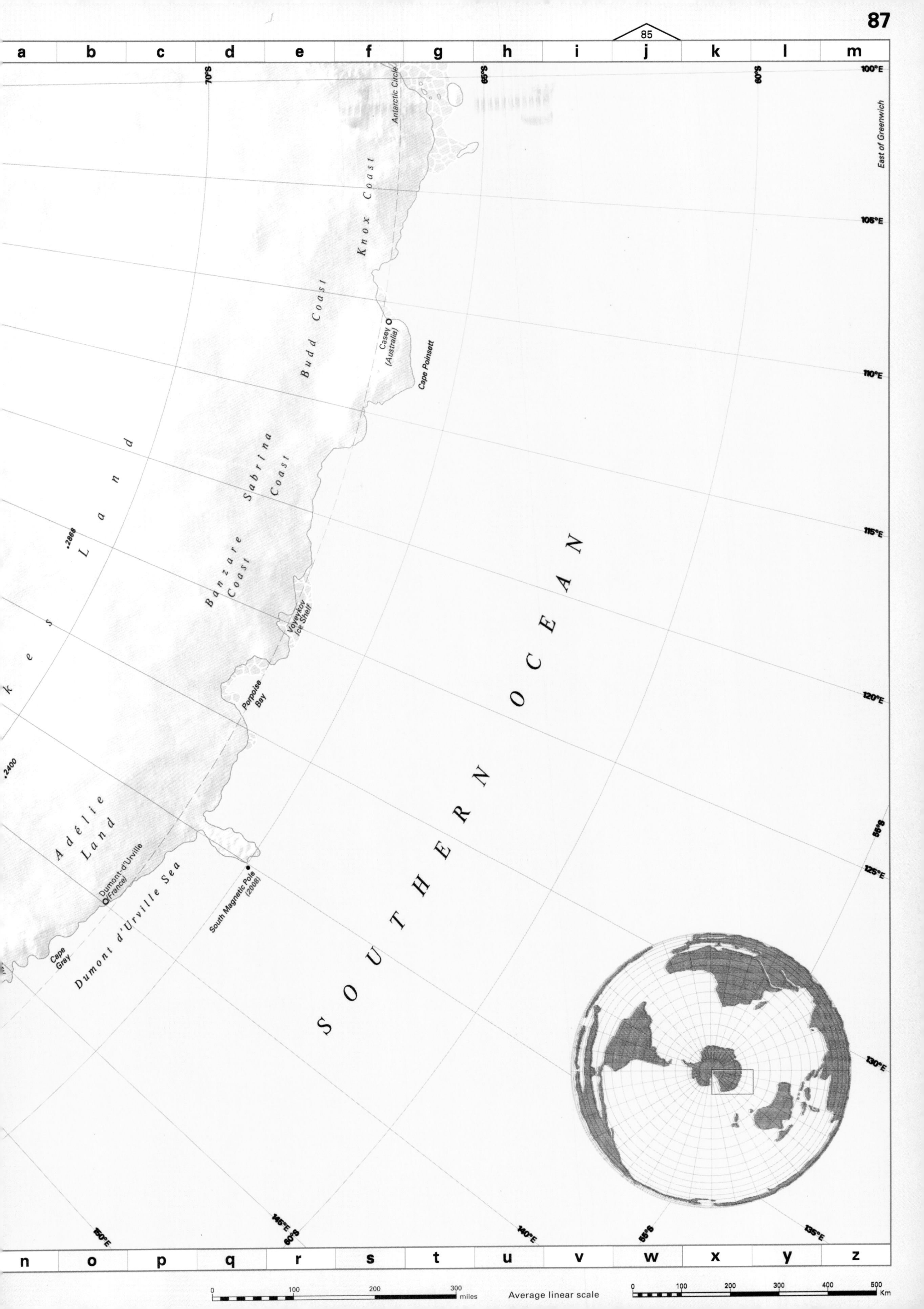
85
a b c d e f g h i j k l m
70°S
65°S
60°S
100°E
East of Greenwich
Antarctic Circle
Knox Coast
105°E
Budd Coast
Casey (Australia)
Cape Poinsett
110°E
Sabrina Coast
Wilkes Land
115°E
.2868
Banzare Coast
Vostok Ice Shelf
SOUTHERN OCEAN
120°E
Porpoise Bay
.2400
Adélie Land
55°S
Dumont-d'Urville (France)
Dumont d'Urville Sea
South Magnetic Pole (2008)
125°E
Cape Gray
130°E
150°E
145°E
60°S
140°E
55°S
135°E
n o p q r s t u v w x y z
0 100 200 300 miles
Average linear scale
0 100 200 300 400 500 Km

Hudson Bay
Foxe Basin
Foxe Channel
Roes Welcome Sound
Committee Bay
Gulf of Boothia
Queen Maud Gulf
McClintock Channel
Dolphin and Union Strait
Great Slave Lake
Great Bear Lake
Lake Athabasca
Reindeer Lake
ALBERTA
SASKATCHEWAN
MANITOBA
NORTHWEST TERRITORIES
NUNAVUT
CANADA
Mackenzie Mountains
Franklin Mountains
Caribou Mountains
King William Island
Southampton Island
Boothia Peninsula
Melville Peninsula
Brodeur Peninsula
Baffin Island
Bylot Island
Rae Isthmus
Edmonton
Churchill
Yellowknife
Eskimo Point (Arviat)
Rankin Inlet
Baker Lake
Chesterfield Inlet (Igluligaarjuk)
Arctic Circle

YUKON TERRITORY
ALASKA
U.S.A.
Brooks Range
Anaktuvik Pass
Arctic Village
Eagle Plain
Porcupine
Old Crow
Mount Chamberlin 2749
Colville
Icy Cape
Wainwright
Chukchi Sea
Deadhorse
Prudhoe Bay
Barrow
Cape Barrow
Kaktovik
Herschel
MacKenzie Bay
Fort McPherson
Arctic Red River
Aklavik
Inuvik
Tuktoyaktuk
Mackenzie
Good Hope
Cape Dalhousie
Cape Bathurst
Cape Perry
Paulatuk
Amundsen Gulf
Sachs Harbour
Holman Island
Minto Inlet
Prince Albert Peninsula
Prince of Wales Strait
Banks Island
Cape Prince Alfred
BEAUFORT SEA
ARCTIC OCEAN
McClure Strait
Mould Bay
Prince Patrick Island
Dundas Peninsula
Melville Island
Melville Sound
Viscount
Stefansson Island
Byam Martin Channel
Hazen Strait
Lougheed Island
Mackenzie King Island
Borden Island
Prince Gustav Adolf Sea
Bathurst Island
Resolute
Cornwallis Island
Table I.
Grinnell Peninsula
Belcher Channel
Cornwall Island
Ellef Ringnes Island
Hassel Sound
Peary Channel
Meighen I.
Amund Ringnes Island
Sverdrup Channel
Sverdrup Islands
Parry Islands
Queen Elizabeth Islands
Graham I.
Norwegian Bay
Axel Heiberg Island
Nansen Sound
Eureka
Devon Island
Jones Sound
Grise Fiord
Sydney Ice Cap
Bjorne Peninsula
North Lincoln Land
Smith Bay
Ellesmere Island
Agassiz Ice Cap
Greely Fiord
United States Range
Cape Alert
Cape Discovery
North Magnetic Pole (2006)
North Pole
65°N 70°N 75°N 80°N 85°N
135°W 140°W 145°W 150°W 155°W 160°W 165°W 170°W
a b c d e f g h i j k l m
n o p q r s t u v w x y z
11 94 91
Average linear scale
0 100 200 300 miles
0 100 200 300 400 500 Km

88

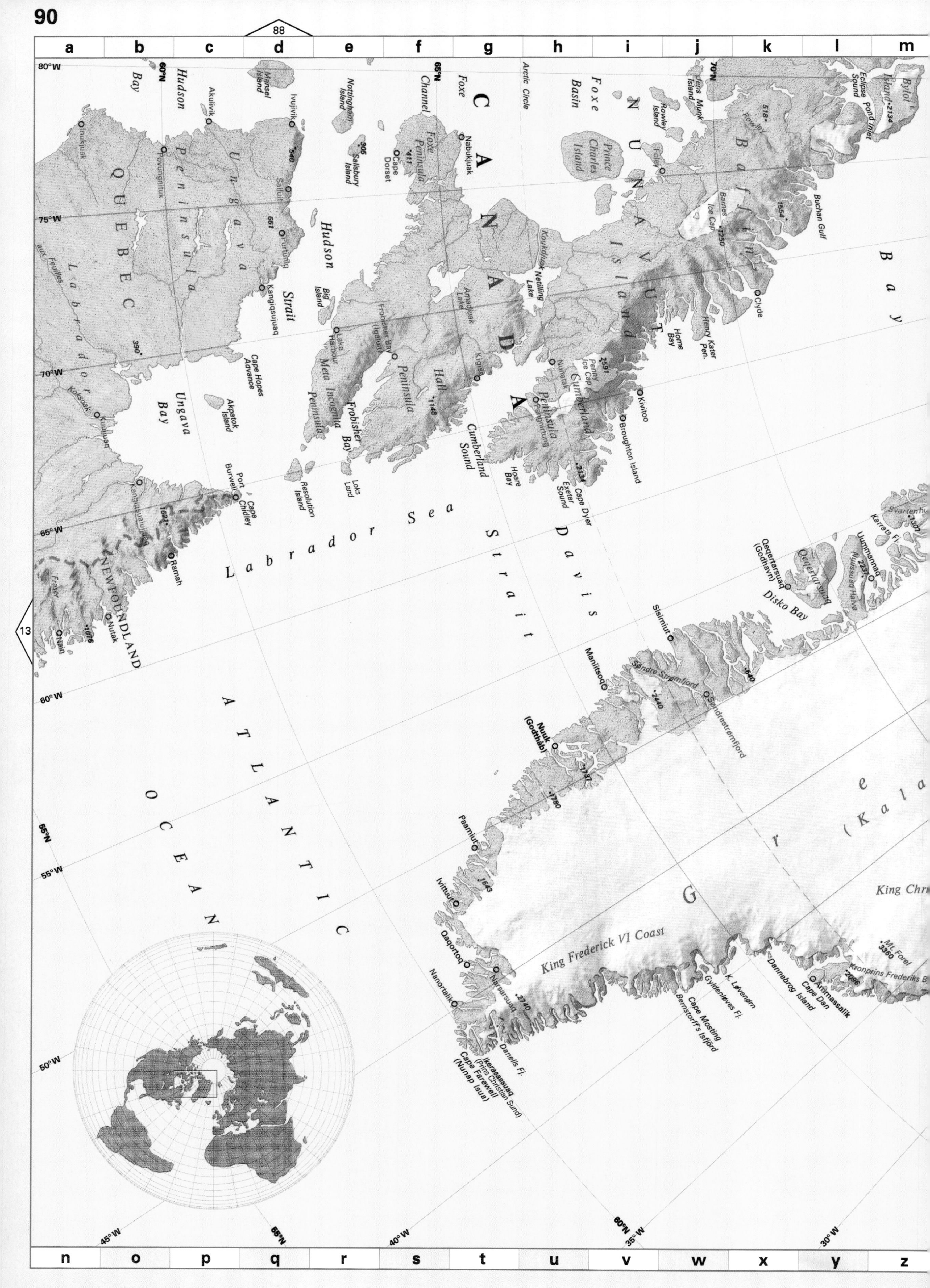

This map shows 1/60 of the earth's surface

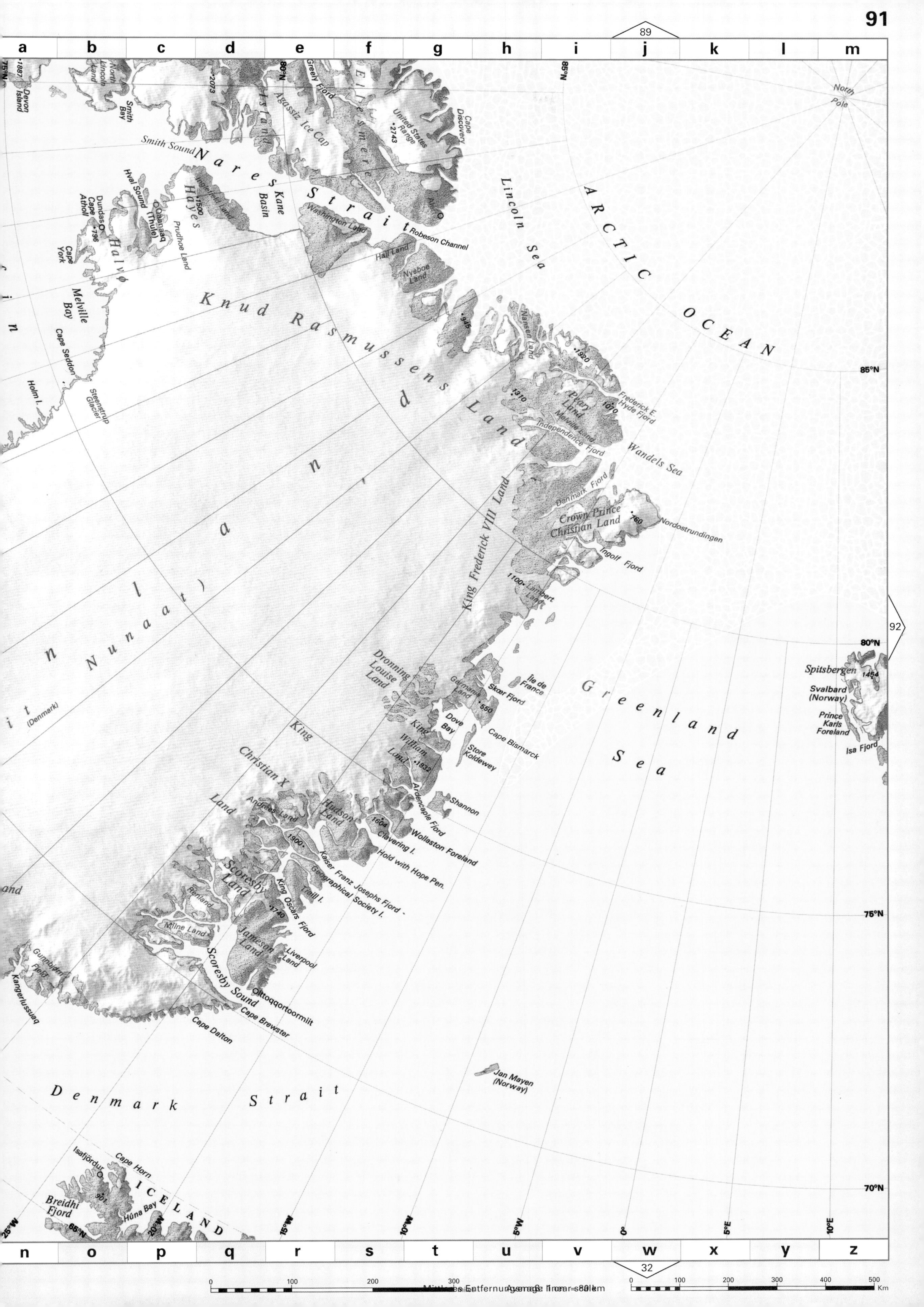
89
a
b
c
d
e
f
g
h
i
j
k
l
m
Devon Island
North Lincoln Land
Smith Bay
Smith Sound
Nares Strait
Kane Basin
Agassiz Ice Cap
Greely Fjord
Ellesmere Island
United States Range
2743
Cape Discovery
Hyal Sound
Dundas
Cape Atholl
796
Qaanaaq (Thule)
Inglefield Land
Hayes Halvø
Prudhoe Land
Cape York
Washington Land
Hall Land
Robeson Channel
Nyeboe Land
Lincoln Sea
ARCTIC OCEAN
North Pole
Melville Bay
Cape Seddon
Holm I.
Steenstrup Glacier
Knud Rasmussens Land
945
Nansen Land
1920
1310
Peary Land
1070
Frederick E. Hyde Fjord
Independence Fjord
Wandels Sea
Danmark Fjord
Crown Prince Christian Land
760
Nordostrundingen
Ingolf Fjord
King Frederick VIII Land
1100
Lambert Land
Greenland (Kalaallit Nunaat) (Denmark)
85°N
80°N
75°N
70°N
92
Spitsbergen
1454
Svalbard (Norway)
Prince Karls Foreland
Isa Fjord
Île de France
Greenland Sea
Dronning Louise Land
Germania Land
Skær Fjord
550
Dove Bay
Cape Bismarck
Store Koldewey
King William Land
1832
King Christian X Land
Shannon
Ardencaple Fjord
Hudson Land
1604
Andrées Land
Wollaston Foreland
Clavering I.
Hold with Hope Pen.
1900
Kaiser Franz Josephs Fjord
Geographical Society I.
Traill I.
Scoresby Land
Renland
King Oscars Fjord
1740
Milne Land
Jameson Land
Liverpool Land
Gunnbjørn Fjeld
Kangerlussuaq
Scoresby Sound
Ittoqqortoormiit
Cape Brewster
Cape Dalton
Jan Mayen (Norway)
Denmark Strait
Isafjördur
Cape Horn
901
Breidhi Fjord
Húna Bay
ICELAND
25°W
20°W
15°W
10°W
5°W
0°
5°E
10°E
65°N
n
o
p
q
r
s
t
u
v
w
x
y
z
32
0
100
200
300
400
500
Km

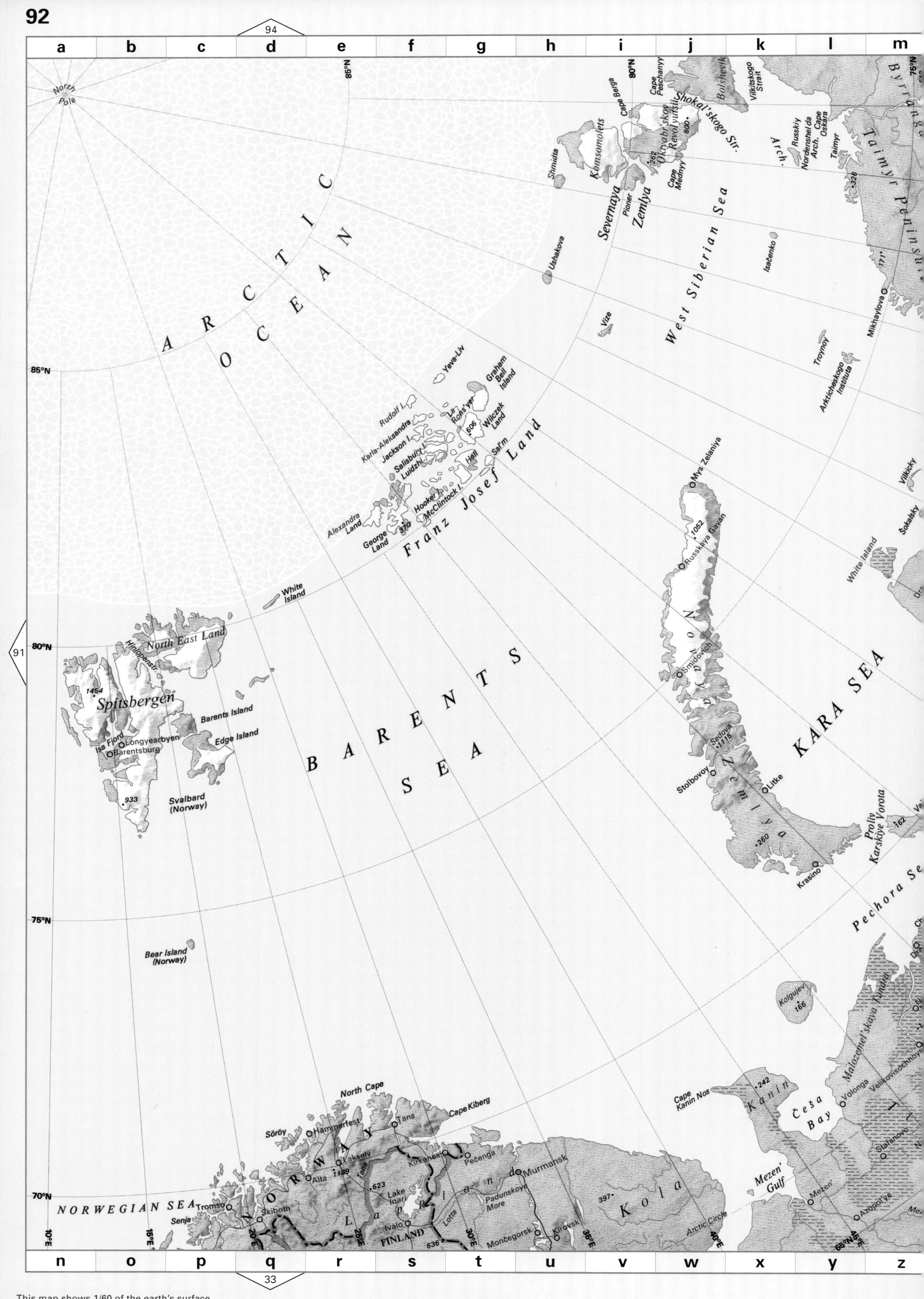
94
a b c d e f g h i j k l m
North Pole
ARCTIC OCEAN
85°N
Cape Berga
Cape Peschanyy
Bolshevik
Vilkitskogo Strait
Shokal'skogo Str.
Oktyabr'skoy Revolyutsii
Komsomolets
Shmidta
Severnaya Zemlya
Pioner
Cape Mednyy
Russkiy
Nordenshel'da Arch.
Cape Oskara
Taimyr
Taimyr Peninsula
West Siberian Sea
Ushakova
Isačenko
Vize
Mikhaylova
Troynoy
Arkticheskogo Instituta
Yeva-Liv
Graham Bell Island
Rudolf I.
La Rons'yer
Wilczek Land
Karla-Aleksandra
Jackson I.
Salisbury I.
Luidzhi
Hall
Sal'm
Franz Josef Land
Alexandra Land
George Land
Hooker I.
McClintock I.
Mys Zelaniya
Russkaya Gavan
Novaya Zemlya
Smidovich
Sedova
Stolbovoy
Litke
Krasino
Vilkicky
Šokalsky
White Island
White Island
KARA SEA
Proliv Karskiye Vorota
Pechora Sea
91
80°N
North East Land
Hinlopenstr.
Spitsbergen
Barents Island
Edge Island
Isa Fjord
Longyearbyen
Barentsburg
Svalbard (Norway)
BARENTS SEA
75°N
Bear Island (Norway)
Kolgujev
Malozemel'skaya Tundra
Velikovisochnoye
Cape Kanin Nos
Kanin
Česa Bay
Volonga
Stafanovo
Mezen' Gulf
Mezen'
Azopol'ye
North Cape
Cape Kiberg
Söröy
Hammerfest
Tana
NORWAY
Lakselv
Kirkenes
Pečenga
Alta
Murmansk
Kola
Lapland
Lake Inari
Padunskoye More
NORWEGIAN SEA
Tromsø
Senja
Skibotn
Ivalo
FINLAND
Lotta
Mončegorsk
Kirovsk
Arctic Circle
70°N
10°E 15°E 20°E 25°E 30°E 35°E 40°E 45°E
n o p q r s t u v w x y z
33

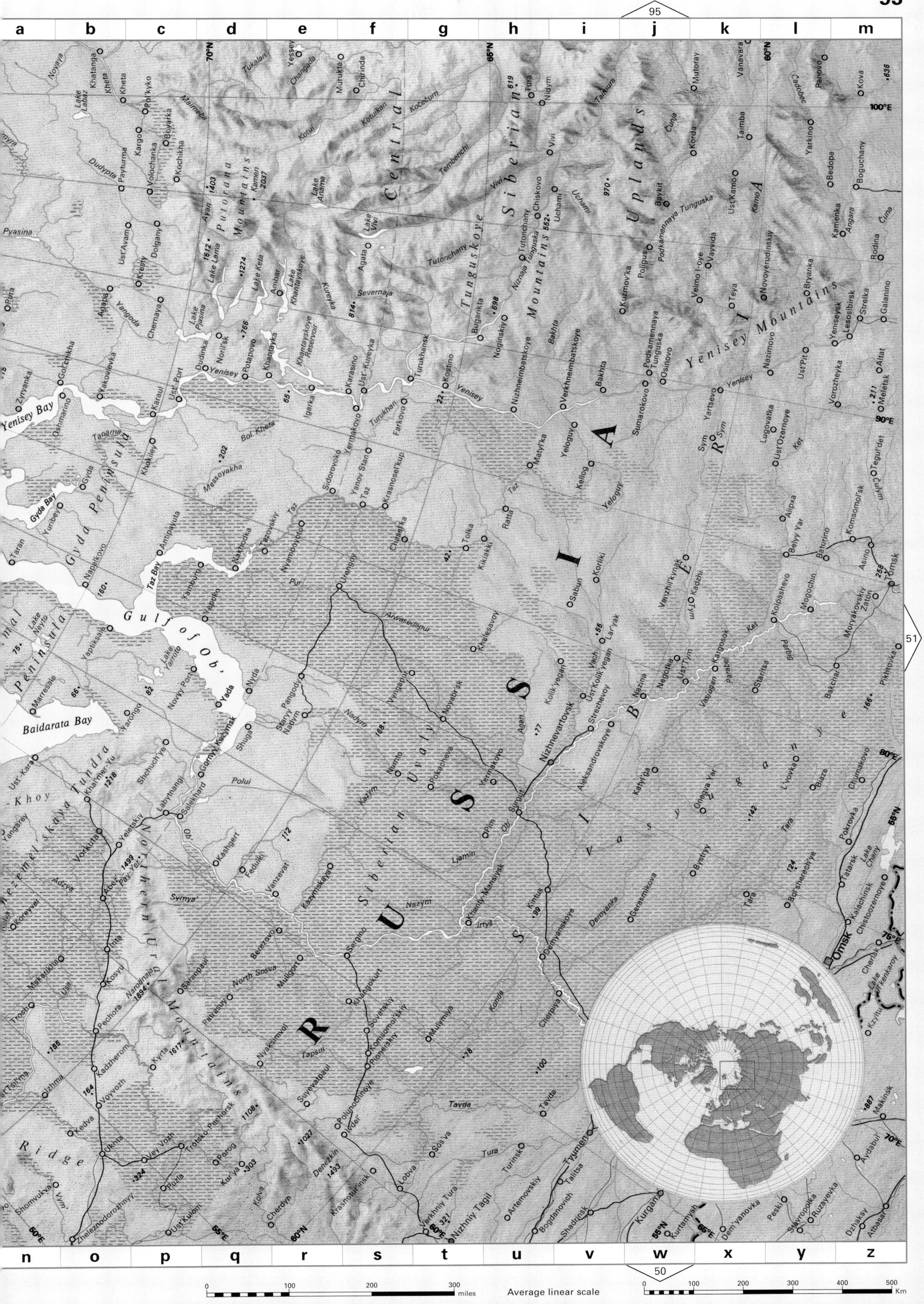
95
a b c d e f g h i j k l m
Khatanga
Lake Labaz
Kheta
Pol'kyko
Maimecha
Tukalan
Yessey
Changada
Murukta
Chirinda
Kotuikan
Kotui
Kočečum
Central
Siberian
Upland
Tura
Nidym
Taimura
Mutoray
Vanavara
Panovo
Kova
100°E
Kargo
Boyarka
Kochikha
Volochanka
Dudypta
Payturma
Putorana Mountains
Kamen 2037
1403
Lake Anama
Tembenchi
Vivi
Chiskovo
Uchami
970
Cunja
Korda
Taimba
Yarkino
Bedopa
Boguchany
Baykit
Podkamennaya Tunguska
Ust'Kamo
Kamo
Kamenka
Angara
Čuna
Pyasina
Ayan
Lake Vivi
Tutonchany
Tunguska Mountains
552
Ust'Avam
Dolgany
Kresty
1612
Lake Lama
1274
Lake Keta
Ambar
Lake Khantayskoye
Agata
Severnaja
Kureyka
614
Poligus
Veimo-I-oye
Vayvida
Rodina
Agapa
Yangoda
Chernaya
Lake Pyasina
766
Noril'sk
Khantayskoye Reservoir
Buganikta
598
Noginskiy
Kuzmovka
Teya
Novoyerudinskiy
Bryanka
Yenisey Mountains
Yeniseysk
Lesosibirsk
Strelka
Galanino
Gol'chikha
Yakovlevka
Dudinka
Yenisey
Potapovo
Khantayka
Karasino
Ust'-Kureyka
Turukhansk
Kostino
Nizhneimbatskoye
Verkhneimbatskoye
Bakhta
Podkamennaya Tunguska
Osinovo
Nazimovo
Ust'Pit
Vorozheyka
Altat
Meletsk
Zyryanka
Yenisey Bay
Oshmarino
Ust'-Port
Karaul
Igarka
Yermakovo
Turukhan
Farkovo
Sumarokovo
Yartsevo
Sym
Lugovatka
Ust'Ozernoye
Ket
90°E
Gyda Peninsula
Tanama
Khakhley
Bol'Kheta
Messoyakha
202
Sidorovsk
Yanov Stan
Krasnosel'kup
Taz
Matyl'ka
Yeloguy
Kellog
Tegul'det
Gyda Bay
Gyda
Yuribey
Antipayuta
Tazovskiy
Nakhodka
Nyamboyoto
Charkha
Tolka
Kikiakki
Ratta
Korliki
Alipka
Komsomol'sk
Belyy Yar
Baturino
Asino
Tomsk
Taran
Napalkovo
Taz Bay
Yamburg
Pur
Urengoy
Sabun
Vanzhil'kynak
Kedzhi
Tym
Kolpashevo
Mogochin
Molchanovo
Zaton
Gulf of Ob'
Yamal Peninsula
Lake Neyto
Yaptiksale
Yepoko
Aivasedapur
Khalesavoy
Lar'yak
Vakh
Kolik'yegan
Ust'Kolik'yegan
Negotka
Ust'Tym
Kargasok
Parabel'
Starica
Bakchar
Parbig
Pikhtovka
51
Marresale
Lake Yarroto
Novyy Port
Nyda
Yada
Pangody
Vyngapur
Noyabr'sk
Nadym
Staryy Nadym
Nazina
Strezhevoy
Vasyugan
Baidarata Bay
Yarongo
Shchuch'ye
Shuga
Gornyy Kazymsk
Numto
Pokachevo
Yermakovo
Nizhnevartovsk
Aleksandrovskoye
Katyl'ga
Onega Yar
L'vovka
Blaza
Chumakovo
80°E
Ust'-Kara
Khalmer-Yu
Bol'shezemel'skaya Tundra
Labytnangi
Salekhard
Polui
Kazym
Surgut
Pim
Tara
Pokrovka
55°N
Vorkuta
Yeletskiy
Kashgert
Fedulki
Ob'
Siberian Uvaly
Ljamin
Vasyugan'ye
Lake Chany
Abez'
Vanzevat
Kazymskaya
Nazym
Khanty-Mansiysk
Kintus
Gerasimkova
Bystryy
Bol'sherech'ye
Tatarsk
Kalachinsk
Chistoozernoye
Adzva
Northern Ural Mountains
Synya
Berezovo
Sergino
Irtys
Demyansk
Demyanskoye
Demyanka
Omsk
75°E
Inta
Makarikha
Saranpaul
Narodnaja 1894
North Sosva
Mulugort
Kharagokurt
Konda
Cherniya
Kenkaroy
Kosyu
Usa
Trosh
Pechora
Paksaduy
Nyaksimvol
Sovetskiy
Komsomol'skiy
Pionerskiy
Mulymya
Tapsui
Kadzherom
Kyrta
1617
Izhma
Vovvozh
Suyevatpaul
Polunochnoye
Ivdel'
Tavda
Makinsk
Kedva
Ukhta
Vey Vozh
Troitsko-Pechorsk
Porog
Ridge
Kur'ya
Denezhkin 1493
Sos'va
Tura
Turinsk
Tyumen
Avdabul
70°E
Shomvukva
Vim'
Zheleznodorozhnyy
Puzla
Ust'Kulom
Kolva
Cherdyn
Krasnoturinsk
Lobva
Verkhnyy Tura
Nizhniy Tagil
Artemovskiy
Bogdanovich
Talitsa
Shadrinsk
Kurgan
Kurtamysh
Dem'yanovka
Peski
Stavropolka
Ruzayevka
Dzhaksy
Atbasar
50°E
55°E
60°E
65°E
55°N
RUSSIA
SIBERIA
n o p q r s t u v w x y z
50
0 100 200 300 miles
Average linear scale
0 100 200 300 400 500 Km

10
a
b
c
d
e
f
g
h
i
j
k
l
m
Chukchi Sea
Long Strait
Wrangel Island
East Siberian Sea
New Siberian Islands
Laptev Sea
ARCTIC OCEAN
Kolymskiy Mountains
Kolymskiy Plain
Northern Anyuskiy Mountains
Southern Anyuskiy Mountains
Byrranga Mountains
Mys Shmidta
Pevek
Krasnoarmeyskiy
Retkucha
Ilirney
Cherskiy
Ambarchik
Mal. Baranikha
Bennetta
Kotel'nyy
Stolbovoy
Novaya Sibir'
Bol'shoye Zimov'ye
Laptev Str.
Vilkicki Str.
Shokal'skogo Str.
Bolshevik
Komsomolets
Oktyabr'skoy Revolyutsii
North Pole
70°N
75°N
80°N
85°N
170°W
175°W
180°
175°E
170°E
165°E
160°E
89
n
o
p
q
r
s
t
u
v
w
x
y
z
92

53
a
b
c
d
e
f
g
h
i
j
k
l
m
R
U
S
S
I
A
Dzhugdzhur Mountains
Cherskogo Mountains
Verkhoyanskiy Mountains
Stanovoy Mountains
Patomskoye Plateau
Vilyuy Mountains
Central Siberian Uplands
Yakutsk
Khandyga
Amga
Aldan
Lena
Vilyuy
Olekminsk
Lensk
Mirnyy
Vilyuysk
Nyurba
Suntar
Zhigansk
Verkhoyansk
Batagay
Ust'-Nera
Oymyakon
Tomtor
Okhotsk
Ayan
Chumikan
Tynda
Neryungri
Chara
Bodaybo
Vitim
Kirensk
Chernyshevskiy
Tura
Khatanga
Olenëk
Anabar
Popigay
Kotuy
Moyero
Angara
Ust'-Ilimsk
Arctic Circle
52
150°E
145°E
140°E
135°E
130°E
125°E
120°E
115°E
110°E
105°E
100°E
60°N
55°N
65°N
70°N
n
o
p
q
r
s
t
u
v
w
x
y
z
93
0
100
200
300
miles
Average linear scale
0
100
200
300
400
500
Km

Principal sources for the thematic maps: Amnesty International Report 2001. * www.ancientscripts.com 2001. * Buch und Buchhandel in Zahlen. Frankfurt 1987. * British Geological Survey, Natural Environment Research Council: World Mineral Statistics 1979-1983. 1995-1999. * Brown, Louise: Sex Slaves: The Trafficking of Women in Asia. London 2000. * CIA World Factbook 2000. * Dathe, Heinrich und Paul Schöps (eds.): Pelztieratlas. Jena 1986. * Deutsche Gesellschaft für Luft und Raumfahrt: Astronautische Start-Verzeichnisse und Raumflugkörper-Statistiken 1957-1987. * Diercke Länderlexikon. Braunschweig 1983. * Durrell, Lee: State of the Ark. London 1986. * www.eia.doe.gov 2001. * Encyclopedia Britannica. 15th ed. 32 vls. 1985. * Encyclopedia Britannica Book of the Year 1986. 1987. 1988. * www.ethnologue.com 2001. * Fischer Weltalmanach 1986. 1987. 1988. 2001. 2002. * Food and Agricultural Organization of the United Nations (FAO). Rome: FAO Production Yearbook 1985. 1986. FAO Food Balance Sheets 1975-1977. 1979-1981. FAO Yearbook of Fishery Statistics 1983. FAO Trade Yearbook 1986. www.fao.org 2001. * Haack. Atlas zur Zeitgeschichte. Gotha 1985. * Herre. Wolf und Manfred Röhrs: Haustiere - zoologisch gesehen. Stuttgart 1973. * www.infoplease.com 2001. * Institut für Seeverkehrwirtschaft und Logistik, Bremen: Shipping Statistics Yearbook 2000. * The International Institute of Strategic Studies (ILSS): The Military Balance 1986-1987. 1995/1996. 2000. * International Labour Organization (ILO). Geneva: Yearbook of Labour Statistics 1978. 1979. 1980. 1981. 1982. 1983. 1984. 1985. 1986. 1987. Income Distribution and Economic Development. An Analytical Survey. Geneva 1984. Sixth African Regional Conference. Application of the Declaration of Principles and Programme of Action of the World Employment Conference. Geneva 1983. STAT Working papers, Bureau of Statistics 1950-2010. Geneva 1997. www.ilo.org 2001. * International Road Transport Union: World Transport Data. Geneva 1985. * International Telecommunication Union: Table of International Telex Relations and Traffic. Geneva 1987. * Inter-Parliamentary Union (IPU): Women in Parliament 1988. Participation of Women in Political Life and in Decision-Making Process. Geneva 1988. Distribution of Seats Between Men and Women in National Assemblies. Geneva 1987. www.ipu.org 2001. * Jain, Shail: Size Distribution of Income. Compilation of Data. World Bank Staff Working Paper No.190. Nov. 1974. Washington 1975. * Kidron, Michael and Ronald Segal: The State of the World Atlas. London 1981. The New State of the World Atlas (revised ed.). London 1987. * Kurian, George Thomas: The New Book of World Rankings. New York 1984. * Länder der Erde. Berlin 1985. * McDowell, Jonathan: Harvard-Smithsonian Center for Astrophysics. * Meyers Enzyklopädie der Erde (8 vls.). Mannheim 1982. * Moroney, John R.: Income Inequality. Trends and International Comparisons. Toronto 1979. * Myers, Norman (ed.): GAIA - Der Öko-Atlas unserer Erde. Frankfurt 1985. * www.nasa.gov 2001. * Nohlen, Dieter and Franz Nuscheler (eds.): Handbuch der Dritten Welt. 8 vls. Hamburg 1981-1983. * Ökumene Lexikon. Edited by Hanfried Krüger, Werner Löser et al. Frankfurt 1983. * Peters, Arno: Synchronoptische Weltgeschichte. 2 vls. München 1980. * Saeger, Joni and Ann Olson: Der Frauenatlas. Frankfurt 1986. * Serryn, Pierre: Le Monde d'aujourd'hui. Atlas économique. social, politique, stratégique. Paris 1981. * South: South Diary 1987. 1988. * Statistisches Bundesamt, Wiesbaden: Statistisches Jahrbuch 1999. Statistik des Auslandes. Vierteljahreshefte zur Auslandsstatistik. 1985-1987. Statistik des Auslandes. Länderberichte. * Stockholm International Peace Research Institute (SIPRI): SIPRI Yearbook 1987. World Armaments and Disarmament. New York 1987. * Taylor, Charles Lewis and David A. Jodice: World Handbook of Political and Social Indicators. New Haven. London 1983. * UNESCO: Statistical Yearbook 1974. 1975. 1976. 1977. 1978. 1979. 1980. 1981. 1982. 1983. 1984. 1985. 1986. 1987. * UNICEF: The State of the World's Children 1987. * The United Nations (UN): UN Statistical Yearbook 1983/84. 1999. UN Demographic Yearbook 1972. 1979. 1984. 1985. 1986. National Accounts Statistics. Compendium of Income Distribution Statistics. New York 1985. UN Energy Statistics Yearbook 1984. UN Yearbook of International Trade Statistics 1982. 1983. 1984. 1986. 1998. Selected Indicators of the Situation of Women 1985. UN Industrial Statistics Yearbook 1983. 1984. World Conference of the United Nations Decade for Women: Equality, Development and Peace. Copenhagen 1980. World Culture Report 2000. World Education Report 2000. World Health Report 2000. The World's Women 2000. Activities for the Advancement of Women: Equality, Development and Peace. Report of Jean Fernand-Laurent. 1983. UNCTAD Handbook of Statistics 2000. UNIDO International Yearbook of Industrial Statistics 1995. 1996. 1997. 1998. 1999. 2000. 2001. www.un.org 2001. * University of Stellenbosch, Department of Development Administration and the Institute for Cartographic Analysis: The Third World in Maps. 1985. * Westermann Lexikon der Geographie. Edited by Wolf Tietze. Braunschweig 1968. * World Almanac & Book of Facts 1985. 1986. 1987. * The World Bank: World Development Report 1980. 1981. 1982. 1983. 1984. 1985. 1986. 1987. 1999/2000. 2000/2001. World Labour Report 1984. World Tables 1984. World Atlas of the Child 1979. Social Indicators of Development 1987. The World Bank Atlas 1987. World Economic and Social Indicators. Document of the World Bank. 1980. World Development Indicators 2001. www.worldbank.org 2001 * World Energy Resources 1985-2020, Renewable Energy Resources. The Full Reports to the Conservation Commission of the World Energy Conference. Published for the WEC by IPC Science and Technology Press 1978. * The World in Figures. Editorial information compiled by the Economist. London 1987. * World Health Organization (WHO), Geneva: World Health Statistics. Annual. * Völker der Erde. Bern 1982. * Voous, K.H.: Atlas of European Birds. New York 1960.

NATURE, HUMANKIND AND SOCIETY IN 212 THEMATIC MAPS

Each map presents a single subject. As a result, it is possible to dispense with symbols and allow the information to be expressed entirely in terms of colour: dark colours for high values, light for low ones. This makes it easy to see and assimilate the content of the maps - an important feature, since up to 16 maps can be dedicated to a single subject.

The individual subject should not be considered in isolation. The mutual interaction between all spheres of life, the intricacies of nature and culture, of economics, nations and society, mean that each of the subjects can be understood only in connection with the other 48 double-page spreads.

This richness and multiplicity of facts and insights is however the minimum which someone of our time must have in mind if they wish to form their own opinion on the current situation in the world and in their own country. Without this effort, their own view of the world can never be clear and reliable.

Over 40,000 individual pieces of factual information have been compiled for these 212 thematic world maps. They were obtained almost exclusively from published materials of the United Nations and other international organisations. There reliability is presumed, and an average of annual data available from 1980 onwards has been calculated. Where official figures were not available, estimates were made in consultation with the leading experts in the various fields concerned. No indication is given of these estimates, since their reliability is no less than that of the official figures.

The names of countries appear also on the small world maps. These can be read with the aid of a magnifying glass or by reference to the large whole-page maps such as that of „States" on pages 112-113.

Brief texts on each subject are intended to aid mental categorisation and to make historical connections plain. In addition, they contain the figures of extreme cases which cannot be extracted from the average values given by the colour-coding.

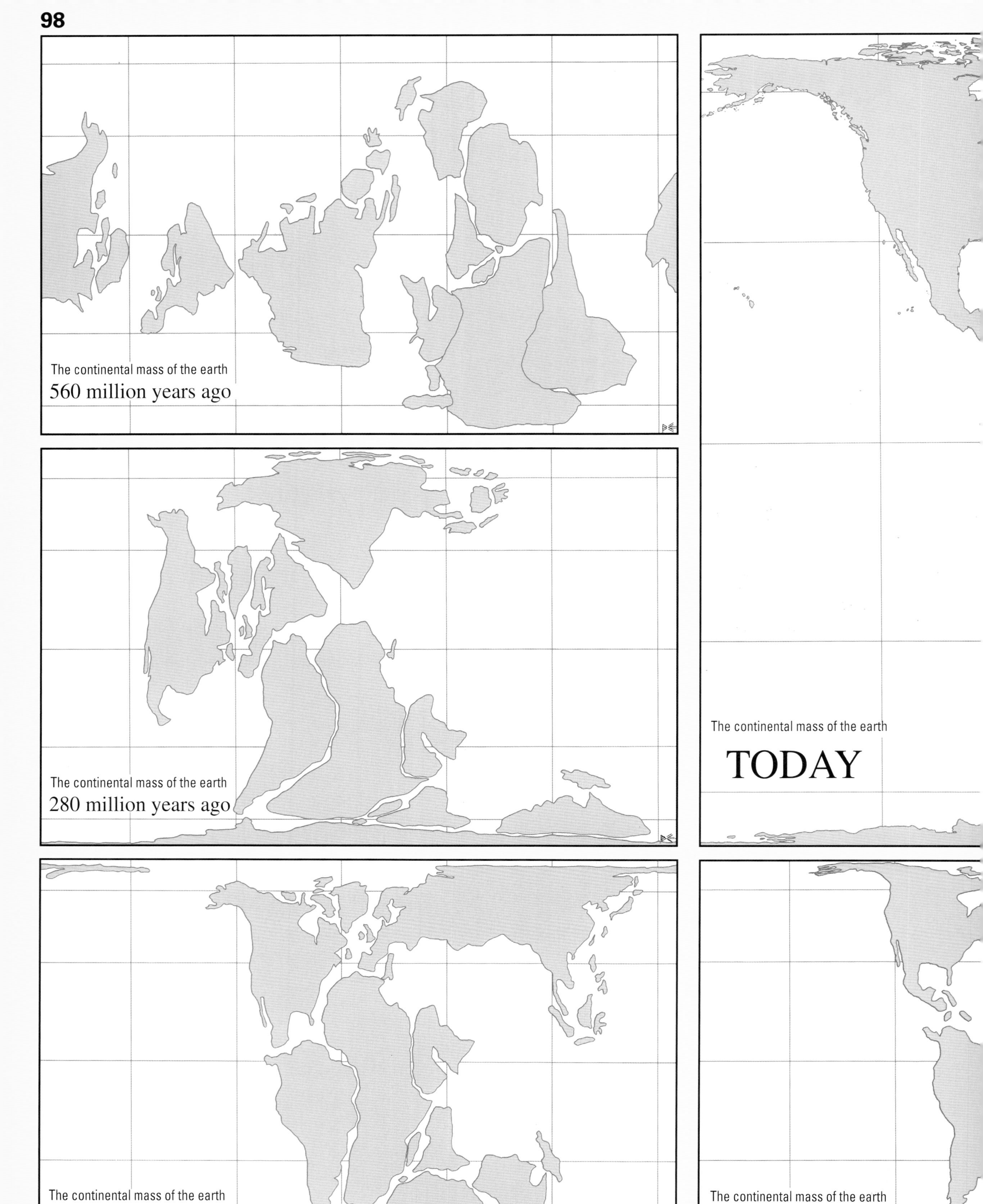

The continents of the earth are in constant movement. The cause of this DRIFT is the displacement of half a dozen gigantic sheets of the earth's crust. These change their position by an average of three metres in the course of one human generation. There are however great differences in the speed of this movement; in the Atlantic area it is only one to two centimetres a year, but in the Pacific it is up to fifty centimetres annually.

THE CON

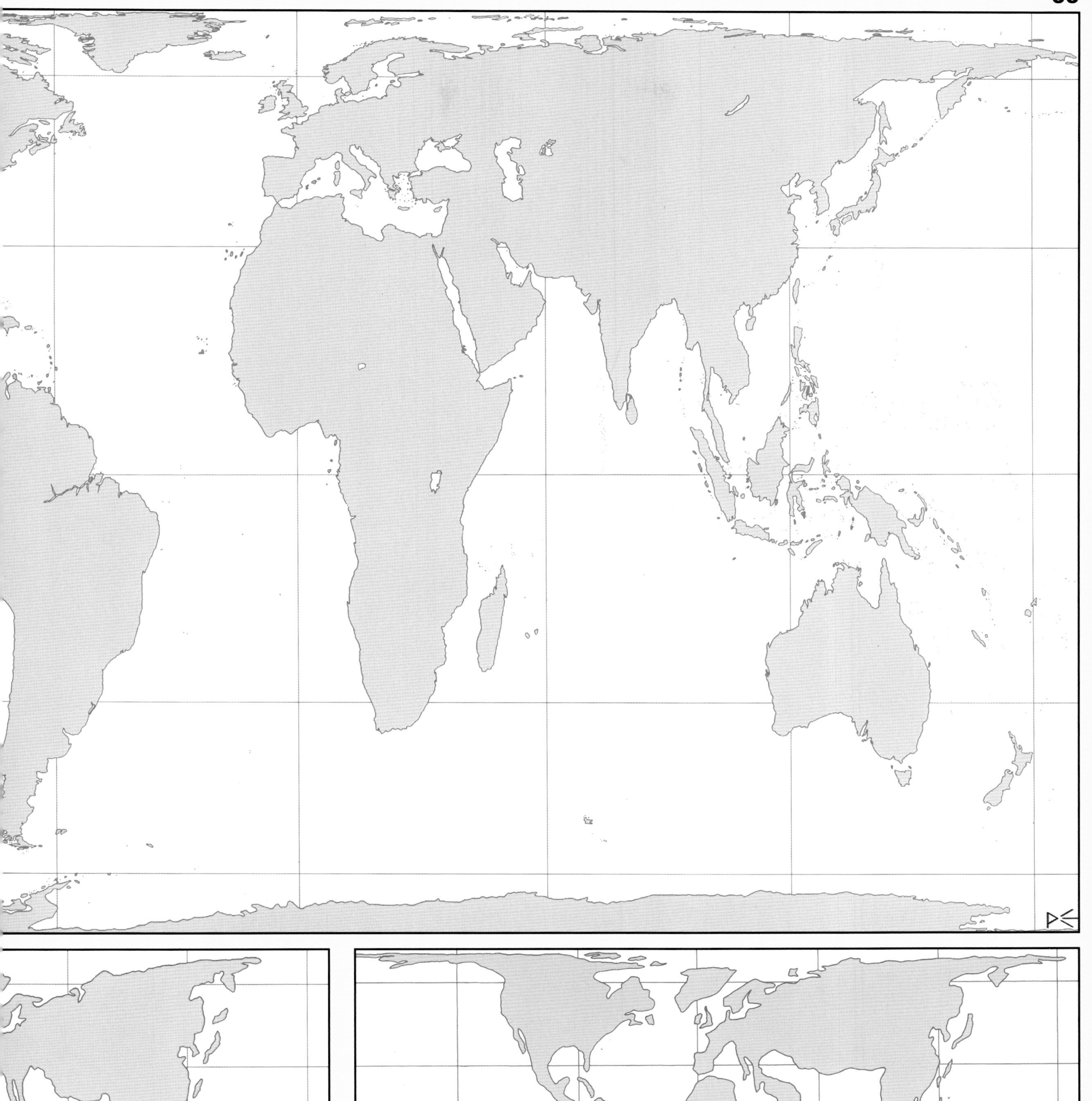

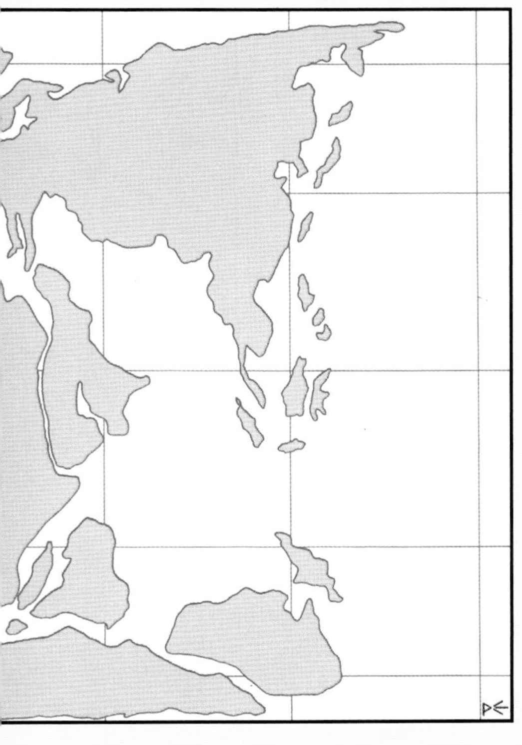

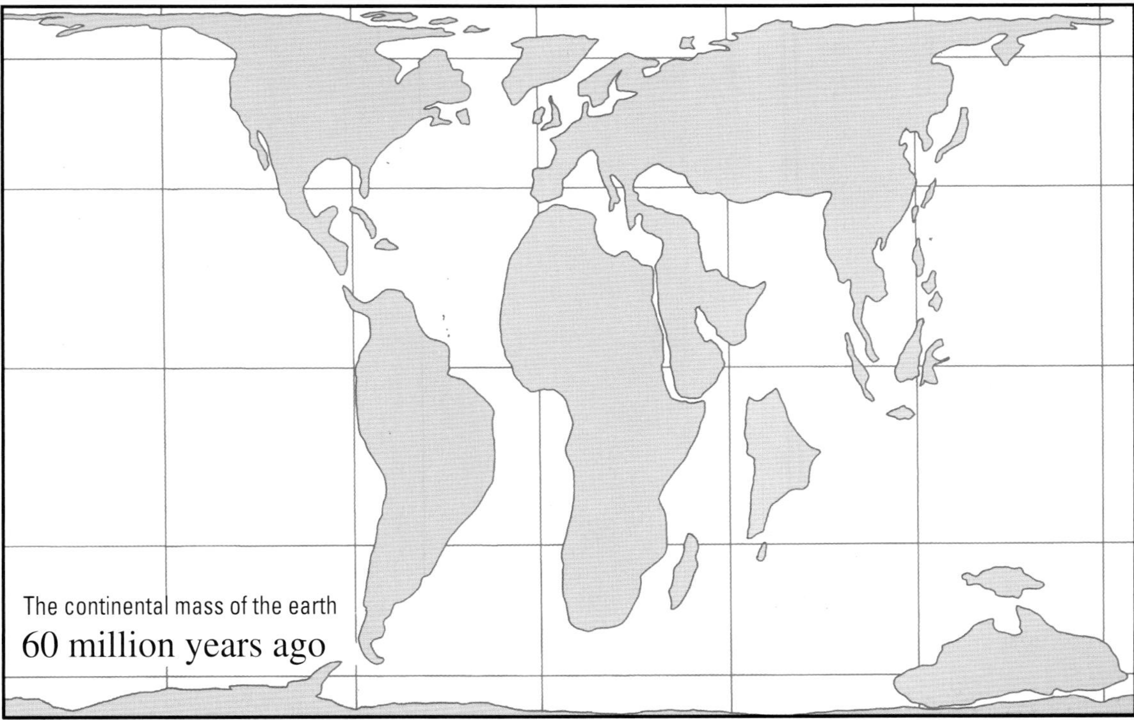

The continental mass of the earth

60 million years ago

TINENTS

Some 280 million years ago Europe was a separate land mass which could be regarded as a continent in its own right. By 200 million years ago Europe had combined with Asia into a single large continent. Later, Arabia, India and Africa joined onto it, while North and South America had also united into a single continent. In comparison with these two real continents of AMERICA and EURASIA, Australia, Greenland and Antarctica are only enormous islands.

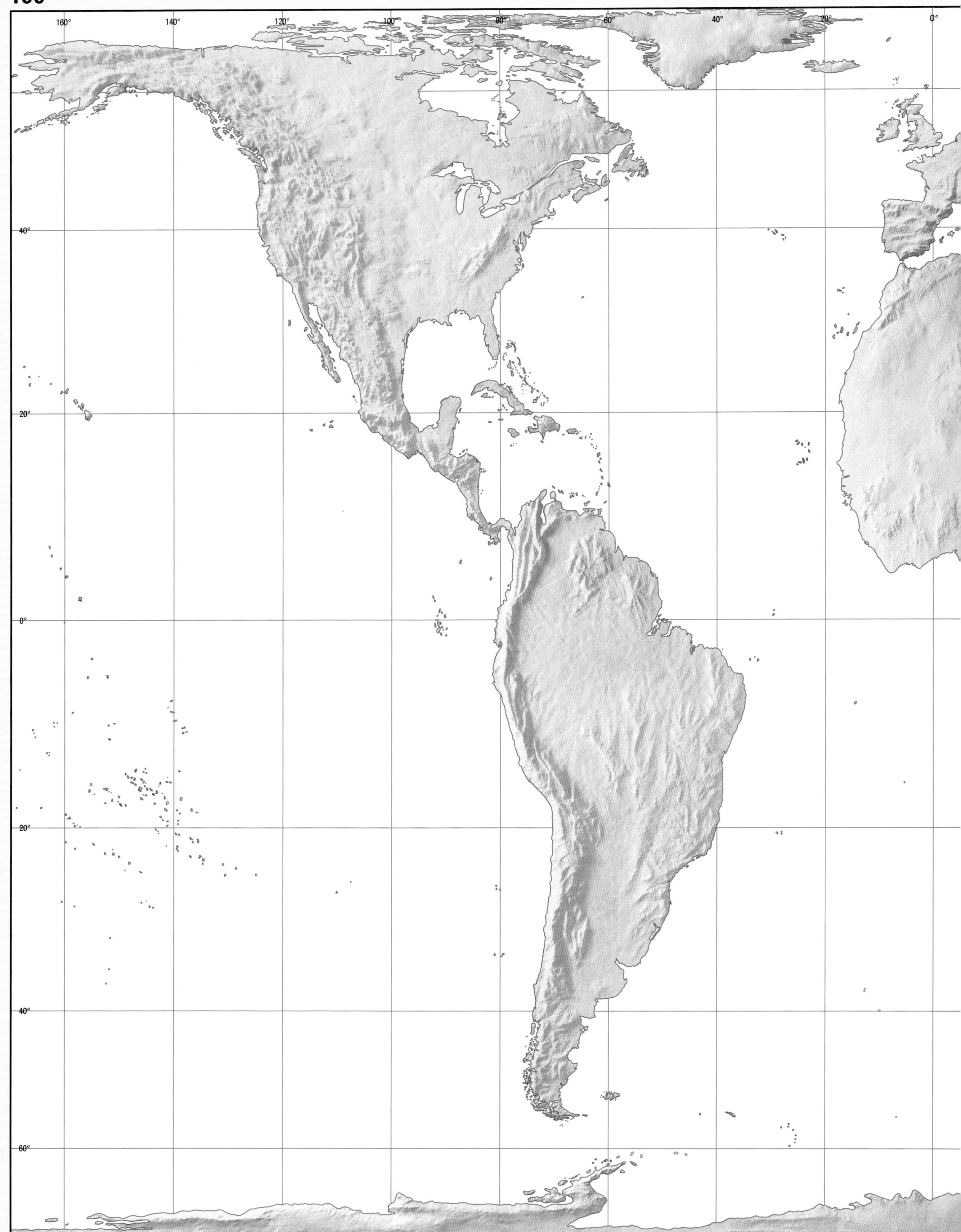

Most mountain ranges were formed by the displacement of the continental land masses. When these collided, they forced up folds which became the high ranges (Alps, Andes, Himalaya). Then, over millions of years, parts of them were worn away by glaciers and erosion into highlands and tablelands. Where mountains are of volcanic origin (Iceland, Java, Hawaii) new lava bursting out of the earth counteracts this natural erosion. Periods of major earth movement alternate with quieter periods, but in general it can be said that the mountains are still constantly changing.

MOUN

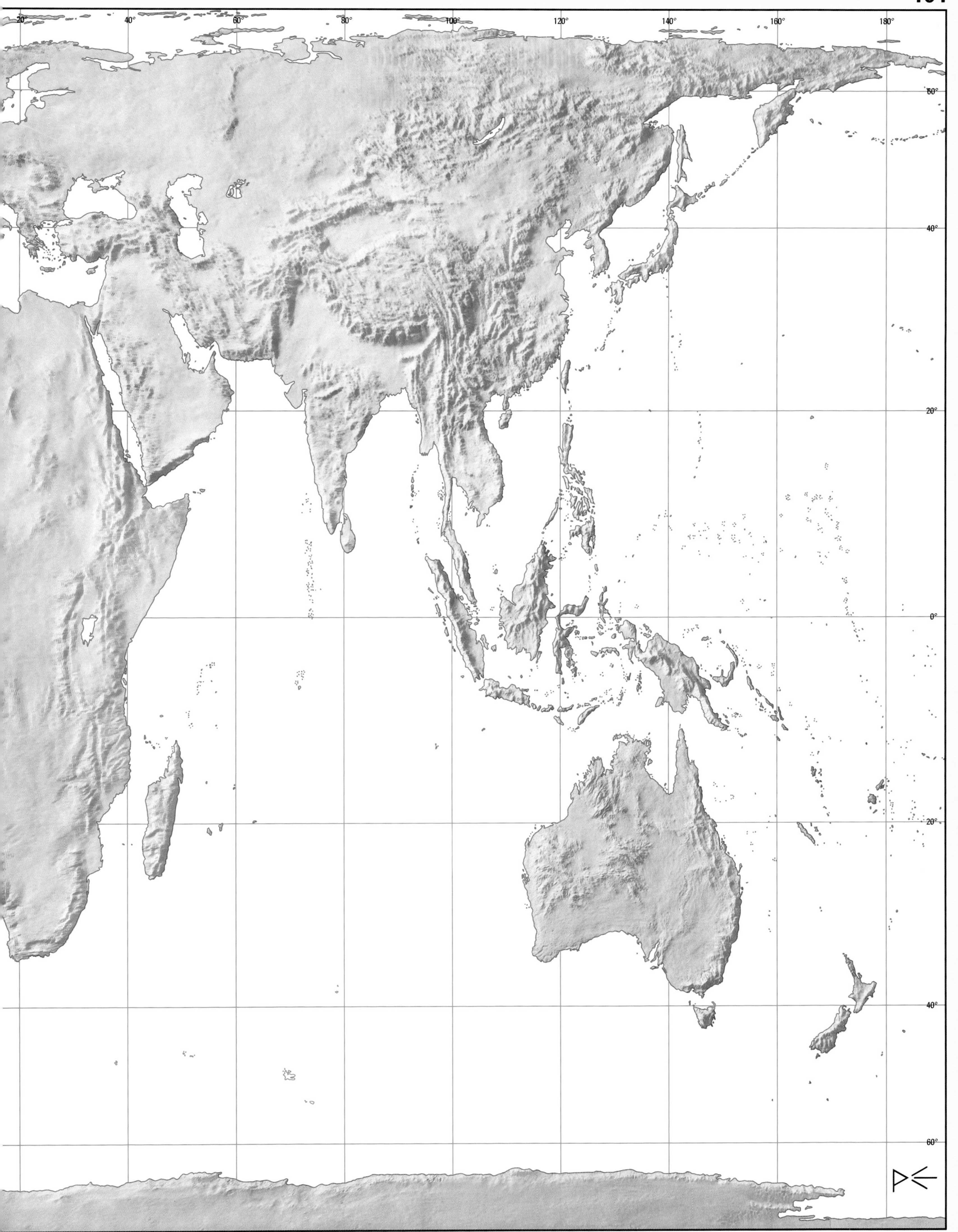

Mountains are natural barriers and therefore often form the boundaries between countries. Up to the 20th century, mountainous regions were often isolated from the rest of the world and people there followed their own traditional ways of life. In recent decades, however, these regions, especially in Europe, have been opened up to tourist traffic. Mountains and valleys have thus lost much of their individual character. Tourism has become a major source of income and ski-lifts, hotels and motorways are increasingly changing their appearance.

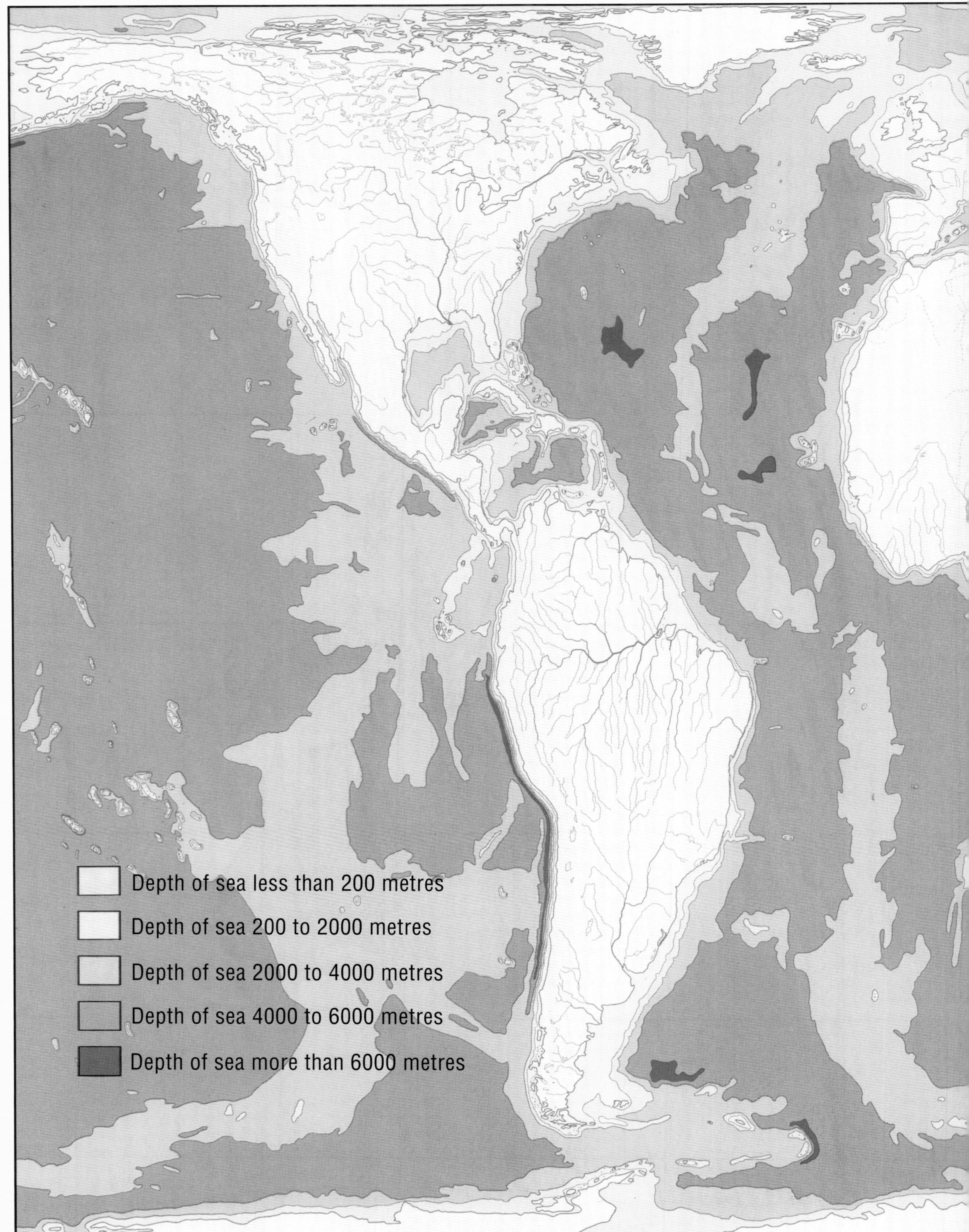

Every second, the earth's rivers carry more than a million cubic metres of water down to the oceans, where it evaporates and returns to the interior of the land masses as rain or snow. It runs into the rivers and is again carried down to the sea. This eternal water cycle, which involves less than one-hundredth of 1% of all the earth's water, is the essential basis for life and human existence. Two-thirds of our body consists of water.

RIVERS A

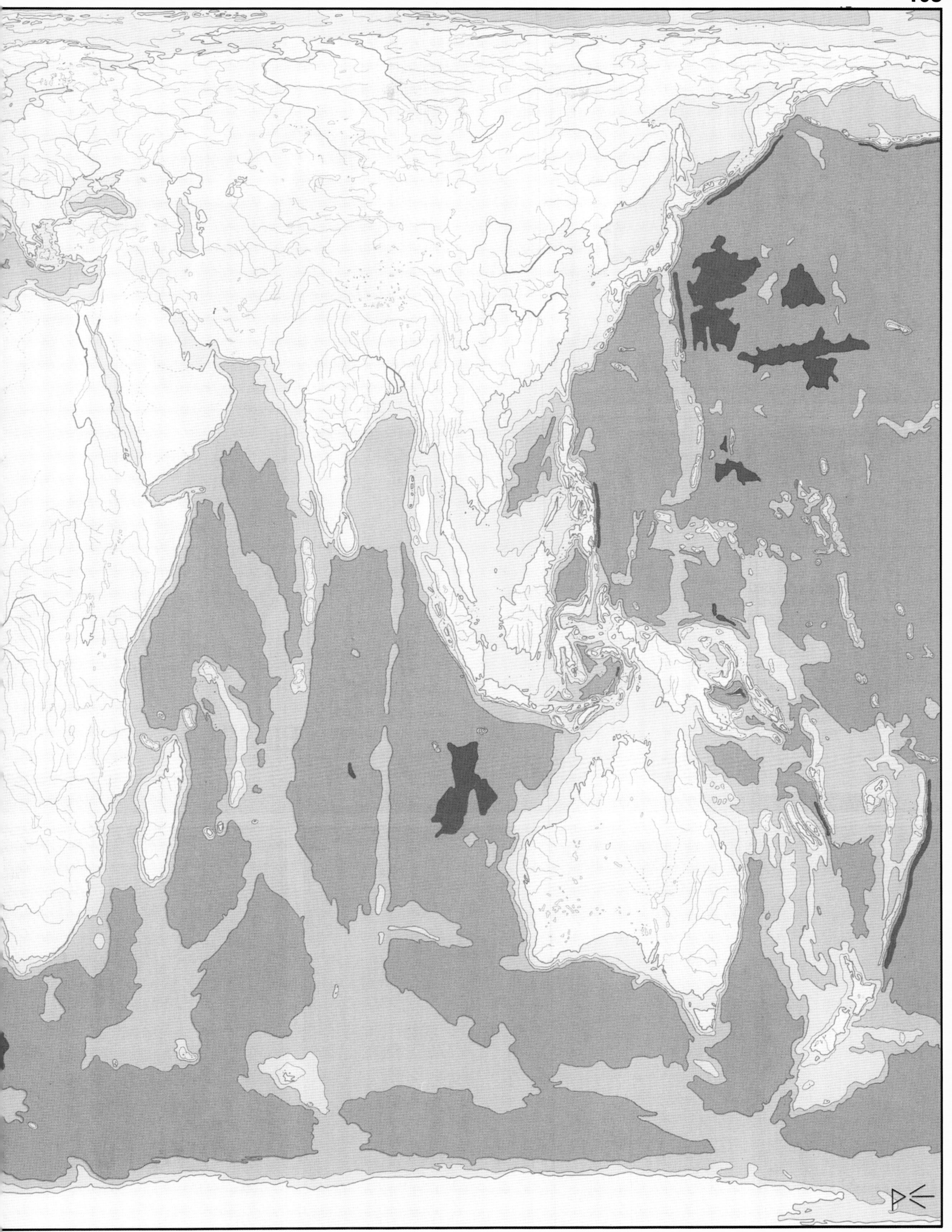

ND SEAS

The first human civilisations rose beside great rivers such as the Euphrates, Tigris, Nile, Indus and Huang He. The first city-states emerged there because people found drinking water and fish to provide nourishment. Rivers became their traffic routes, irrigated their fields and provided energy. The oceans, which covered more than 70% of the Earth's surface, limited the movement of early people until they learnt to conquer them with ships and thus make the Earth a single living space.

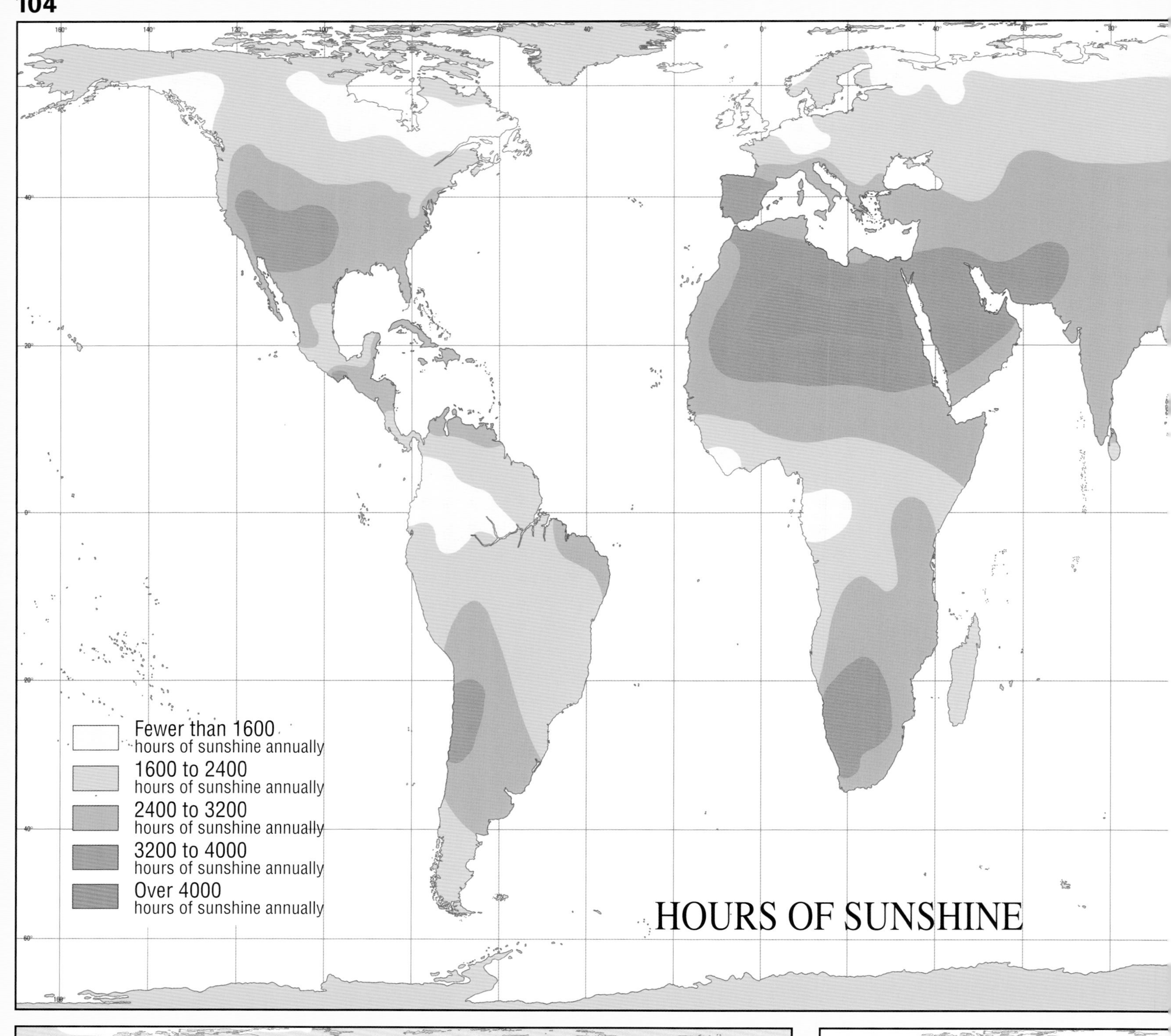

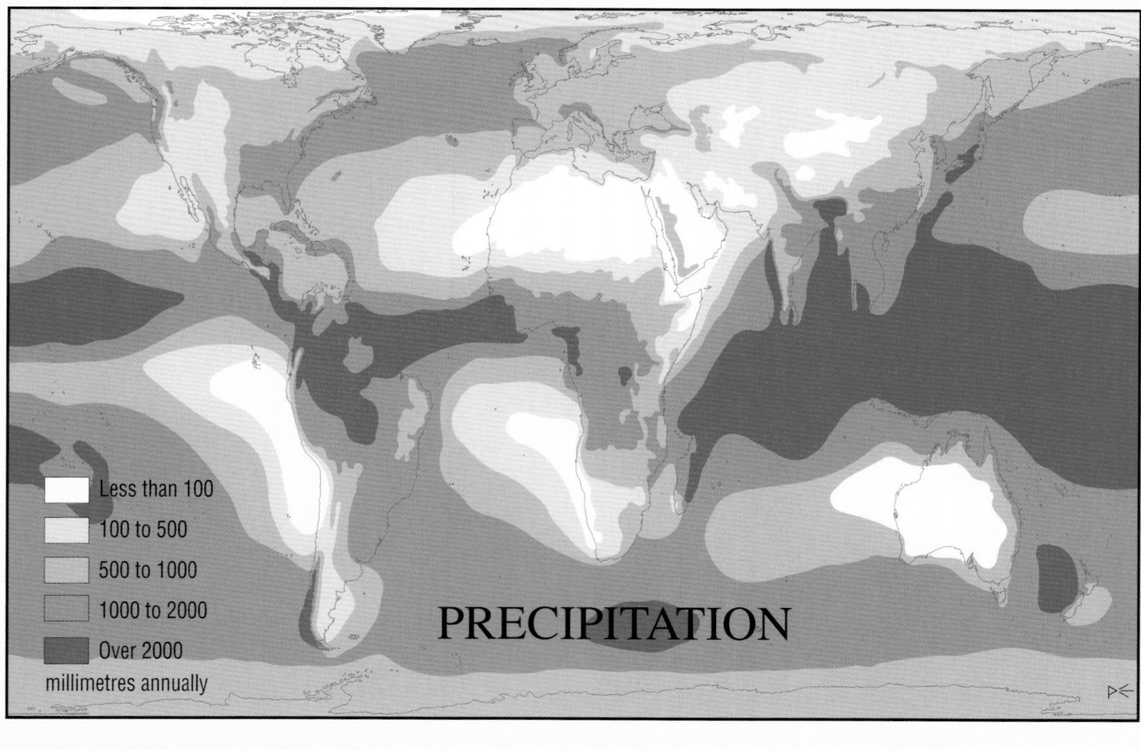

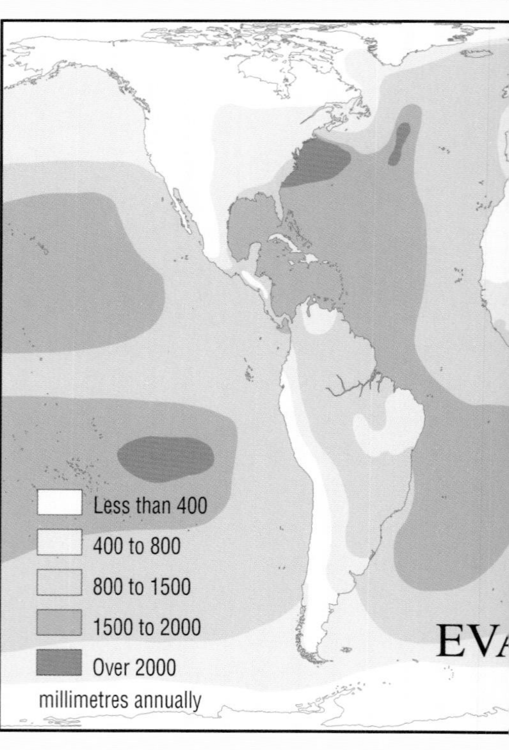

Like the rest of nature, people depend for their existence on weather factors, which together make up the climate. No matter how adaptable we may be, we cannot live everywhere on the earth's surface. 20% of it is covered by snow and ice and is uninhabitable. Another 20% is inhospitable desert, and a further 25% either consists of steep mountains, has insufficient soil, or is marshy or flooded land. Thus only about one-third of the earth's surface possesses suitable conditions for human habitation.

SUN AND

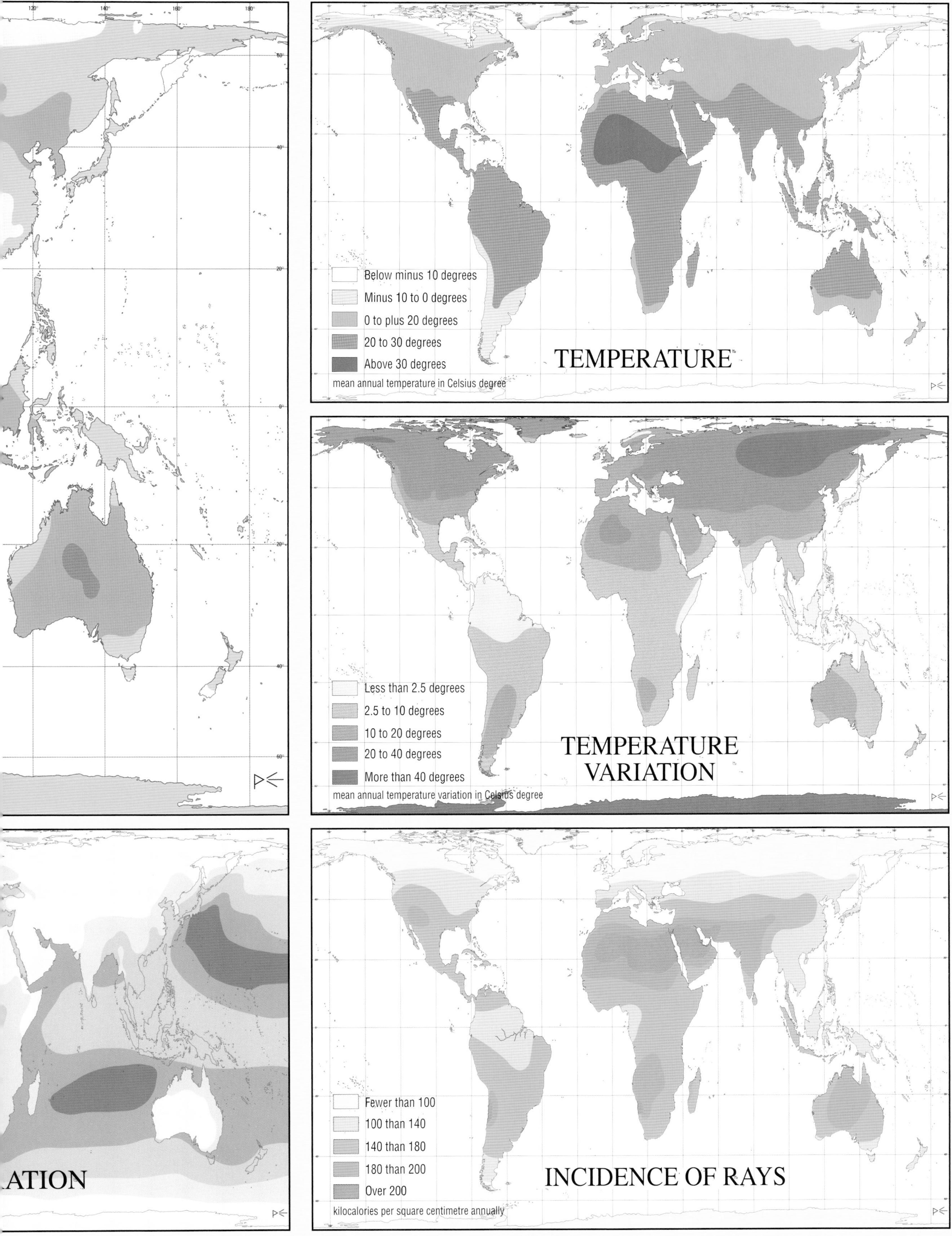

CLIMATE

A hundred years ago people began scientific research into weather in all its aspects—atmospheric pressure, temperature, humidity, hours of sunshine, amount of precipitation, cloud formation and wind. The origin of all weather conditions is the sun. This ball of fiery gas around which we travel once a year and whose diameter is about one hundred times greater than that of the earth, is one of the smallest of the over 200 billion fixed stars in the Milky Way. But it is our sun's light and warmth which make life on earth possible.

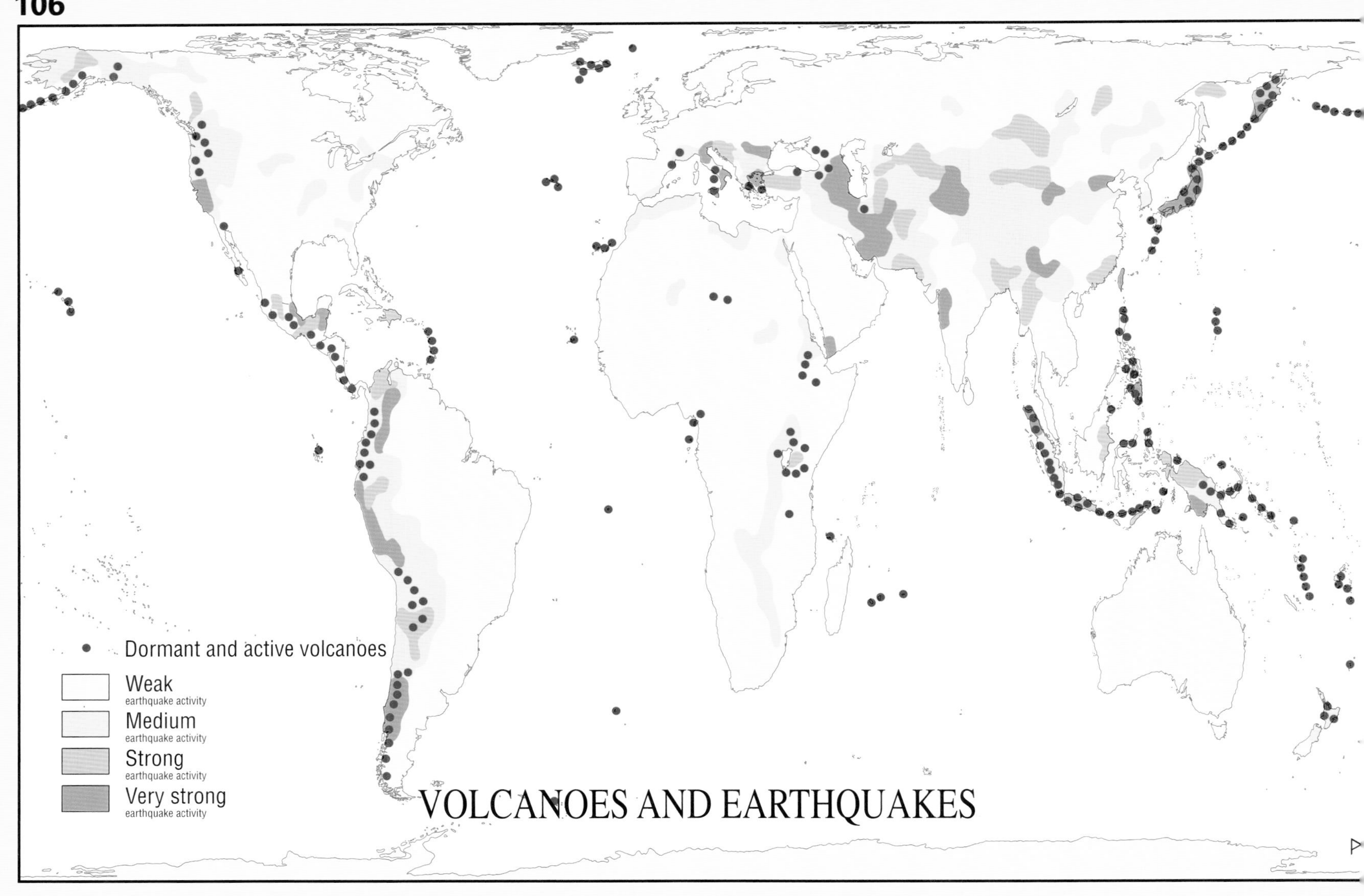

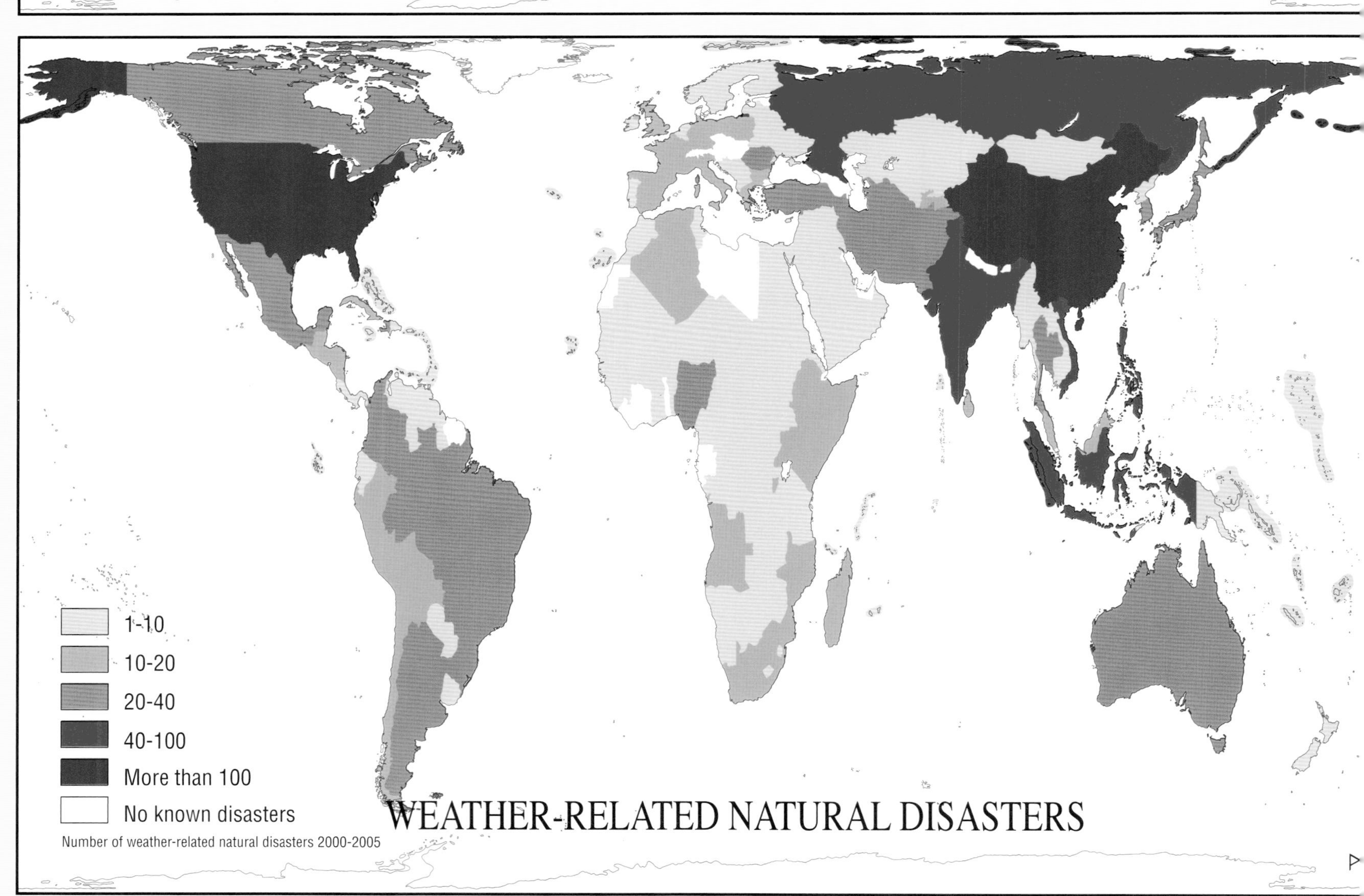

We have learned to control some natural dangers—wild animals, potentially fatal infections and famine. But fires, floods, earthquakes, hurricanes and volcanic eruptions are still a threat. However, now that these have been researched, the ability to predict them means that they claim fewer lives than those lost through traffic accidents. Earthquake research, new building methods, the prediction of volcanic eruptions and worldwide hurricane and flood warning systems have made natural occurences less perilous.

NATURA

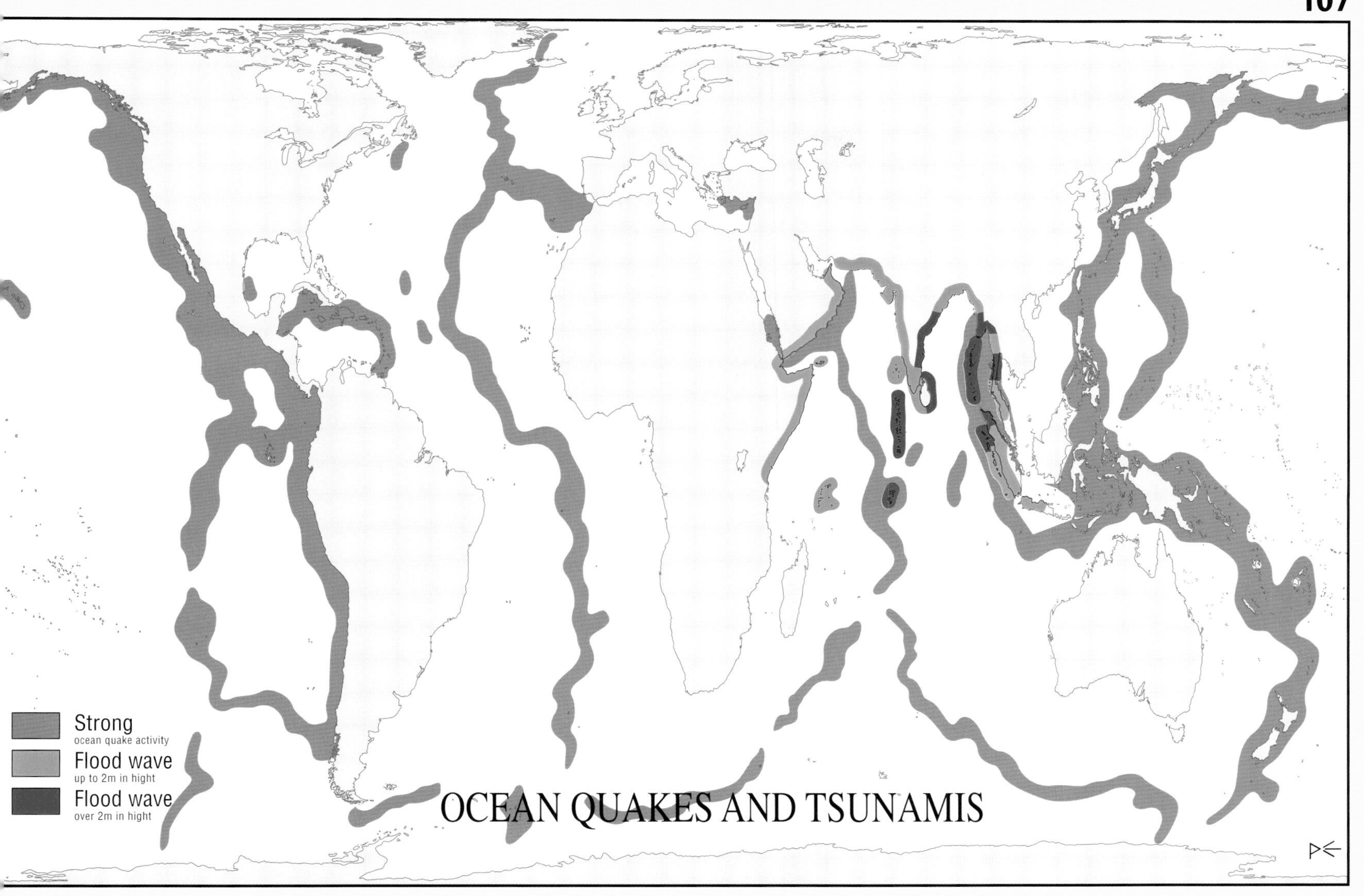

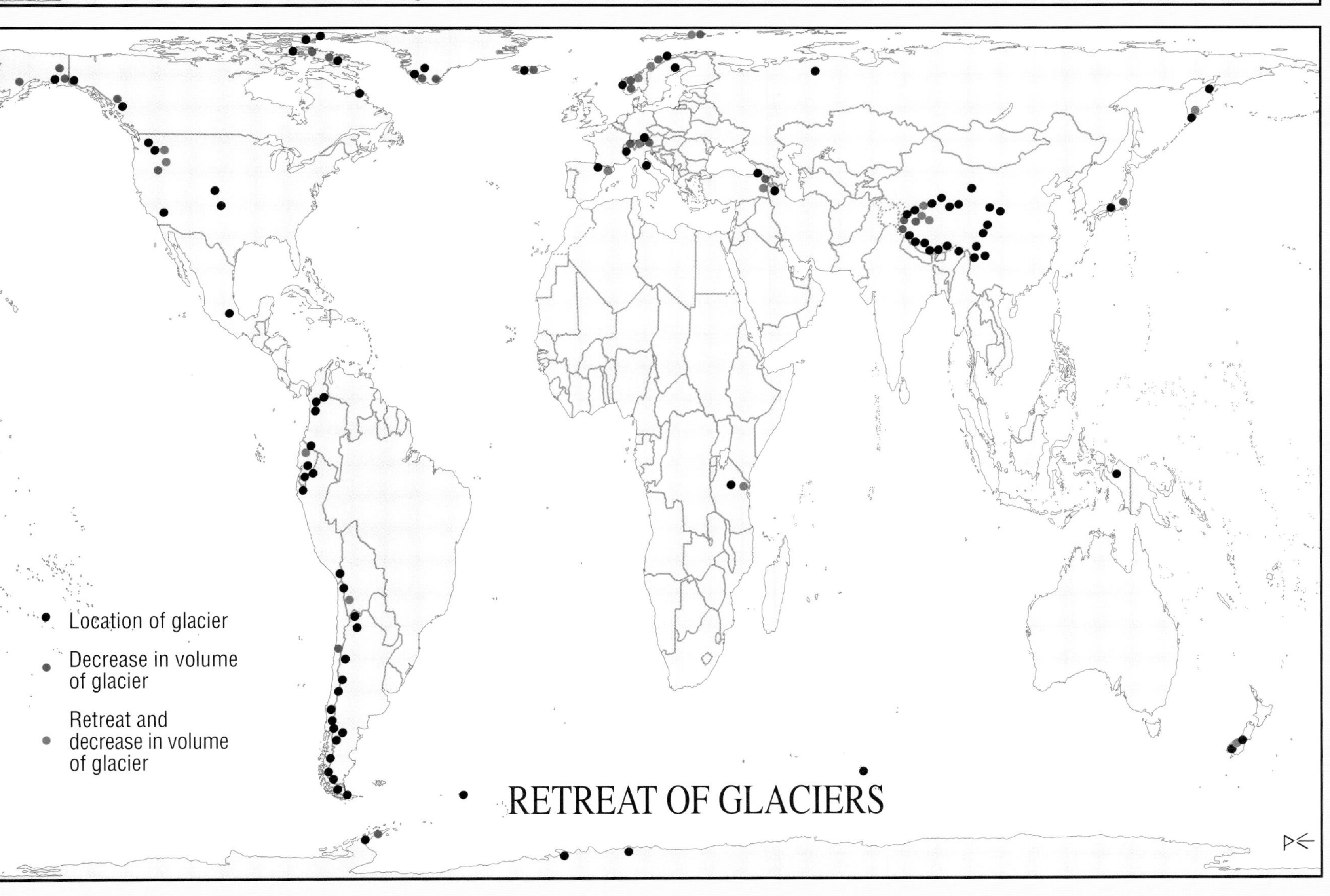

DANGERS

Greed for profit has led to the ruthless exploitation of natural resources and resulted in the emergence of new dangers in our day—deforestation, exhaustion of the soil, the pollution of air and water, the rapid spread of deserts, the poisoning of rivers and seas, the extermination of plant and animal species and the destruction of the ozone layer. These changes in the natural environment can threaten life and today they pose a greater danger to our existence than all the natural perils confronting us.

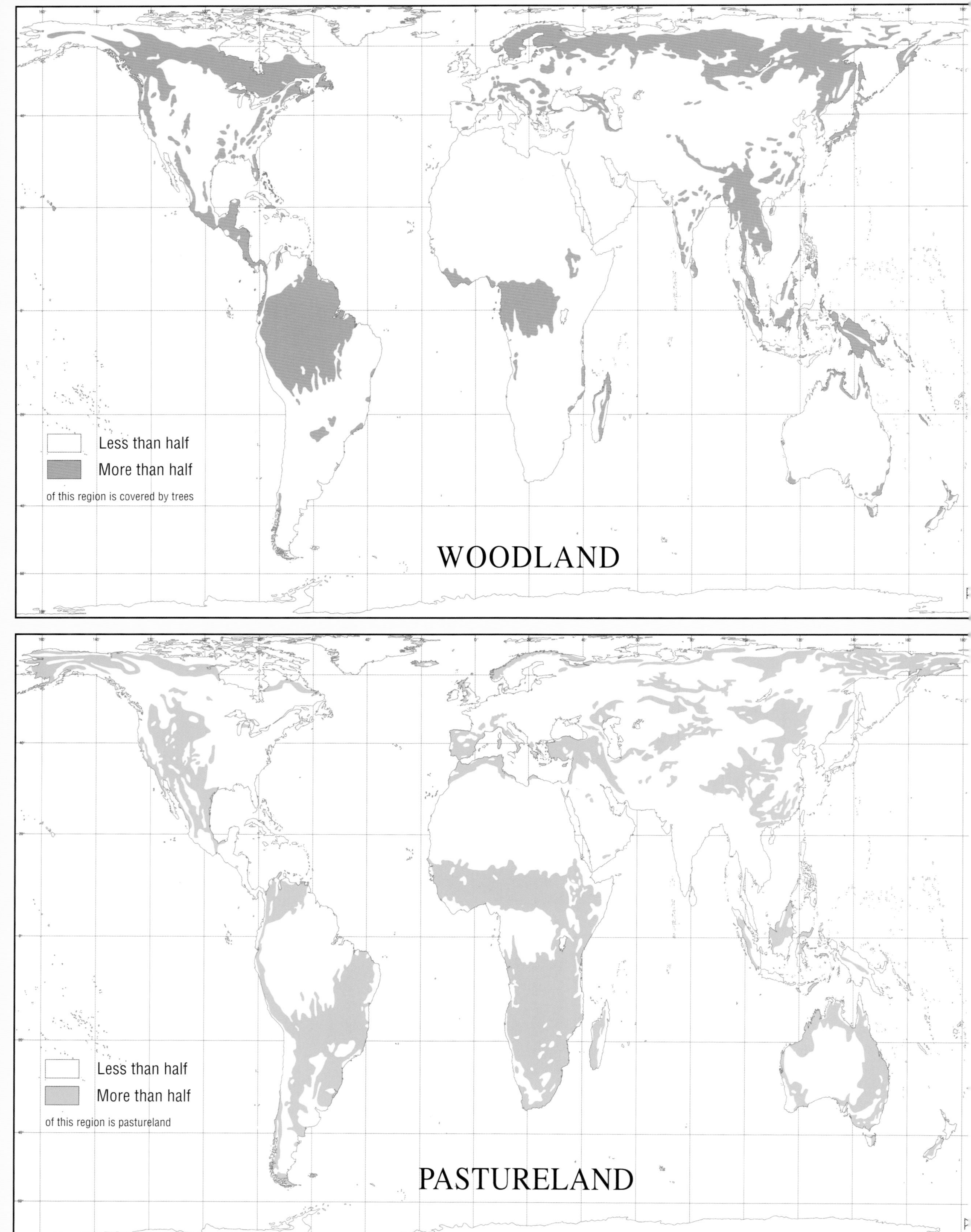

When life first appeared on earth two billion years ago, there was no oxygen. Only through plant life did the atmosphere become enriched with oxygen and thus make animal life possible. All higher life forms today, exist only because of the oxygen continually produced by plants. In addition, a sufficient quantity and quality of plant life are necessary as the basis to feed both people and animals. Yet every day some of the 25,000-odd higher types of plant life on the earth are exterminated by humans.

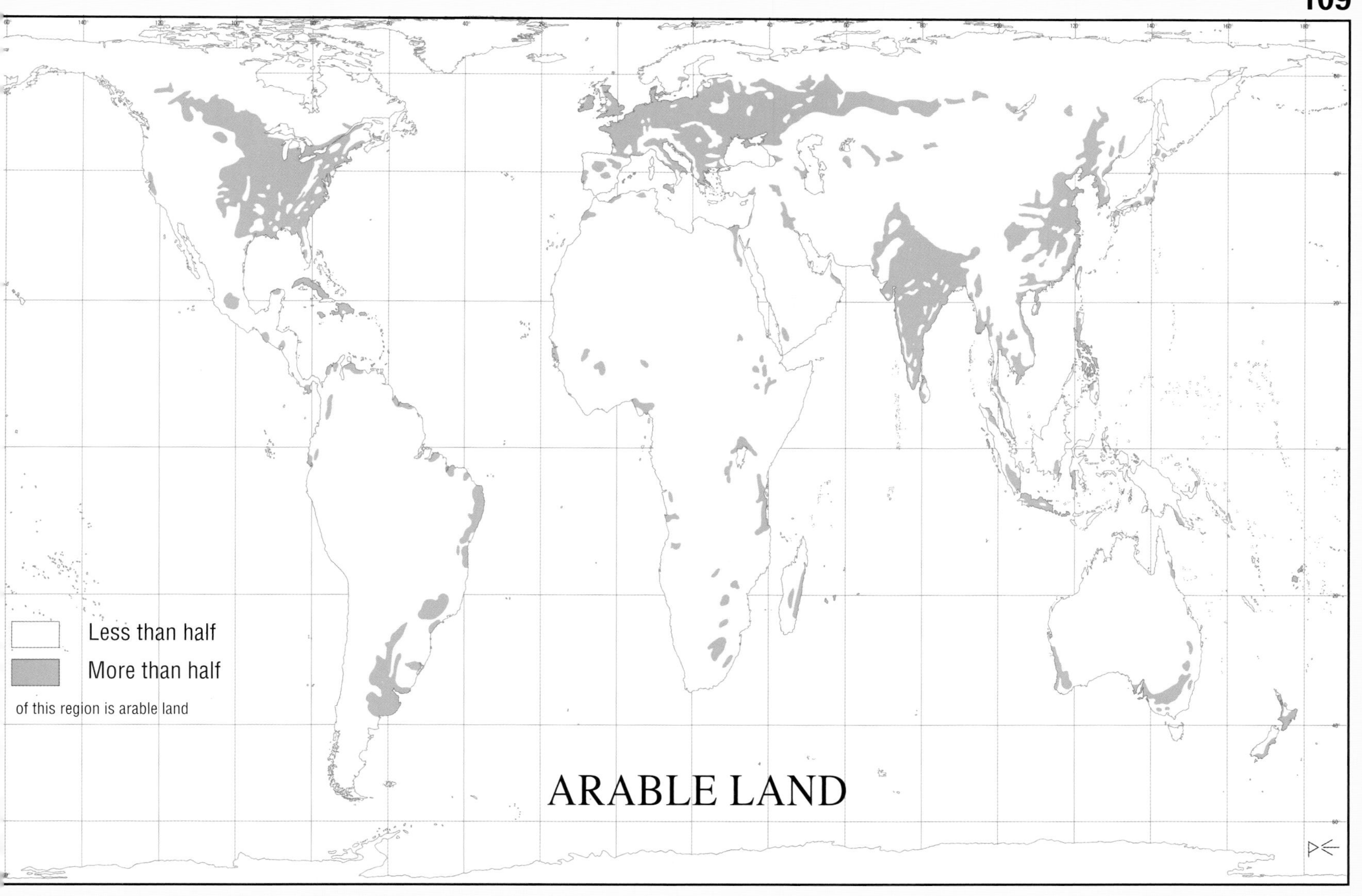

Plants are less demanding than people. They flourish in regions where permanent human settlement is impossible—steppes, tundra, prairies, karst, savannahs, taiga. Grass and moss grow almost everywhere in these regions and provide food for most animals. If we include the fertile fields and meadows of the temperate zones, about 50% of the land surface of the earth is pastureland. The other half consits of woodland (26%), arable (11%) and desert (13%).

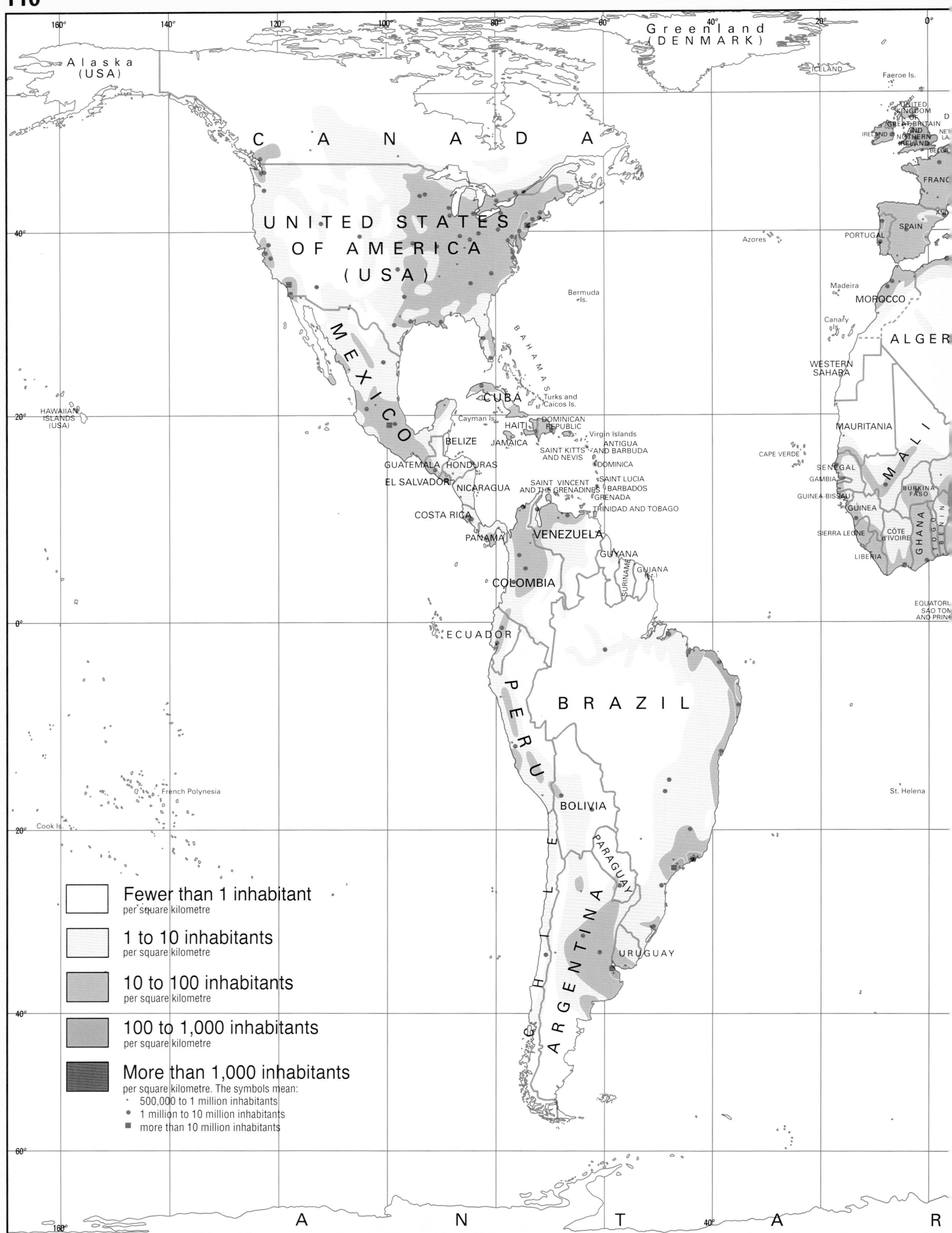

Some 50,000 years ago humans began to build shelters. At first they made constructions of branches and skins, which could be transported. Soon mud and clay began to be used to cover the framework of branches. This was the beginning of our settled existence. People started to tame animals and cultivate the soil. The first towns were built about 6,000 years ago, and the first city of a million inhabitants appeared 2,000 years ago.

PEOPLE A

In about 1840 the world's population was one billion. One hundred years later it had doubled. Since then it has been growing by a further billion every 10 to 20 years. Today the average population density is about 25 people per square kilometre. Some countries however have over 500 inhabitants per square kilometre and in two hundred major cities there are over 10,000 to a square kilometre. In North Africa, Siberia, Canada, South America and Australia, however, vast areas are completely unpopulated.

With the earliest city-states 5,000 years ago came the first communities that saw themselves as political entities within territorially defined areas. As a result of the drive of the rich and powerful to subjugate and exploit other peoples these states grew into great empires that expanded and asserted themselves by force. The latest expression of this conception of the state was the Europeanisation of the world in the past five centuries. Its culmination in our epoch opened the way for all peoples to live together in solidarity in a world state.

STA

Most of today's states have come into existence in the past 60 years. After the Second World War there were 70 independent states in the world; today there are nearly 200. And there are still a few colonies struggling for their political independence. In many independent states, dependence on the new superpowers has intensified to such an extent that their sovereignty is only nominal. In addition, an ever-growing number of linguistic, ethnic and cultural groups are seeking independence from the states in which they find themselves.

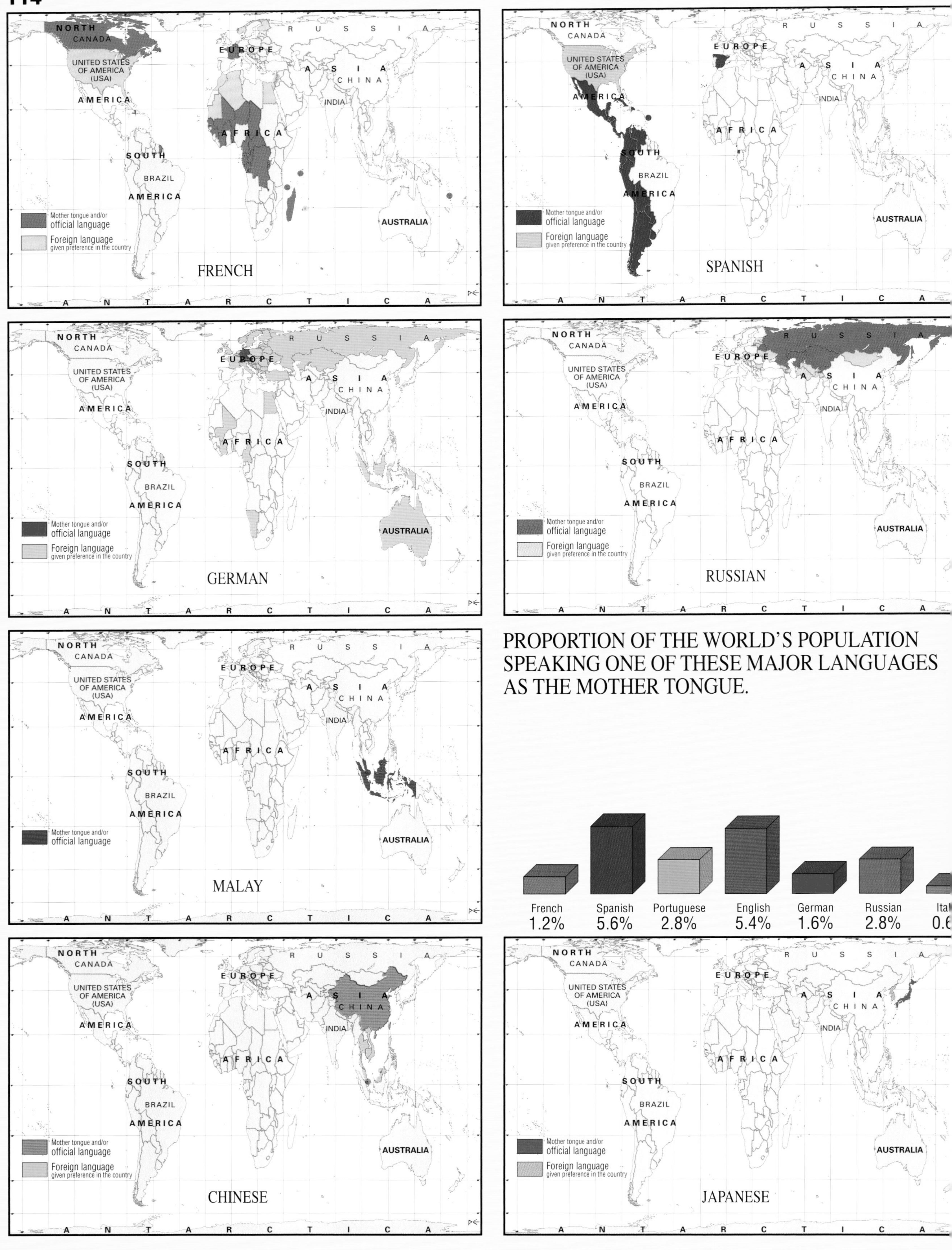

After attaining an upright stance some 20 million years ago, humans began to improve their verbal expression. But only after the creation and use of tools less than 1 million years ago did development towards articulate speech begin. In constant interaction with the growth of our mental ability, human language has since become more and more elaborate. The process is still going on, and every human being is contributing to it. By increasing our own vocabulary we contribute to the general growth of comprehension. Exactitude of the spoken language is both the expression and the origin of clear thinking, and makes possible a true and comprehensive view of the world.

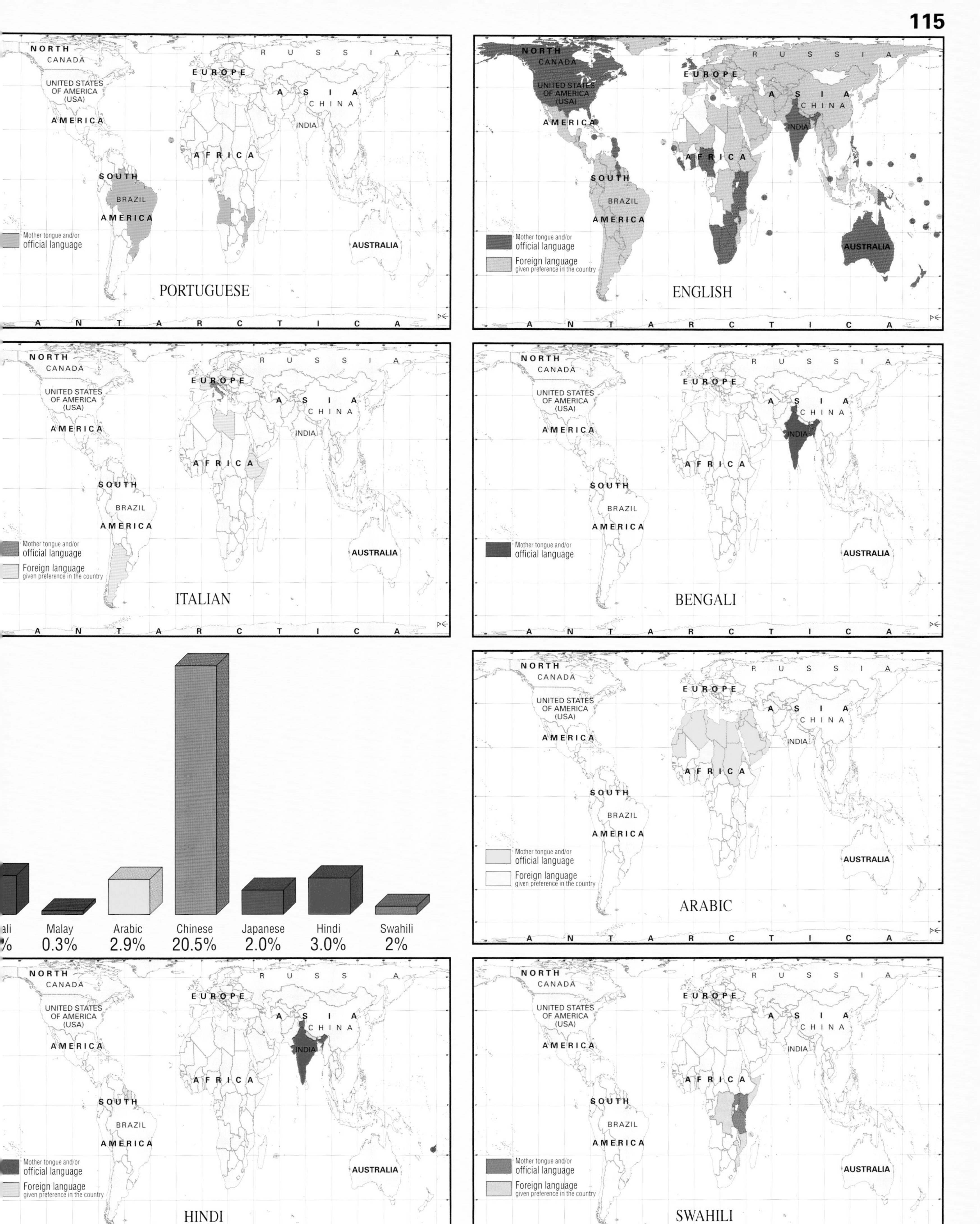

UAGES

Although the world's various cultures are drawing closer together, there are still thousands of different languages in existence. But only fourteen of these are spoken as a mother tongue by more than 1% of humankind. After the 500 years of Europeanisation of the world, half of these major languages are European in origin. It also seems probable that English, spoken 500 years by only 1% of the world's population, will become the world language used by all peoples of the world. But at the same time the multiplicity of languages will remain as long as the peoples ot the earth retain their cultural identities.

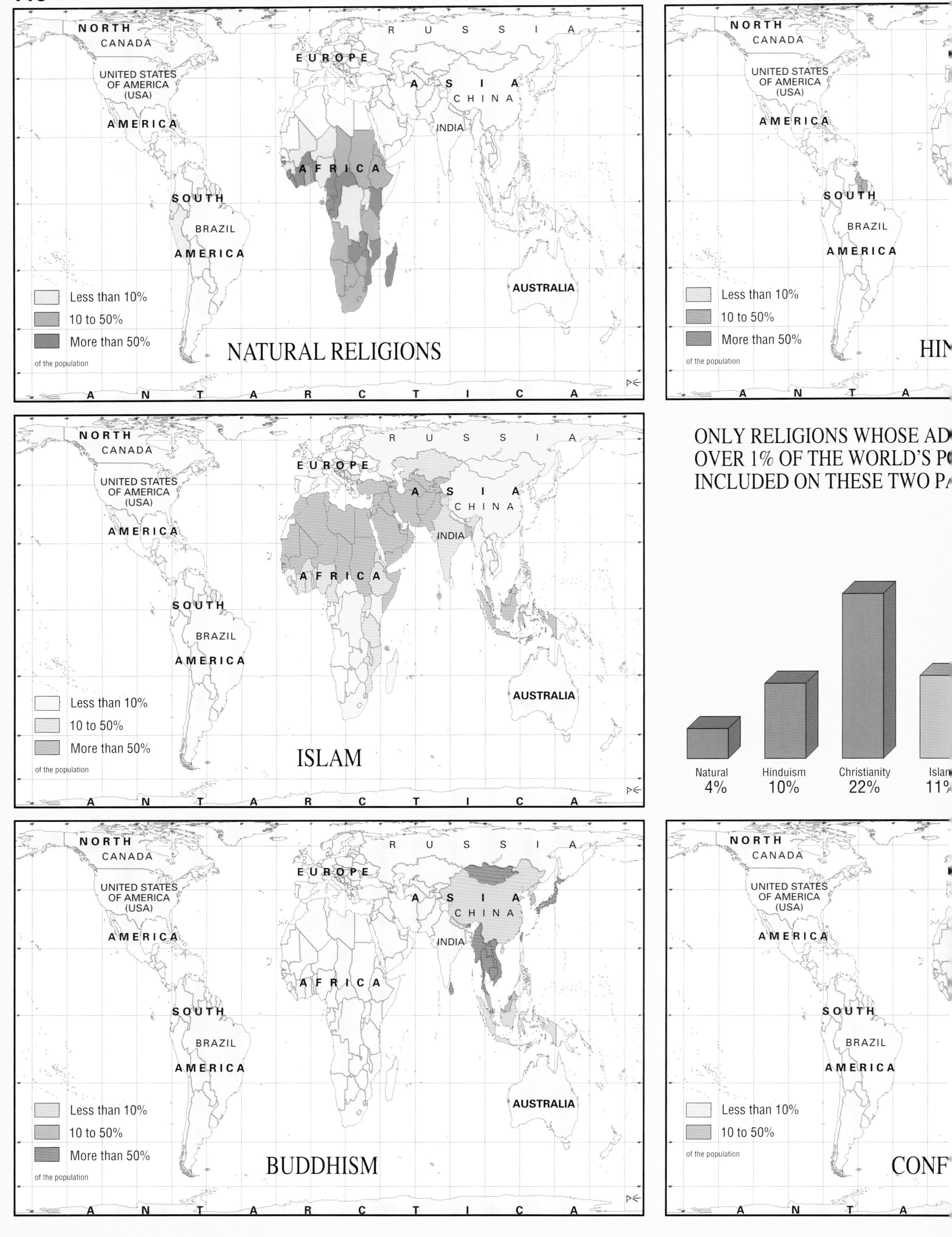

For over two millions years, humans lived without religion. A few thousand years ago people started to try to influence the powers of nature by means of prayers and sacrifices. These nature-oriented religions were followed by world religions that tried to remove the fear of mortality by promising the continuation of life after death. In the past few hundred years science has harnessed the powers of nature and brought humanity back to nature, reconciling us to our mortality. As a result, only 40% of the world's population now subscribe to one of the world religions.

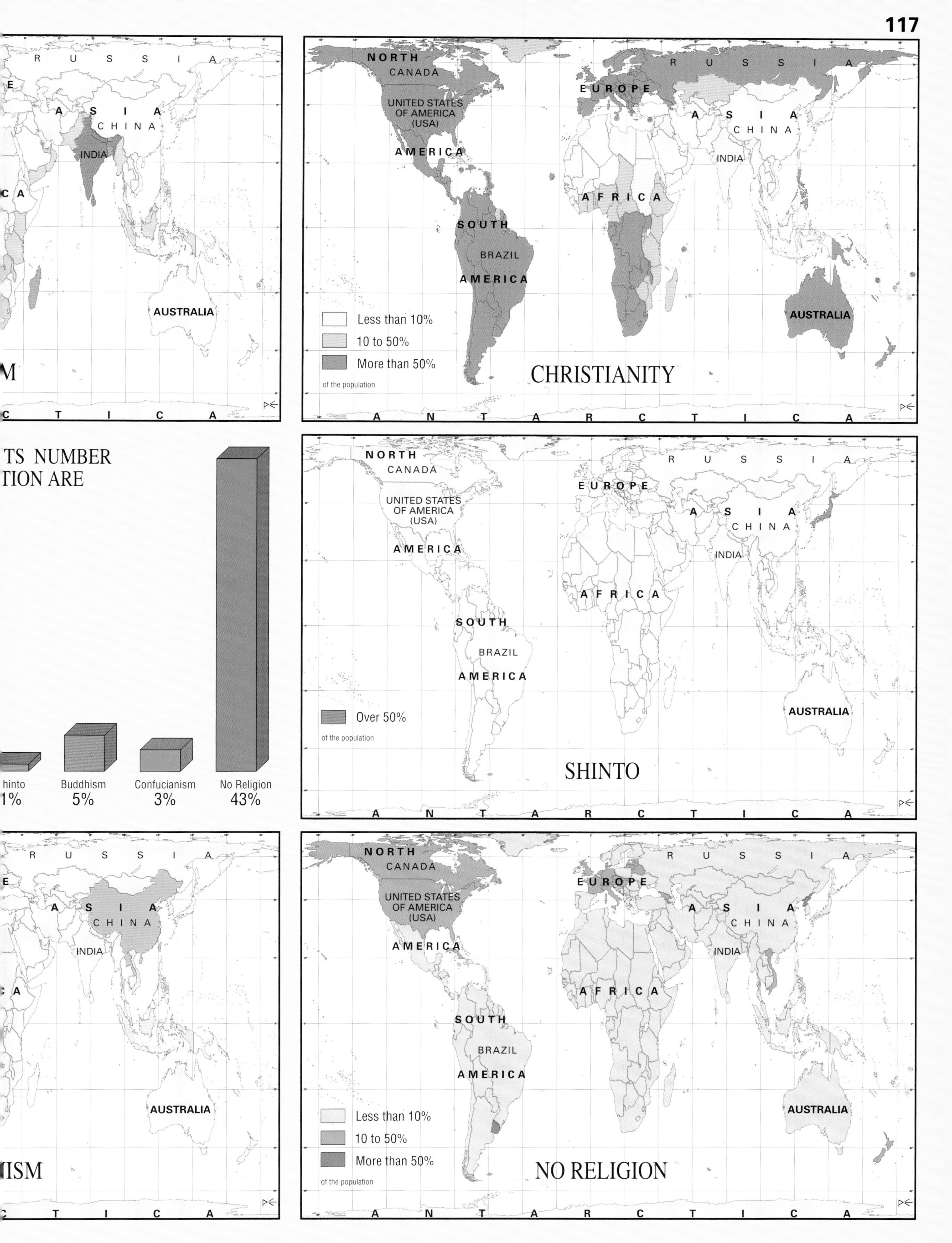

GIONS

If we define religion as belief in a God, only three of the six world religions qualify. These are Hinduism, Christianity and Islam. Shintoism is about the veneration of nature, Confucianism teaches ethics without a hereafter and Buddhism also has no concept of God. A significant element of the three major world religions is about domination. Hinduism divides the classes into castes as God's commandment, Islam has been used to justify Arab and Ottoman Turkish military and political expansion, and Christianity became the ideological basis for the Europeanisation of the world in the age of discovery.

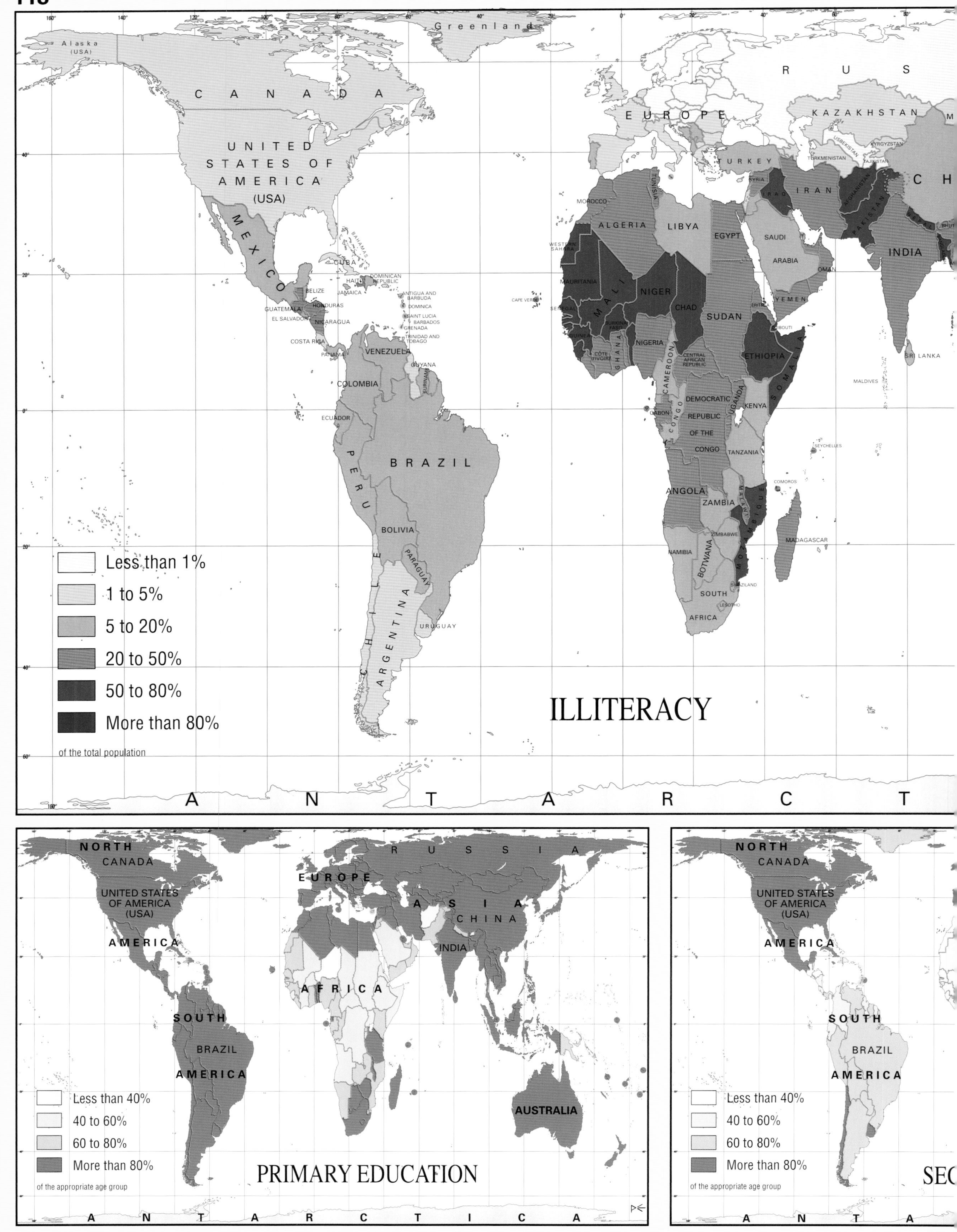

In the world of today a human needs almost two decades to acquire the knowledge necessary for coping with modern life. Without this knowledge a person would know nothing of their rights, be unable to realise their potential and could not comprehend the dangers which confront them. Without education they have no chance of controlling their own lives, but become mere pawns in the hands of those who possess education and will use such people for their own ends. The acquisition of education is therefore a means of participating in today's world.

EDUC

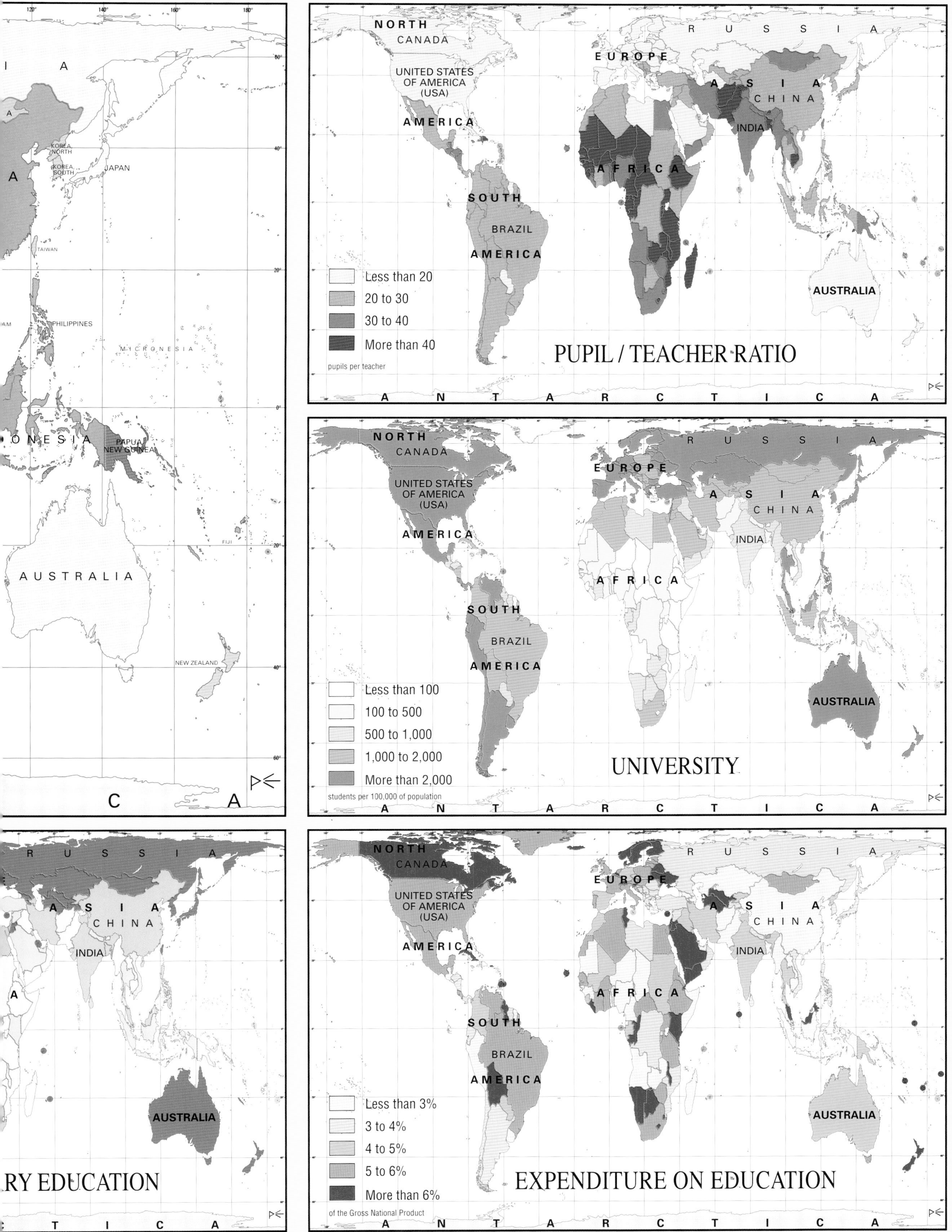

TION

More than a quarter of the world's population is illiterate and therefore excluded from general education. In Africa half the population is illiterate, in Asia more than a quarter, in India nearly half and in Latin America nearly a fifth. Even in Europe and North America there are still millions of people who cannot read or write. Although in most developing countries more than 80% of children attend a elementary school, the number of illiterate people in the world declines only slowly. Three-quarters of these are women. Higher education is and remains largely a privilege of rich peoples and individuals.

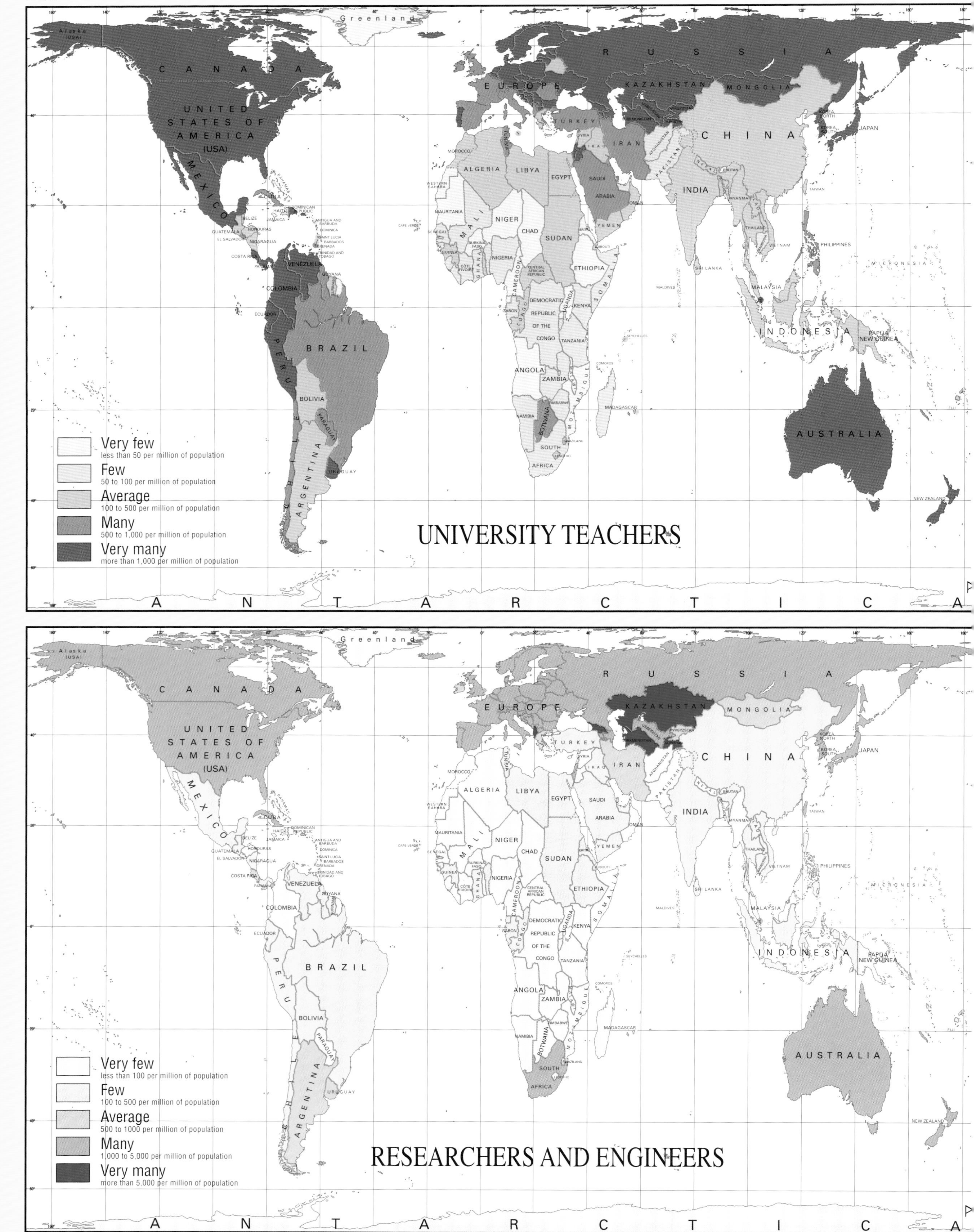

Today we regard ourselves more and more as living in the Age of Science. Detailed research into nature has given impetus to technical progress, and each day hundreds of new inventions add new dimensions to our lives. This technical revolution, which was set in train by science, has increased production on all levels to such an extent that it would be possible to assure sufficient and humane living standards for everyone in the world.

THE S

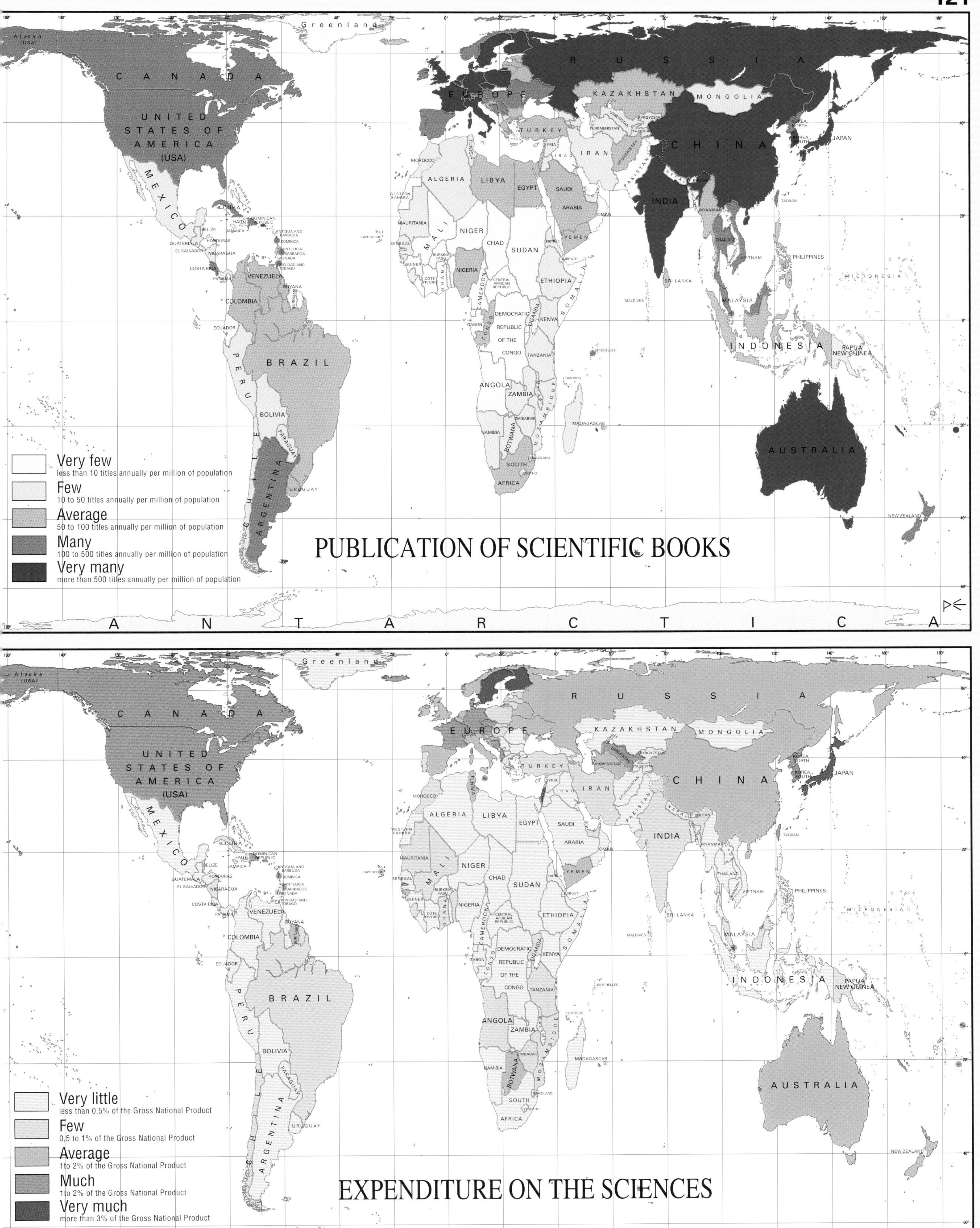

Recent advances in science and mathematics have resulted in the development of computers. Scientific progress has furthered co-operation and revolutionised the production and distribution of goods. People's ability to share in scientific progress however still varies greatly. The poorer peoples of Africa, Asia and Latin America do not yet have the means to participate in the university-orientated scientific process.

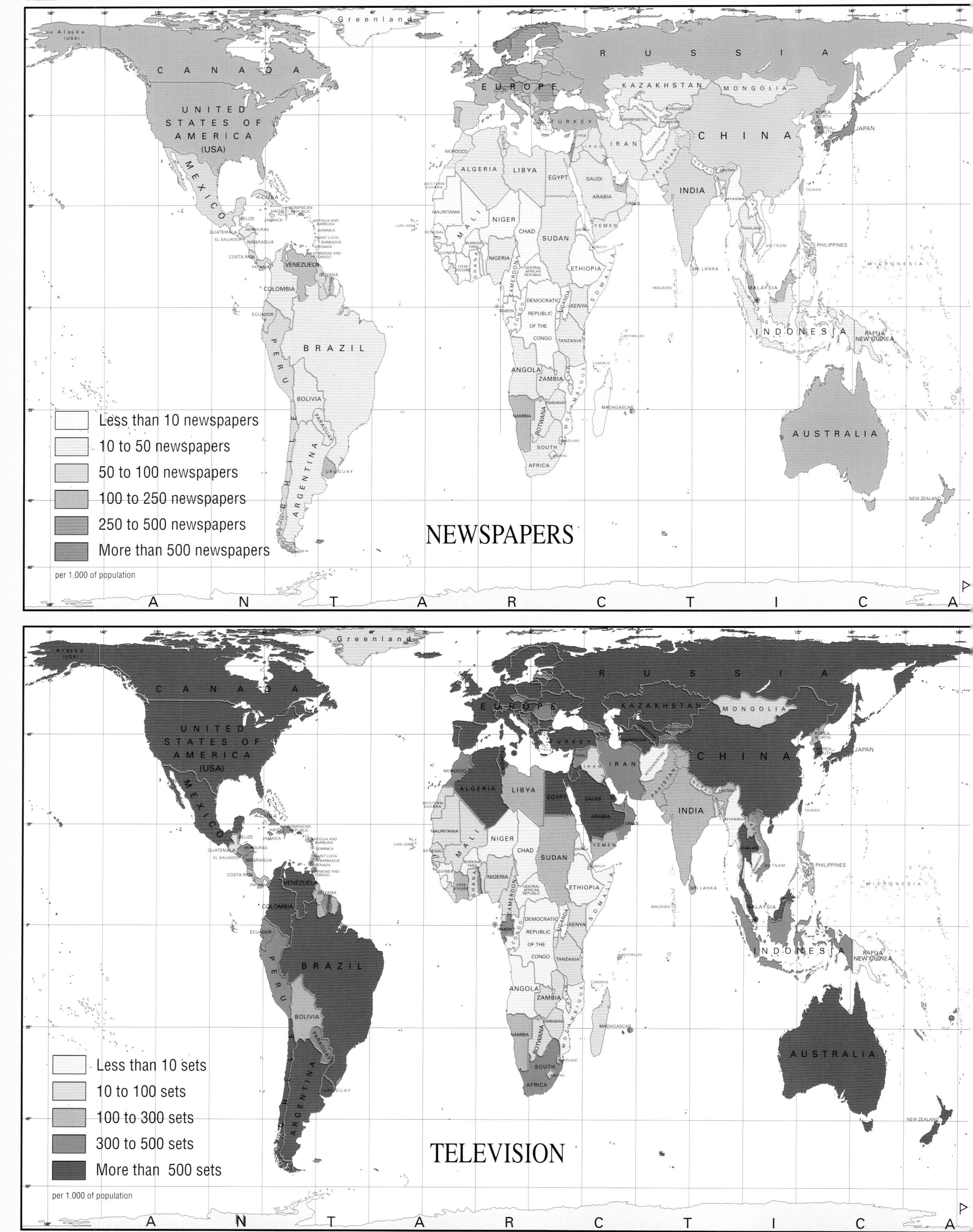

Access to information about events of general significance is increasingly a vital precondition for people's work and their ability to cope with their own ever more complicated personal lives. Often the information media are the only source for the lifelong process of further education for adults, which has become essential because of the rapidly changing world. The northern hemisphere, which has only one-quarter of the world's population, uses up three-quarters of the paper available for newspapers and books, that is, ten times as much per inhabitant as in the south.

INFOR

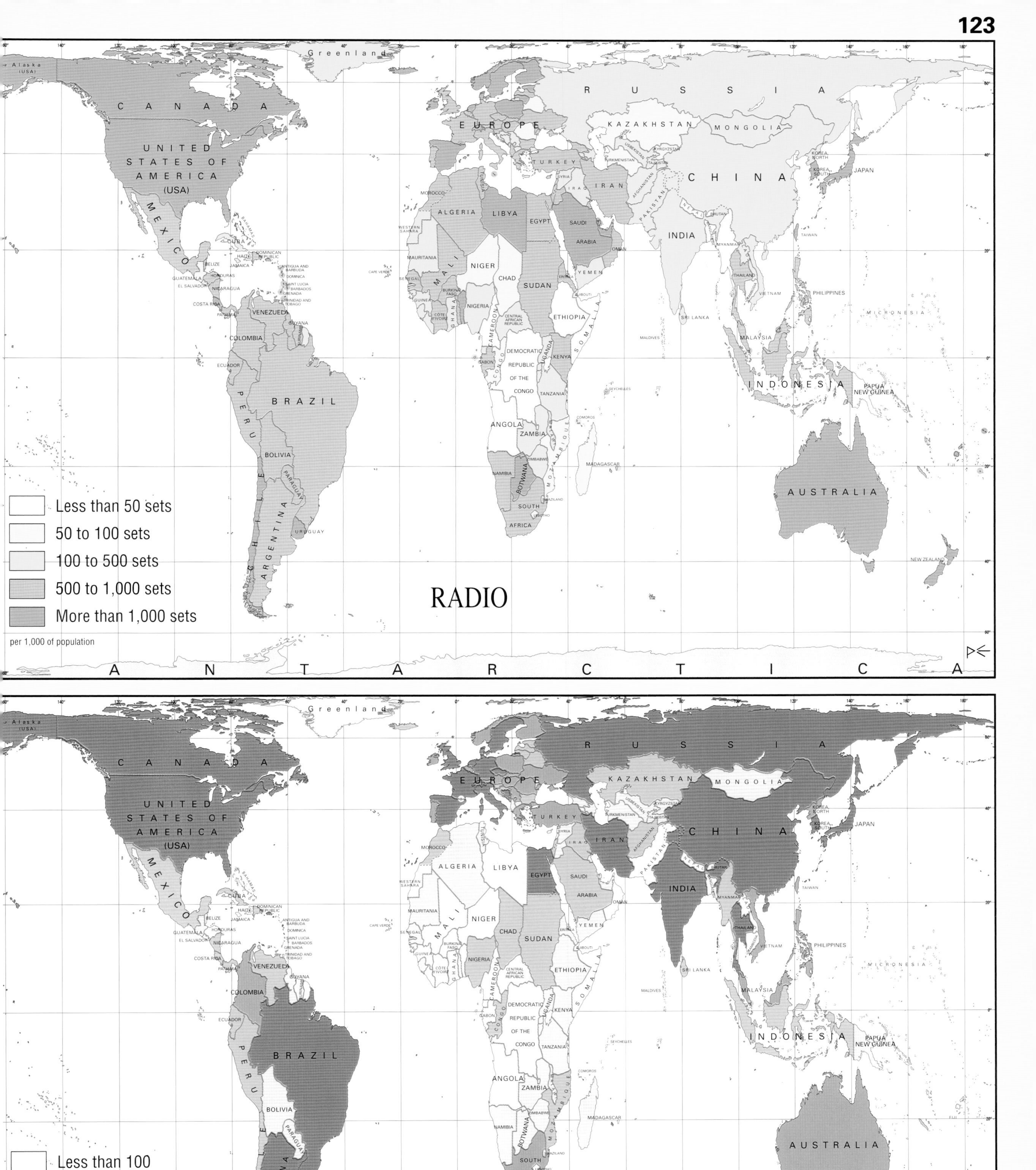

Radio and television are particularly important in the southern hemisphere, because with their help even illiterate people can gain access to information. But while in the northern hemisphere at least every other person has either television or radio, in the southern hemisphere only 10% are in this position. There is only one newspaper per thousand of the population in Mali, Niger and Burkina Faso while in Norway and Japan there is one for every two people. In Malawi 500 people effectively share one television set, while in France and the United States there is one for every two people.

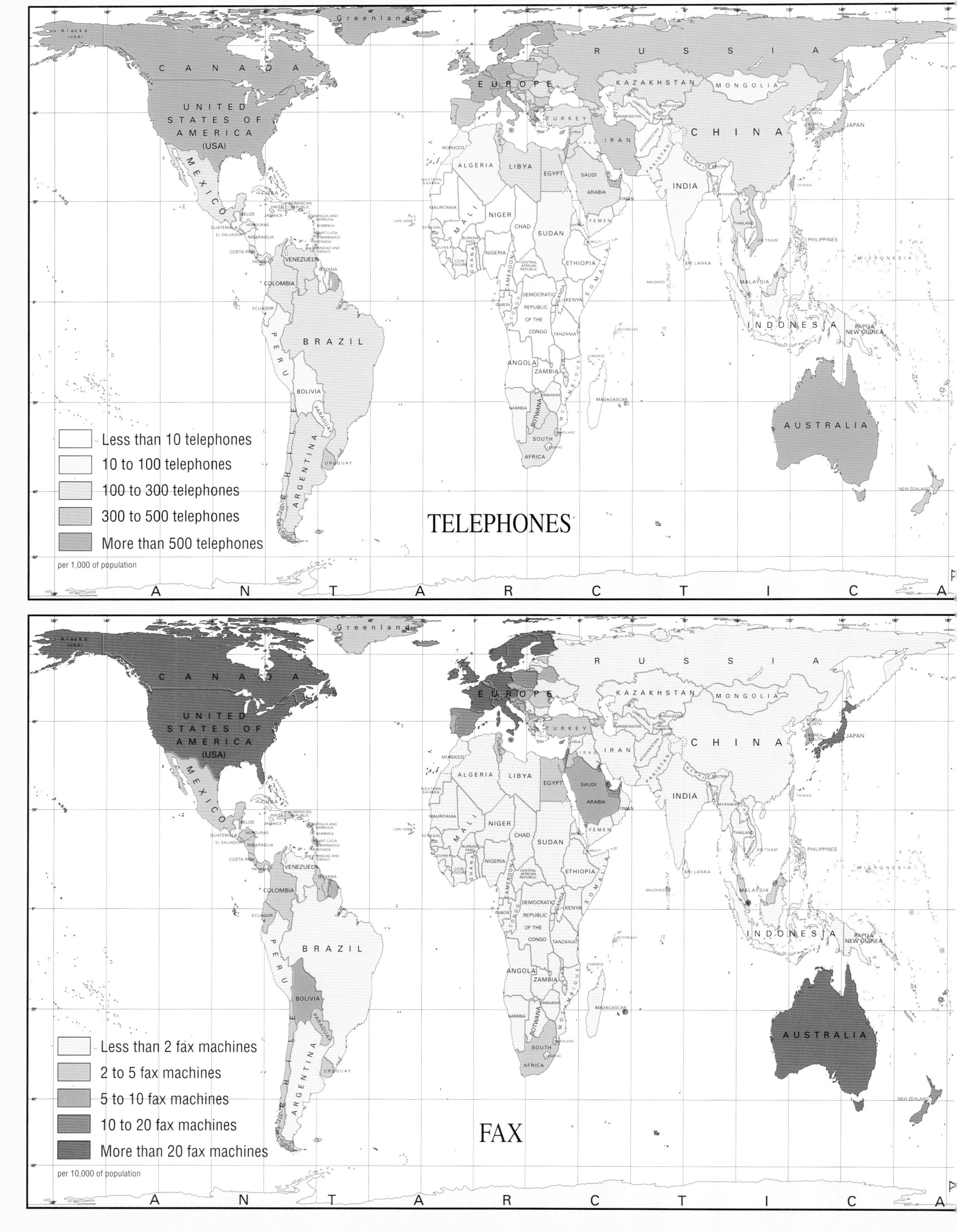

Today an essential part of world communication is the sending and receiving of personal information. Every person in every country could, theoretically, connect directly with everyone in the world. But the ability to do so is still a privilege which the rich states and peoples can use to bolster their position because communicating with text presupposes knowledge of reading and writing, which means it is denied to the poorest third of the world's people, those who are illiterate.

COMMUN

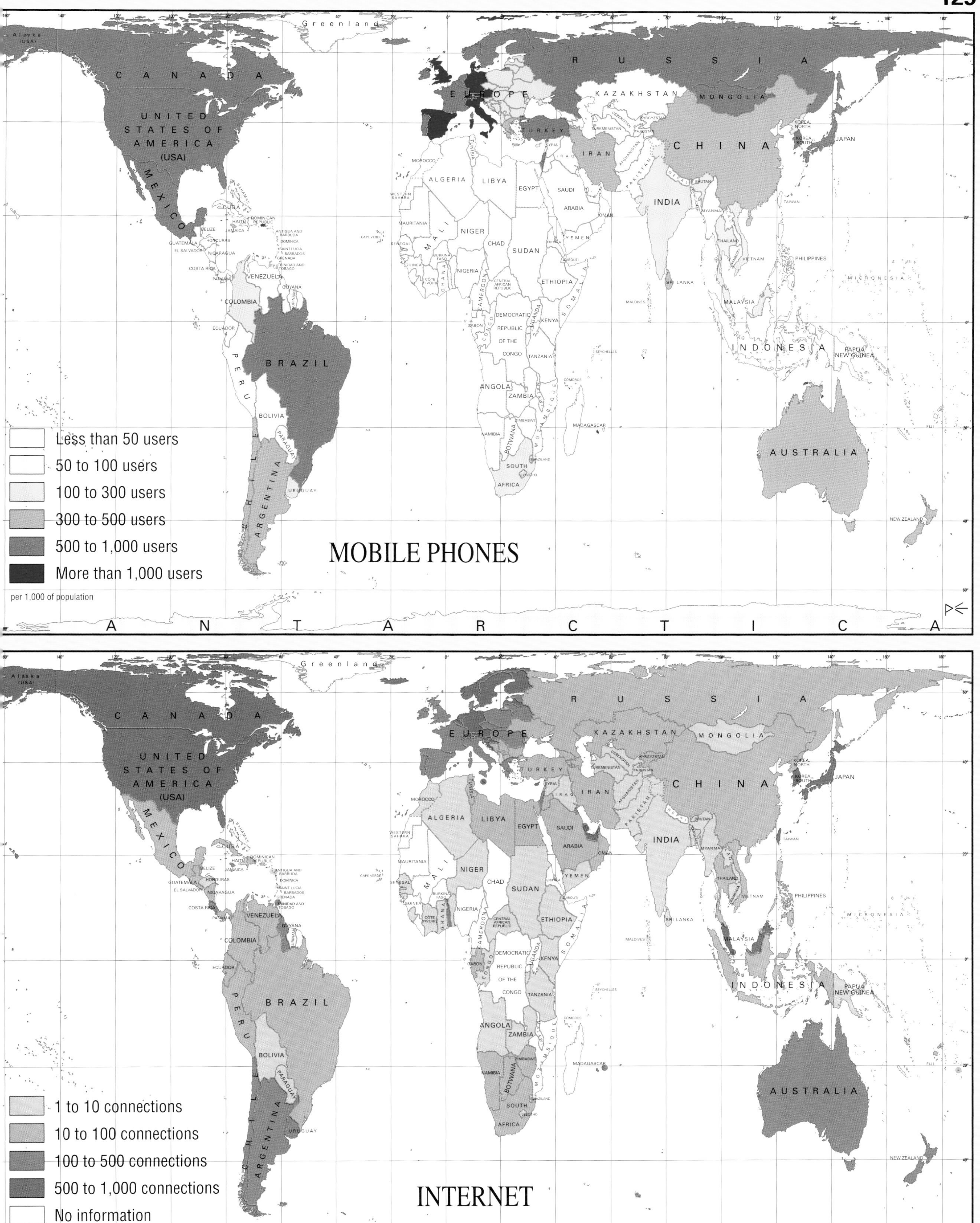

CATIONS

Telephone communication is not available world-wide. In the rich industrial states there are 500 telephones per 1,000 inhabitants (US 630, Canada 610, France 570, Monaco 1015). Poor developing countries like Mauritania have only one telephone per 1,000 of the population. The transmission of documents (fax) is also still a privilege of the rich industrial states, as is the computer, which is revolutionising communication.

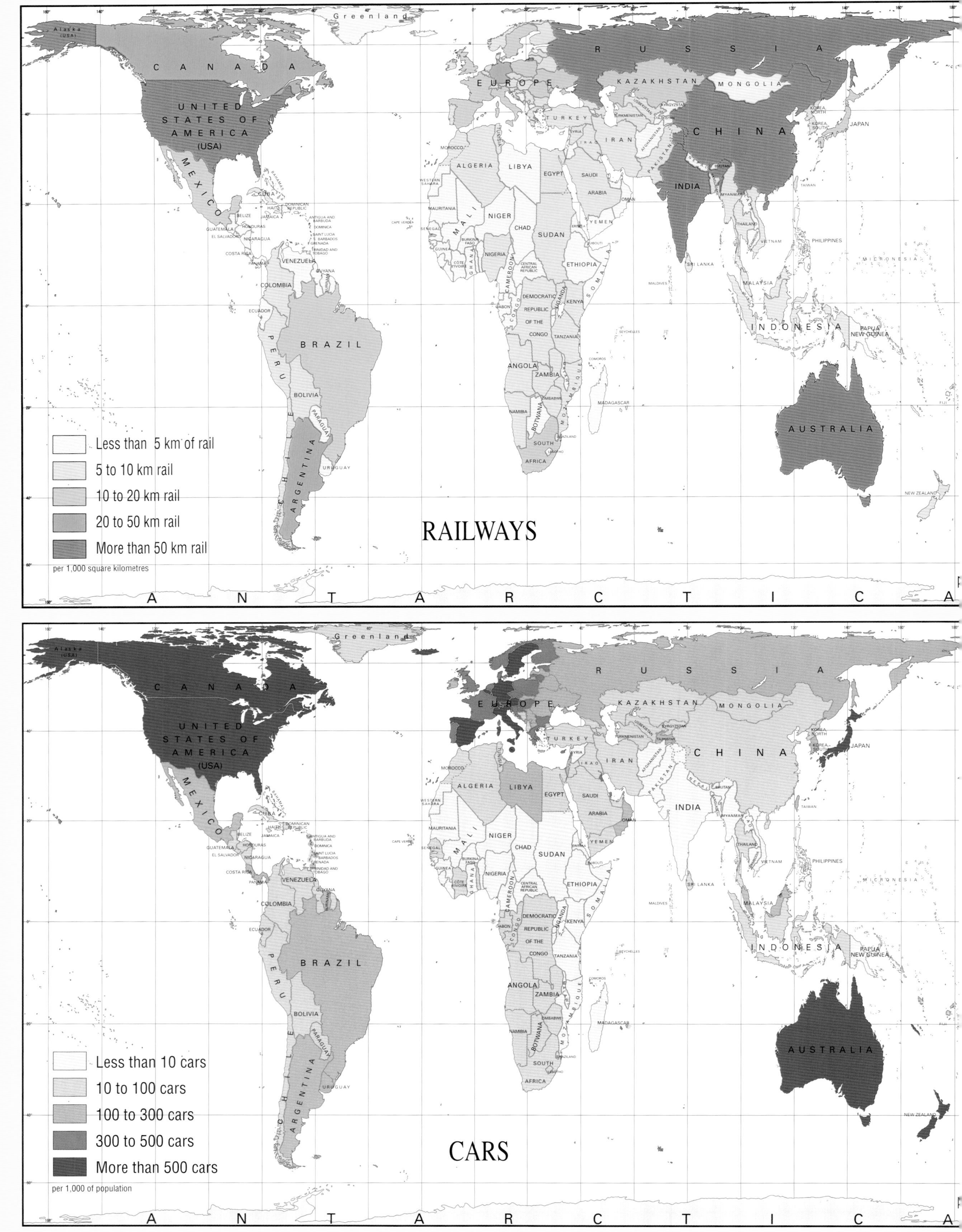

For centuries people were limited to travel within their own localities because only horses, donkeys, mules, ox-carts and horse-drawn vehicles were available. The use of ships and later, cars and planes meant that the movement of people and goods rapidly developed into global traffic. In our age of closely woven contacts it is possible to participate in world events at all levels.

TRAFFIC

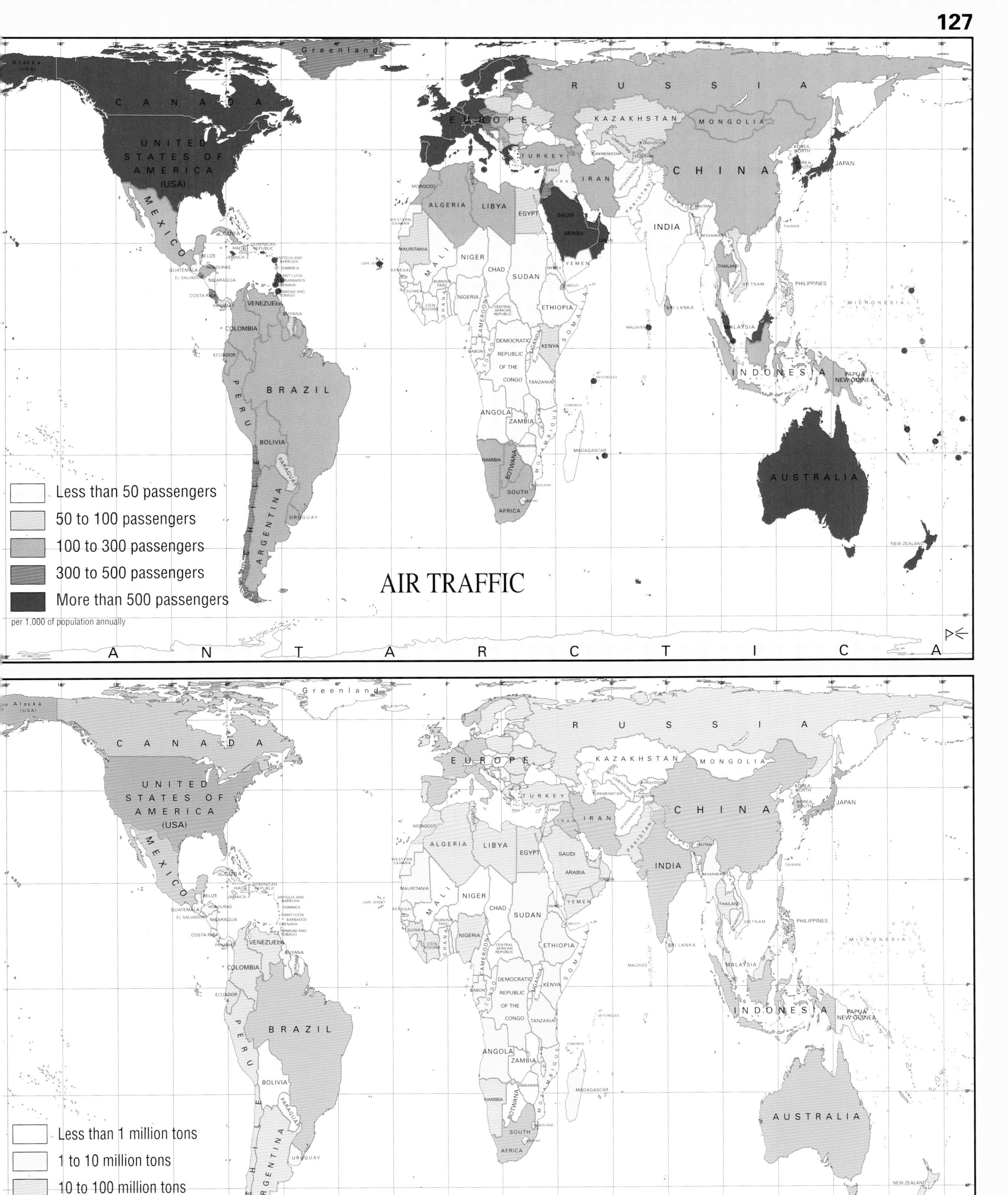

DENSITY

Half of the world's people live in countries with fewer than one car per 1,000 inhabitants (for example in Burma, Bangladesh, Mozambique, Afghanistan, Somalia). In most of these developing countries railways, air travel and shipping are still only slightly developed. While Japan ships nearly 700 million tons of goods a year, much bigger developing countries in general, ship well under one million tons each.

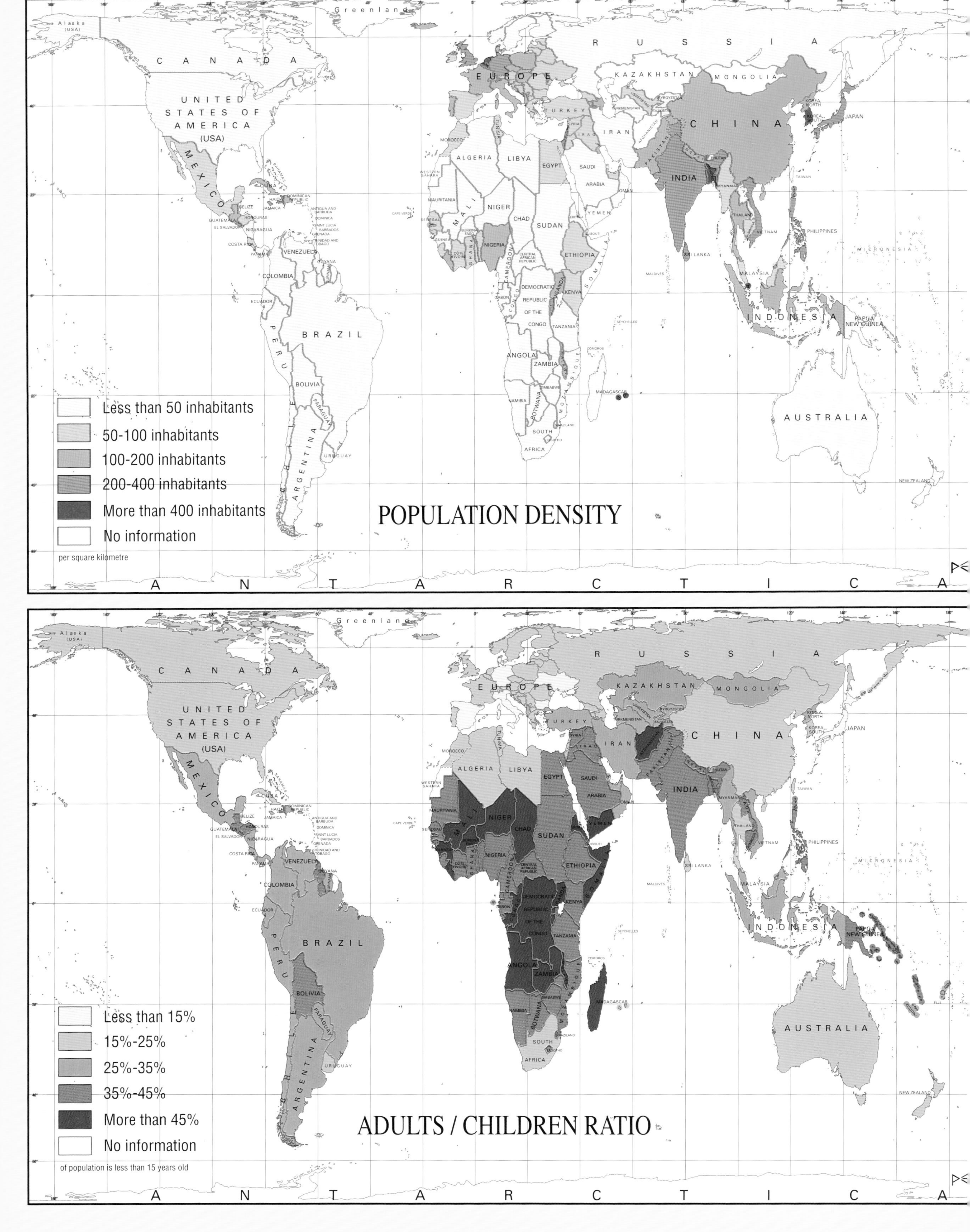

Young nations possess the future but the old nations have a firm hold on the present and deny the younger ones their rightful place in the family of peoples. In Yemen 48% of the population is under 15; in Germany 15%.

POPULATION

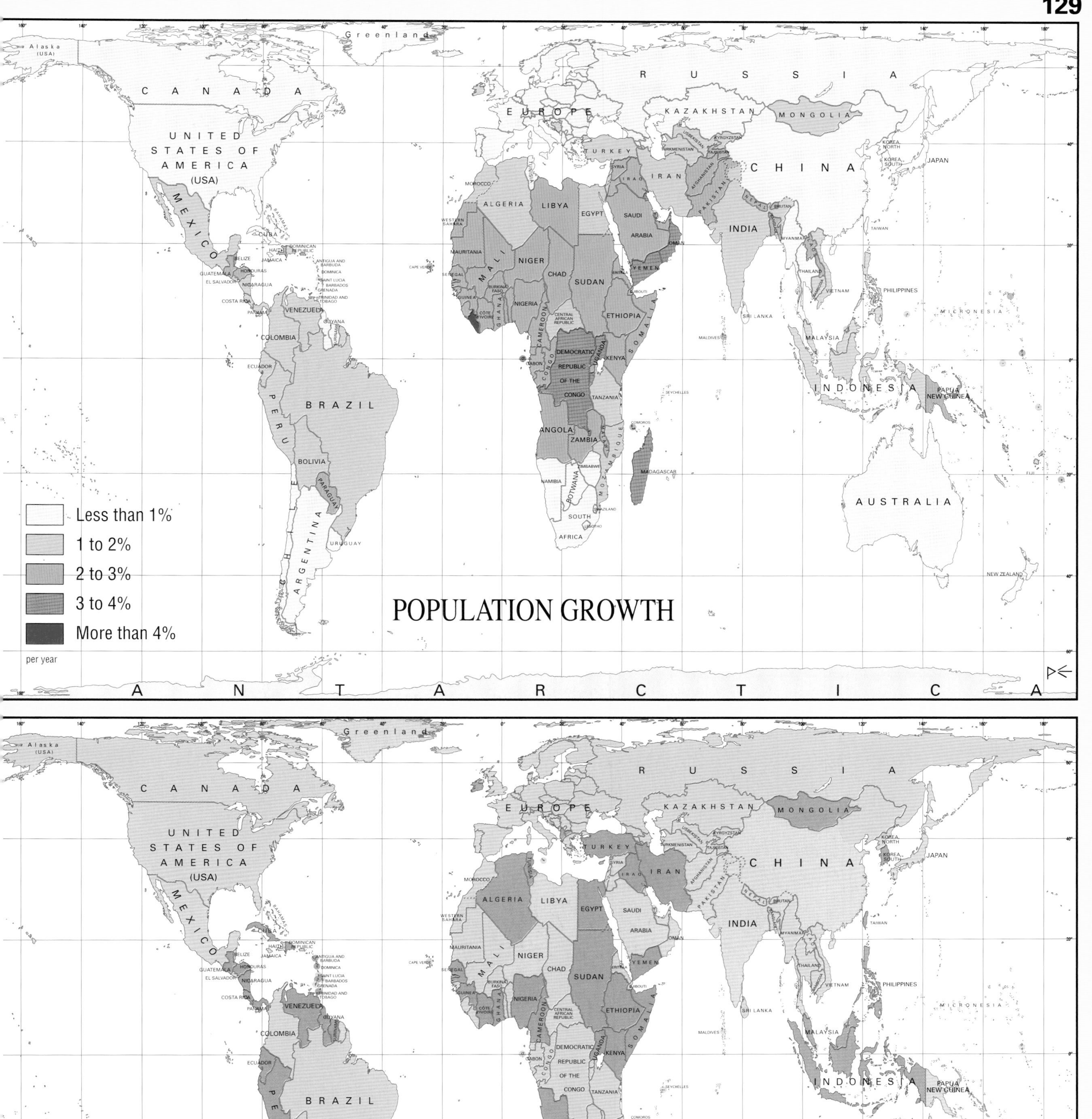

STRUCTURE

The increase in population amongst poor nations and recent population decline in rich nations also reflect the uneven distribution of wealth inherited from 500 years of colonial exploitation. As the affluence of the rich industrial nations in the North has increased, so has the population in the poor countries of the South.

Advances in civilisation on all levels are expressed by an increase in life expectancy. Five thousand years ago the average length of life was about 20 years, 500 years ago people lived on average to 30 and 100 years ago to barely 40. Nowadays life expectancy in the rich world is around 70 and in some countries almost 80.

LIFE EXP

ECTANCY

Life expectancy varies greatly in different countries of the world. A person born in Uganda or Ethiopia has a life expectancy today of about 40 years, while in rich industrial countries like France, England, Germany, Canada and Japan life expectancy is about twice as long.

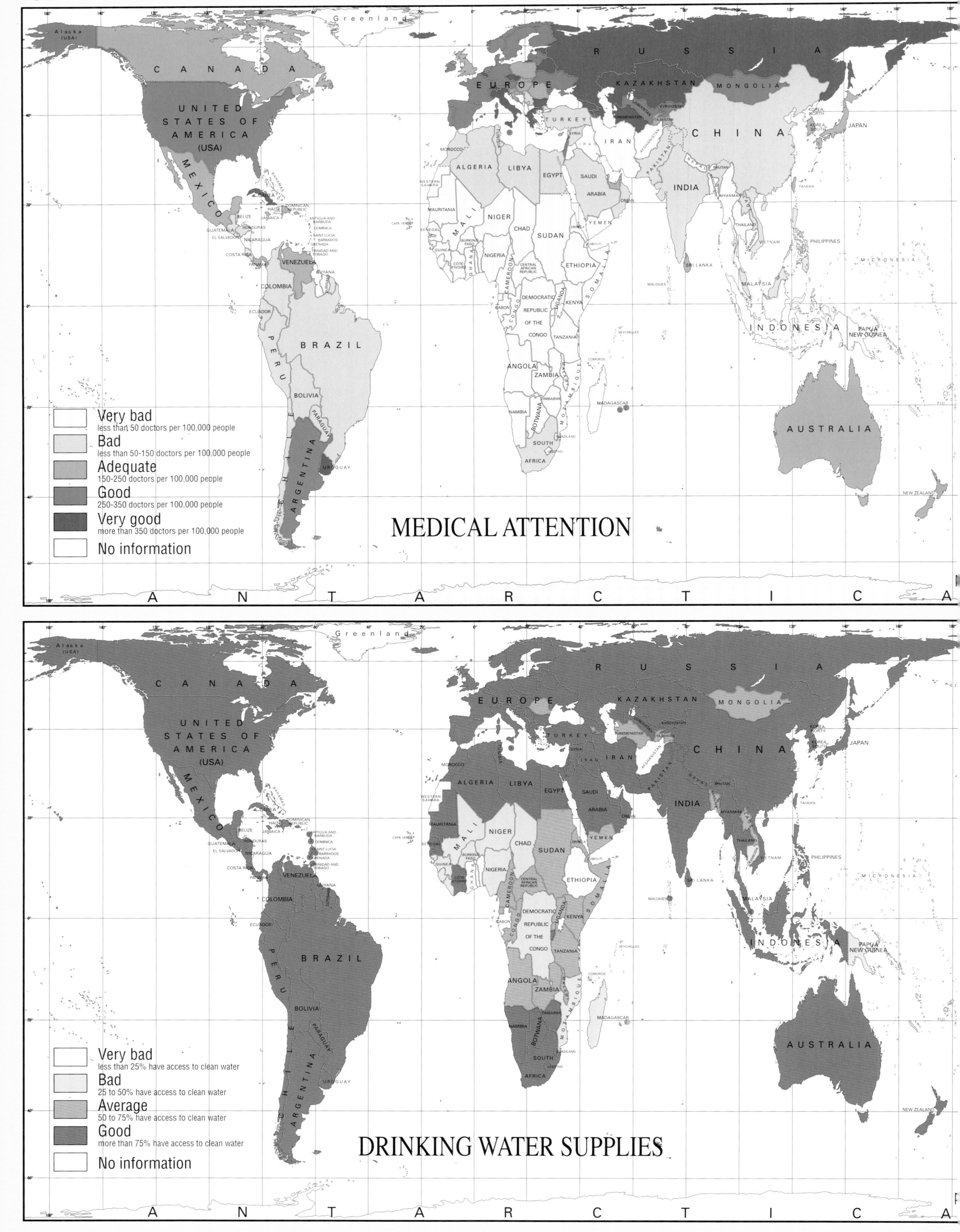

To remain healthy a person needs sufficient food and clean drinking water. Today both could be secured for everyone. But health depends on other factors: pollution-free air to breathe, regular and sufficient sleep, enough exercise, avoidance of toxins and toxic stimulants, bodily cleanliness, adequate living space, natural diet and clothing, sunny and calm living conditions, and harmony with the rhythms of nature.

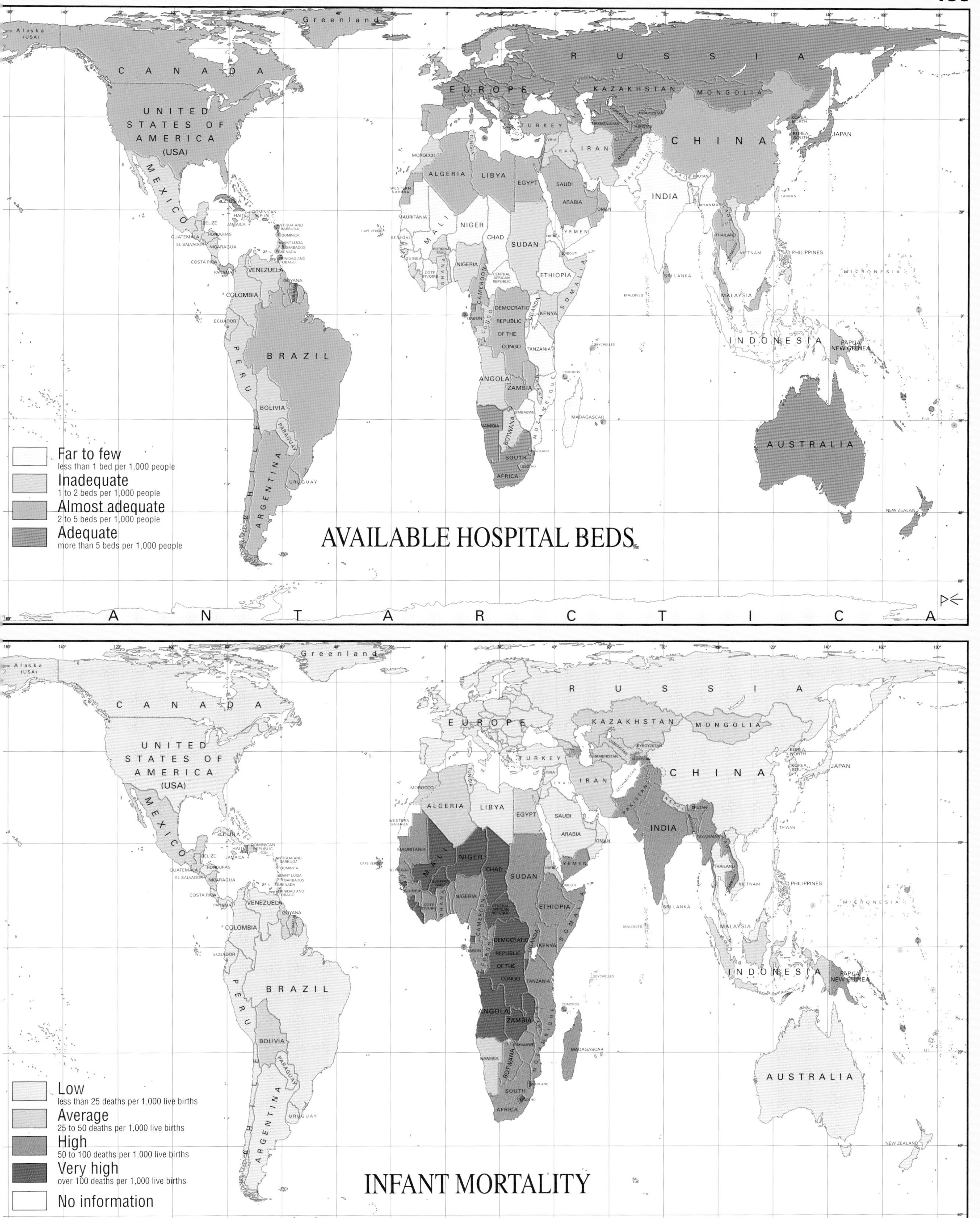

LTH

When illness strikes the availability of a doctor can mean the difference between life and death. Where there is one doctor for every 300 people, as in Belgium, Bulgaria or Austria, the provision of medical care is assured. Where there are a hundred times more people per doctor (as in Chad with 30,000 or Burkina Faso with 32,000) people mostly have to rely on themselves. The same applies to the provision of hospitals: in Switzerland there are 55 inhabitants per hospital bed, in Niger 10,000. Thus health is today the privilege of the wealthier peoples.

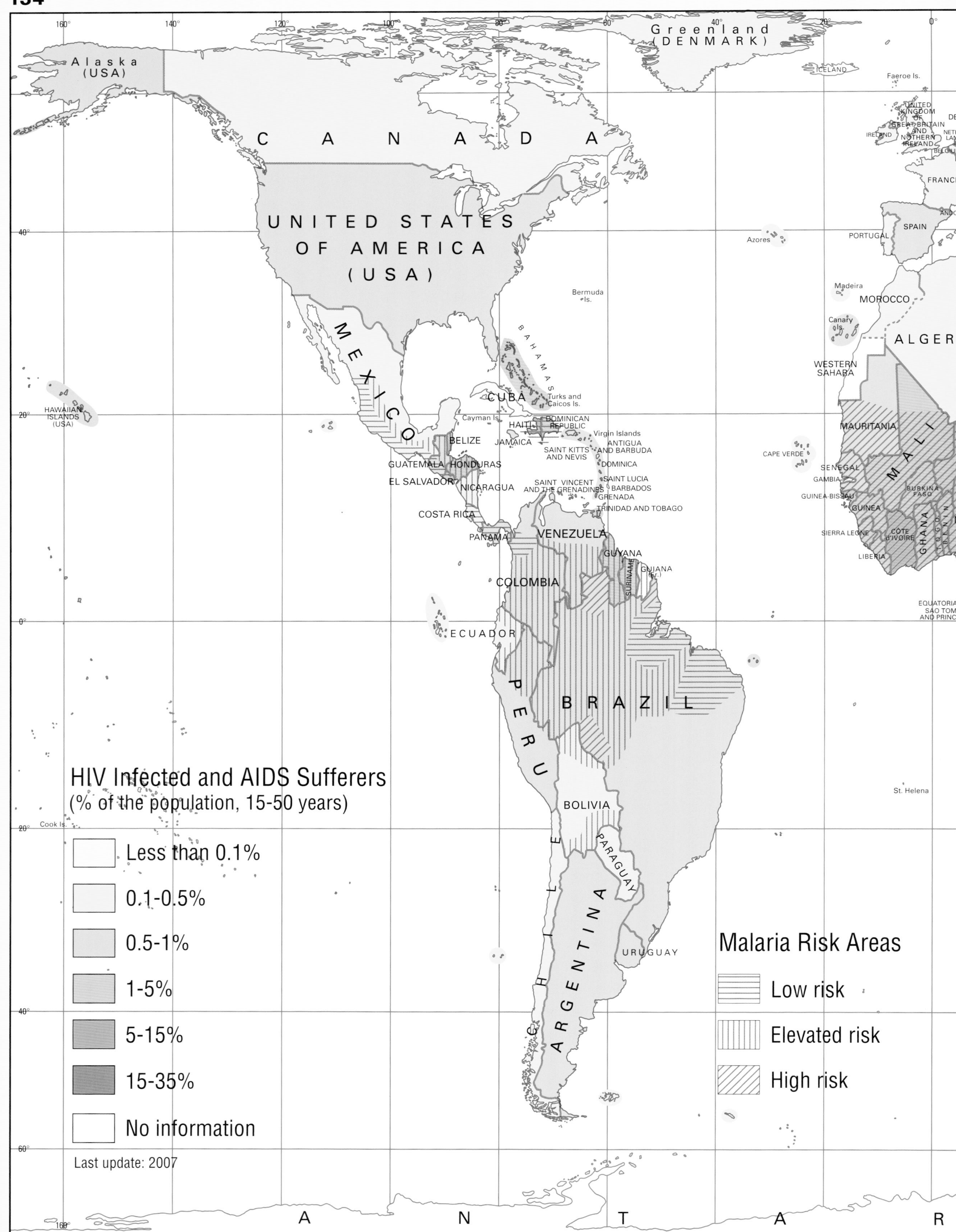

Since 1985, more than 25 million people have died from the after-effects of acquired immune deficiency syndrome around the world. In the meantime, approx. 33 million people live with HIV all over the world. AIDS not only constitutes a threat for each individual, but is also an outstanding problem in the globalized world. Roughly tens of thousands of HIV infections and millions of attacks of malaria can be traced to the interplay of the two diseases. Competence networks have formed on an international level, which have set up data pools, forming the basis of large-scale international studies that are of paramount importance to HIV research and thus for the treatment of the patients.

HIV/AIDS - MAL

RIA RISK AREAS

About 10 % of the world's population become infected with malaria every year, and over one million people die of it. It is feared that the situation could worsen still on account of environmental changes, unrest among the population, growth in population, travel behaviour und resistances against drugs and pesticides. The new approaches to getting malaria under control as well as growing conscience all over the world have led to an increase in funds for research and monitoring measures. In order to buck the dispersal trend, a permanent political and financial commitment is necessary. It is only in this way that successful programmes for the reversal of the current development can be established. This will also boost research.

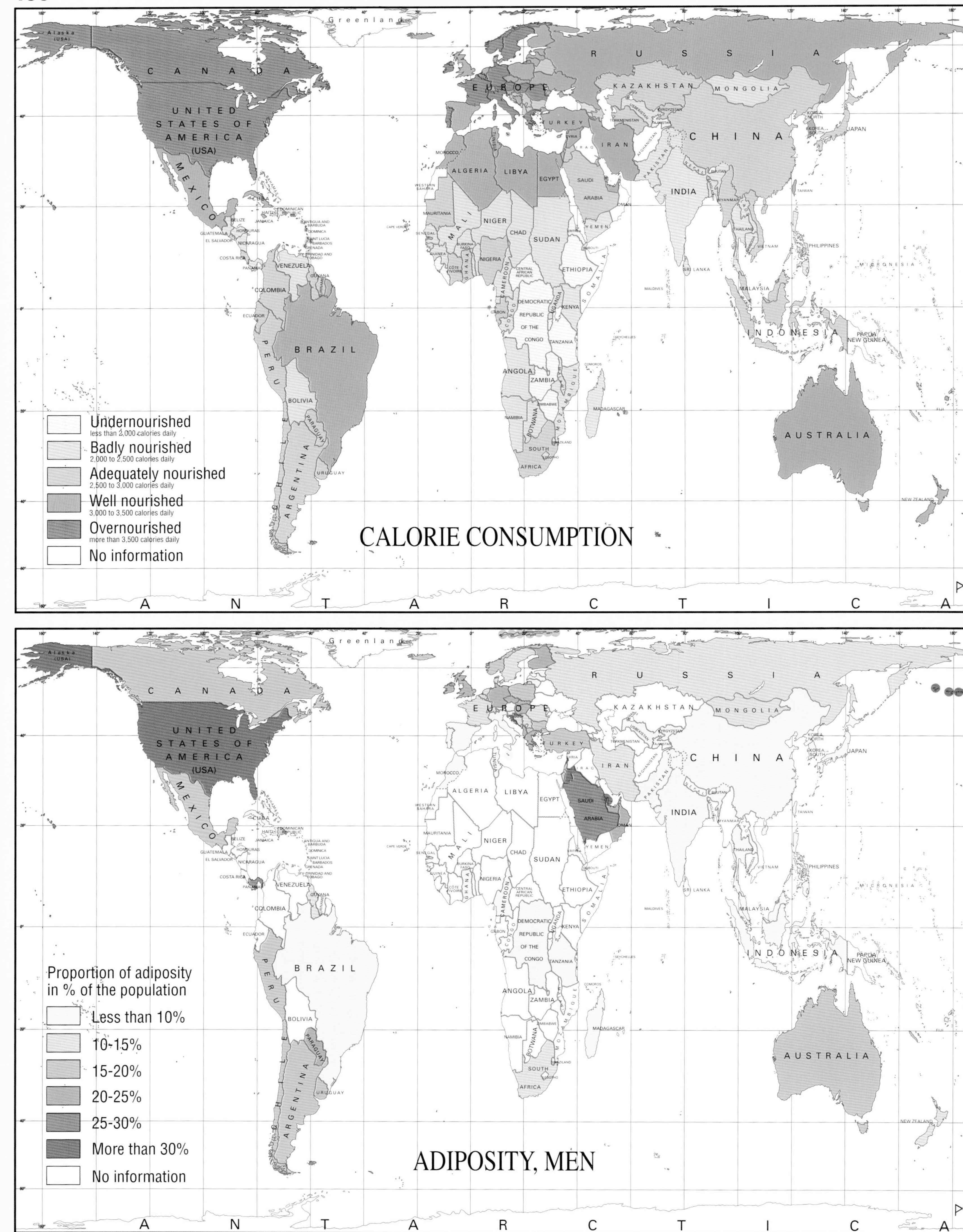

Life processes require constant energy, which all animals must renew through food. After people had begun to produce their own food through agriculture and animal husbandry, the human diet became better and more regular. Most of the world's people normally have enough to eat. Food production is increasing faster than the rise in the world's population. Today therefore, nutrition is only a question of distribution. The superfluous food of richer people and nations could supply the needs of the poor and hungry.

NUTR

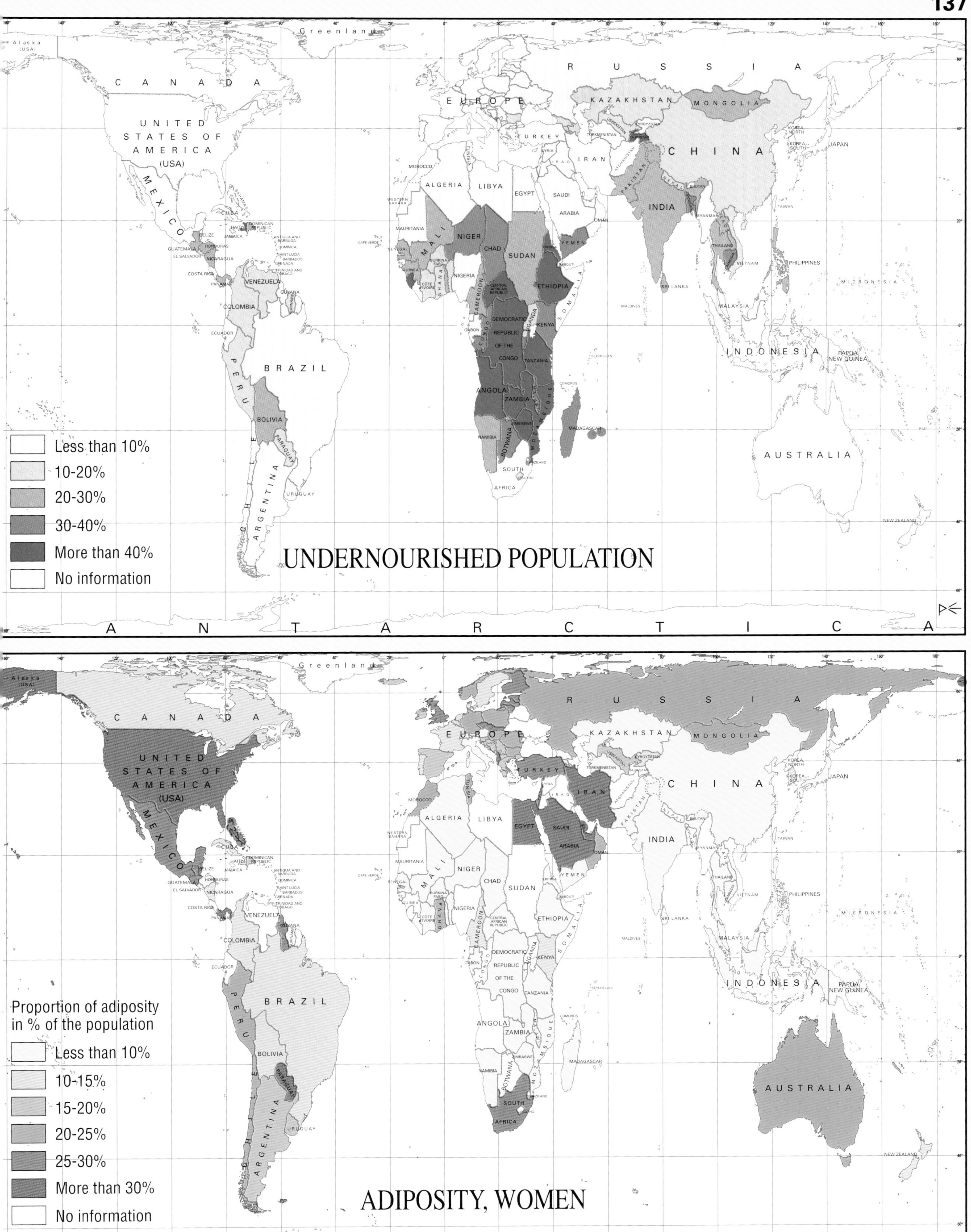

TION

According to studies conducted by the World Health Organization in the year 2000, more than 1 billion people were overweight and at least 300 million obese. Of the latter, 115 million lived in developing and emerging countries. This means there are more overweight than underweight people in the world. According to estimates, 31% of the population of the USA have a body-mass index of more than 30 and are thus considered to be obese. Statistics there also demonstrate that the socially disadvantaged (the less educated and relatively poor) as well as disadvantaged minorities (people of colour) are more likely to be overweight than more privileged groups of society.

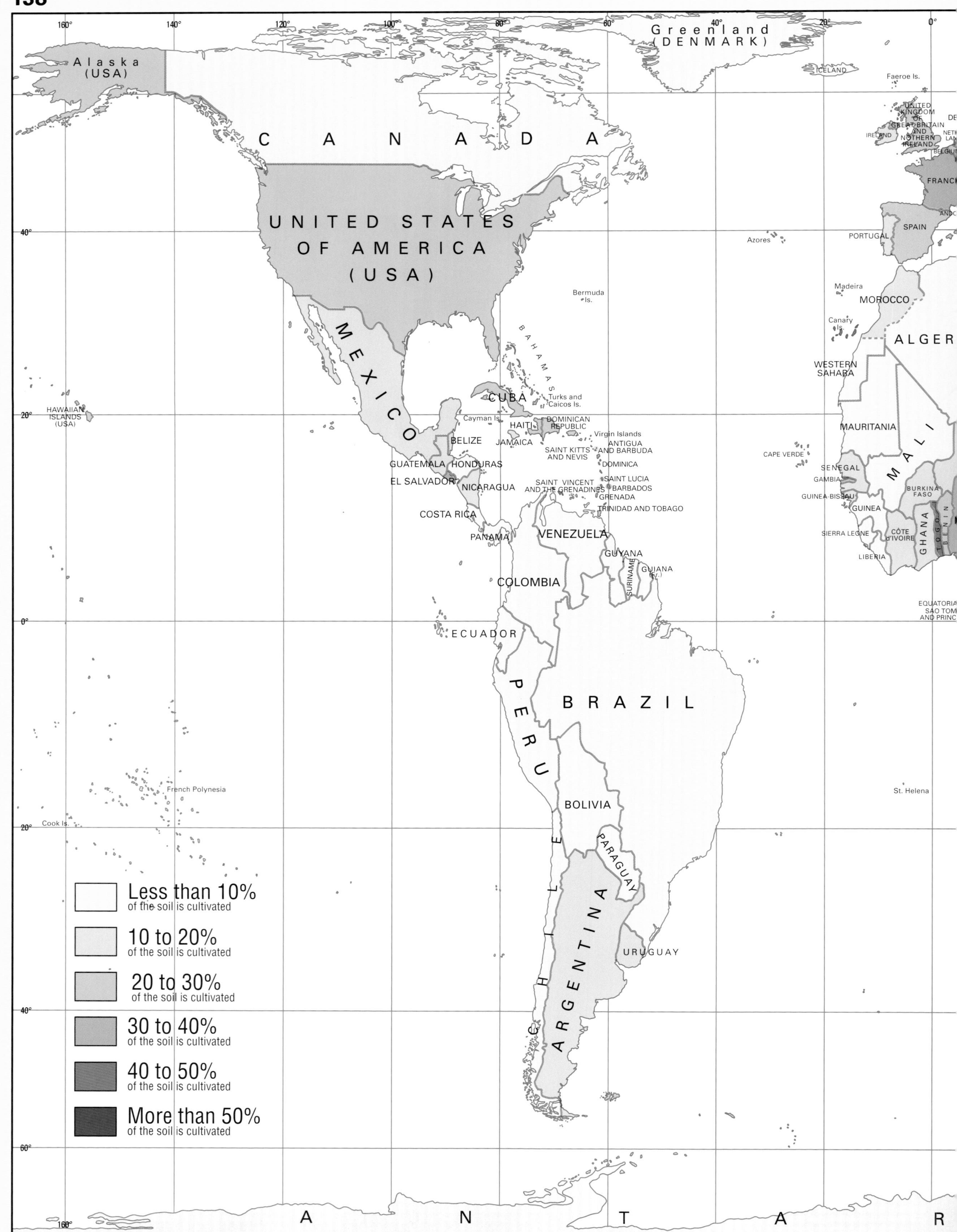

For thousands of years people lived without cultivating the soil. Even today large groups of the world's population do not practise agriculture but live on wild plants, fruit and roots plus meat from animals which they hunt or breed. But cultivation of the soil was precondition for human settlement and therefore for the development of culture. Today the agricultural use of land is a safeguard for our existence.

SOIL CUL

TIVATION

Less than 10% of land is used for agriculture. Many conditions have to be met before cultivation can be undertaken – not only human ability but also sufficient sunshine, an adequate water supply and good drainage, moderate temperatures and of course, suitable soil. The proportion of land used for cultivation varies between 1% (Libya, Angola) and 63% (Denmark).

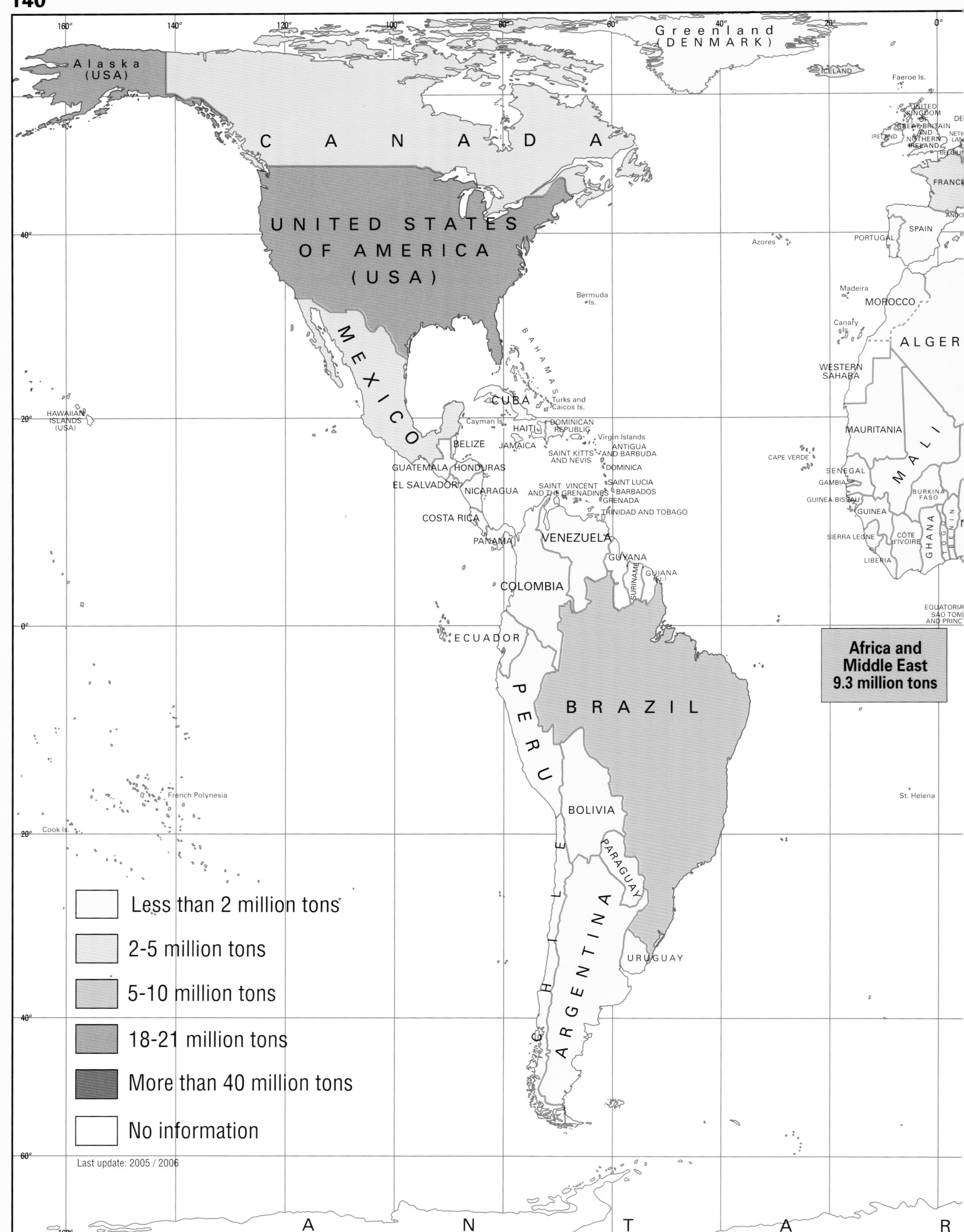

Since the Minoan era (around 1,700 BC), agricultural acreage has been strewn with animal and human faeces in order to enhance the crop. In the 19th century, people also began to use ash, chalk, and marl for manuring. Around 1840, the growth-stimulating effect of nitrogen, phosphates, and potassium was proved. Nitrogen was obtained in the form of nitrates, recovered from guano, excrements of sea birds. Since these supplies were limited, however, a method to produce nitrates synthetically was contemplated.

US
CHEMICAL

OF FERTILIZERS

Between 1905 and 1908, catalytic ammonia synthesis was developed chemically in Germany. The mass production of ammonia forms the basis of the production of synthetic ammonia fertilizers, called “fertilizer” for short. Since World War II, the industry has introduced ever more effective and selectively usable fertilizers to the market. In the last quarter of the 20th century, chemical fertilizer came increasingly under fire because its excessive use was held responsible for different types of ecological damage. Since around 1985, the use of fertilizers has been decreasing in Germany.

With the breakthrough of industrialisation in the mid-19th century, agriculture too benefited from improvements. Before this Green Revolution, three farmers were required to produce enough to feed one town-dweller over and above their own requirements. Nowadays one farmer can produce enough food for thirty town-dwellers. Agricultural production has thus increased almost a hundredfold over the past hundred years.

CROP

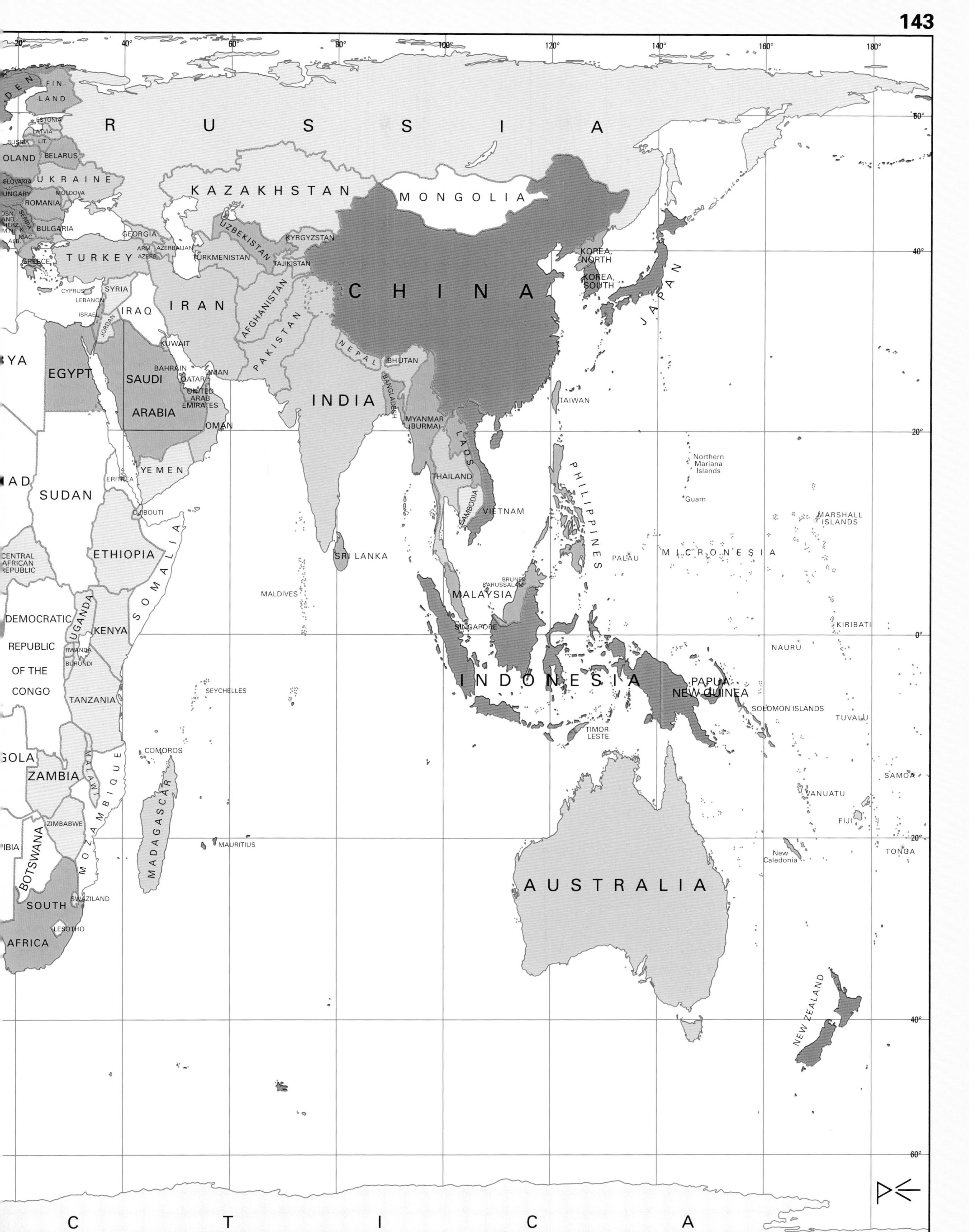

YIELD

The enormous increase in farming yields was achieved by the mechanisation of farming, artificial fertilizers and pesticides as well as the breeding of more productive animals. Because of the high costs involved, the poor countries can make only limited use of these methods. Thus farmers in Niger produce only 38 tons per square kilometre while French farmers produce 730 tons and the Japanese 600 – that is, over fifteen times as much.

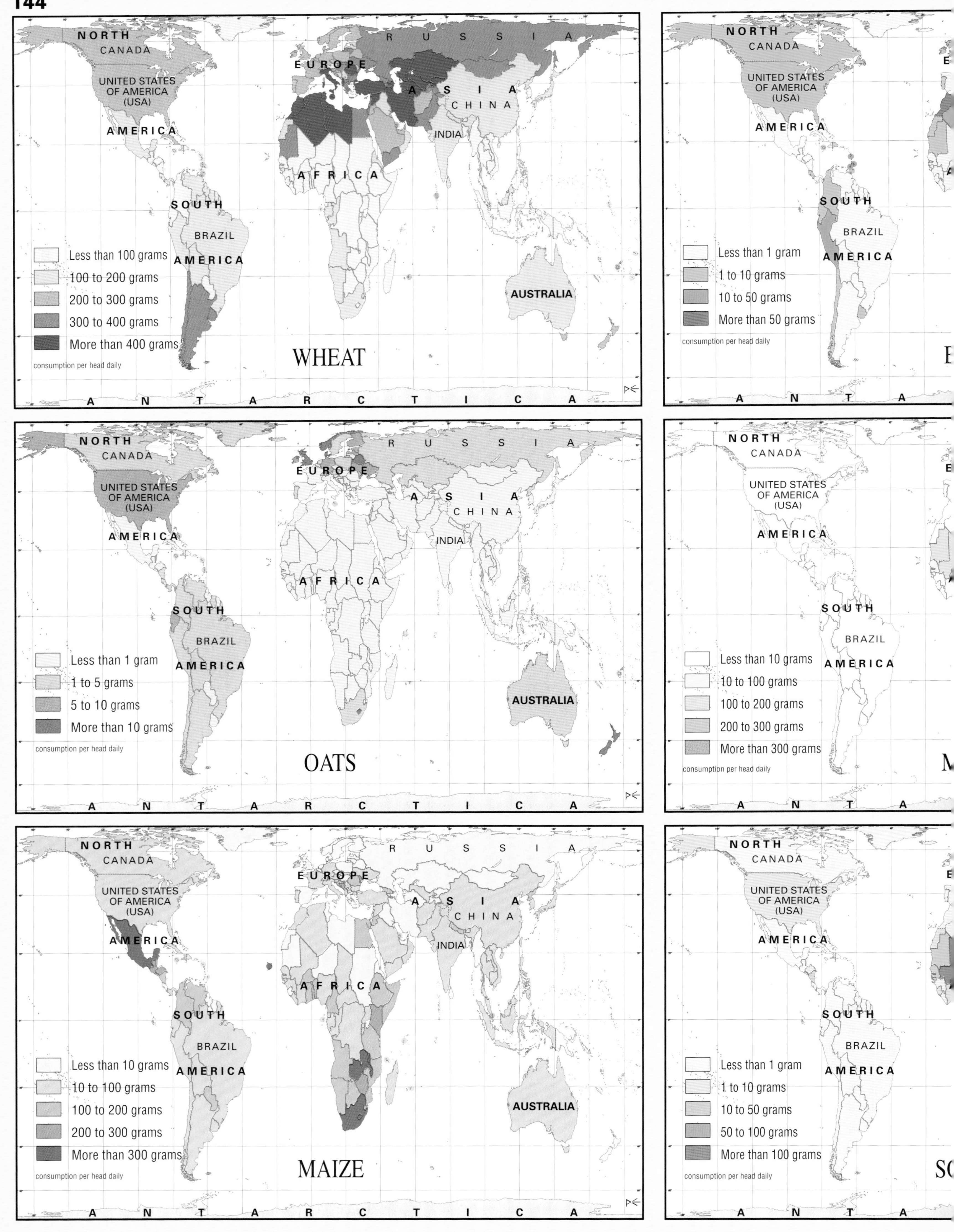

More than half of arable land worldwide is used to grow grain, which represents 56% of the calories consumed by people. Grain is a complete foodstuff containing not only carbohydrates, protein and fat but also the most important minerals and vitamins. In various forms it is still the staple food of the human race.

STAPLE FO

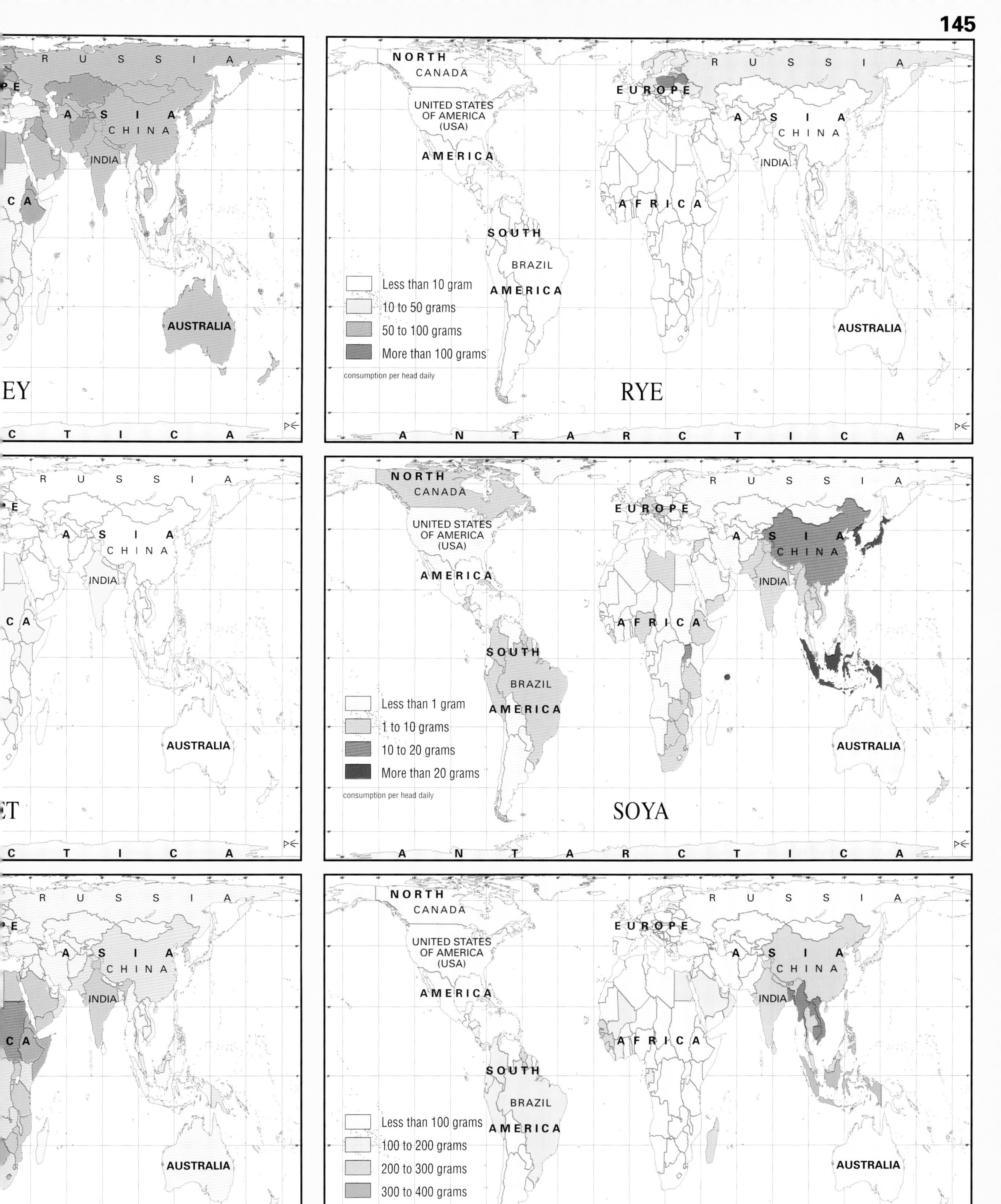

ODSTUFFS

The percentage of grain in the diets of different countries varies considerably. In the developing countries of Asia, Africa and Latin America nearly three-quarters of calories consumed come from grain, compared with a little over a quarter in the rich industrial nations. In the rich world, most of the grain is used as fodder for animals.

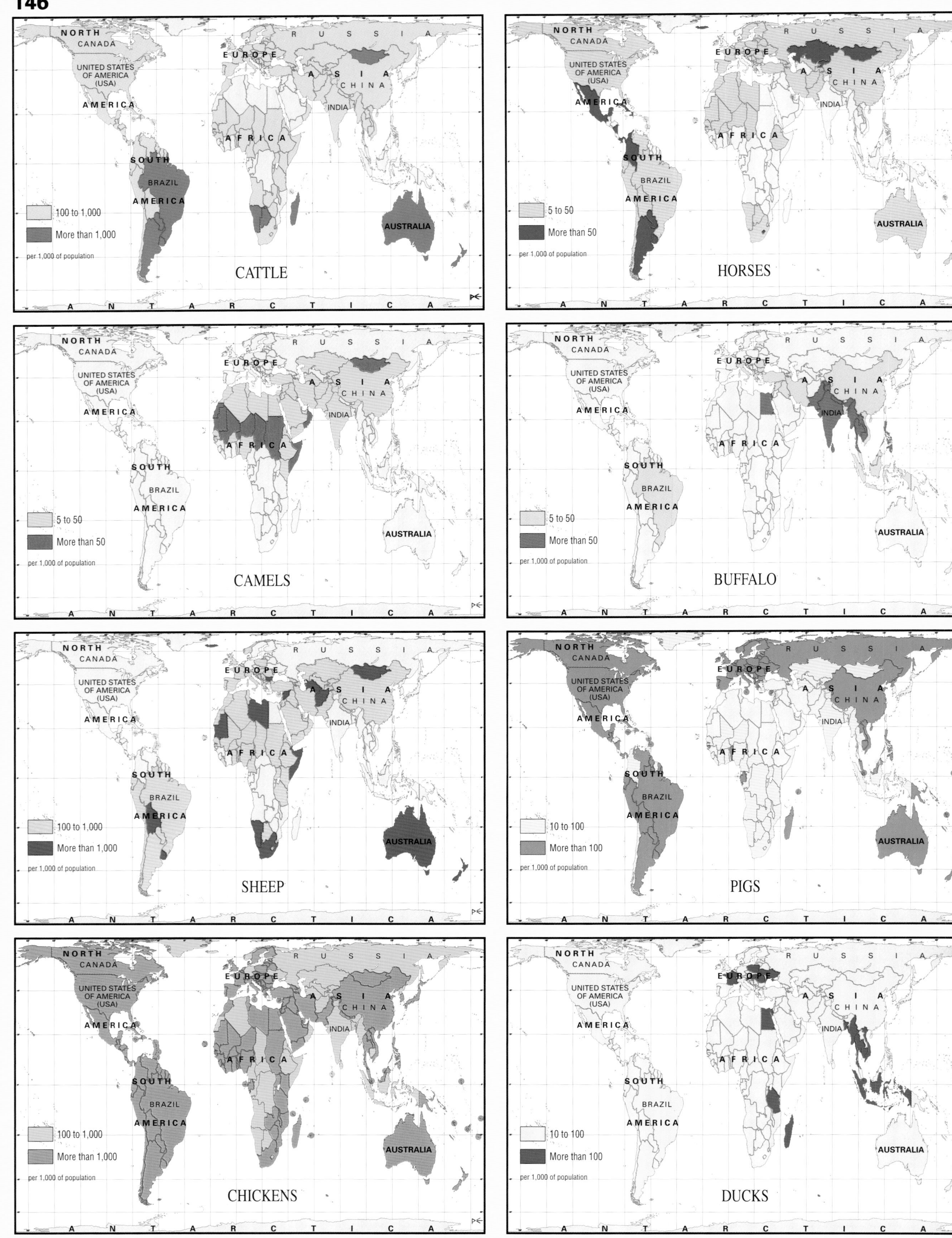

Only a dozen of the 6,000 species of wild mammals and a few birds and insects have been domesticated by humans. With these they developed a close link; they kept them in or near their own home, provided for their needs, and protected them from predators. With the rapid industrialisation of animal husbandry, this relationship has however deteriorated into a merely financial interest.

ANIMAL H

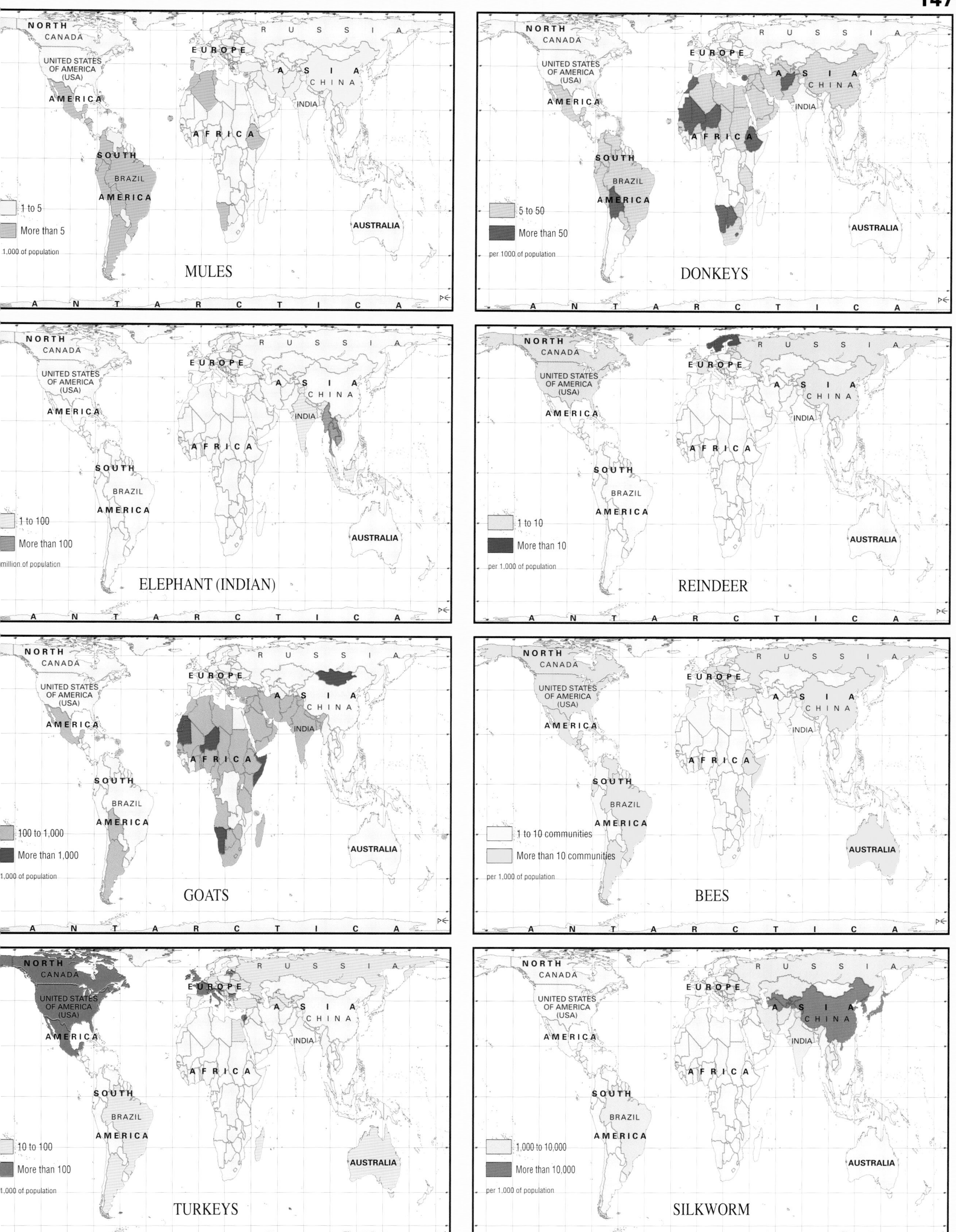

SBANDRY

The successful subjugation of animals almost 10,000 years ago served as a model for human exploitation of other people. Slavery was first limited to people of foreign tribes but 5,000 years ago it was extended to members of the same people as serfdom and wage work. At the same time the gap between rich and poor widened immeasurably. But this polarisation also had its origin in animal husbandry because increased ownership of animals in the hands of a few made a big social difference even 5,000 years ago.

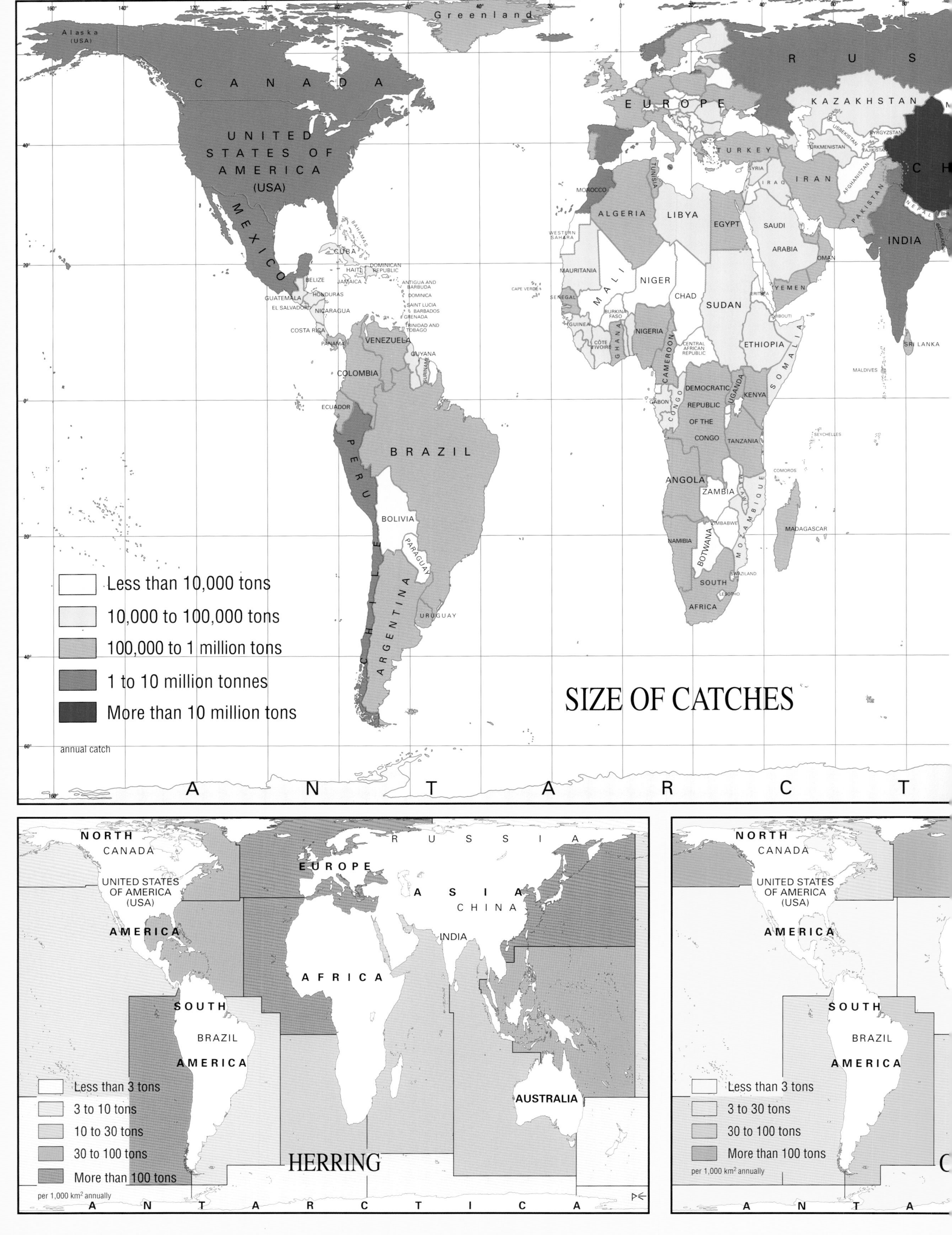

Before people learnt about farming and breeding animals they lived by hunting, gathering and fishing. Fish may originally have been caught by hand but later hunted by spear and finally with nets or hooks and lines. Fishing was limited initially to inland and coastal waters but was extended to the ocean with the introduction of factory ships and trawl nets. The size of the catch trebled in the second half of the 20th century as a result of onboard processing, refrigerated transport and commercialisation. Scientific research, international agreements and increased fish breeding have made for a further growth in production and lower prices so that fishing has become even more important.

FIS

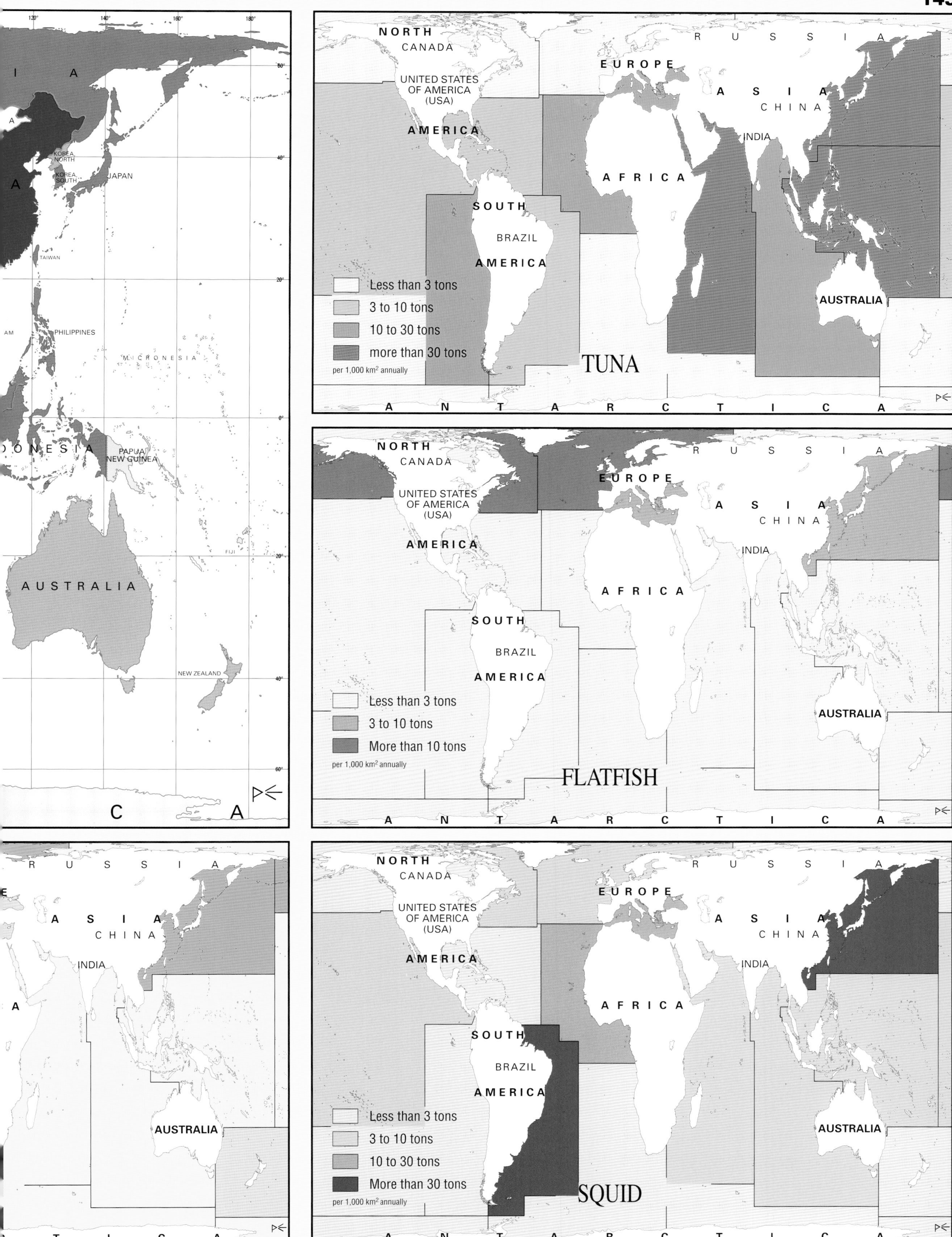

ING

Barely one-third of the global catch of fish is eaten as fresh fish. 40% are frozen, smoked or preserved by other methods. 30% are processed into fish oil, margarine and fish meal or delivered to industry for use in cosmetics, medicaments and soap. A larger number of species of both fresh-water and sea fish are eaten in the hotter countries than in regions nearer the poles, where consumption is increasingly limited to fewer kinds. Almost half the fish caught today are varieties of herring or cod. Fish are cold-blooded and are being eaten more and more in preference to the more expensive flesh of warm-blooded animals by health and money-conscious consumers.

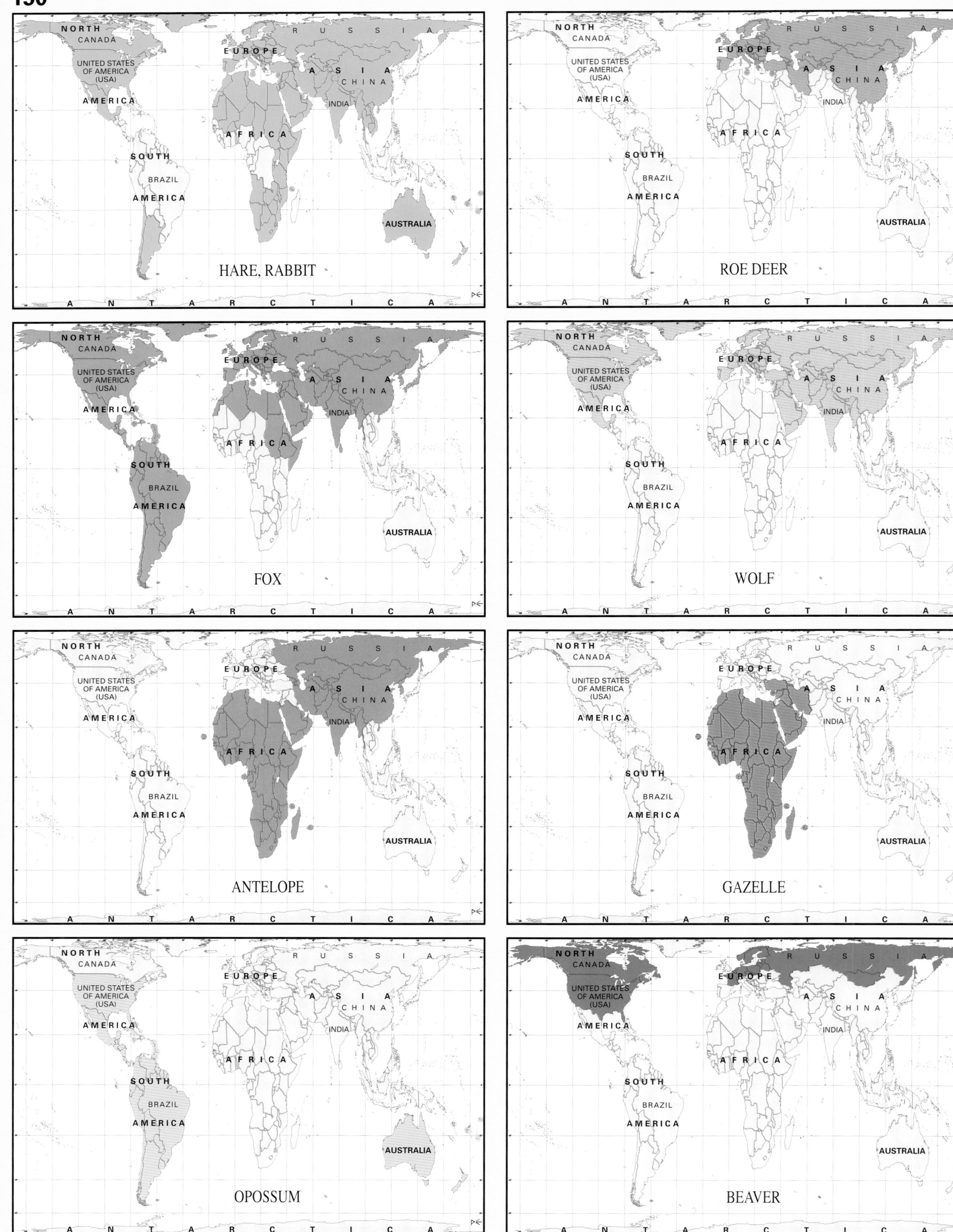

From the very beginning humans killed the animals in their surroundings. Meat from these and from those they kept, together with fish, plants and fruit, made up their diet. They also hunted those predatory animals whose prey they themselves would have become had they not killed them. They provided them too with skins which were used for shelter and clothing. Hunting required courage, strength and skill, and people developed these qualities. They also strengthened their sense of community, for hunting was undertaken as a co-operative activity. For all peoples at all times hunting was the responsibility of the men, whose character had essentially been influenced by it.

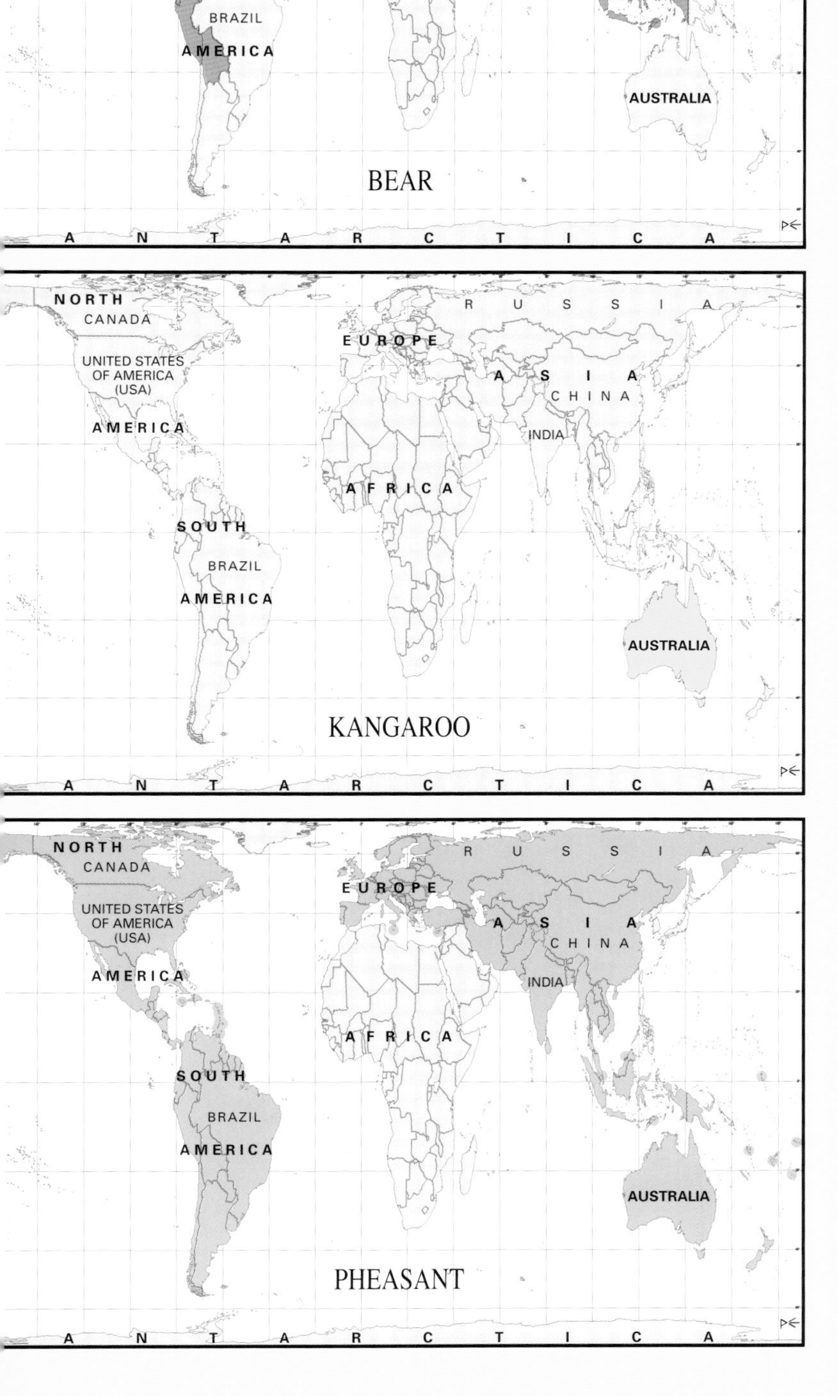

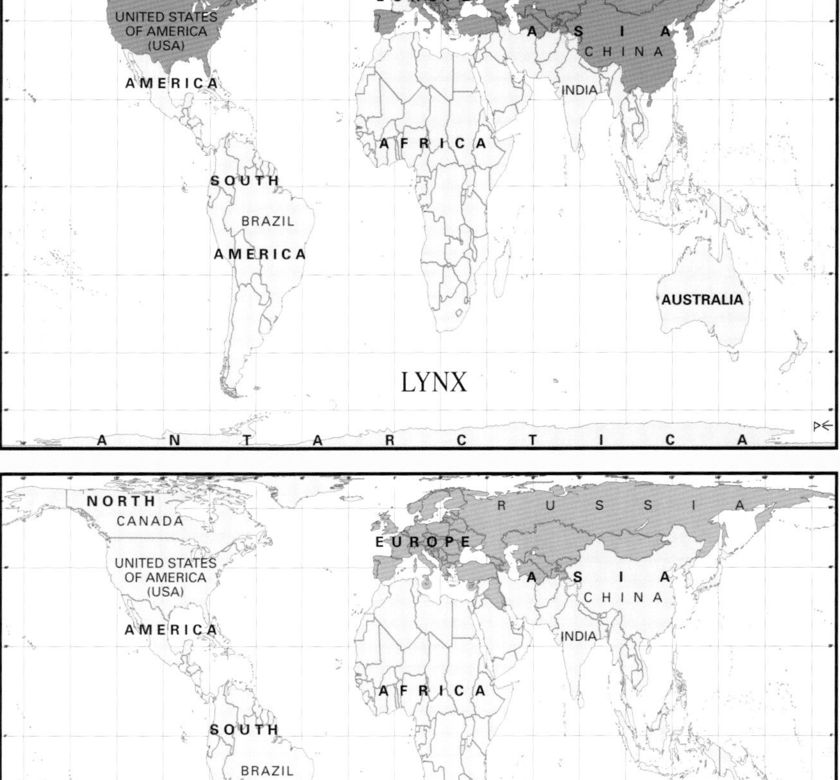

TING

With the emergence of farming and animal husbandry and because of the elimination of predators, hunting lost its original importance in most parts of the world. Food and clothes were available without it. With the invention of firearms hunting lost its danger. Therefore nearly everywhere it became just a sport for the wealthy. Parallel to this was the increasingly commercialised slaughter of wild animals whose skins or horns commanded high prices. In this way the equilibrium between the last hunting peoples and their natural environment has been destroyed.

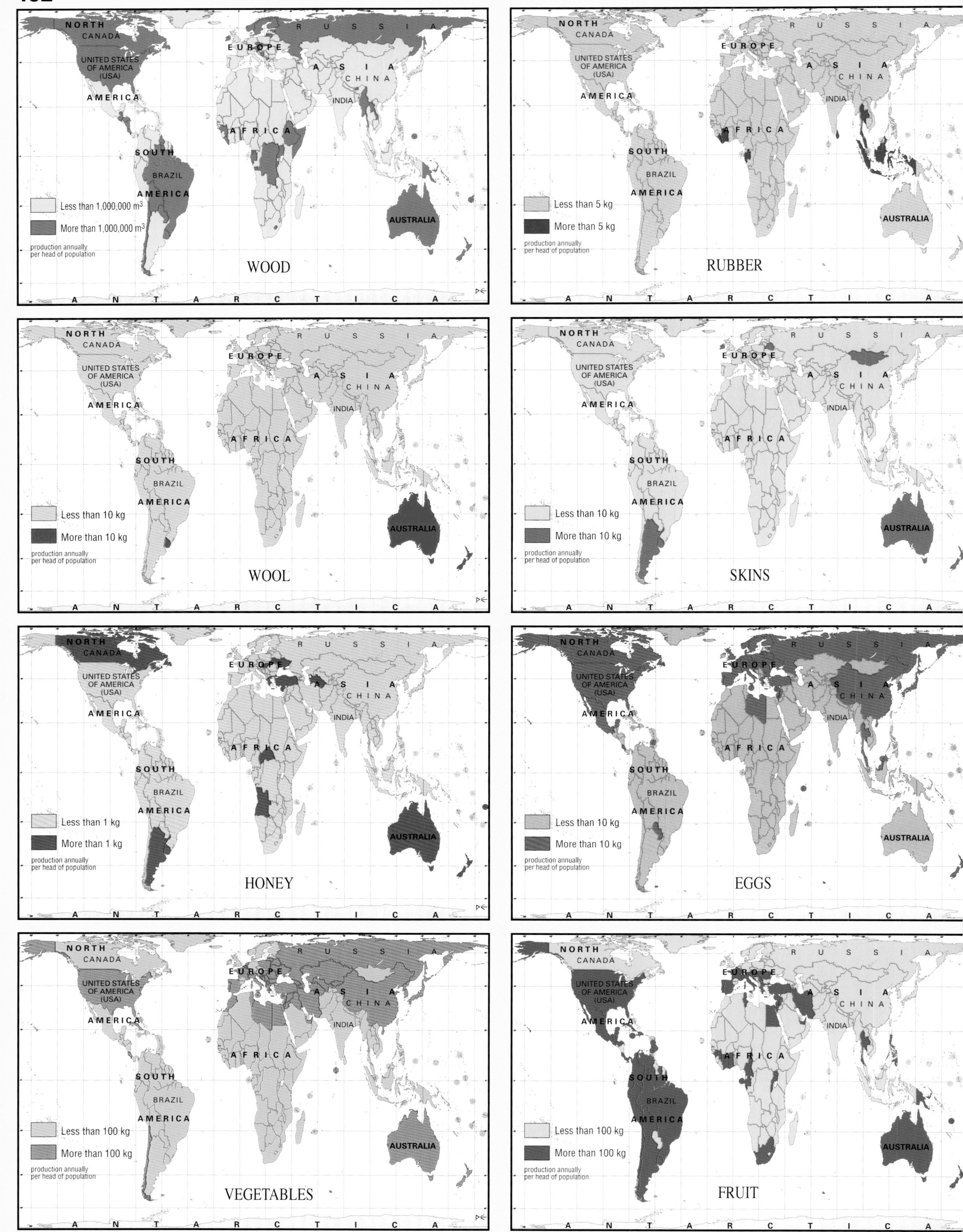

Like all other creatures, humans were for millions of years quite content with the plants and fruits provided by nature and the meat of the animals they kept or hunted. A few thousand years ago we began to improve upon nature's work, in order to satisfy our needs for food, clothing or shelter. Natural products are the essential basis for life and are therefore produced almost everywhere in the world. In addition they are also being produced in ever greater quantities to serve as raw materials for industry to process further.

NATURAL

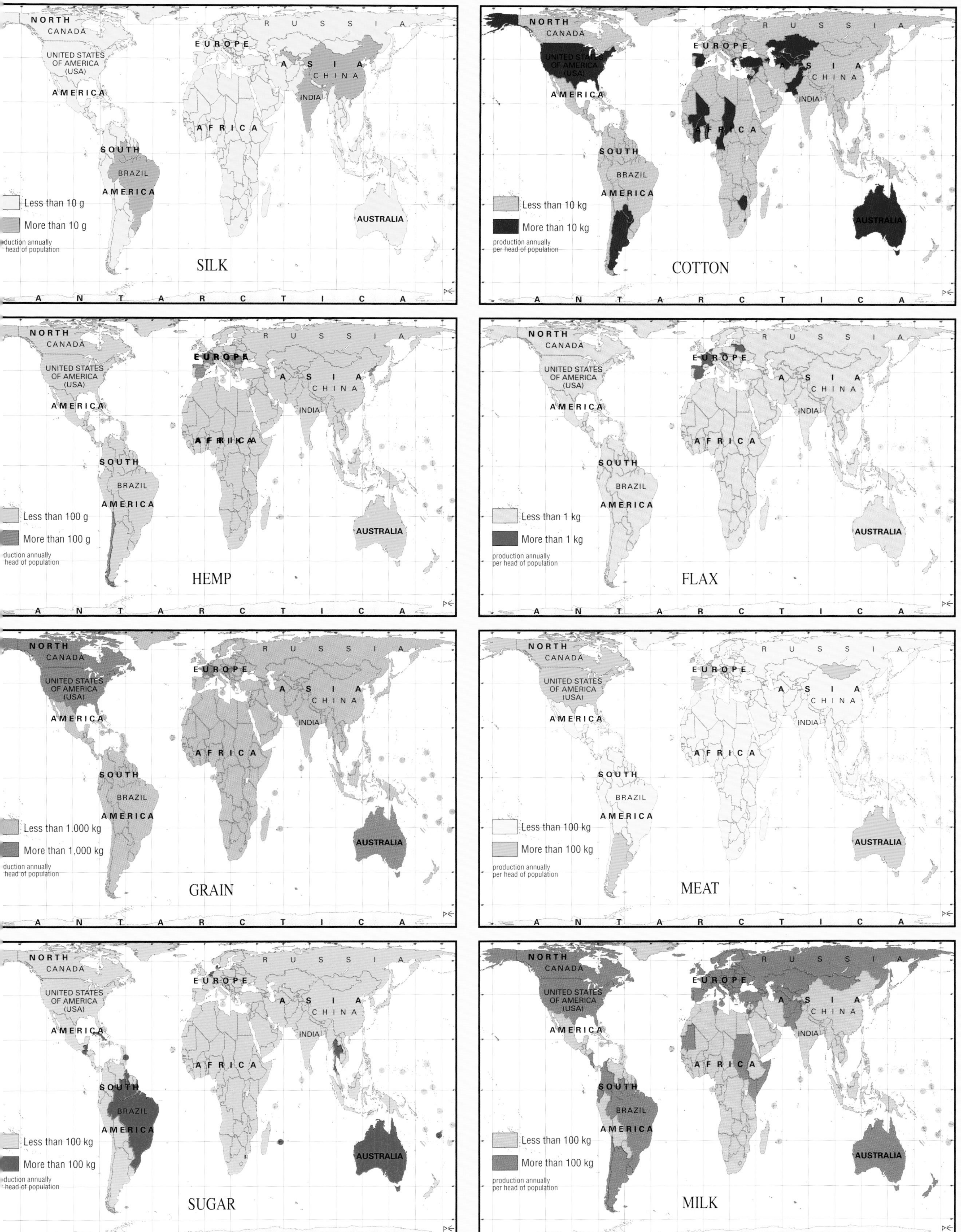

RODUCTS

Food, furs and wood are produced nearly everywhere in the world, and increasingly wool and cotton too. Rarer goods like silk, rubber, flax and hemp are produced in only a dozen countries. So much silk is produced in a single country, China, that it meets more than half the world's demand. Since natural products are available nearly everywhere in abundance their world market prices are low, particularly now that the rich industrial states, with their superior technology, biology and chemistry, dominate their production.

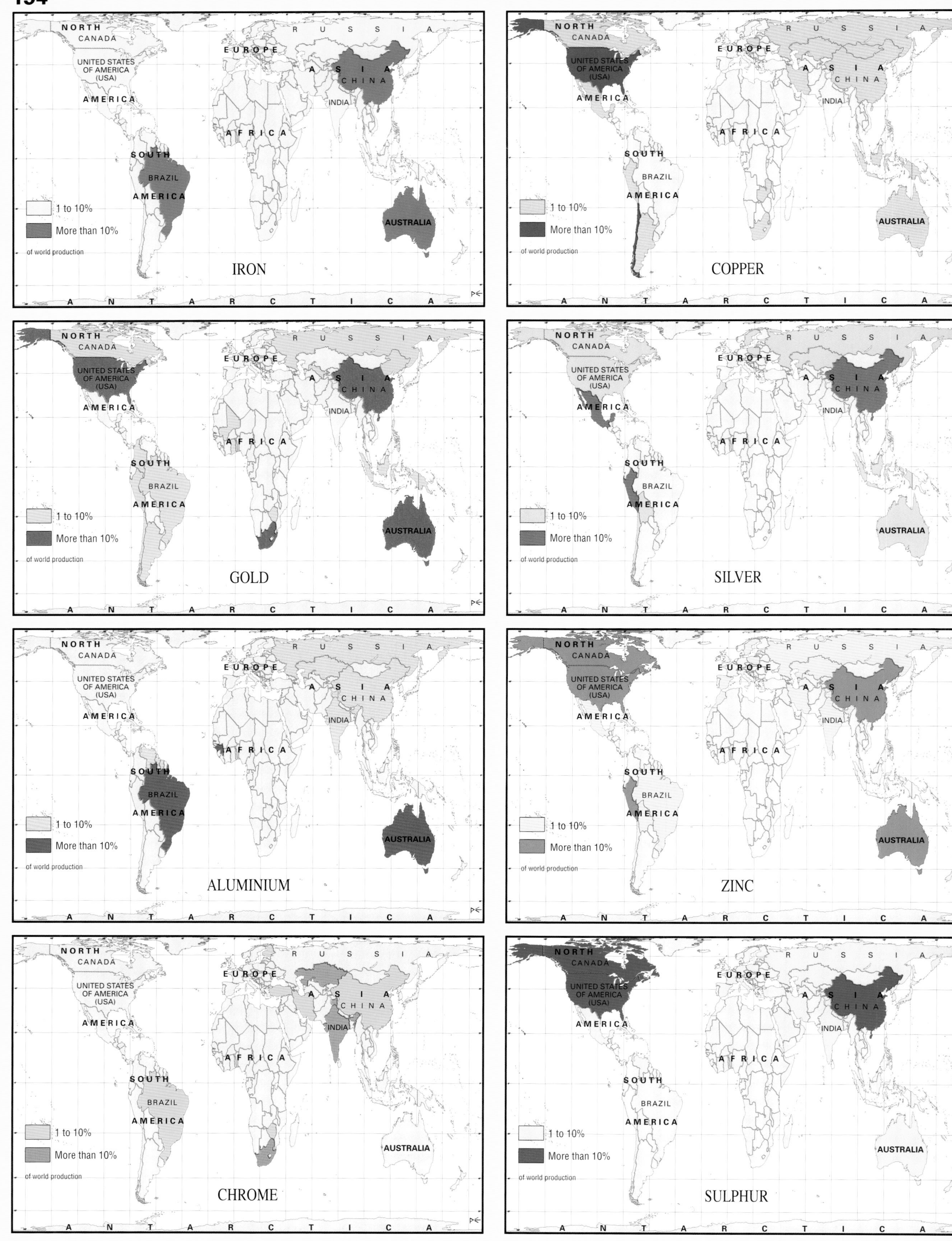

Of the inorganic components found everywhere in the earth's crust, minerals have been used since time immemorial for the manufacture of tools (copper, iron) and the possession of gold, silver and diamonds has always been regarded as a sign of wealth. Throughout the course of history the desire to possess the source of these raw materials has often been the reason for invading and conquering another country. Even today each country is concerned to secure its own access to the minerals it needs for its economy and the production of weapons.

MINERAL

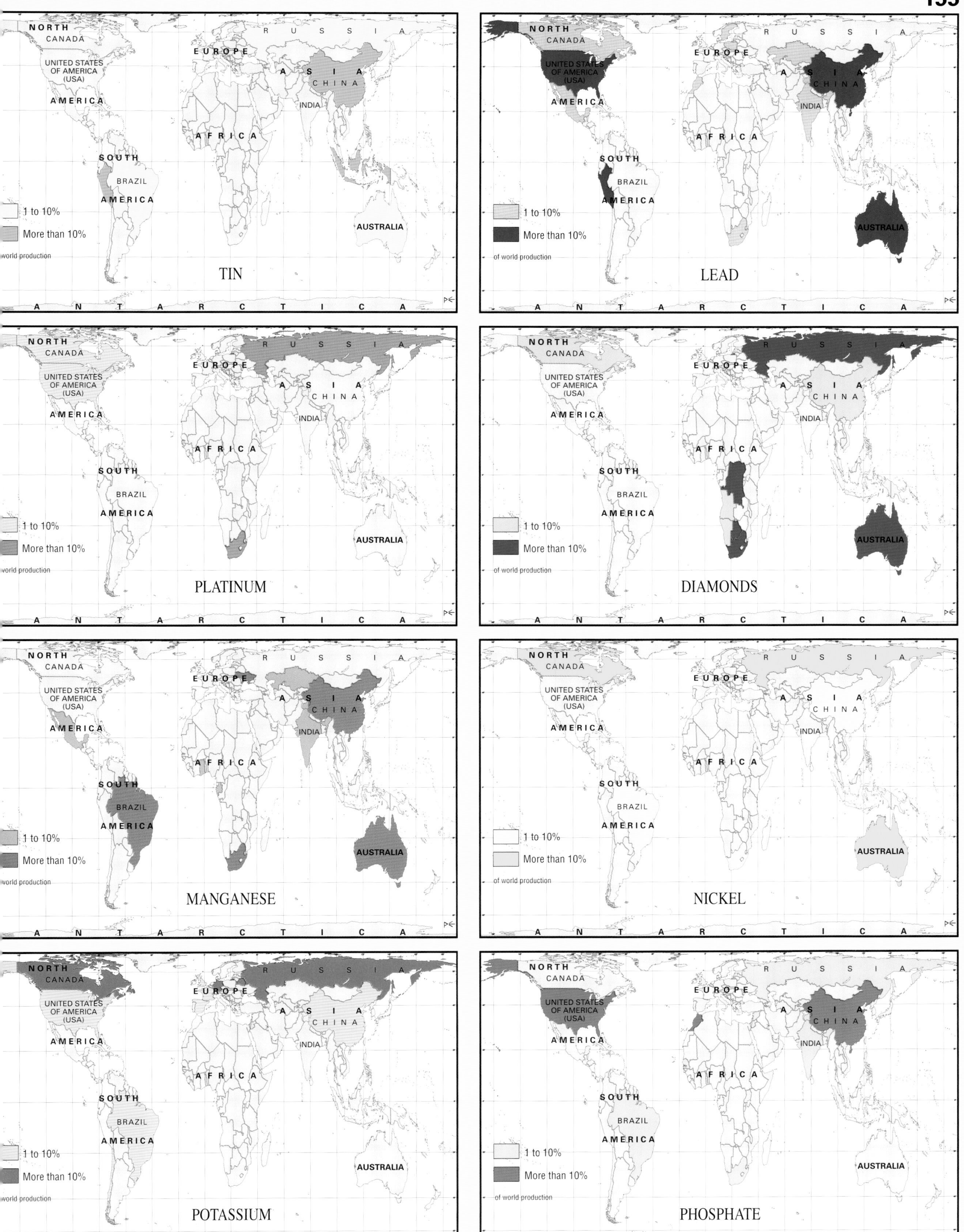

ESOURCES

The countries of the North seek to decrease their dependence on the mineral resources of the poor countries of the South for economic and strategic reasons. They buy or occupy large areas of foreign states and stockpile large amounts of minerals that they do not have themselves. Increasingly, the rich countries are looking to recover minerals from scrap or to replace with synthetic products. Despite this, most industrial nations are still heavily dependent on the South's mineral resources.

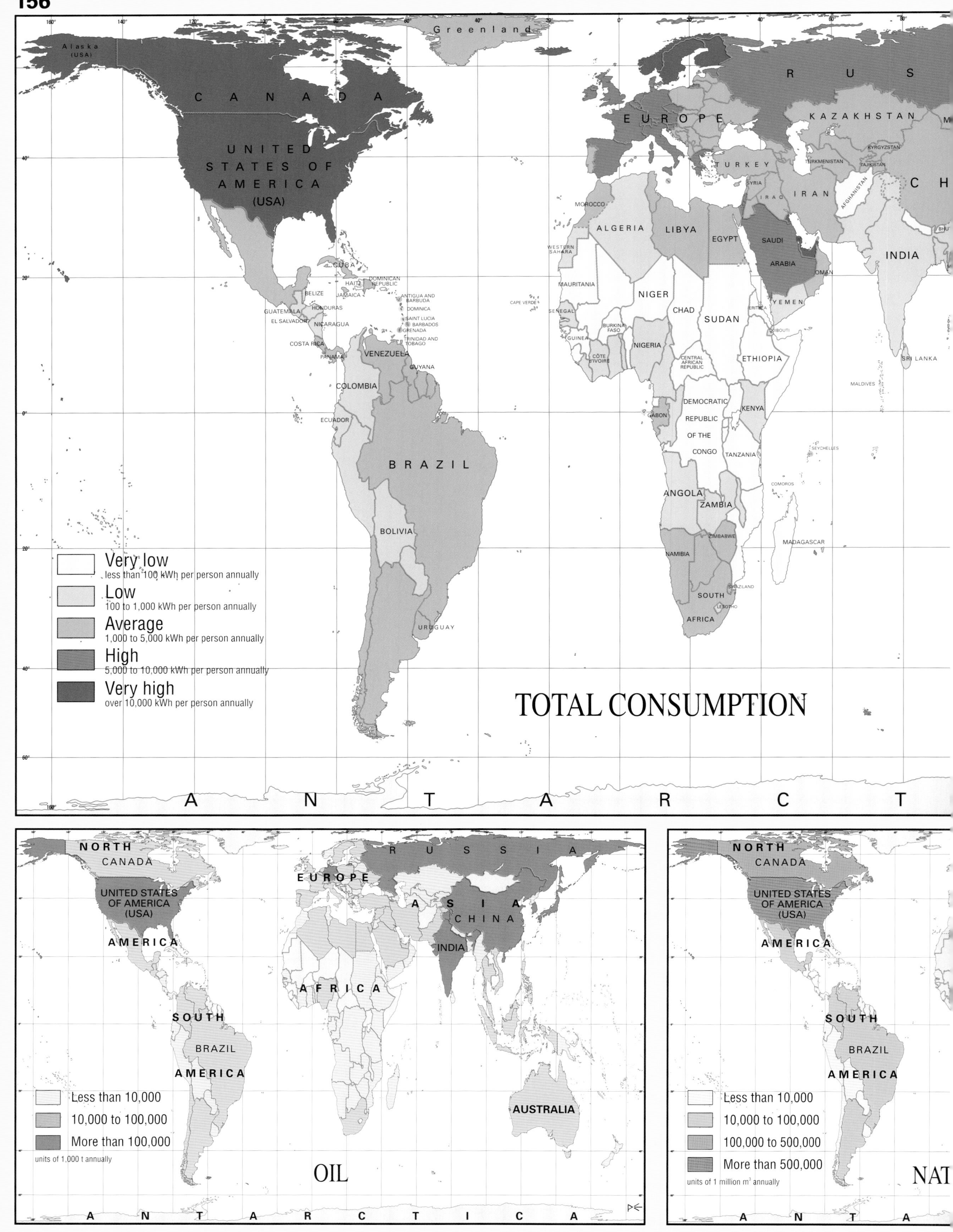

In contrast to the rest of the natural world, humans use more energy than they actually need for living. This surplus defines the level of a person's living standard. World history is thus an expression of people's striving for more energy. After domesticating animals men increased the energy available to them by using people (salves, women, children, workers, employees). In our epoch humans can use the energy contained in matter to meet their needs without exploiting animals or people.

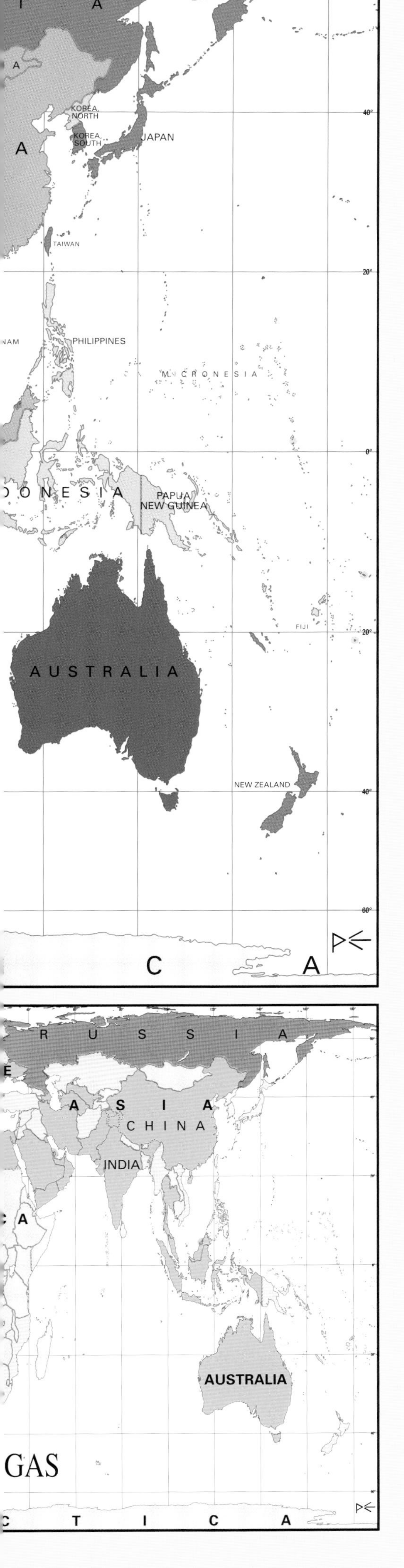

NORTH CANADA
UNITED STATES OF AMERICA (USA)
AMERICA
EUROPE
RUSSIA
ASIA
CHINA
INDIA
AFRICA
SOUTH AMERICA
BRAZIL
AUSTRALIA
ANTARCTICA

Less than 10,000
10,000 to 100,000
More than 100,000
units of 1 million kWh annually

WATER POWER

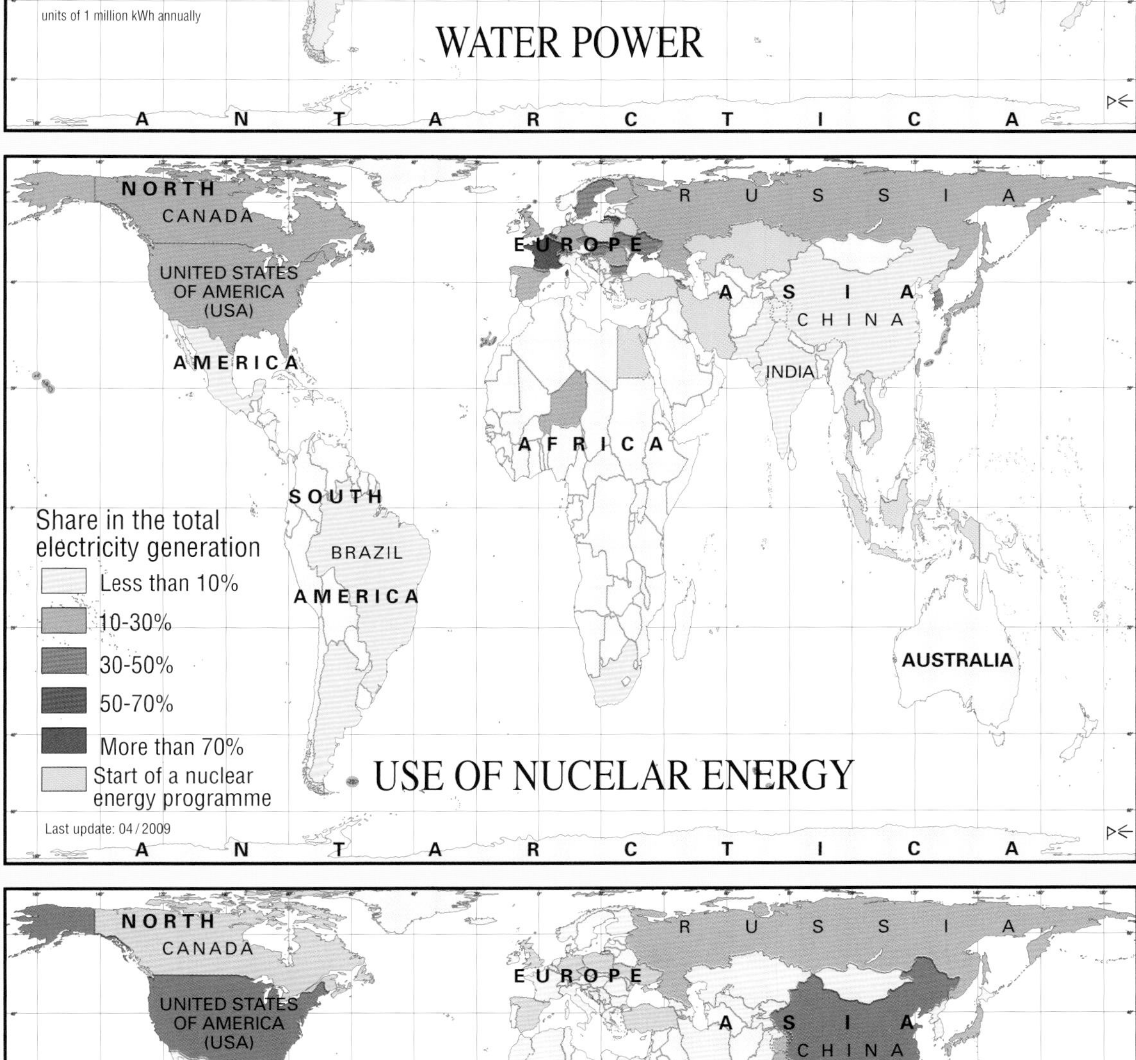

NORTH CANADA
UNITED STATES OF AMERICA (USA)
AMERICA
EUROPE
RUSSIA
ASIA
CHINA
INDIA
AFRICA
SOUTH AMERICA
BRAZIL
AUSTRALIA
ANTARCTICA

Less than 10,000
10,000 to 100,000
100,000 to 500,000
More than 500,000
units of 1,000 t annually

COAL

RGY

All our energy sources except water can be used only once. Water renews itself endlessly but covers only 3% of our needs; oil meets 43% of our energy needs, coal 32%, natural gas 18%, nuclear power 2%. Our increasing demand will exhaust natural energy reserves within a few decades, apart from hydro power, whose use could be increased six-fold. But in solar power we have inexhaustible reserves at our disposal. Eight minutes of sunshine could supply the entire energy needs of the world for one year.

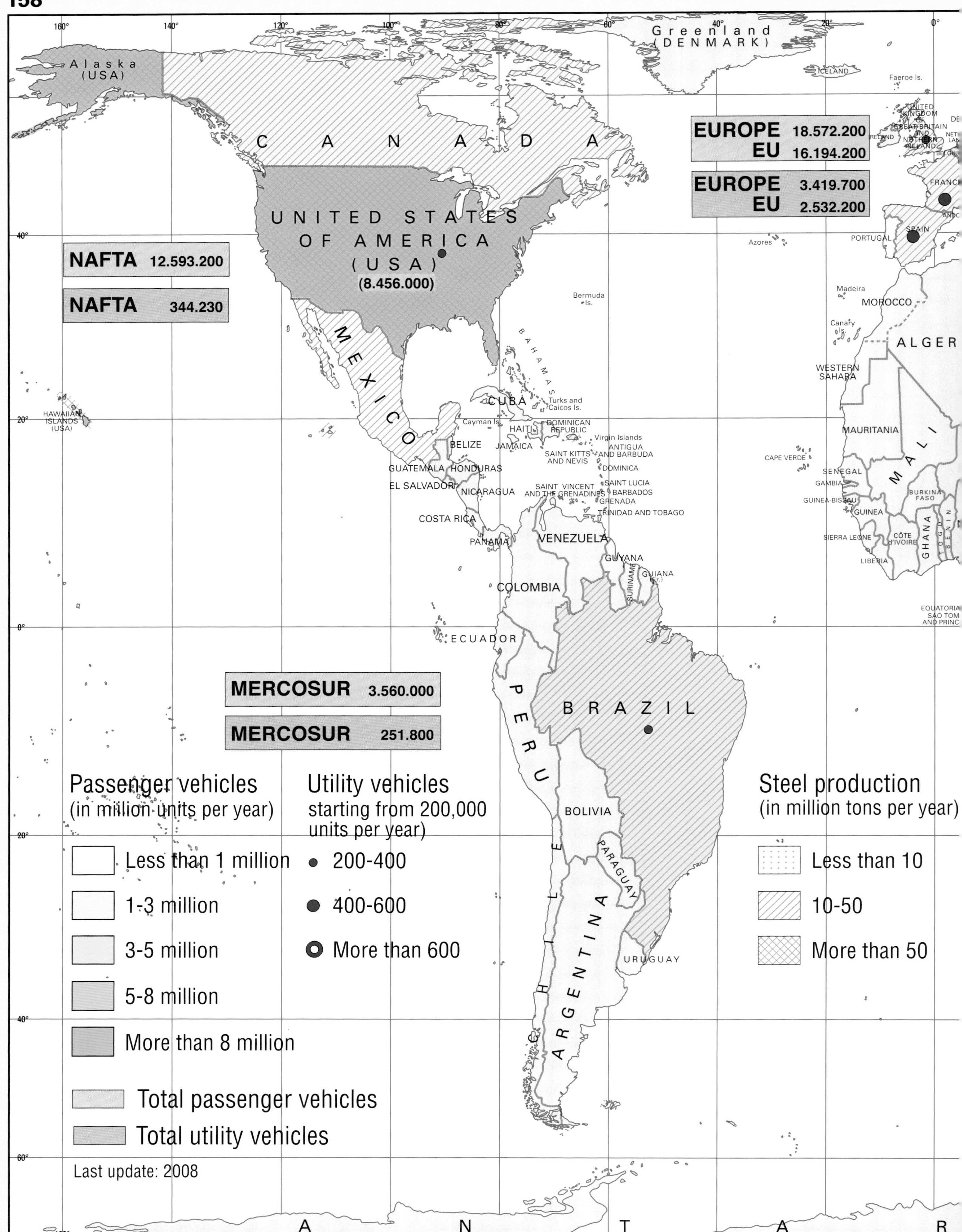

One of the most important products in industrial nations is steel. During the last decades, the shares of some states in the worldwide steel production have varied considerably. Whereas the quantities in the traditional steel locations like Great Britain and Luxembourg have diminished, emerging countries like Mexico and South Korea have notched up strong increases. It is China, however, which has recorded the biggest increases. Its share constituted only 3% or so in 1970 and increased to 26% in 2004, at an average annual rate of almost 12%. Automobile production is also mainly located in the big industrialised nations, but for reasons of personnel costs, the big car makers outsource production to some emerging countries.

STEEL AND AUTOM

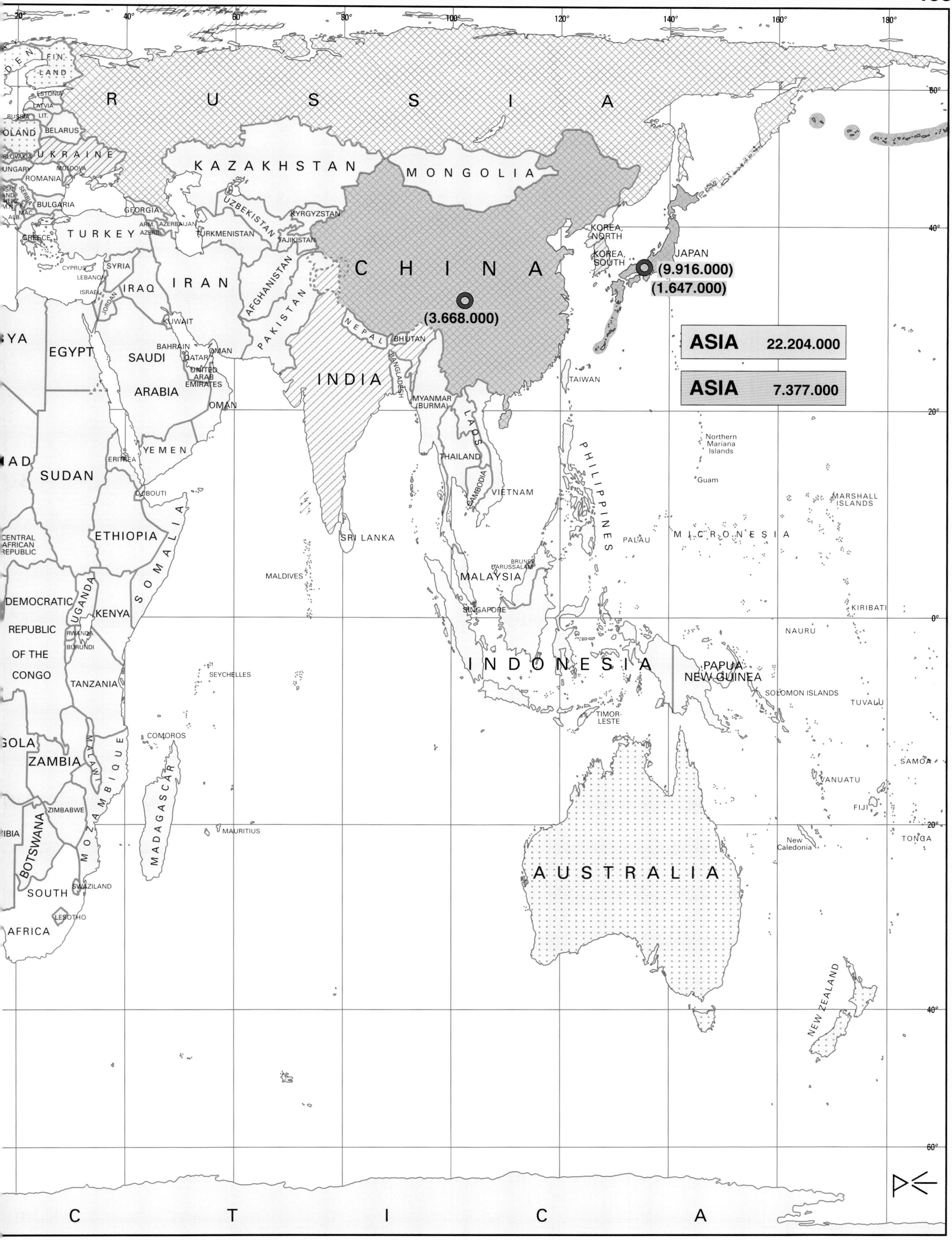

ASIA 22.204.000

ASIA 7.377.000

JAPAN (9.916.000) (1.647.000)

CHINA (3.668.000)

BILE PRODUCTION

If all trades dealing with the processing of natural products rank among the branches of industry, handcraft is also part of them. It holds its own in large parts of the world alongside real industry based on the division of labour through the use of machines and the fact that, to a large extent, it manufactures the same goods. More than a dozen countries have reached the highest performance level in almost all areas of production: the industrial nations. They reach a total industrial output of more than $5,000 per inhabitant and per year. Less than half manage over $10,000, and only the Federal Republic of Germany achieves more than $14,000. Three quarters of all the countries in the world attain an industrial annual production amounting to less than $1,000 per inhabitant, and almost half of them not even $100.

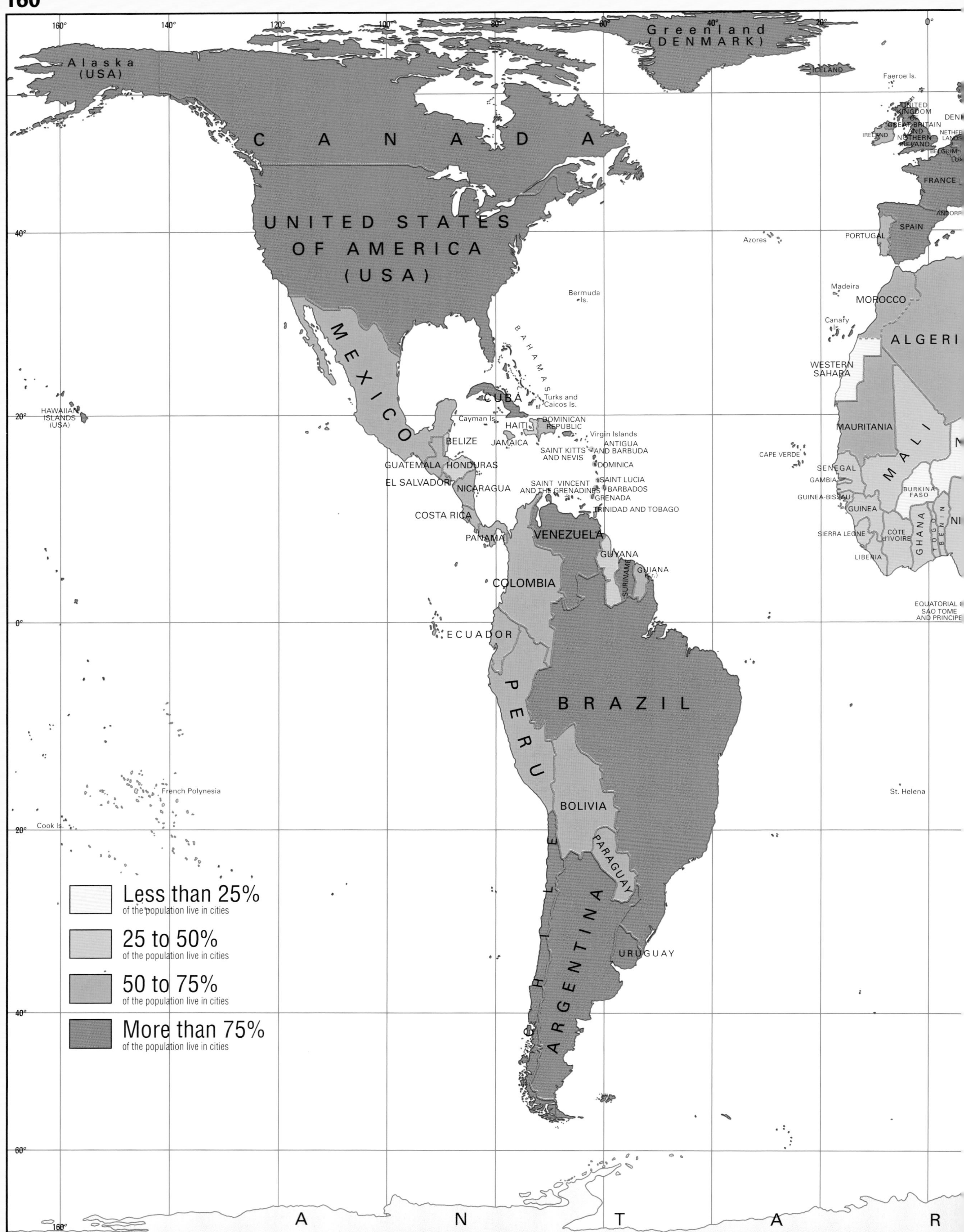

Around 6,000 years ago the first cities appeared in Asia and Africa. About 3,000 years ago cities grew up in Europe. Initially they were only ruling foci that had to be supported by the surrounding rural populations, but they soon became centres of human culture. Arts, sciences and education all developed in cities. The new refined living conditions led to urban ways of life.

URBAN

SATION

Over 100 years ago industry enabled cities to develop into centres of production. This led to a rapid increase in the number of their inhabitants. Nowadays more than three-quarters of the population of industrial countries live in cities (Britain 90%) but the same is also true of some developing countries (Venezuela 87%). In more than 50 countries of the world less than a quarter of the population lives in cities (Burundi 9%).

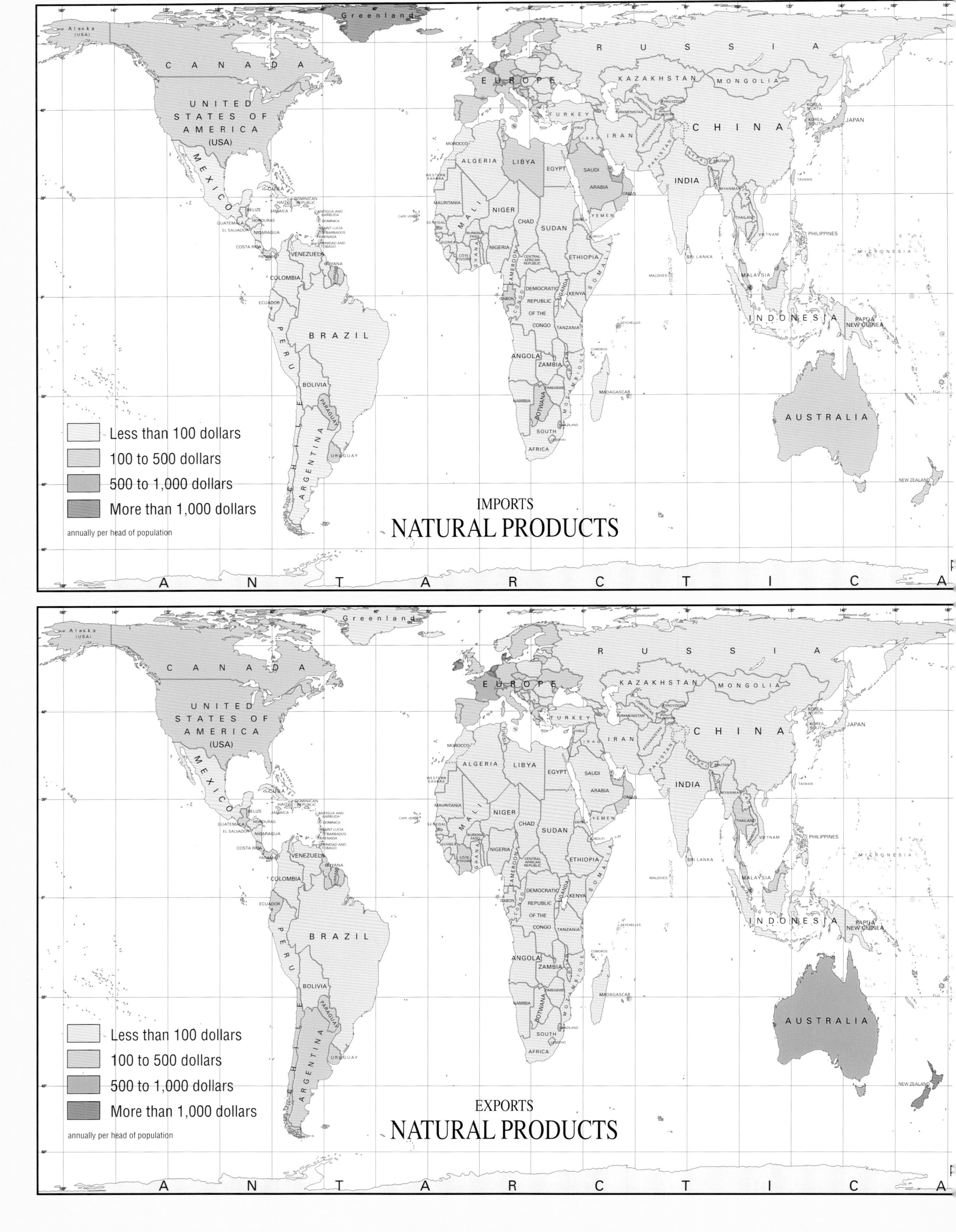

For 500 years the Europeans took the labour and goods they wanted from the rest of the world by force. Slowly, cunning replaced force and trade became the instrument of exploitation. It was an unequal exchange. The Europeans set the prices for which goods would be traded. This method of accumulating wealth was developed to such perfection that the Europeans could end their colonial domination in the second half of the twentieth century without losing is benefits.

WORLD

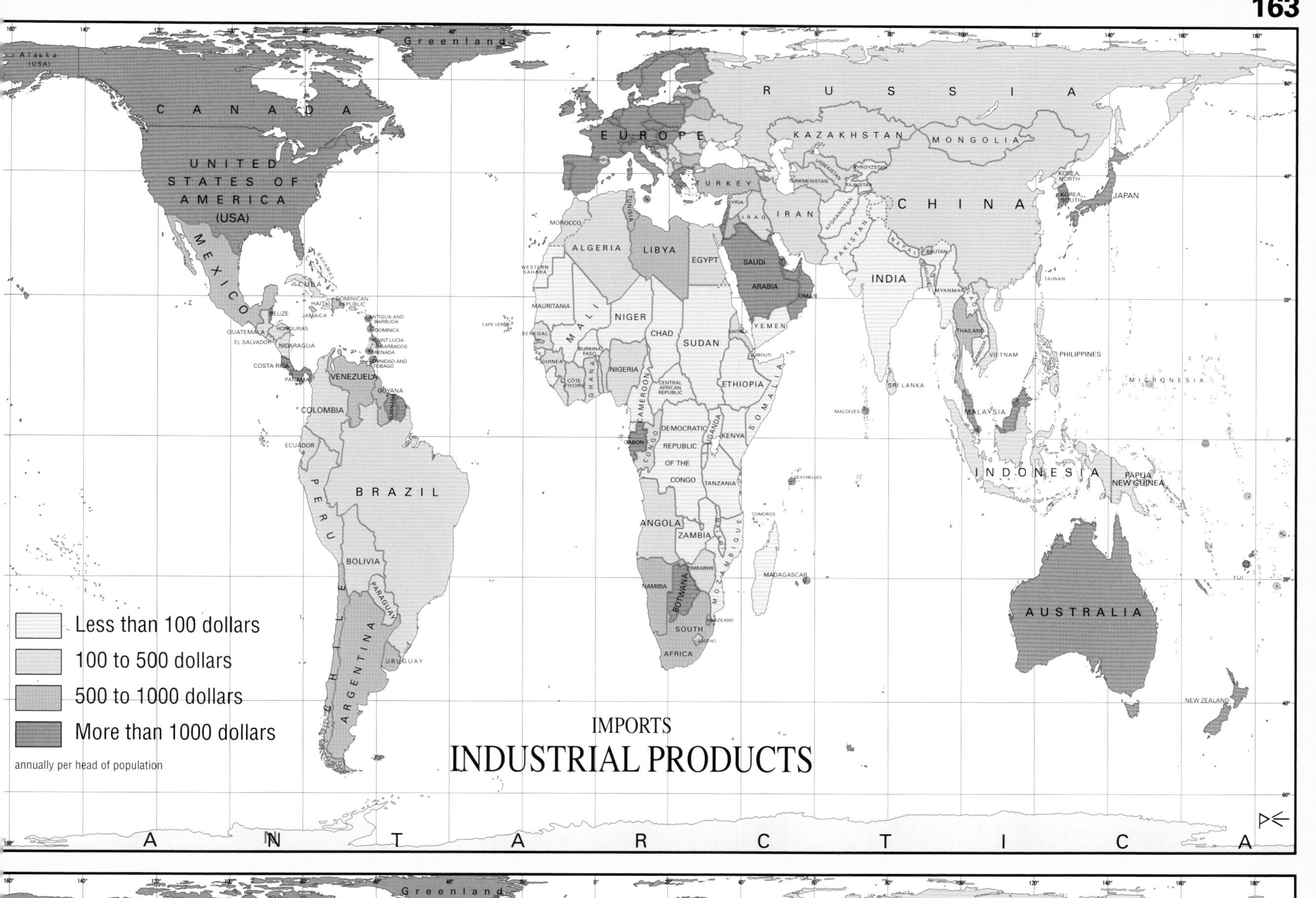

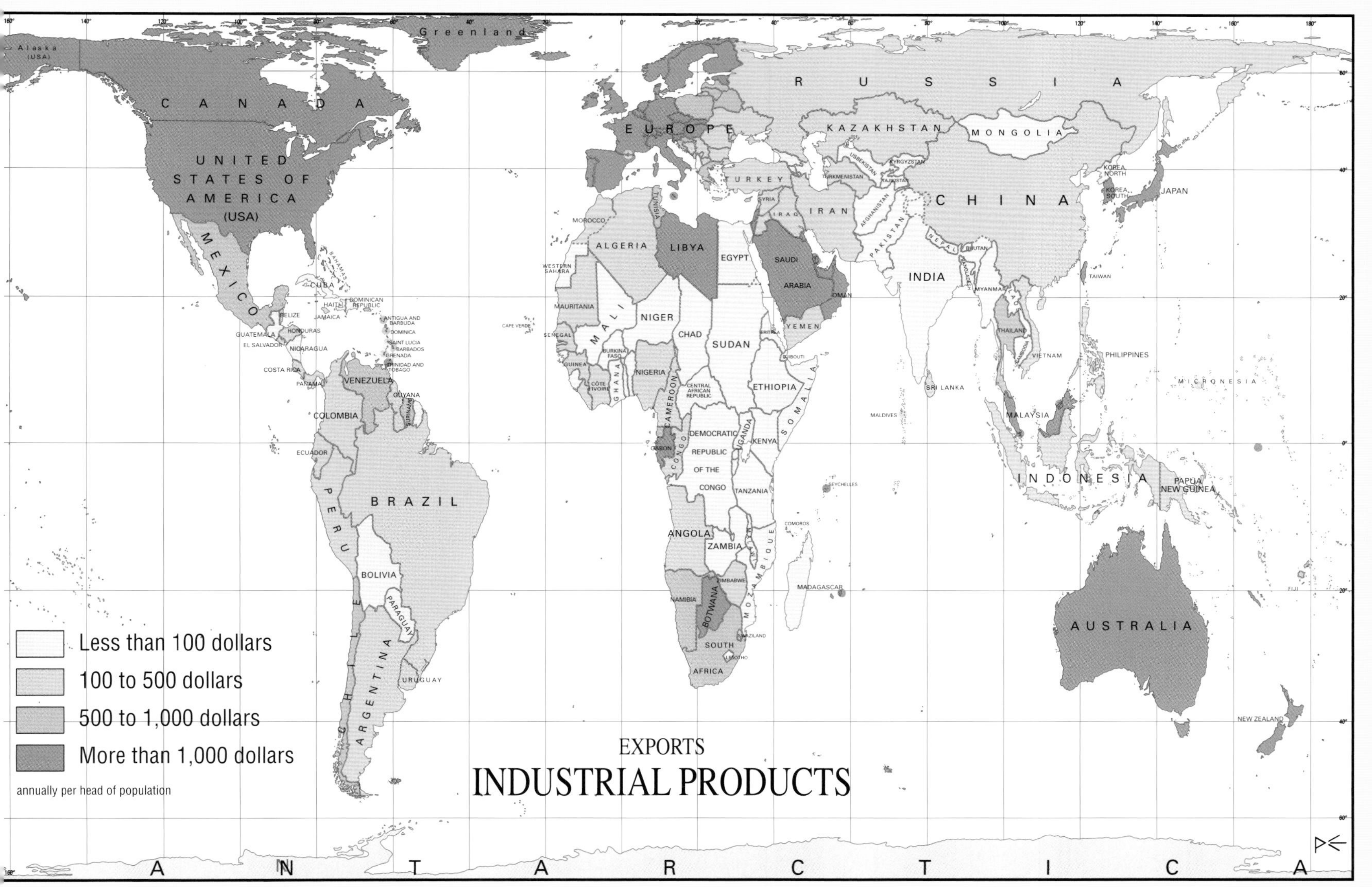

TRADE

Since the end of European colonial domination in the 1970s, prices for raw materials, which had already been far too low, fell by a third, whereas prices for industrial products increased fivefold. Within three decades this manipulation of world market prices caused the purchasing power of the developing countries to sink to an all-time low. Developing countries are to a large extent excluded from the worldwide exchange of goods because they have to hand over their products at a fraction of their real value.

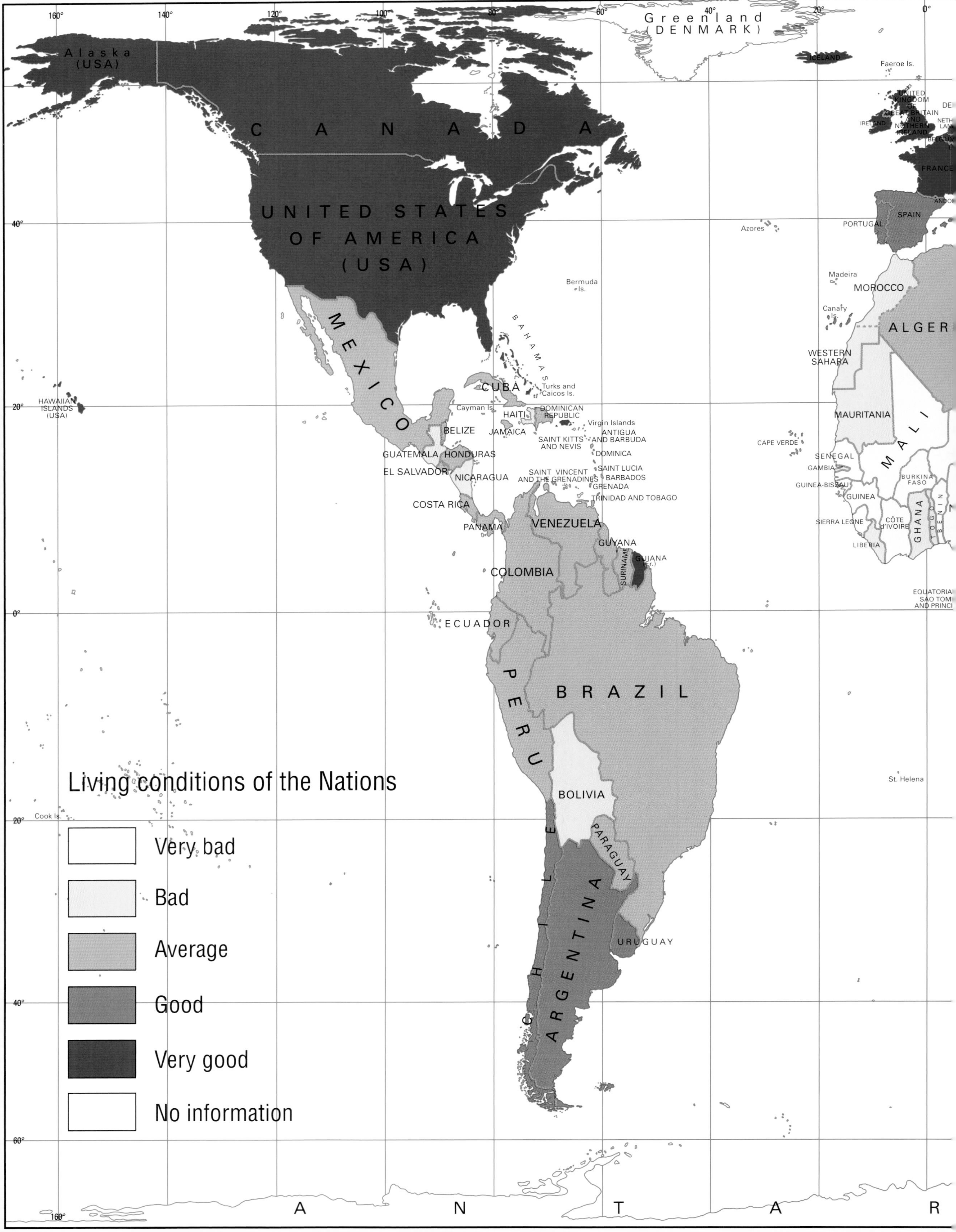

Living conditions of the Nations

Very bad

Bad

Average

Good

Very good

No information

THE LIVING CONDI
(HUMAN DEV

The Human Development Index, abbreviated HDI, has been published since 1990 in the annual "Human Development Report" of the UNDP (United Nations Development Programme). The HDI was developed by Pakistani economist Mahbub ul Hag, Indian Nobel laureate Amartya Sen as well as British economist and politician Mghnad Desai. It takes into account the gross national income of a country per inhabitant in purchasing power parity,

NS OF THE NATIONS
PMENT INDEX)

life expectancy and the level of education by means of the population's literacy rate and percentage of children enrolled in grade school. The life expectancy factor is considered as an indicator of health care, nutrition, and hygiene. The level of education, as well as income, represents acquired knowledge and participation in public and political life and an adequate standard of living. Since the year 2000, the countries Canada, Norway, and Iceland have alternatingly held first place in the HDI ranking.

Average income:

- Less than 500 dollars per person annually
- 500 to 1,000 dollars per person annually
- 1,000 to 5,000 dollars per person annually
- 5,000 to 10,000 dollars per person annually
- More than 10,000 dollars per person annually

24% of the world's population consumes 83% of the world's income. That leaves just 17% of world income for the remaining 76%. Therefore each inhabitant of the rich countries has on average 16 times more of world income.

POOR NATIONS

RICH NATIONS

In the poorest countries the average income is $100-300 a year (Burundi $120, Ethiopia $100, Sierra Leone $130). In the richest nations the annual average income is more than one hundred times greater (Norway $33,000, Switzerland $38,000, Japan $32,000).

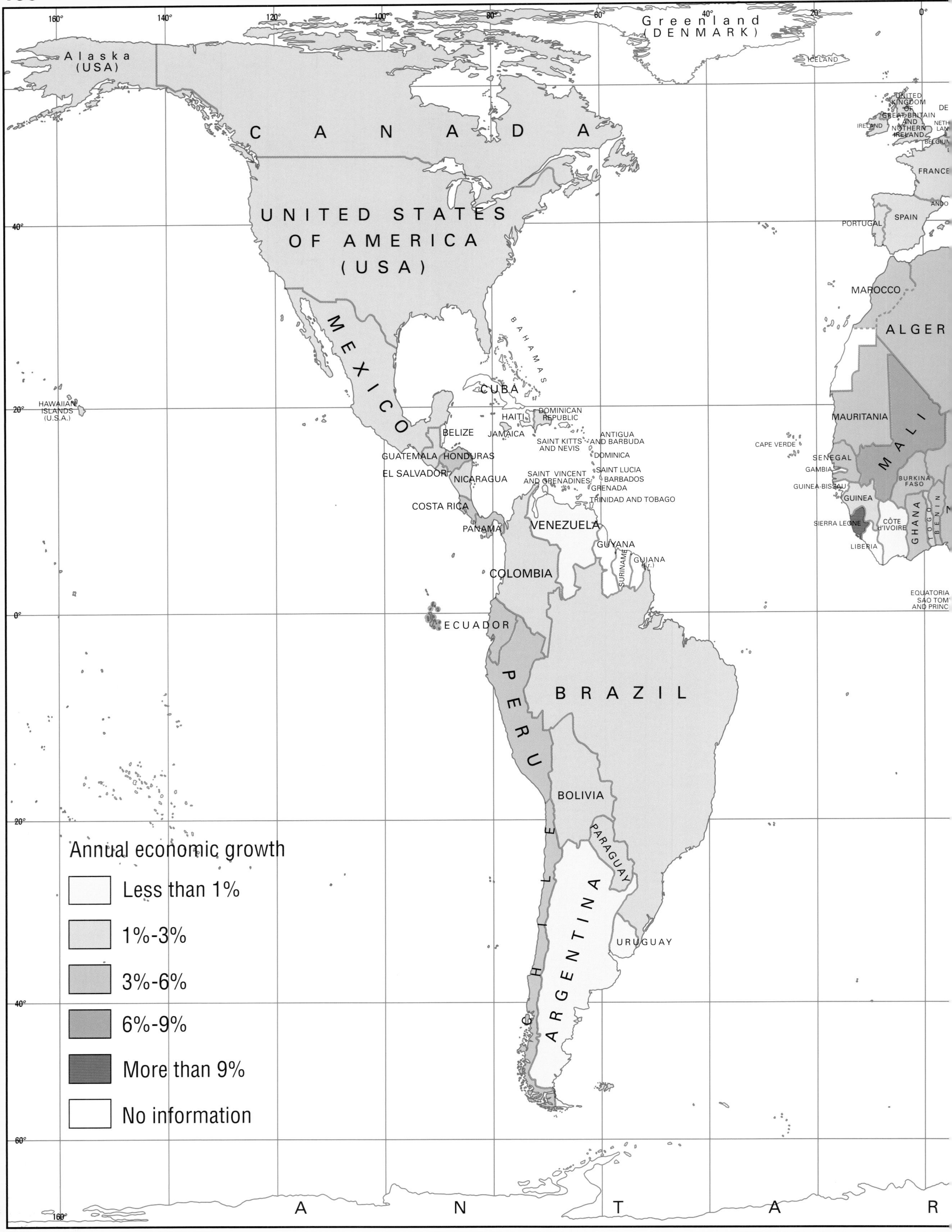

Alaska (USA)
Greenland (DENMARK)
ICELAND
CANADA
UNITED STATES OF AMERICA (USA)
MEXICO
BAHAMAS
CUBA
HAITI
DOMINICAN REPUBLIC
JAMAICA
BELIZE
GUATEMALA
HONDURAS
EL SALVADOR
NICARAGUA
COSTA RICA
PANAMA
SAINT KITTS AND NEVIS
ANTIGUA AND BARBUDA
DOMINICA
SAINT LUCIA
SAINT VINCENT AND GRENADINES
BARBADOS
GRENADA
TRINIDAD AND TOBAGO
VENEZUELA
GUYANA
SURINAME
GUIANA (Fr.)
COLOMBIA
ECUADOR
PERU
BRAZIL
BOLIVIA
PARAGUAY
CHILE
ARGENTINA
URUGUAY
HAWAIIAN ISLANDS (U.S.A.)
IRELAND
UNITED KINGDOM OF GREAT BRITAIN AND NORTHERN IRELAND
FRANCE
PORTUGAL
SPAIN
MAROCCO
ALGER
MAURITANIA
MALI
CAPE VERDE
SENEGAL
GAMBIA
GUINEA-BISSAU
GUINEA
SIERRA LEONE
LIBERIA
CÔTE d'IVOIRE
BURKINA FASO
GHANA
TOGO
BENIN
EQUATORIA
SAO TOM
AND PRINC
ANTAR

Annual economic growth

- Less than 1%
- 1%-3%
- 3%-6%
- 6%-9%
- More than 9%
- No information

In the age of globalisation the economic growth of a country does not depend just on its own economic achievements. What is decisive today is the terms on which it exchanges its own products with the rest of the world. This relationship is defined by the rich industrial nations through world market prices.

ECONOMIC

GROWTH

In the last ten years economic growth in Western countries has slowed: France 1.5%, Germany 1.7%, Britain 2.4%, Italy 1.2%, Switzerland 0.4%. Socialist countries have stabilised their economic growth: China 6.3%, Vietnam 8.4%, Laos 6.6%.

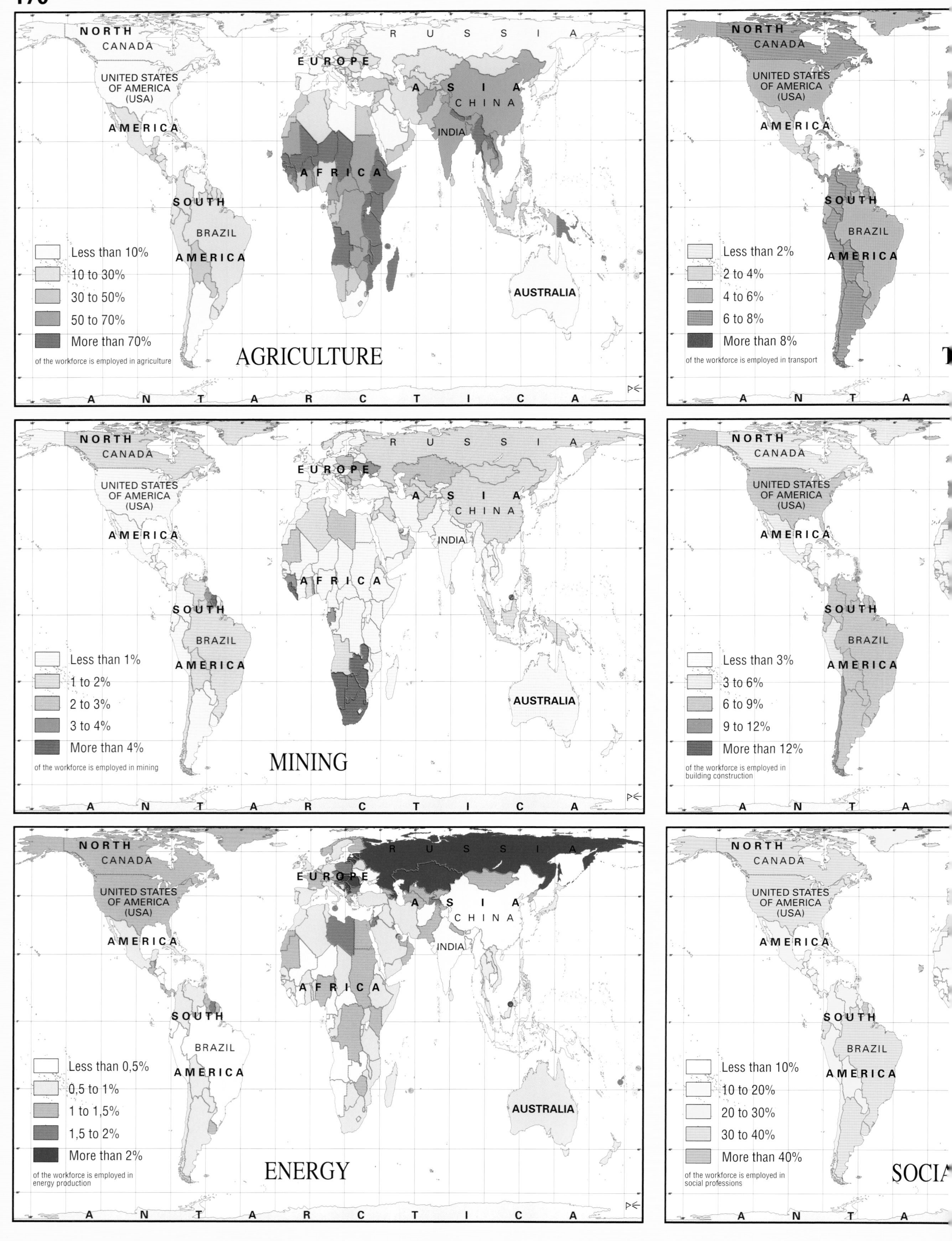

The distribution of employment among the various sectors of an economy does not say anything about each sector's relative strength or productivity. The US for example, where only 3% of its workers are in agriculture today, is the greatest wheat supplier in the world, whereas Tanzania cannot even feed its own population, although 82% of its people work in agriculture.

EMPLOYMEN

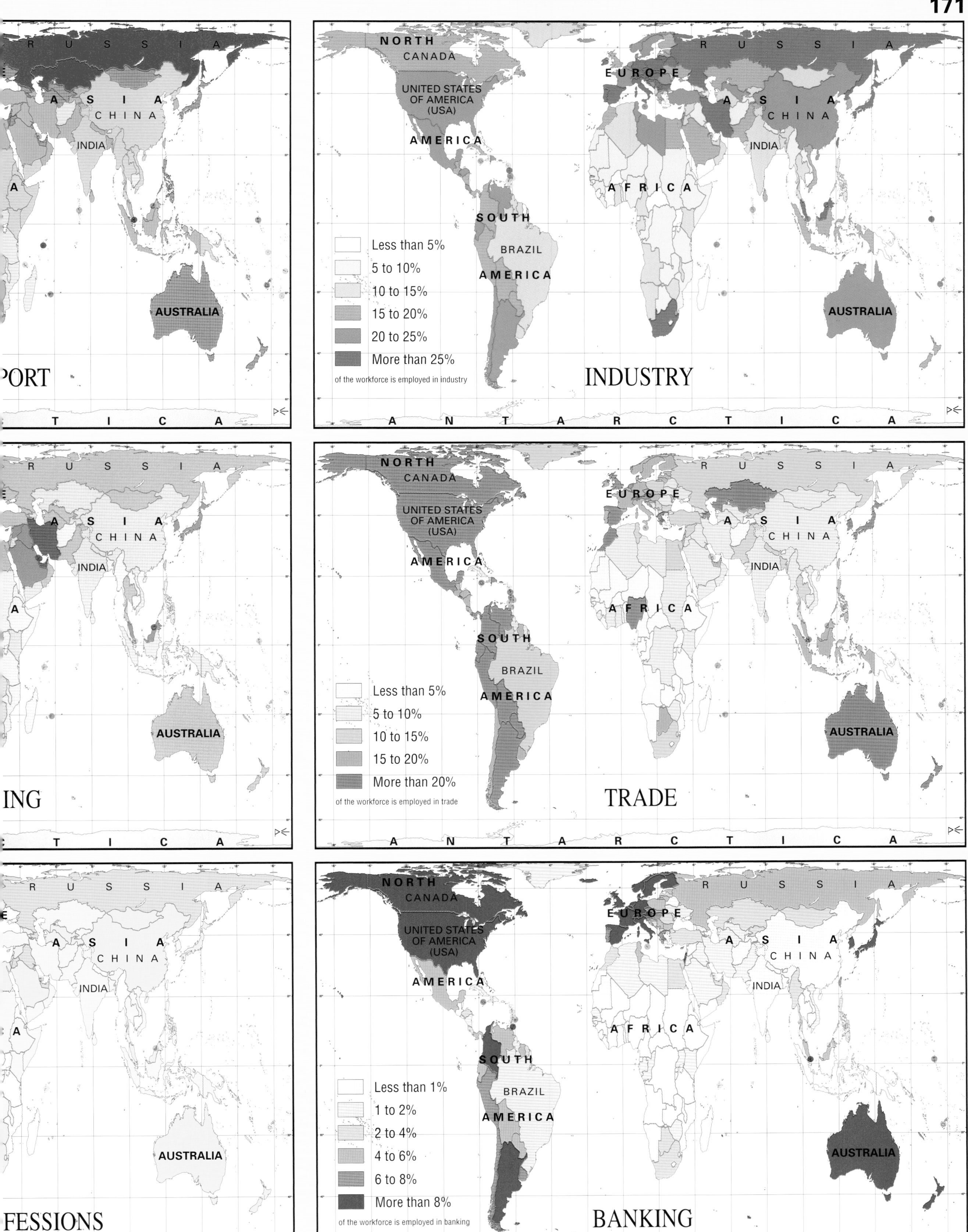

T STRUCTURE

As a result of increasing automation in the industrial countries, in most cases fewer than a quarter of the workforce is now employed in industrial production, for example Germany and Italy. In such countries the emphasis of employment has shifted towards trade and services, (which has already overtaken industry in Canada and the US), and towards banking, which is increasingly shaping economies and states.

It is nearly a million years ago since the first people began to exchange products of equal value. With the beginning of farming and the domestication of animals about 10,000 years ago this economy of meeting immediate needs was slowly pushed aside by the market economy. Trade replaced barter, and striving for profit became the engine of economics. Rich and poor emerged, and with them domination and subservience, robbery and war.

ECONOM

C ORDER

The market economy reached its highest form in capitalism which has developed with the machine age over the past 250 years. With capitalism, the gulf between rich and poor grew immeasurably throughout the world. Over the past 100 years some countries have been trying, by abolishing private ownership of the means of production, to bring back the old economic system of meeting only real needs but at a higher level in the form of planned economies. Nearly a quarter of human kind, or 1.4 billion people, live in those socialist countries today.

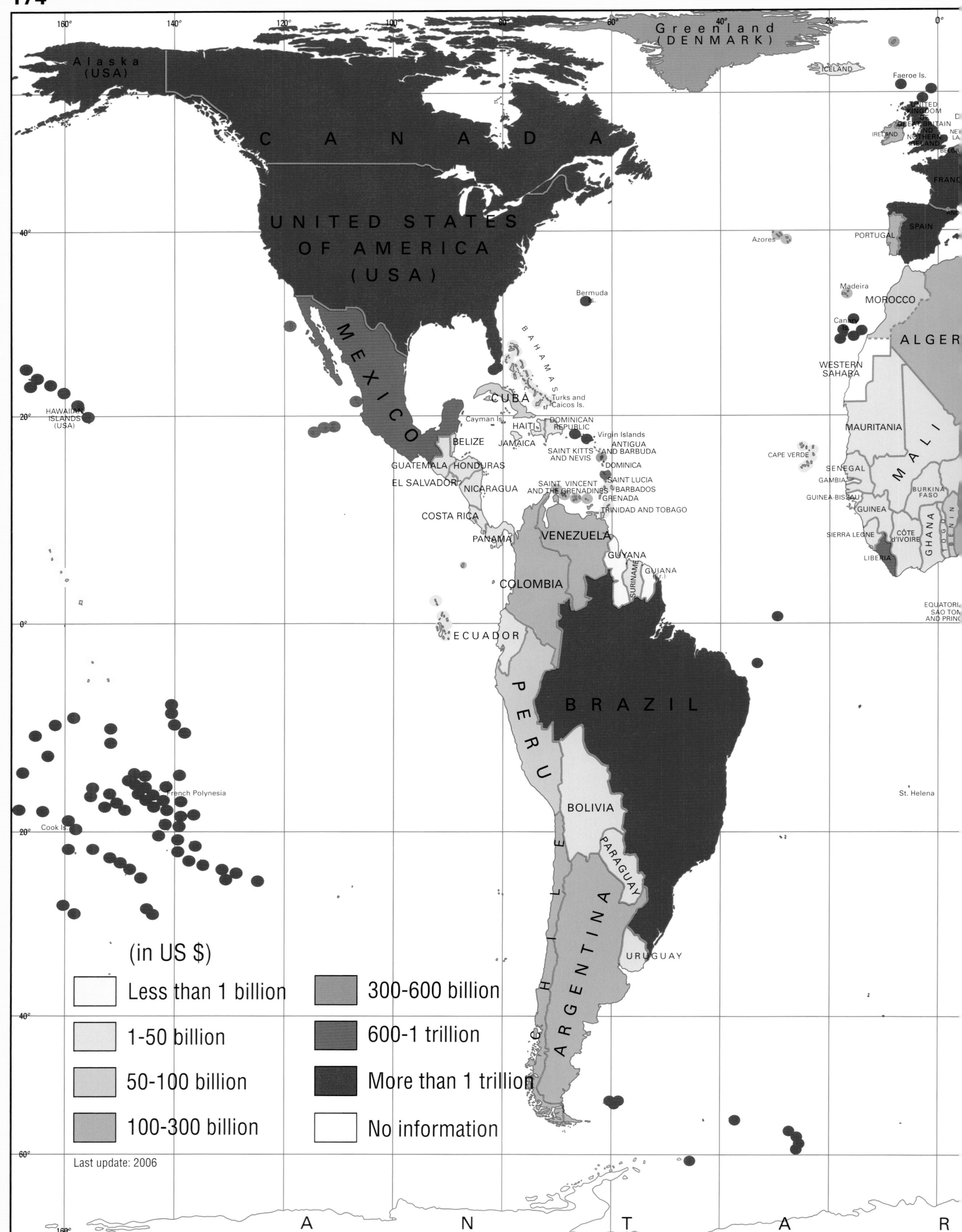

The indicator of growth of a state's overall politico-economic performance, an important basis of public finances, is the gross domestic product (GDP). It includes the value of all goods and services produced and provided, respectively, within the borders of a national economy in one year. It does not include reductions for the depreciation of capital in kind or the decrease in and the exhaustion of resources. It is problematic, however, to compare the GDPs of different nations in case of varying economic systems,

PUBLIC
GROSS DOMEST

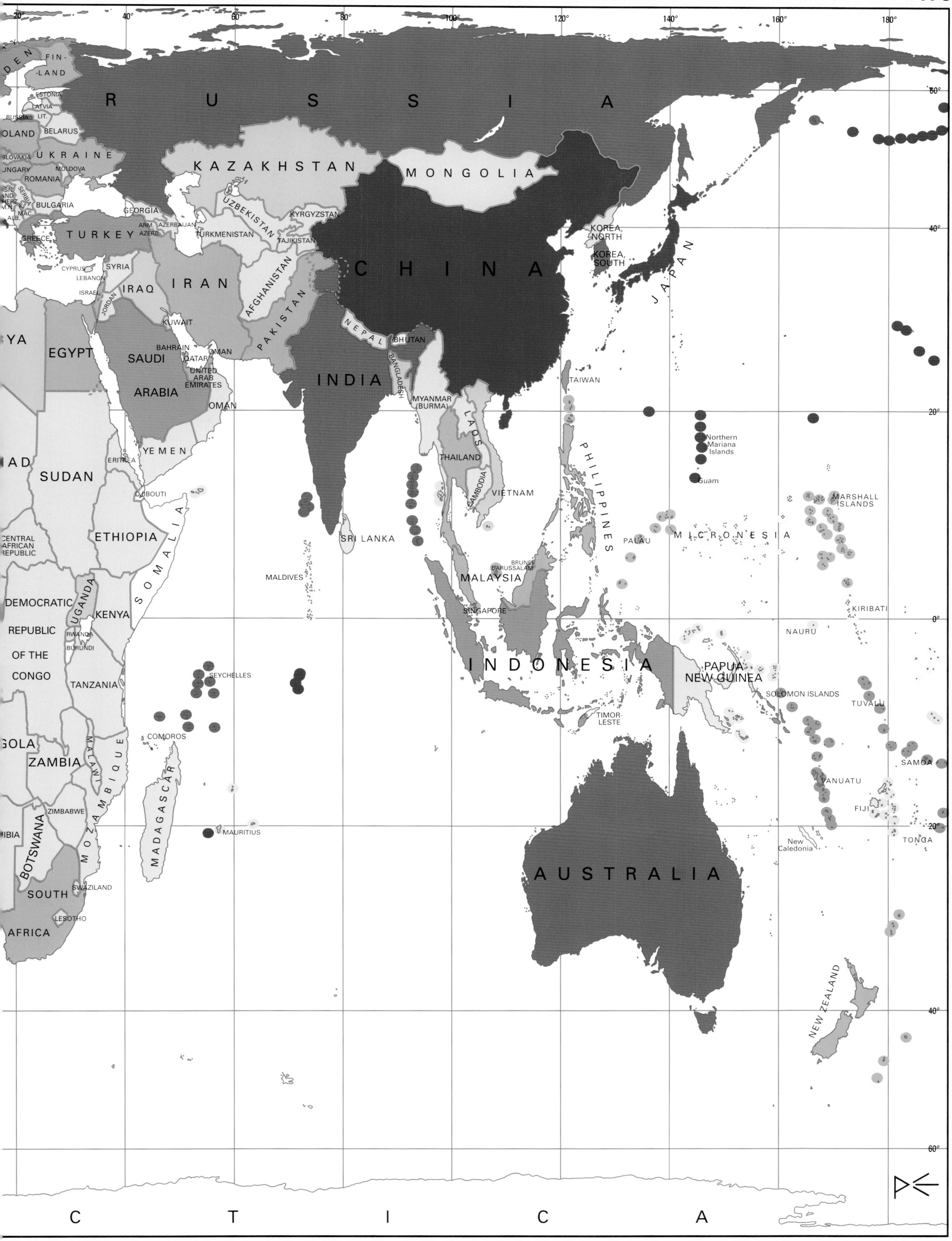

NANCES
PRODUCT (GDP)

just as the conversion into US $ is disadvantageous because it does not take into account changes in currency exchange rates, the varying nature of purchase power and price differences relating to domestic economy and foreign trade. For instance in Germany, the GDP increased by 2.5% in 2007 in real terms compared to the previous year. The following economic areas are significantly involved in the constitution of the GDP in Germany: financing, renting and enterprise service providers, agriculture and forestry, the fishing industry, the building trade, commerce, the catering trade, traffic, public and private service providers, and industry.

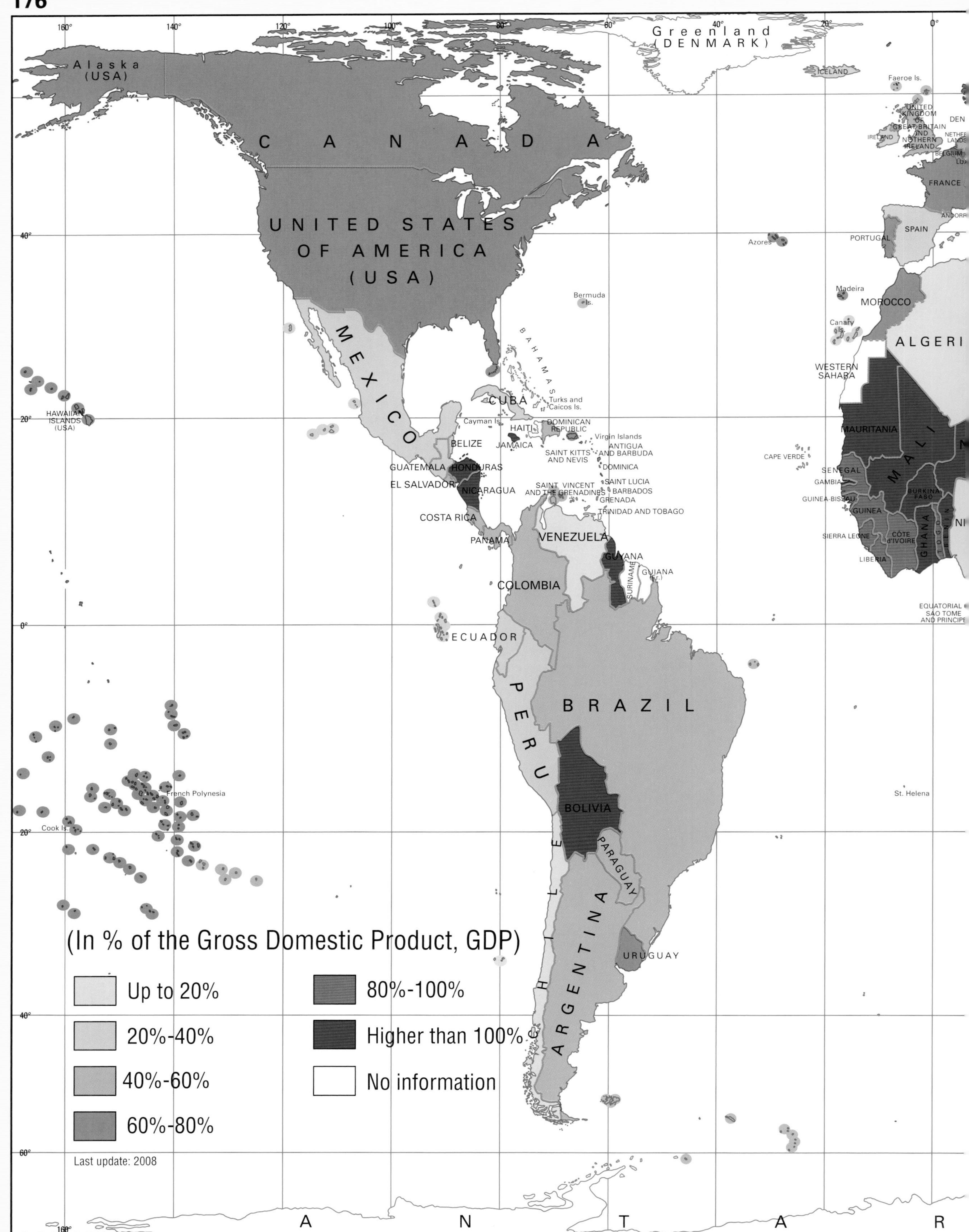

The national debt was also an important subject of global economics and monetary policy in the years 2007 and 2008. It is the ratio of debt to economic and financial performance rather than the amount of debt which is crucial. The main problem of developing countries is the incapability of many debtor nations to pay off their foreign debt in time. Certainly many Western European states and the USA are highly indebted, but this is largely a matter of domestic debt.

NATION

AL DEBT

Moreover, these states have no difficulty in paying off their debt on grounds of the economic power and prosperity of their citizens. Debt servicing can only become dangerous with regard to fiscal and financial policy, and it restricts the states' plans for future investment. For developing countries, their high foreign indebtedness often means excessive demands on their economic productivity. According to the World Bank, the total foreign indebtedness of all developing countries has increased by more than 33% in recent years.

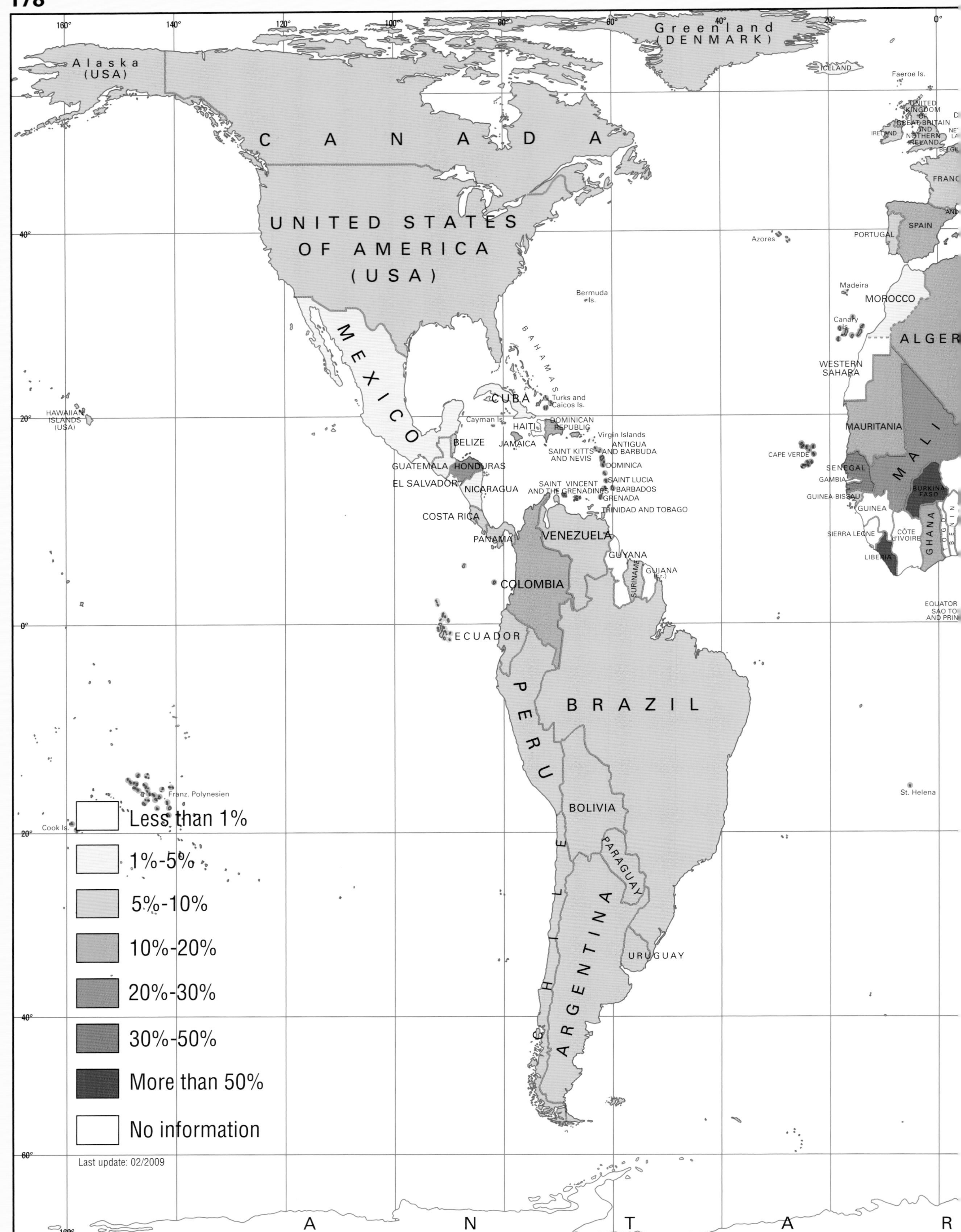

Work shapes the face of the earth and of man; it defines the way man lives, his language, behaviour and thinking; work brings about the progress of mankind and is the source of education, science and morals. To be excluded from work means the loss of personal fulfilment and exclusion from the community. A society which is unable to secure work for all its members on a lasting basis bears responsibility for hopelessness, self-neglect, flight from reality and criminality.

UNEMPL

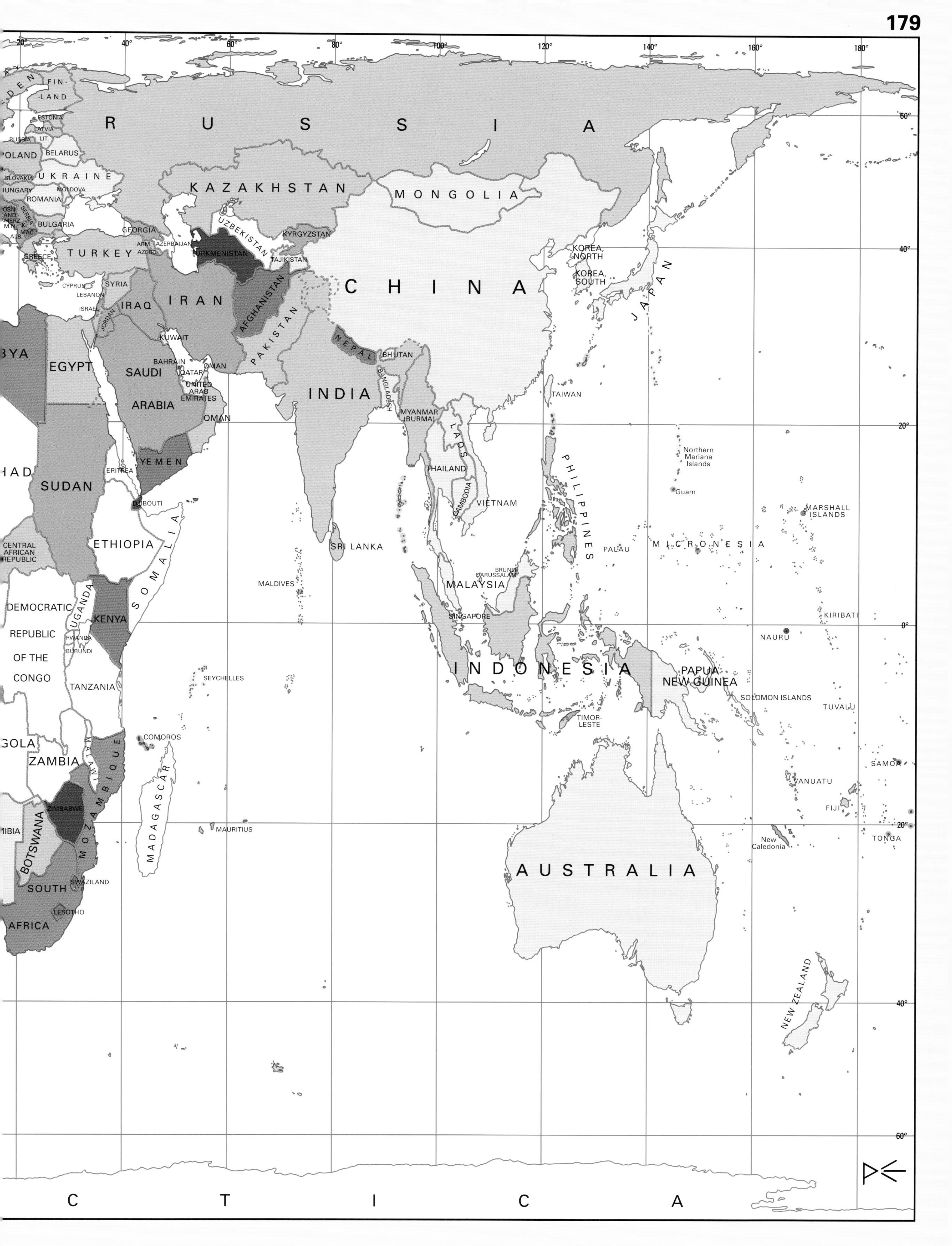

OYMENT

According to official figures worldwide there are, today, about 150 million unemployed, of whom about 50m are in Western Europe. But in many countries only those entitled to unemployment benefit are counted so the real figure is higher. Unemployment was seen at the beginning of the 20th century as the worst form of economic crisis; today it is an integral part of the capitalist market economy. In socialist countries the right to work, which was in part established as a constitutional right, has been abandoned as a result of the partial restoration of the market economy.

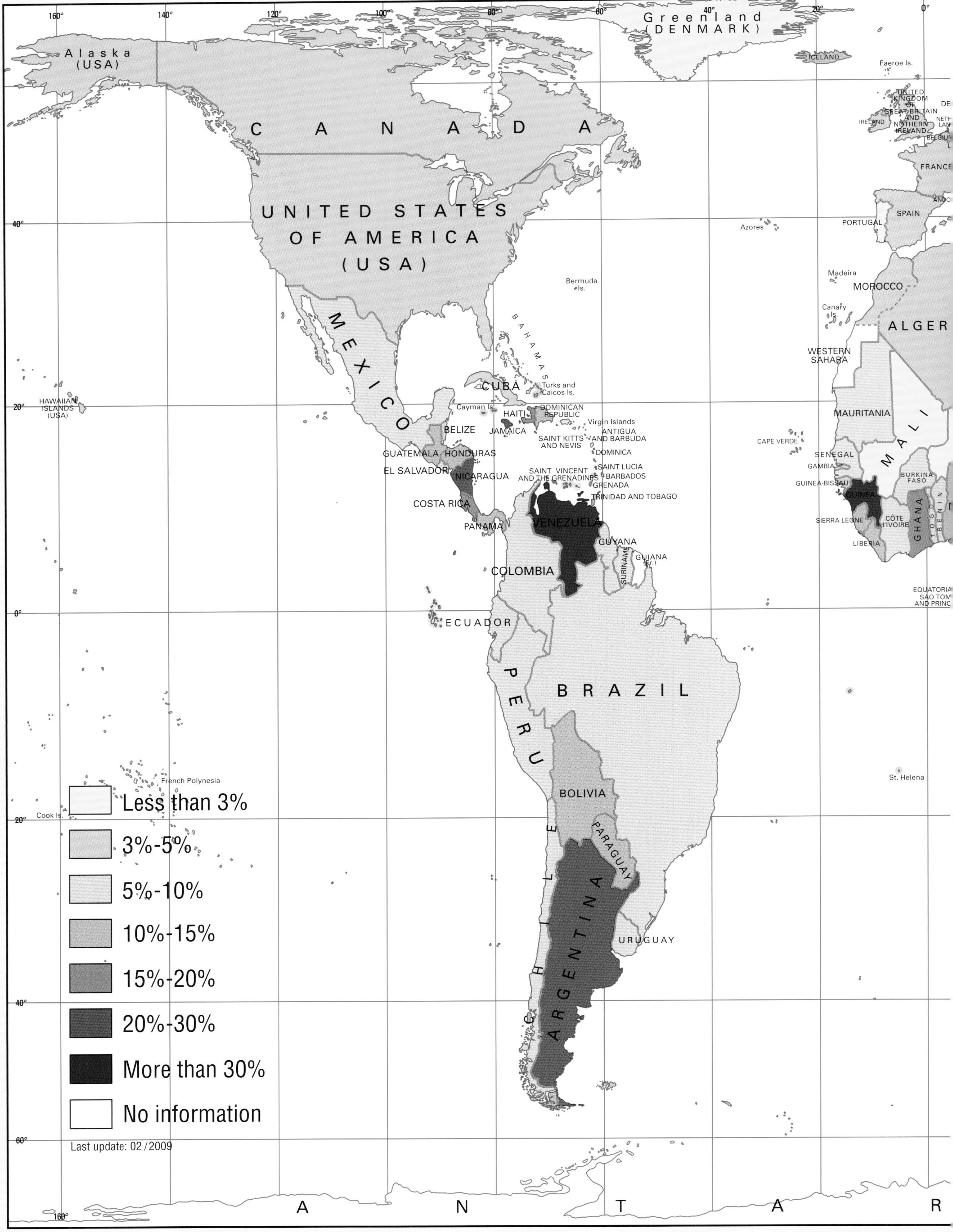

Less than 3%
3%-5%
5%-10%
10%-15%
15%-20%
20%-30%
More than 30%
No information

Last update: 02 / 2009

INFL

Devaluation of money, which many governments bring about by increasing the money supply, disturbs the balance between the supply of money and that of goods. It increases the prices of consumer goods without simultaneously increasing pay and is therefore at the expense of the waged and salary workers whose real income decreases. Those who profit are national governments, whose debts sink, and banks, who earn from the reduced value of their deposits, and the owners of capital assets which rise in value. If devaluation is not followed by increased production, inflation brings crises and unemployment.

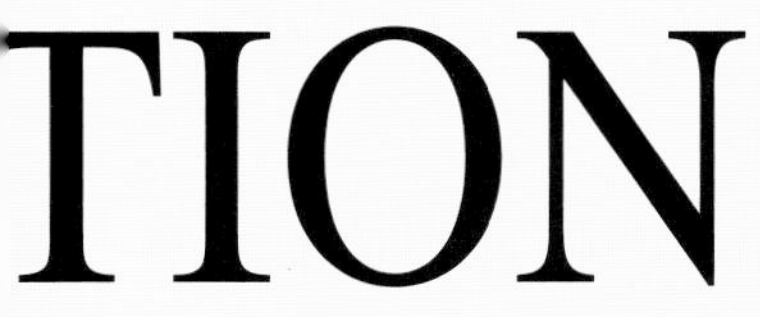

Paper money, which was introduced in Europe about 200 years ago to replace coins of precious metals, made possible the arbitrary increase of money supply and hence inflation. For 150 years inflation remained the last resort to overcome breakdowns after wars and revolutions. But from the middle of the 20th century doses of inflation have been continually used by the rich industrial nations of the world as an instrument of monetary policy. Their inflation roughly keeps pace with rising incomes achieved by wage demands, whereas the inflation rate of developing countries adds to the poverty of their peoples (Zambia 11%, Somalia 75%)

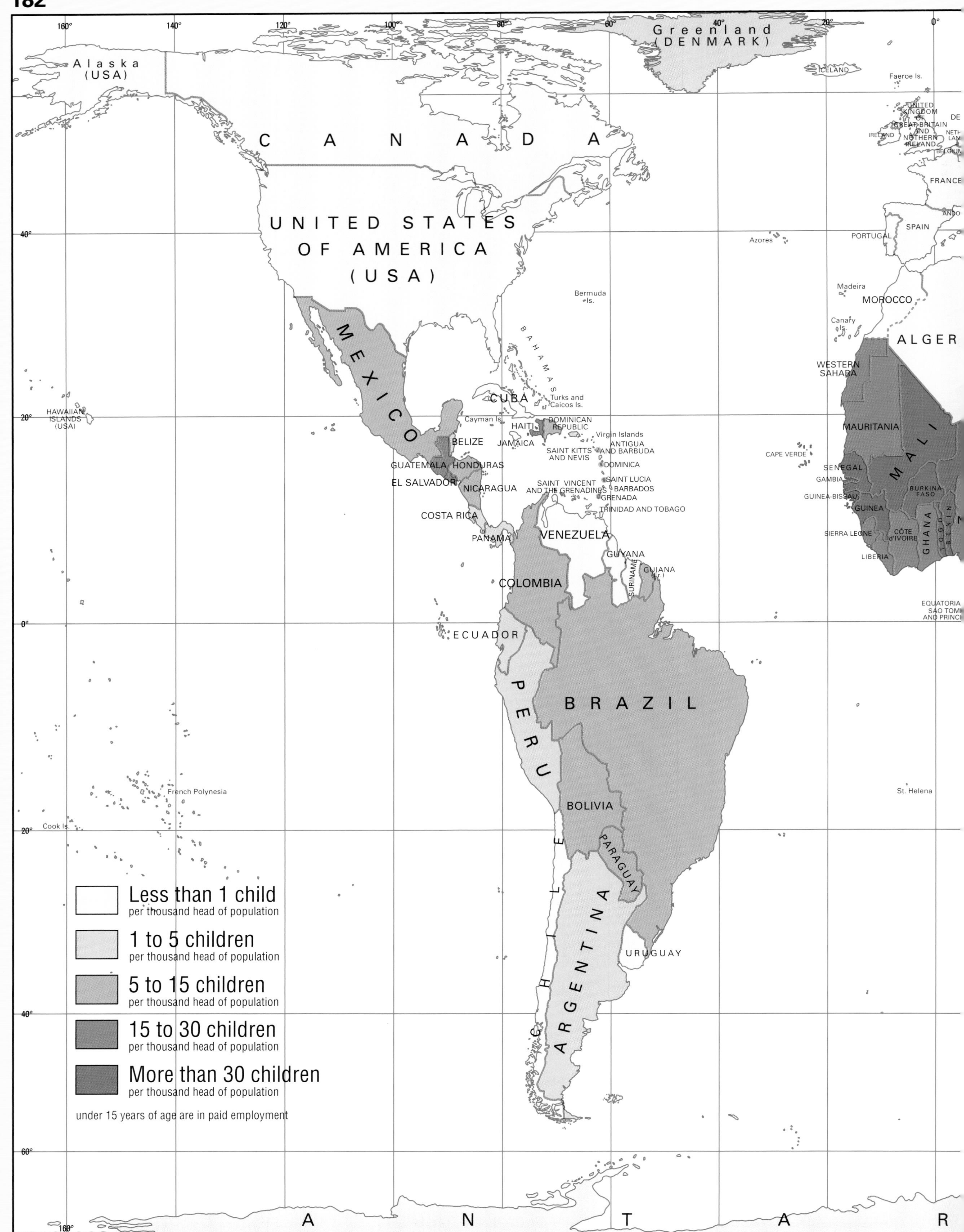

The use of children to help in the house and in the fields is as old as human culture. This help served a useful educational purpose when it did not overwork the children. During industrialisation child labour throughout Europe shifted from the parental house to workshops in which children were merely cheap labour. This paid child labour, the worst form of exploitation, was suffered only by children of the poorest people, whose families were hungry and deprived. Today child labour is incompatible with regular attendance at school or learning a profession so it leads to lifelong social disadvantage.

CHILD

ABOUR

As the most acute forms of exploitation shifted to the colonial areas so did most child labour. Political independence of colonies did not end exploitation because economic dependence continued. In the rich industrial countries, increasing condemnation of child labour led to a loss of educational opportunities at work but did not wholly eradicate child labour. Socialist countries are increasingly trying to bring work into education in polytechnic schools.

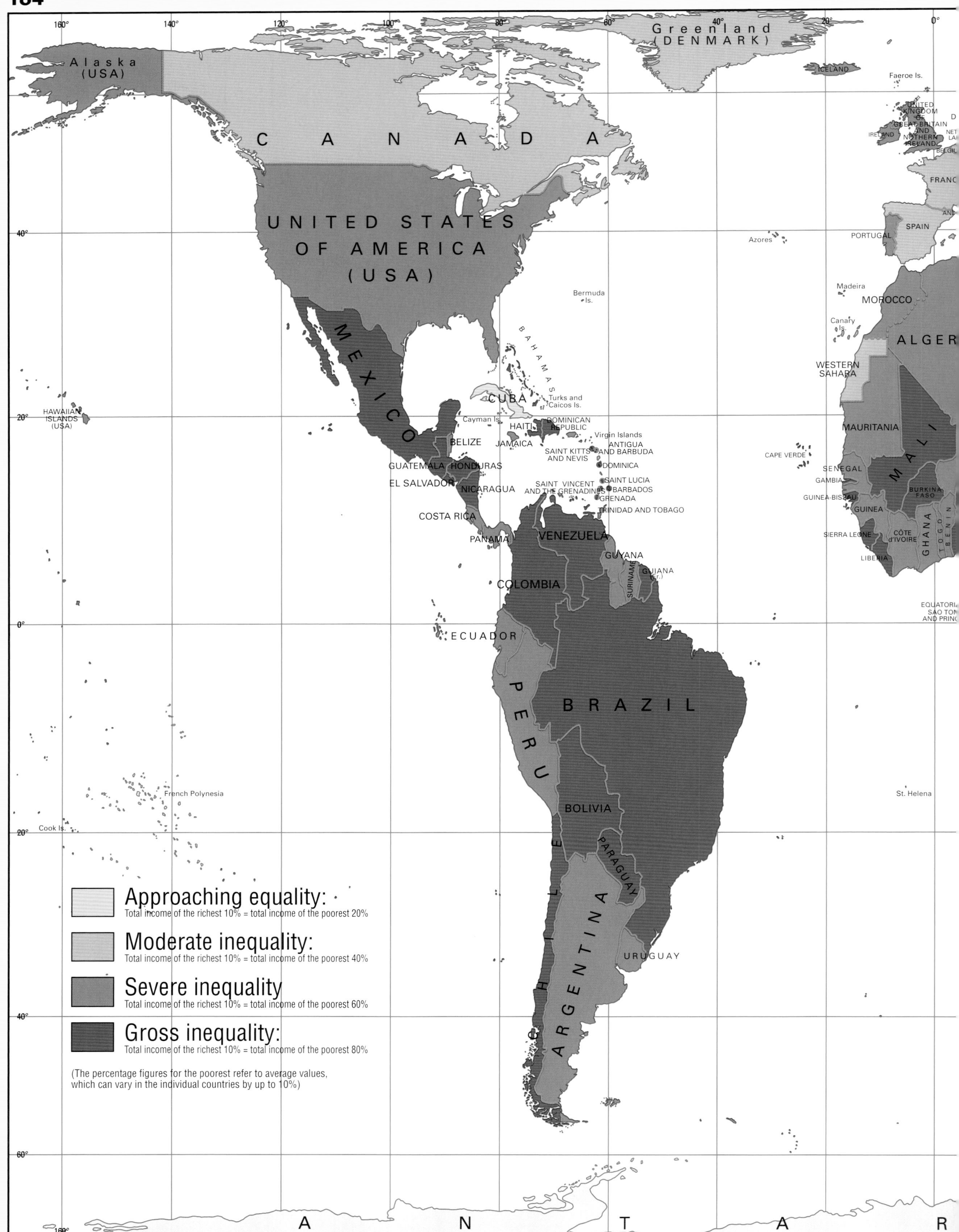

Increasing inequality which evolved with the beginnings of the market economy 6000 years ago has today reached a high point. Directors of big companies earn many millions yearly. Heads of big American concerns have annual salaries of $20-30m; thus in one minute they earn as much as the poorest people in the developing countries earn in their whole lives. The owners of big companies have annual incomes running into billions while every day worldwide more than 100,000 people starve to death.

INEQU

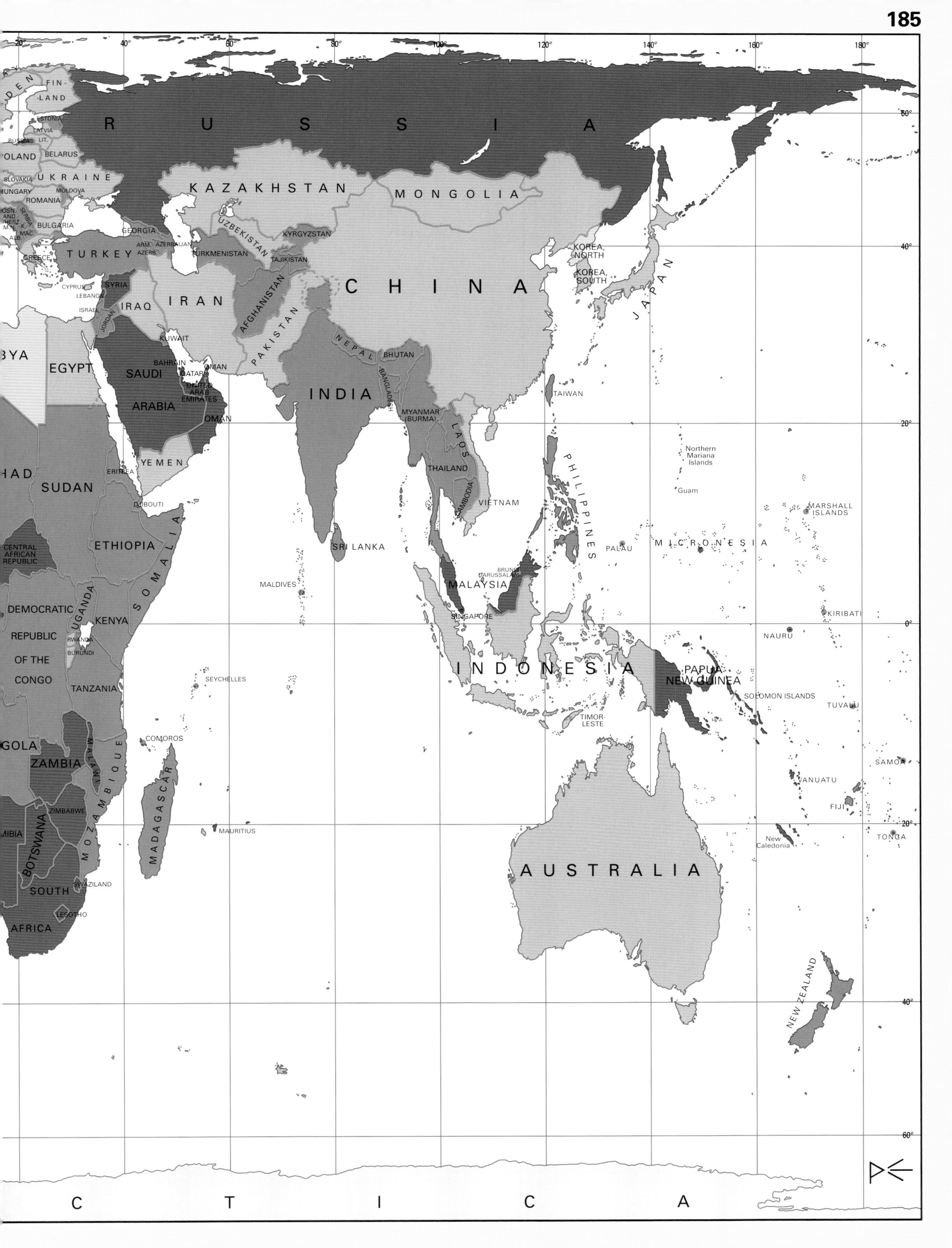

Income distribution in different countries does not depend on the size of average income. Inequality in poor developing countries is often greater than in the rich industrial nations. In South Africa the income of the richest mine owner is about $2 billion a year, three times as much as all five million inhabitants of Chad earn together in a year. The extreme inequality which comes from private ownership of the big companies has been overcome by a fifth of mankind in the socialist countries without, however, achieving real equality anywhere.

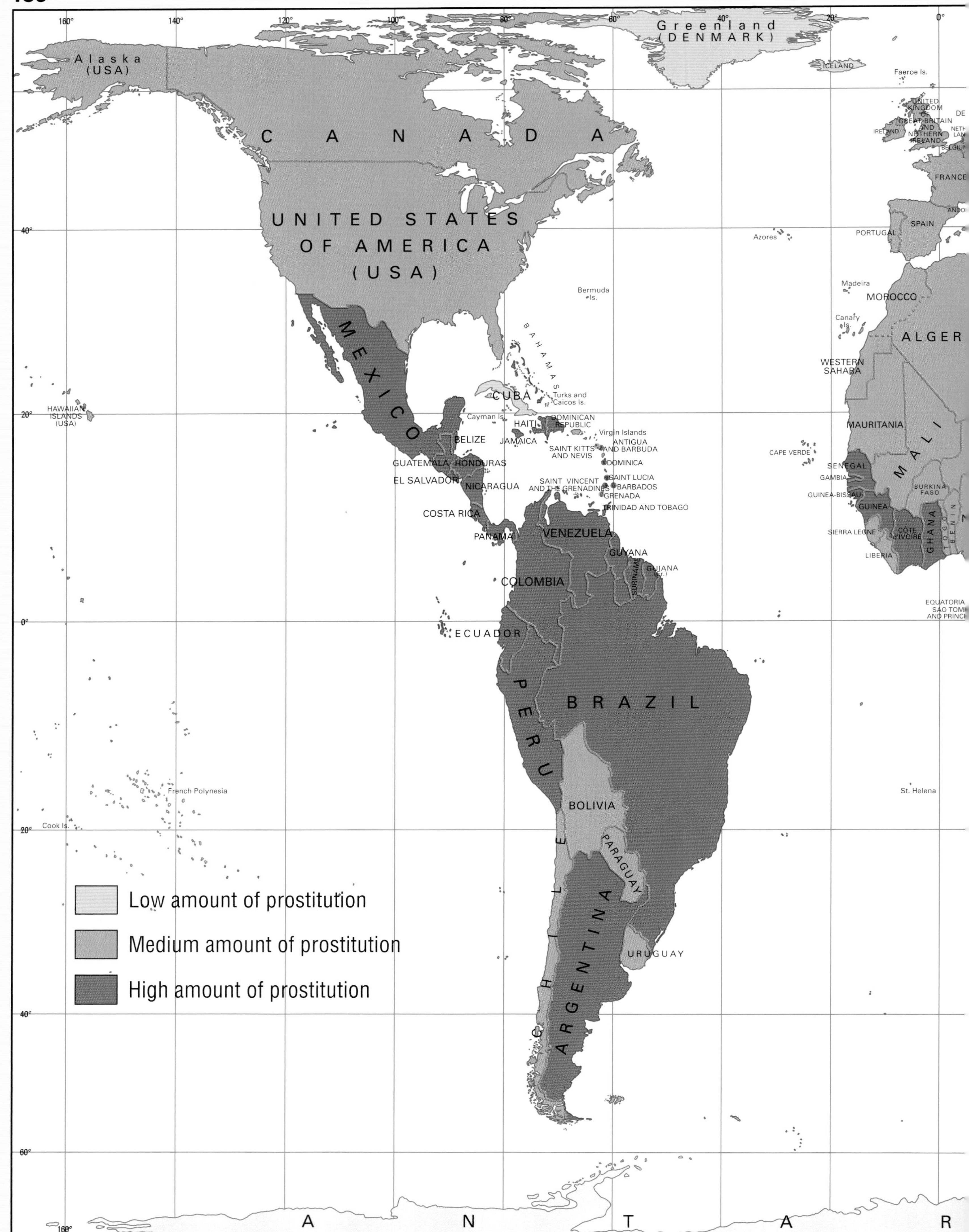

The most horrible form of exploitation of women, girls and children is prostitution. It developed with the emergence of the profit motive in market economies about 6,000 years ago. It was introduced by priests who made money from sacrificing young girls to the temple goddesses (Ishtar, Aphrodite, Venus). Prostitution was later adopted by city states (Athens and Corinth), then by slave dealers and finally by free entrepreneurs. The few people who still live in a state of pure nature do not know prostitution. The United Nations has declared prostitution a form of slavery and demanded its abolition.

PROST

TUTION

Because the worst exploitation is now in non-European countries the emphasis of prostitution has shifted there too. More and more men fly from Western Europe, Japan and North America to those countries in East Asia, Africa and Latin America in which bordello cities have developed (sex tourism). Apart from this more and more women, girls and children from Eastern Europe and the Third World are brought by organised trade to Western Europe, Japan and North America and are there forced into prostitution.

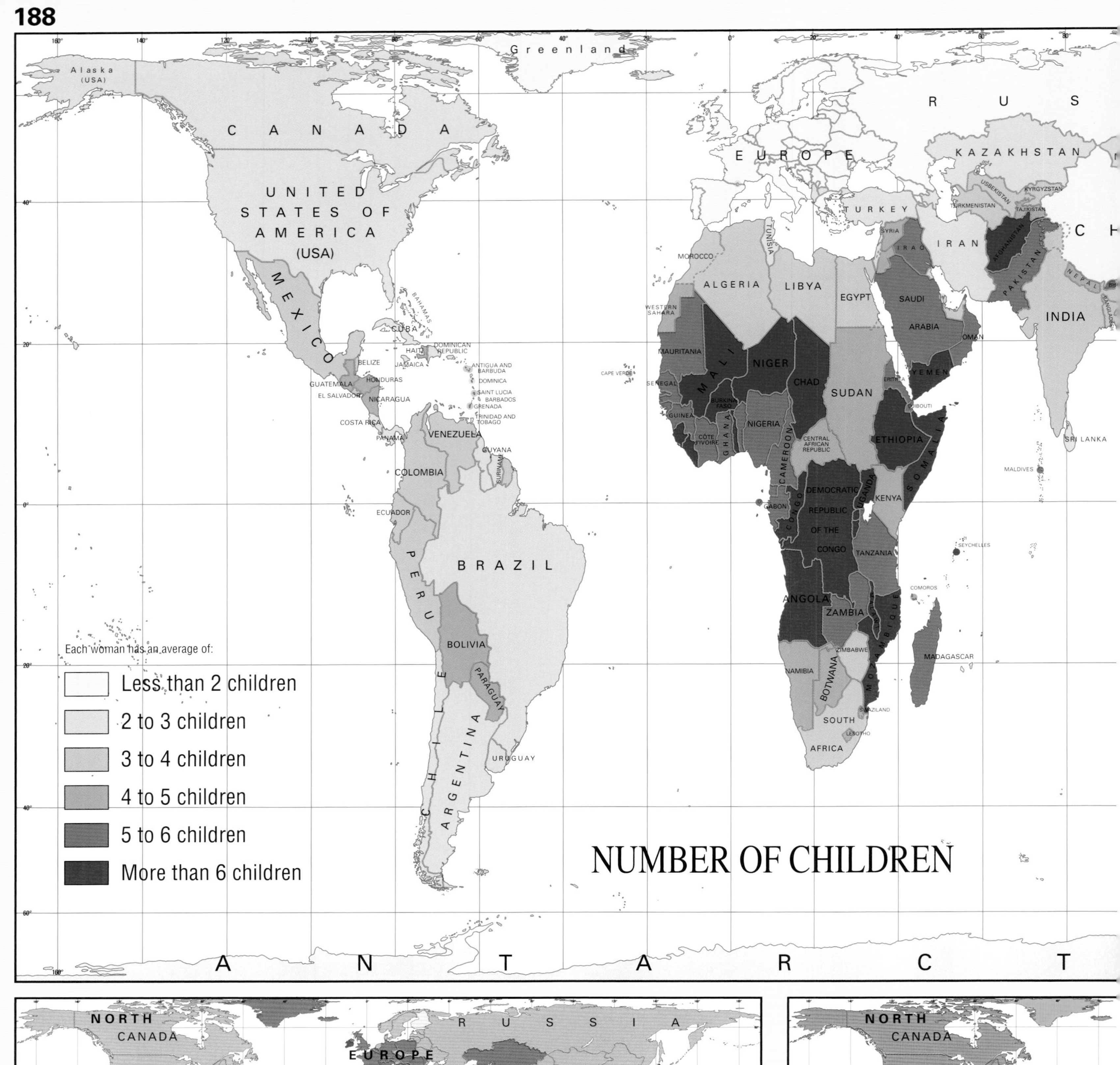

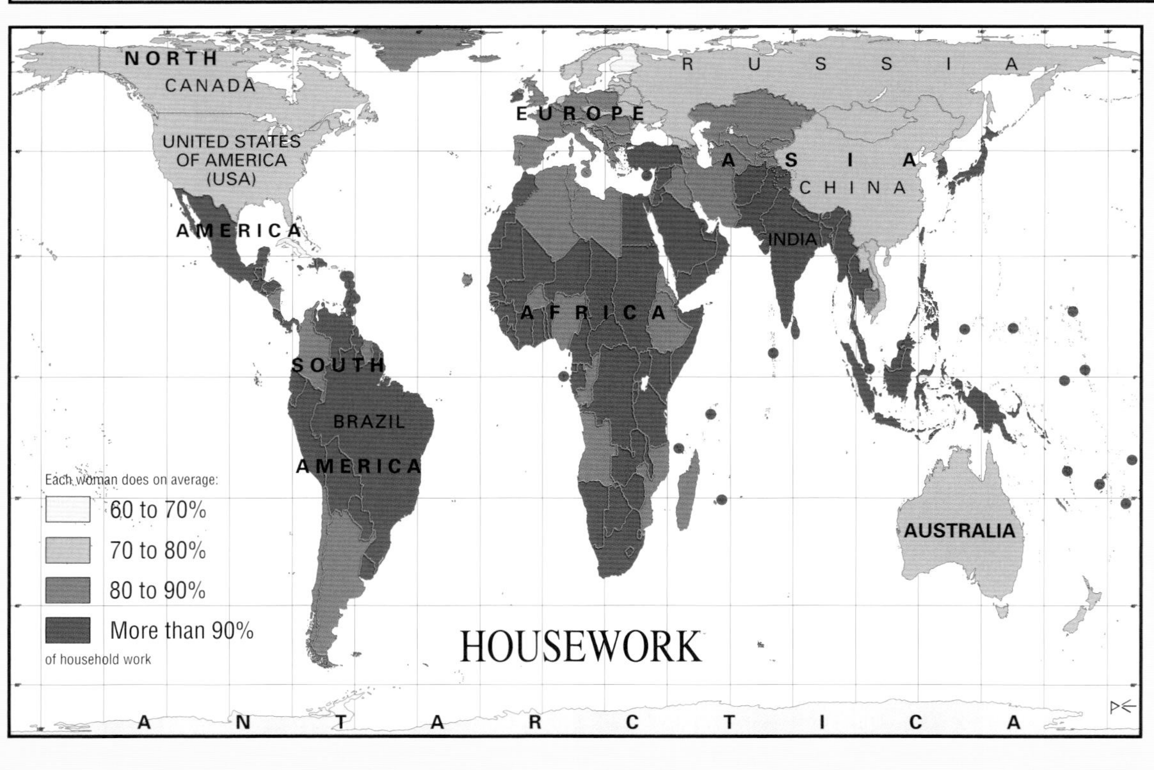

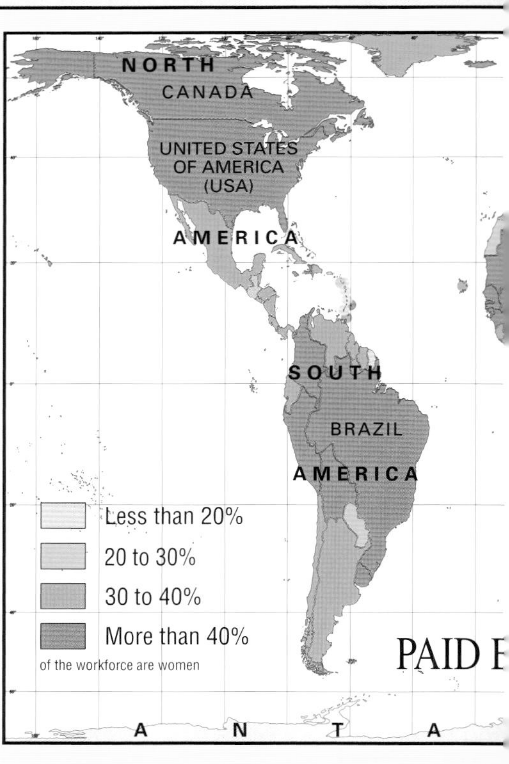

Half the world's people are women. They do two-thirds of the work, earn a tenth of the world's income and own a hundredth of the world's wealth. This statement by the United Nations in 1980 is still valid. The position of women in rich industrial nations has somewhat improved in the meantime but not the position of four times as many women in the poor developing countries. Even in the socialist countries, equality of women is also not yet complete.

THE STATU

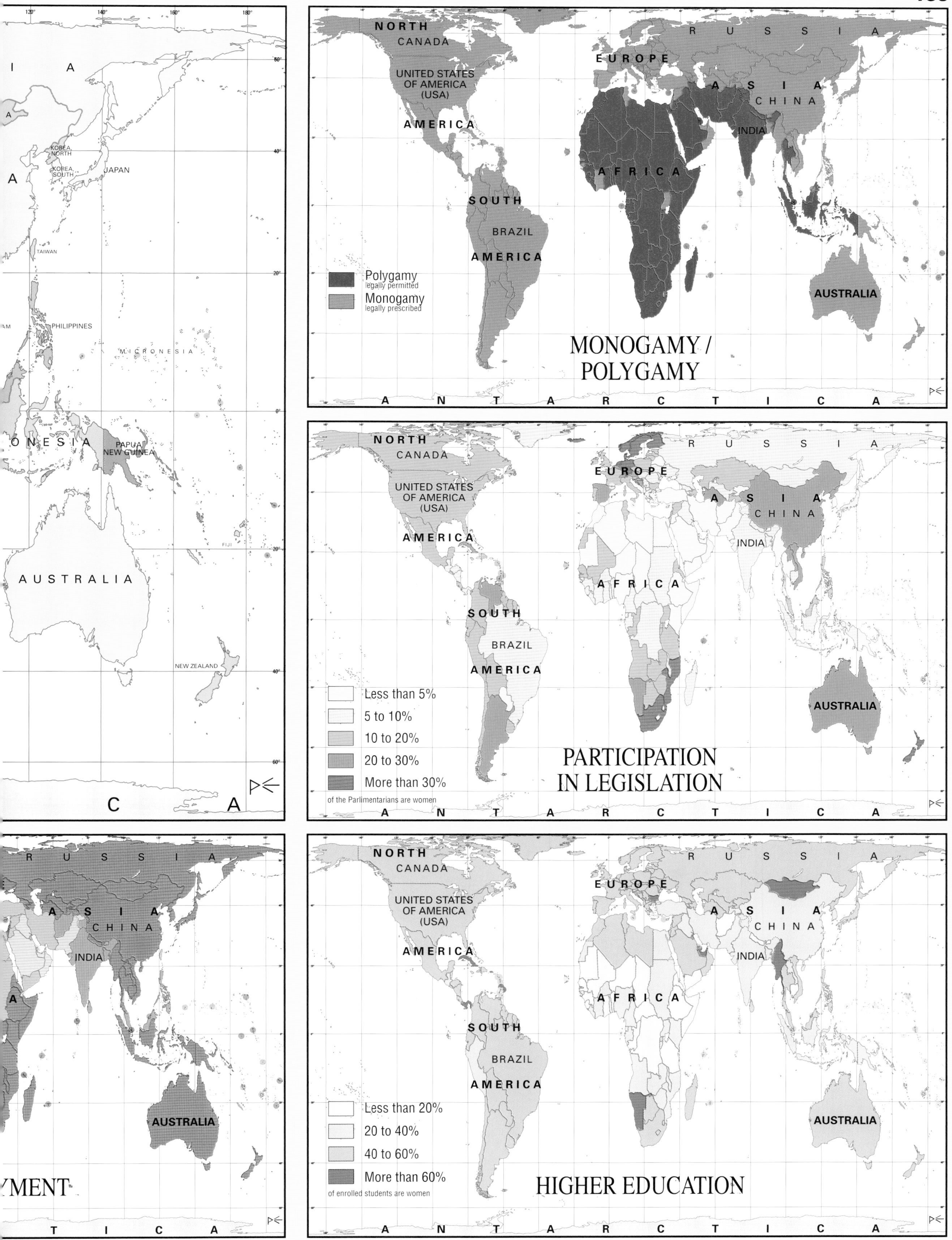

OF WOMEN

Besides bringing up children and doing the housework, more and more women are taking up employment. Worldwide this threefold burden is the price of growing equality. Women seek paid employment in order to be liberated form the narrowness of domestic existence. Their hope that men will share domestic duties is only slowly being realised. And in only a few countries have women achieved full rights over their own bodies (the right to abortion) and equal participation in lawmaking.

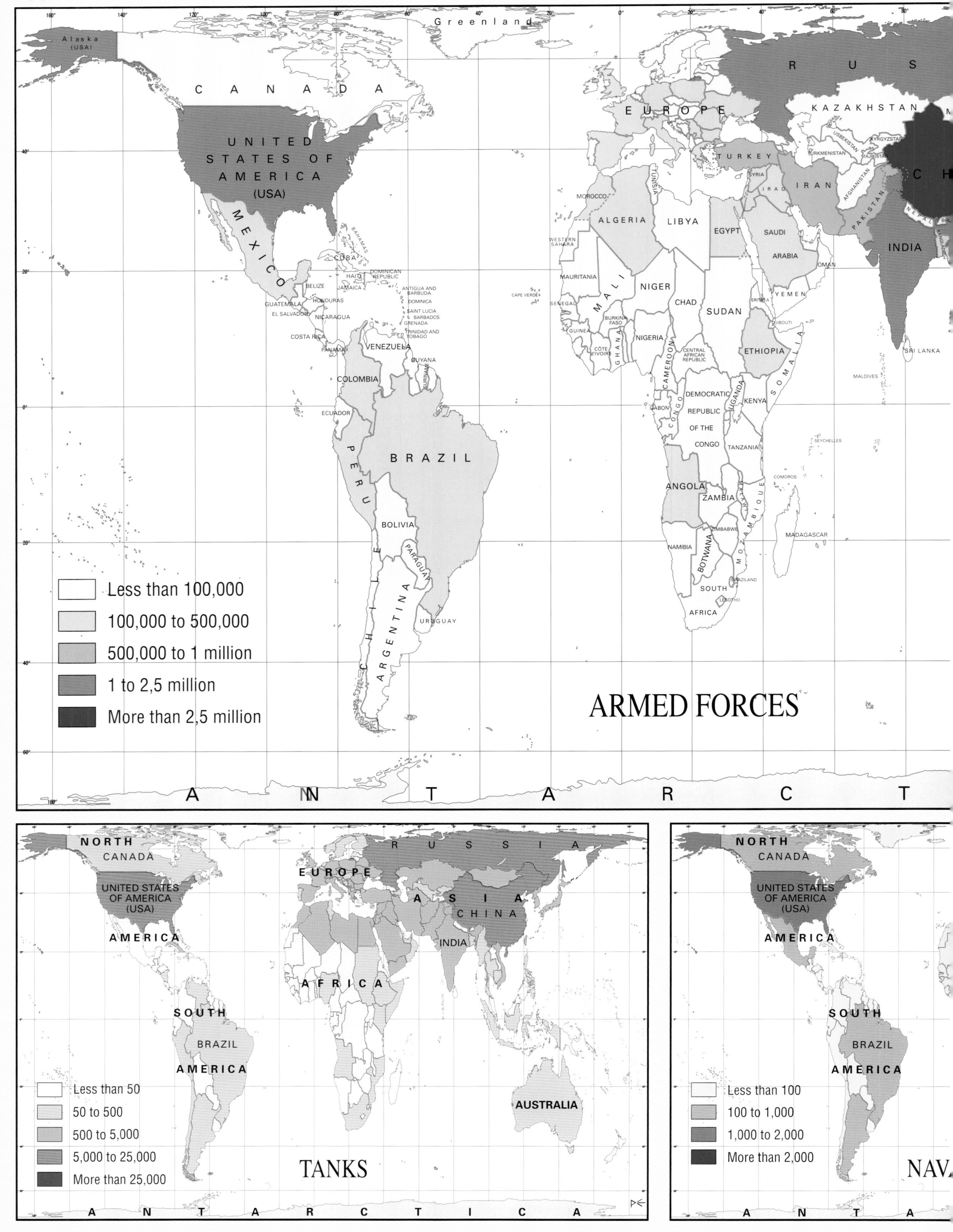

After the Second World War the Soviet Union deferred higher living standards to achieve military balance with the rich industrial nations so it could help the developing countries in their liberation struggles against their white colonial masters as well as giving military help to the spread of the socialist revolution.

RELATIVE MILI

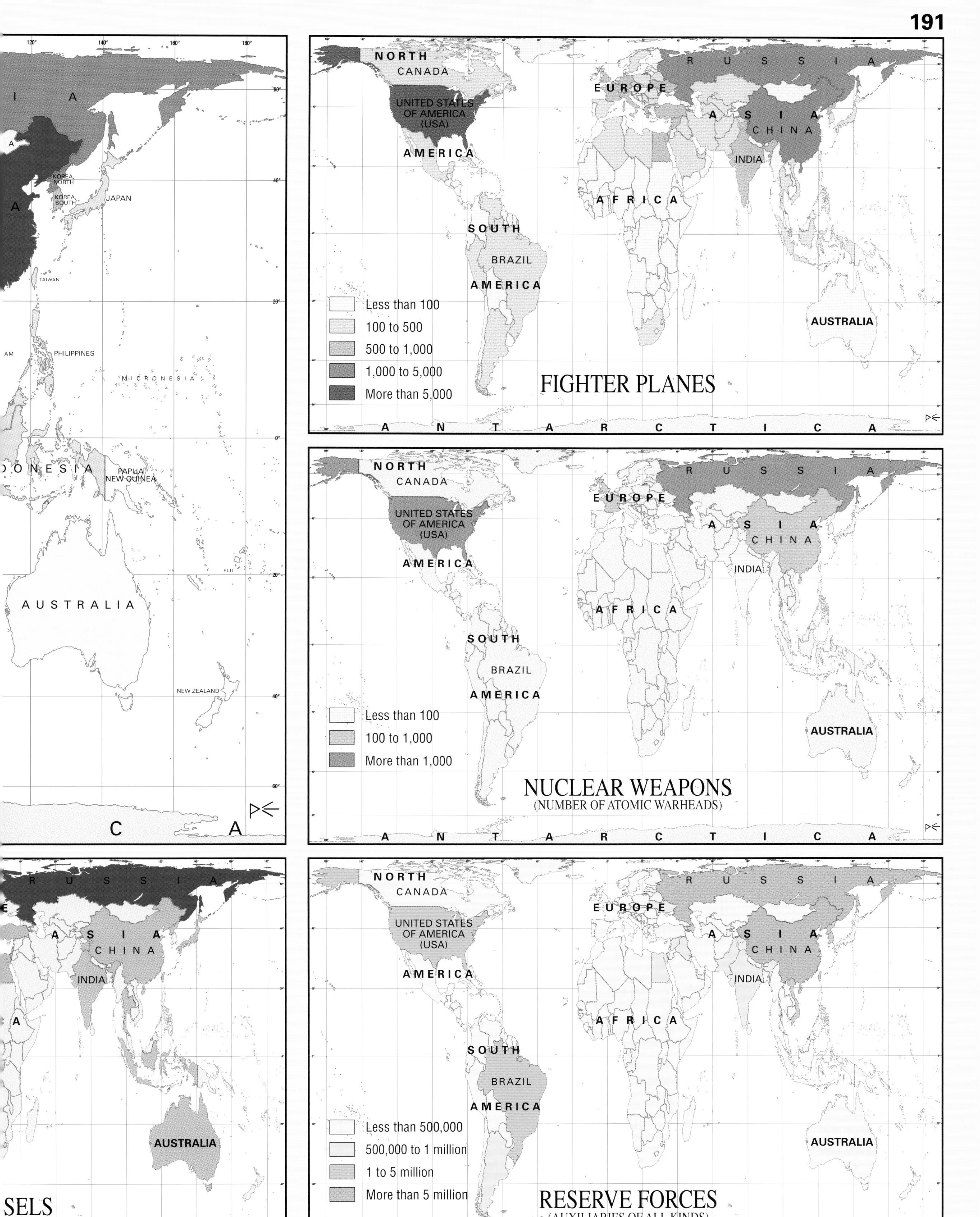

ARY STRENGTH

Since the disintegration of the Soviet Union the United States and the rich industrial nations with just under 0.7 billion people confront the socialist countries with 1.3 billion people and the developing countries with 4 billion people. Thus the predominance of the rich industrial nations rests on their technological/military superiority.

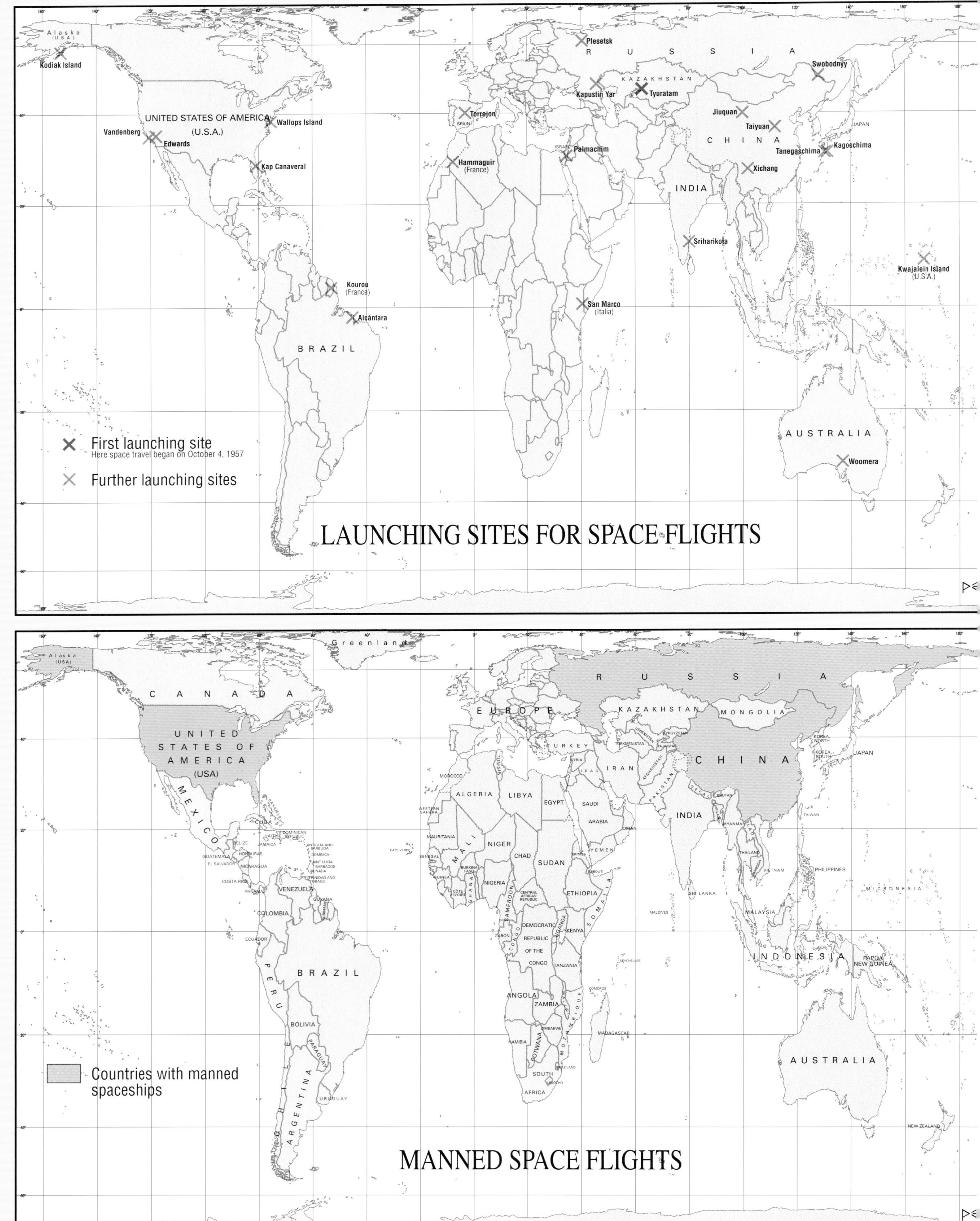

On 4 October 1957 the Soviet Union launched the first spacecraft (Sputnik). Four years later on 12 April 1961 the Russian, Yuri Gagarin, was the first person to fly into space. Thus began a new epoch in human history, opening unlimited new possibilities for mankind's urge to explore and act.Since then hundreds of manned spacecraft and many thousands of

THE CONQU

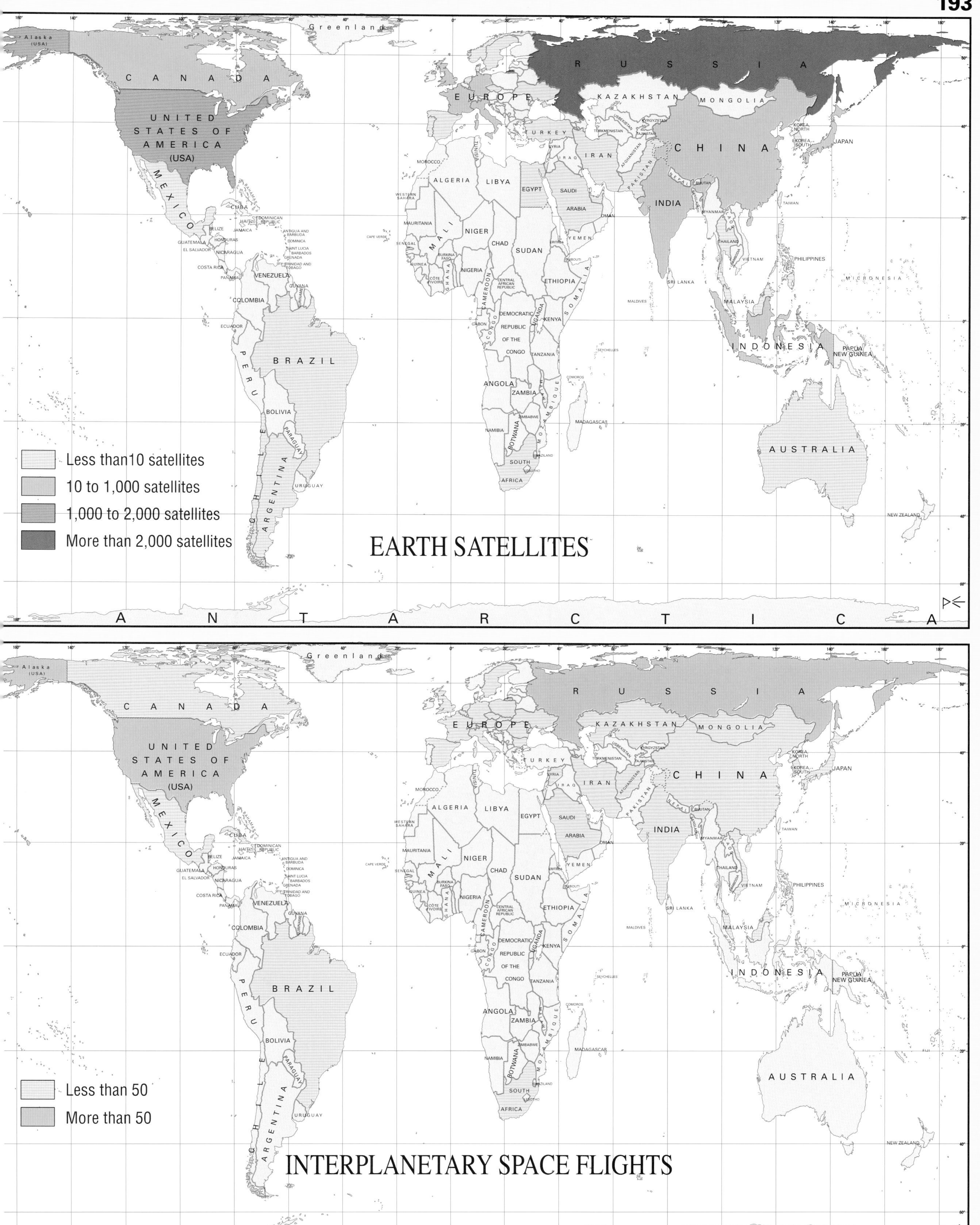

ST OF SPACE

probes and satellites have been sent into space from a good dozen launching sites. Men have landed on the moon and returned to earth (Armstrong and Aldrin). Pictures and data from Mars and other celestial bodies have been sent back to earth. Weather forecasting, navigation, telephone traffic, television and the internet have all been improved by satellites.

Each name in the index is followed by a page number and a letter. On the page referred to, the letter can be found either at the top or at the bottom of the map frame. In the first case, the place is in the upper half of the map vertically below the letter; otherwise it is on the lower half of the map vertically above the letter. If a name extends over several letters, the given letter indicates its beginning.

Names such as countries or oceans which cover a large area on the map are listed with their page number only. However, if they extend over two pages, two page numbers are shown – the left-hand and right-hand page numbers being linked with a dash. Names of countries, oceans, rivers and mountains that extend over more than a double page are listed under each separate page. A dash between two nonconsecutive page numbers means that the place appears on all maps between and including those two pages.

The headwords are in alphabetical order. Names with prefixes like „Saint" or „Bad" can be looked up under the initial letter of the prefix. Place names appear on the maps in their widely-used Anglicised form, or in their local spelling or a standard transliteration of that local spelling. The index also includes local forms of names where the Anglicised form has been used on the map. In these cases the local name is followed by the Anglicised name in brackets. This indicates that the place name appears on the map, at the reference given, in the form shown in brackets, not in its local form.

Aachen 33 n
Aaro 25 v
Aargub 36 o
Aba (China) 62 h
Aba (Nigeria) 41 s
Aba (Dem. Rep. Congo) 42 w
Abadan 39 g
Abadeh 39 j
Abadla 36 k
Abaeté 29 g
Abaetetuba 26 l
Abag Qi 57 v
Abakan 56 i
Abakan (River) 56 h
Abakan Mountains 56 h
Abala 41 c
Abalessa 37 o
Aban 56 l
Abancay 24 z
Abarqu 39 k
Abay 55 j
Abay (River) 43 b
Abaya, Lake 43 e
Abbabis 48 h
Abbe, Lake 43 o
Abbeville 32 y
Abbotabad 60 m
Abbot Ice Shelf 80 m
Abd al Kuri 43 m
Abdulino 54 m
Abéché 42 c
Abengourou 40 x
Abeokuta 41 p
Aberdeen Lake 12 y
Aberdeen (South Africa) 49 n
Aberdeen (South Dakota) 15 h
Aberdeen (UK) 32 x
Aberdeen (Washington) 14 g
Aberfoyle 74 o
Aberystwyth 32 w
Abez' 50 l
Abha 39 p
Abidjan 40 x
Abilene 19 b
Abingdon Downs 73 l
Abitibi 16 i
Abitibi Lake 16 j
Abiyata, Lake 43 c
Abkit 53 k
Abminga 77 d
Abomey 41 o
Abong Mbang 41 w
Abou Deïa 42 b
Abqaiq 39 h
Abra Pampa 28 f
Abrolhos, Archipelago dos 29 m
Abu al Abyad 39 x
Abu Dara, Ras 38 z
Abu Dhabi 39 x
Abu Habi 42 k
Abu Hadriyah 39 h
Abu Hamed 38 w
Abuja 41 f
Abu Kamal 39 b
Abu Madd, Ras 38 z
Abumonbazi 42 r
Abun 25 r
Abun (River) 25 q
Abune Yosef 43 c
Abu Shagara, Ras 38 z
Abu Simbel 38 v
Abu Tig 38 i
Abuye Meda 43 d
Abuyog 67 j
Abu Zabad 42 i
Acailandia 27 n
Acámbaro 19 n
Acampamento 26 i
Acampamento de Indios 25 w
Acaponeta 18 x
Acapulco 19 n
Acarai, Serra 25 l
Acaraú 27 f
Acarigua 23 d
Acatlán 19 p
Acayucan 19 r
Accra 40 z
Achacachi 25 p
Achénoumma 37 v
Achinsk 56 i
Achsu 54 w
Acklins 21 q
Acomayo 25 n
Aconcagua 28 p
Aconguija, Nev. de 28 s
Acre 25 p
Açu 27 u
Açu (River) 27 u
Acude Araras 27 s
Acude Orós 27 t
Ada 15 v
Adafer 36 s
Adak 10 v
Adale 43 v
Adam 60 o
Adam's Br. 64 u
Adamaoua, Highlands 41 u
Adan 66 c
Adana 35 z
Adanac 15 b
Adapazari 35 j
Adarama 38 x
Adare, Cape 86 s
Adavale 74 o
Ad Da'ain 42 g
Ad Darb 39 p
Ad Dawadimi 39 r
Ad Dibdibah 39 f
Addis Ababa 43 c
Addis Zemen 43 b
Ad Diwaniyah 39 e
Ad Dueim 42 k
Ad Duwaid 39 c
Adelaide 77 g
Adelaide Island 82 u
Adelaide Peninsula 12 m
Adelaide River 73 c
Adélieland 87 o
Aden 43 h
Aden, Gulf of 43 h
Aderbissinat 41 g
Adi Arkay 43 b
Adigrat 43 d
Adi Keyih 43 d
Adilabad 61 q
Adi Quala 43 c
Adirondack Mountains 16 z
Admiralty Inlet 13 d
Admiralty Islands 71 n
Ado 41 q
Adoni 64 f
Adour 34 f
Adrar 36 l
Adrar of the Iforas 36 z
Adré 42 d
Adriatic Sea 35 a
Adwa 43 c
Adycha 53 c
Adygalakh 53 f
Adzhima 59 b
Adzva 50 k
Aegean Sea 35 t
Affolé, Montagnes de l' 40 g
Afghanistan 55/60
Afgoi 43 t
Afif 39 p
Afikpo 41 s
Aflou 34 t
Afmadu 43 s
Aftout 36 w
Afuá 26 l
Afyon 35 w
Agadem 41 j
Agadez 37 r
Agadir 36 f
Agadyr' 55 k
Agailas 36 p
Agamor 41 a
Agan 51 q
Agapa 51 i
Agaro 43 a
Agartalá 61 z
Agassiz, Cape 82 w
Agassiz Ice Cap 13 h
Agata 51 j
Agboville 40 x
Agdam 54 v
Agdz 36 h
Agen 34 f
Agnew 76 g
Agnibilekrou 40 x
Agoare 41 c
Agra 61 c
Ağri 54 t
Agrigento 37 h
Agrínion 35 r
Agryz 54 m
Aguadas 22 z
Agua de Chale 18 e
Aguapei, Serra 25 w
Agua Prieta 18 i
Aguarico 22 y
Aguas Belas 27 u
Aguascalientes 18 z
Aguas Calientes, Sierra de 28 r
Aguaytía 24 x
Aguaytia (River) 24 x
Aguelock 36 z
Aguelt el Melah 36 s
Agueraktem 36 u
Aguilas 36 l
Agulhas 48 x
Agulhas, Cape 48 x
Aha, Mount 48 l
Ahaggar 37 p
Ahar 54 v
Ahmadi 39 g
Ahmadnagar 61 n
Ahmar Mountains 43 e
Ahmedabad 60 y
Ahnet 37 n
Ahome 18 h
Ahoskie 16 y
Ahuachapan 22 c
Ahwar 43 i
Ahwaz 39 g
Aibak 60 i
Aileron 73 r
Ailigandi 22 k
Aim 53 o
Aim (River) 53 n
Aimorés 29 k
Aïn Beïda 34 x
Ain Benimathar 36 k
Aïn ben Tili 36 s
Ain Oussera 34 u
Ain Selra 34 s
Ain Sukhna 38 i
Aioun Abd el Malek 36 u
Aiquile 28 f
Aïr 37 r
Airo 25 h
Airy, Mount 16 w
Aitape 70 y
Aiwasedapur 51 r

Aix-en-Provence **34** j
Aizawl **62** o
Ajaccio **34** k
Ajaguz **56** c
Ajaguz (River) **56** b
Ajdabiyah **38** a
Ajmer **61** a
Ajni **55** u
Ajryk **55** m
Akabli **37** n
Akchar **36** o
Ak-Dovurak **56** i
Akelamo **69** m
Akespe **55** c
Aketi **42** r
Akhisar **35** v
Akhmîn **38** i
Akhtyrka **35** l
Akimiski Island **16** i
Akita **59** o
Akjoujt Faye **36** p
Akkabak **55** d
Akkani **10** k
Akkarga **55** d
Aklavik **11** k
Aknoul **34** q
Akobo **42** l
Akola **61** o
Akordat **43** b
Akoupé **40** x
Akpatok Island **13** x
Akqi **56** o
Akranes **32** b
Akreijit **36** s
Akrerèb **41** g
Akritas, Cape **35** r
Aksakovo **54** m
Aksaray **35** y
Aksay **54** m
Akşehir **35** w
Aksu **55** j
Aksuat **55** f
Aksum **43** c
Aksumbe **55** u
Aktag **56** r
Aktau (Kazakstan) **54** y
Aktau (Kazakstan) **55** k
Aktaz **56** s
Aktogaj (Kyrgyzstan) **56** n
Aktogaj (Kazakstan) **56** c
Aktyubinsk **55** a
Akulivik **13** s
Akure **41** q
Akureyri **32** e
Akwanga **41** f
Alabama (River) **19** j
Alabama (State) **19** k
Al Abyar **35** r
Alachakh **53** p
Al Adam **35** s
Alagoinhas **27** u
Al Ain **39** y
Alajuela **22** g
Alakanuk **10** m
Alakol', Lake **56** d
Al Alamein **38** g
Al Amarah **39** g
Alamogordo **18** k
Alamos **18** i
Aland Islands **33** h
Alanya **35** x
Al Aqaylah **37** m
Al Artawiyah **39** e
Alaşehir **35** v
Alashan Desert **57** p
Al Ashkhirah **60** q
Alaska **11**
Alaska, Gulf of **11**
Alaska Peninsula **11** o
Alaska Range **11** e
Al Aswaa, Ras **39** n
Alat **55** r
Ala Tau **55** z
Alatyr **54** i
Alausí **22** x
Al Awaynat **37** t
Al Aziziyah (Iraq) **39** e
Al Aziziyah (Libya) **34** z
Albacete **34** r
Al Bad **38** k
al Baida **43** h
Al Balyana **38** i
Albanel Lake **17** a
Albania **35** d
Albany (Australia) **76** e
Albany (Georgia) **20** j
Albany (New York) **16** z
Albany (Oregon) **14** g
Albany (River) **16** h
Albany Downs **74** q
Al Bardi **35** t
Al Batinah **39** z
Al Bayda **35** r
Alberga **77** d
Alberta **14** l
Albert, Lake **42** w
Albert Lea **16** o
Albert Nile **42** x
Alberton **17** e
Al Bir **38** l
Ålborg **33** p
Albuquerque **15** p
Albury **78** o
Al Busaiyah **39** e
Alcantara **27** b
Alcova **15** p
Alcudia **34** u
Aldabra Island **47** d
Aldama **18** k
Aldan **52** x
Aldan (River) **53** a
Aldar **56** l
Aleg **40** d
Alej **56** d
Alejandro Sekirk **30** c
Alejsk **56** d
Aleksandrovskoye **51** q
Aleksandrovsko-Sakhalinskiy **53** r
Alekseyevka (Kazakstan) **55** j
Alekseyevka (Ukraine) **54** d
Alekseyevo **53** h
Além Paraiba **29** j
Alençon **34** f
Alenquer **26** h
Aleppo **54** q
Alert **91** g
Alert, Cape **89** t
Alerta **25** n
Alessandria **34** k
Ålesund **33** a
Aleutian Islands **10** u
Alevina, Cape **53** v
Alexander Archipelago **11** x
Alexander Bay **48** u
Alexander Island **82** w
Alexandra **79** r
Alexandra, Lake **77** t
Alexandra, Land **50** f
Alexandria (Australia) **73** s
Alexandria (Egypt) **38** g
Alexandria (Louisiana) **19** g
Alexandria (Minnesota) **15** j
Alexandria (Virginia) **16** y
Al Faiyûm **38** h
Al Farcia **36** f
Al Faw **39** g
al Fúla **42** i
Alga **55** a
Algarroho del Aguila **30** j
Algarve **34** n
Algatart **55** w
Al Gaydah **43** m
Algeciras **34** p
Alger (Algiers) **34** u
Algeria **34/36-37**
Al Ghaydah **39** w
Algiers **34** u
al Hadd, Ras **60** q
Al Hadithah (Iraq) **39** c
Al Hadithah (Saudi Arabia) **38** m
Al Hajar al Gharbi **39** z
Al Hajar ash Sharqi **60** p
Al Hammâm **38** h
Al Hanakiyah **39** o
Al Hasakah **54** r
Al Hazm **39** r
Al Hijaz **38** m
Al Hillah **39** s
Al Hoceima **34** q
Aliabad **39** l
Aliança **25** f
Ali-Bagata **53** a
Alicante **34** s
Alice Springs **73** r
Aligarh **61** c
Alijos, Rocas **18** d
Alindao **42** q
Alipxa **51** u
Ali-Sabieh **43** e
Al Isawiyah **38** m
Aliwal North **49** o
Al Jabal al Akhdar **35** r
Al Jaghbub **38** d
Al Jalamid **39** b
Al Jawf (Libya) **38** p
Al Jawf (Saudi Arabia) **39** a
Al Jubayl **39** h
al Junayna **42** e
Aljustrel **36** g
Al Kamil **60** p
Al Karak **38** l
Al Khaburah **60** o
Al Khamasin **39** q
Al Khandaq **38** u
Al Khârga **38** u
Al Khârga Oasis **38** u
Al Khartum **42** k
Al Khaznah **39** y
Al Kuf **39** f
al Kú, Wadi **42** g
Allahabad **61** r
Allakh-Yun' **53** c
Allegheny **16** w
Allegheny Mountains **16** w
Allende **19** a
Allentown **16** y
Alleppey **64** e
Alliance **15** r
Allier **34** h
Al Lith **39** n
Alma **17** a
Almaden **74** b
al Mado **43** j
al Mahdad **43** g
Al Maiyah **39** d
Al Majm'ah **39** e
Al Mansura **38** i
Al Marj **35** r
Almaty **55** z
Almaznyy **52** f
Almeirim **26** j
Al Mekhily **38** e
Almenara **29** k
Almería **34** r
Al'metyevsk **54** m
al Milk **38** u
Al Minyâ **38** h
Almirante **22** h
Almora **61** d
Al Mudauwara **38** l
Al Mughairah **38** m
Al Muglad **42** i
al Mukalla **43** k
al Mukha **43** f
Alor **69** w
Alor Setar **65** w
Alota **28** e
Alotau **74** f
al Ouaday **42** a
Alpena **16** h
Alpercatas, Serra das **27** p
Alpha **74** q
Alpine **18** l
Alps **34** j
Al Qahirah **38** i
Al Qaisumah **39** e
Al Qalibah **38** m
Al Qariyat **37** i
Al Qunfudhah **39** o
Al Qurain **39** q
Al Qurnah **39** f
Alroy Downs **73** t
Alta **33** i
Alta Gracia **28** t
Altah **56** h
Altai (Mountains) **56** e
Altaj **56** l
Altaj (High Altai) **56** y
Altamaha **20** l
Altamira (Brazil) **26** j
Altamira (Mexico) **20** t
Altamonte Springs **20** l
Altar Desert **18** f
Altares **18** m
Altat **51** w
Altay **56** h
Altejskij **56** f
Altiplano **28** d
Alto do Araguaia **29** a
Alto Hama **44** t
Alto Molócuè **49** k
Alton **16** q
Alton Downs **77** f
Altoona **16** x
Alto Paraiso de Goiás **29** f
Alto Parnaíba **27** n
Alto Purús **25** n
Altun Shan **56** u
Alturas **14** v
Altus **19** b
Alu **71** u
Al Udailiyah **39** t
al Udaiya **42** i
Al Ulaya **39** o
Al Uruq al Mutaridah **39** x
Alva **15** u
Alvarado **19** r
Alvares **25** e
Alvorada **26** y
Al Wafra **39** g
Al Wajh **38** l
Alwar **61** b
Alxa Zuoqi **57** q
Alyaskitovyy **53** d
Alygdzer **56** m
Amadeus, Lake **73** p
Amadjuak Lake **13** j
Amakinskiy **52** f
Amapá **26** k
Amara **43** b
Amarante **27** q
Amarillo **15** s
Amasin **36** p
Amata **77** a
Amatenango **19** t
Amazar **52** v
Amazon **25-26**
Amazon (Solimões) **25** b
Ambala **61** b
Ambalavao **47** q
Ambam **41** v
Ambar **51** i
Ambarchik **10** a
Ambardakh **53** c
Ambato **22** z
Ambikapur **61** t
Ambilobe **47** e
Ambo **24** x
Ambohitralanana **47** g
Ambon **69** z
Ambositra **47** q
Ambovombe **47** q
Ambriz **44** e
Amchitka **10** u
Amderma **50** l
Amdo **61** l
Ameca **18** y
Americana **29** f
American Highlands **84** z
American Samoa **234** n
Amery Basin **85** p
Amery Ice Shelf **84** n
Ames **16** o
Amga **53** a
Amga (River) **53** a
Amguid **37** p
Amherst **17** f
Amiens **34** h
Amindivi Islands **64** b
Amistad Reservoir **19** a
Amizmiz **36** g
Amka **53** q
Amman **38** l
Ammassalik **90** y
Amo **52** b
Amol **54** y
Amolar **28** l
Amourj **40** h
Amoy **63** s
Ampanihy **47** o
Ampato **25** n
Amravati **61** p
Am Raya **41** m
Amritsar **61** a
Amsel **37** p
Amsterdam **33** n
Amsterdam Island **235** r
Amudarja **55** q
Amudat **42** z
Amund Ringnes Island **12** m
Amundsen Bay **84** m
Amundsen Gulf **12** d
Amundsen Scott **83** z
Amundsen Sea **81** n
Amur (River) **58** k
Amur (Sudan) **38** x
Amursk **58** m
Amysakh **52** h
Anabar **52** f
Anaco **23** h
Anaconda **14** m
Anadyr' **10** g
Anadyr, Gulf of **10** h
Anadyr Mountains **10** f
Anadyr' (River) **10** f
Anah **39** c
Anaheim **18** c
Anai **37** t
Anai Mudi **64** f
Anajás **26** l
Anaka **42** y
Anaktuvuk Pass **11** e
Anama **25** i
Anama, Lake **51** k
Anambas Islands **68** i
Anamur **35** x
Anamur, Cape **35** x
Ananjevo **55** z
Anantapur **64** f
Anantnag **61** a
Anápolis **29** e
Anar **39** l
Anarak **39** k
Anatolia **54** p
Añatuya **28** v
Anauá **26** d
Ancha **53** d
Anchorage **11** e
Ancohuma **25** p
Ancona **35** a
Ancud **30** g
Anda **58** f
Andahuaylas **24** z
Andalgalá **28** s
Åndalsnes **33** b
Andalusia **19** k
Andaman Sea **65** f
Andaman Islands **65** d
Andapa **47** f
Anderson (Canada) **89** b
Anderson (Indiana) **16** t
Anderson (South Carolina) **20** l
Andes **22-25/28/30**
Andirlangar **56** r
Andižan **55** w
Andkhoy **55** s
Andoas **24** k
Andong **58** v
Andoom **73** j
Andorra **34** 9
Andöya **33** f
Andradina **29** c
Andranovory **47** o
Andreanof Islands **10** v
Andreas, Cape **35** y
Andrées Land **91** r
Andreyevka **54** k
Andreyevskiy **52** t
Andriamena **47** q
Andros **35** t
Andros Island **21** n
Andulo **44** u
Andyngda **52** g
Anéfis **36** z
Aney **37** v
Angara **51** x
Angarsk **57** c
Angas Downs **73** q
Angel de la Guarda **18** f
Angeleri **42** w
Angeles **67** f
Angel Falls **23** v
Ångerman **33** g
Angers **34** f
Angkor **66** g
Anglesey **32** w
Ango **42** t
Angoche **46** l
Angola **44**
Angostura Reservoir **19** t

Angoulême **34** g
Angra Dos Reis **29** i
Angren **55** v
Angu **42** s
Anguilla **21** y
Angurugu **73** f
Angwa **45** r
Anhua **63** a
Aniak **11** b
Anjouan **47** b
Anju **58** t
Anka **41** e
Ankacho **52** d
Ankang **62** m
Ankara **35** y
Ankazoabo **47** o
Ankazobe **47** q
Anking **63** e
Anlong **62** x
Annaba **34** x
An'Nahûd **42** i
Anna Plains **72** w
Annapolis **16** y
Annapurna **61** g
Ann Arbor **16** u
An Nashash **39** y
An Nasiriyah **39** e
Annecy **34** j
An Nimas **39** p
Anning **62** u
Annitowa **73** t
An Nukhaib **39** b
Anou Mellene **36** z
Ansab **39** d
Anschan **58** e
Anschluss **48** h
Anse d'Hainault **21** q
Anshun **62** x
Anson Bay **73** b
Ansongo **41** a
Antakya **35** z
Antalaha **47** f
Antalya **35** w
Antalya, Gulf of **38** h
Antananarivo **47** q
Antanimora **47** p
Antarctica **84**
Antarctic Peninsula **82** u
Antelope Mine **49** d
Antequera **34** p
Anthonys Lagoon **73** s
Anti Atlas **36** g
Anticosti Island **17** f
Antigua and Barbuda **21** z
Antipayuta **51** d
Antipinskiy **52** m
Antipodes Islands **235** z
Antofagasta **28** b
Antofagasta de la Sierra **28** r
Antofalla **28** q
Antong (Dandong) **58** r
Antonovo **54** l
Antsalova **47** o
Antsirabe **47** q
Antsiranana **47** f
Antsohihy **47** d
An Tuc **66** k
Antuerpia **25** t
Antwerp **33** n
Antwerpen (Antwerp) **33** n
Anuradhapura **64** u
Anvers Island **82** s
Anxi **56** y
Anxious Bay **77** c
Anyang **63** d
Anzhero-Sudzhensk **56** f
Aomori **59** c
Aoraki (Mt. Cook) **79** r
Aosta **34** k
Aouderas **37** r
Aouk **42** c
Aoukr **40** g
Aoukarr **36** w
Aouker **36** s
Aoulef el Arab **36** m
Aozi **37** y
Apalachee Bay **20** j
Apan **19** p
Apaporis **25** a
Aparri **67** g
Apatou **26** i
Apatzingán **18** z
Apennines **34** m
Apere **25** r
Api **42** t
Api (River) **42** t
Apiacás, Serra dos **25** x
Apaiú, Serrado **23** v
Apizaco **19** p
Aplao **24** z
Apo **69** k
Apodi **27** u
Apoera **26** g
Apollo Bay **77** w
Apolo **25** p
Aporema **26** l
Apostle Islands **16** d
Apóstoles **28** z
Apoteri **23** y
Appalachian Mountains **16** v
Appleton **16** r
Apucarana **29** c
Apuka Chukotskaya **10** d
Apure **23** e
Apurimac **24** y
Aqaba **38** k
Aqaba, Gulf of **38** k
Aqcha **60** h
Aqqikkol, Hu **61** j
Aqua Clara **29** b
Aquidabán **28** m
Aquidauana **28** m
Aquidauana (River) **28** m
Araara **43** s
Arabelo **23** u
Arabian Desert **38** j
Arabian Gulf **39** i
Arabian Sea **60** s
Araca **25** g
Aracaju **27** u
Aracati **27** u
Araçatuba **29** d
Aracena **34** p
Aracui **29** j
Arad **35** e
Arafat **39** o
Arafura Sea **73** c
Aragarças **29** c
Araguacema **26** y
Araguaçu **26** y
Araguaia **26** z
Araguaína **26** z
Araguari **26** j
Arak **39** h
Arakan Mountains **62** o
Arak Bordj **37** n
Araks **54** v
Aral **56** q
Aral Sea **55**
Aral'sk **55** d
Aralsor, Lake **54** j
Aramac **74** p
Aramberri **19** o
Aranos **48** k
Arapari **25** y
Arapey **28** y
Arapiraca **27** u
Ar'ar **39** b
Araracuara **24** m
Araranguá **29** r
Araraquara **29** f
Araras **26** v
Araras, Serra das **29** o
Ararat **77** v
Ararat, Mount **54** t
Arari **27** b
Ar Arish **38** j
Arauca **23** c
Arauca (River) **23** f
Aravalli Range **60** z
Araxá **29** f
Arba Minch **43** o
Arbil **39** d
Arcas **19** u
Archangel **50** q
Archei **38** o
Archer's Post **43** p
Arčman **55** n
Arco **14** z
Arcoverde **27** u
Arctic Bay **13** d
Arctic Circle **53** l
Arctic Ocean **10-11/32-33/89**
Arctic Red River **11** m
Arctic Village **11** h
Arctowski **82** r
Ar Dăg **54** q
Ardahan **54** s
Ardakan **39** k
Arde **43** h
Ardebil **54** w
Ardencaple Fjord **91** t
Ardestan **39** j
Ar Dzargala **57** j
Arenas **19** u
Arendal **33** e
Arequipa **25** n
Arere **26** i
Areyonga **73** q
Arezzo **34** m
Arga **53** h
Argadargada **73** t
Arga Sala **52** d
Argentina **28-33**
Argentino Lago **30** u
Arghandab **60** g
Argun **58** c
Argungu **41** d
Argun Youqi **58** c
Argun Zuoqi **58** d
Argut **56** g
Argyle, Lake **73** a
Århus **33** p
Ariamsvlei **48** x
Ariari **23** n
Arica (Bolivia) **28** c
Arica (Colombia) **25** a
Arica (Peru) **24** k
Arinos **29** g
Arinos (River) **25** x
Aripuan **25** w
Aripuan (River) **25** w
Ariquemes **25** t
Arizona **18** f
Arizpe **18** h
Arjeplog **33** g
Arjo **43** a
Arjona **22** m
Arka **53** r
Arkadelphia **19** f
Arkagala **53** g
Arkalyk **55** g
Arkansas **15** w
Arkansas City **15** v
Arkansas (River) **19** g
Arkansas (State) **19** f
Arkbajtal **55** x
Arkhangelsk (Archangel) **50** q
Arkhara **58** i
Arklow **32** v
Arkona, Cape **33** r
Arktichesgo Instituta **51** f
Arles **34** i
Arlington (Texas) **19** c
Arlington (Washington) **16** x
Arman **53** v
Armavir **54** s
Armenia **22** z
Armenia **54** u
Armidale **78** f
Armstrong (British Columbia) **14** i
Armstrong (Ontario) **16** e
Arnauti, Cape **35** x
Arnhem **33** n
Arnhem, Cape **73** g
Arnhem Land **73** d
Aroab **48** x
Arrah **61** u
Ar Ramadi **39** d
Ar Randa **39** p
Ar Rawda **43** i
Arrecifal **23** q
Arrecife **36** c
Arriaga **19** s
Ar Rifa **39** f
Arrilalah **73** y
Arrowsmith **79** s
Ar Rutbah **39** b
Ar Ruwais **39** i
Arsen'yev **58** k
Arsk **50** t
Arta **43** f
Artém **58** k
Artemovskiy **50** y
Artesia **18** l
Arthur **77** x
Arthurs Pass **79** t
Artigas **28** z
Artillery Lake **12** u
Art'omosk **56** k
Arturo Prat **82** s
Artyk (Russia) **53** f
Artyk (Turkmenistan) **55** o
Aru **42** w
Aru Islands **70** r
Arua **42** x
Aruaña **29** c
Aruba **23** c
Arum **25** h
Arun Oi **58** e
Arusha **45** i
Aruwimi **42** s
Arvaj Cheer **57** e
Arviat (Eskimo Point) **13** n
Ary **52** j
Arys' **55** u
Arys, Lake **55** g
Arzamas **54** g
Asab **48** v
Asadabad **60** k
Asahi-dake **59** d
Asahikawa **59** d
Asanib **48** j
Asansk **51** x
Asayta **43** e
Asbe Tafari **43** d
Ascensión Bay **20** v
Ascensión (Bolivia) **25** t
Ascensión (Mexico) **18** j
Ascension Island **234** y
Ascoli **35** a
Asedjrad **37** n
Asela **43** d
Asha **55** a
Ashburton **79** t
Ashburton (River) **72** t
Ashburton Downs **72** s
Asheville **16** u
Ashgabat **55** o
Ashland (Montana) **15** c
Ashland (Wisconsin) **15** l
Ash Sha'am **39** l
Ash Sharmah **38** l
Ash Shihr **43** k
Ash Shubaiyai **39** d
Ash Shumlul **39** g
Ash Shwayrif **37** i
Ashtabula **16** v
Ashton **15** a
Asi **38** l
Asilah **34** p
Asimba **43** d
Asino **51** u
Asir **39** p
Askale **54** s
Asmara **43** c
Asosa **42** m
Aspiring **79** q
Assab **43** f
Assad Reservoir **39** a
Assale, Lake **43** d
As Salman **39** e
Assamakka **37** q
As Samawah **39** e
As Sawadah **39** r
As Sib **60** o
As Sidr **39** n
As Sila **39** v
Assiniboia **15** c
As Sua **39** p
As Sulaiyil **39** r
As Suq **39** p
As Suwaiq **60** o
Astakh **53** c
Astana **55** j
Astara **54** w
Astillero **25** o
Astoria **14** g
Astrakhan **54** i
Asuka **84** d
Asunción **28** y
Aswân **38** w
Asyût **38** h
Atacama Desert **28** p
Atakh-Yuryakh **52** i
Atakor **37** p
Atakpamé **41** n
Atalanka **57** b
Atalaya **24** y
Atambua **69** w
Atar **36** p
Atasu **55** i
Atbara **38** w
Atbara (River) **38** x
Atbasar **55** h
Atchison **15** w
Atebubu **40** z
Athabasca **14** m
Athabasca (River) **14** l
Athabasca, Lake **12** v
Athens (Georgia) **20** k
Athens (Greeee) **35** t
Athens (Ohio) **16** v
Atherton **74** c
Athínai (Athens) **35** t
Athlone **32** u
Atholl, Cape **91** b
Ati **42** b
Atiak **42** x
Atico **24** y
Atikokan **16** c
Atikonak Lake **17** f
Atkarsk **54** h
Atka (USA) **10** w
Atka (Russia) **53** i
Atlanta **20** j
Atlantic City **16** z
Atlantic Ocean **234**
Atlas Mountains **36/37**
Atlasovo **53** y
Atlin, Lake **11** y
Atlixco **19** p
Atouat **66** j
Atoyac **19** o
Atrato **22** l
Attawapiskat **16** g
Attawapiskat Lake **16** e
Attikamagen Lake **17** d
At Timan **42** e
Attopeu **66** j
Attu **10** r
Atuel **30** k
Atyrau **54** l
Auckland **79** u
Auckland Island **235** x
Augathella **74** p
Augrabies Falls **48** x
Augsburg **34** l
Augusta (Georgia) **20** l
Augusta (Maine) **17** o
Augusta Victoria **28** d
Augustus **72** s
Augustus Downs **73** v
Augustus, Mount **72** t
Auld, Lake **72** x
Aurangabad **61** n
Aurillac **34** g
Aurora (Suriname) **26** h
Aurora (USA) **15** x
Aurukun **73** j
Aus **48** u
Aust. Capital Territory **78** d
Austin, Lake **76** e
Austin (Minnesota) **16** p
Austin (Nevada) **14** x
Austin (Texas) **19** d
Austral Downs **73** t
Australia **72-74/76-78**
Austria **33-35**
Autlán **18** y
Auxerre **34** h
aux Feuilles, Rivière **13** v
Auzangate **25** n
Auzu **37** x
Avaré **29** e
Aveiro **34** a
Avenal **14** v
Avezzano **35** a
Avignon **34** i
Avila **34** c
Avispas **18** g
Avon **76** e
Avon Downs **73** t
Avon Park **20** l
Awaran **60** h
Awasa **43** p
Awash **43** d
Awash (River) **43** d
Awaso **40** y
Awat **56** p

Awaynat Wanin **37** h
Aweil **42** h
Awetu **43** r
Awjilah **38** b
Axel Heiberg Island **13** b
Ayacucho (Argentina) **31** c
Ayacucho (Peru) **24** y
Ayakkum Hu **56** u
Ayan **53** p
Ayan (Ilimpeya) **52** b
Ayan (Nepa) **52** o
Ayan (River) **51** k
Ayangba **41** s
Ayaviri **25** o
Ayaya **53** o
Aydabu **55** h
Aydin **35** v
Ayers Roek **73** p
Aykhal **52** e
Aylmer Lake **12** v
Ayoûn el Atroûs **40** g
Ayr (Australia) **74** q
Ayr (UK) **32** w
Ayshirak **55** i
Ayu Islands **70** e
Ayvalik **35** u
Azamgarh **61** s
Azángara **25** n
Azaouad **36** x
Azaouak **41** d
Azaouak (River) **37** o
Azaouak, Vallée de l' **41** e
Azefal **36** o
Azemmour **36** g
Azerbaijan **54** v
Azhbulat, Lake **55** m
Azna **39** h
Azogues **24** i
Azopol'ye **50** s
Azores **234** k
Azoum **42** d
Azov, Sea of **54** c
Azrou **36** i
Azuero Peninsula **22** i
Azúl **31** b
Azum **42** e
Azurduy **28** g
Az Zaydiyah **43** f
Az Zilli **39** e
Az Zuqar **43** f

B

Baacagaan **57** n
Baalbek **38** l
Babaçulândia **26** z
Babahoyo **22** w
Bab al Mandab **43** f
Babanusa **42** i
Babar **70** o
Babati **45** i
Babel Thuap **70** f
Babile **43** f
Babo **70** r
Baboua **41** y
Babuyan Islands **67** l
Bababal **27** o
Bacajá **26** j
Bacan Islands **69** l
Bacău **35** h
Bachardok **55** p
Bachinivas **18** j
Bachu Liuchang **56** o
Back **12** m
Bae Ninh **62** x
Baco **67** g
Bacolod **67** i
Badain Jaran **57** o
Badajós **25** g
Badajós, Lago **25** g
Badajoz **34** o
Badgastein **35** a
Bad Hersfeld **33** p
Badillo **23** a
Badin **60** w
Badr Hunain **39** n
Badu **70** x
Badu (Island) **73** j
Baduen **43** w
Badulla **64** v
Baeza **22** y
Bafang **41** u
Bafang (River) **40** f
Bafatá **40** c
Baffin Bay **13** g
Baffin Island **13** g
Bafia **41** u
Bafoulabé **40** f
Bafoussam **41** u
Bafq **39** l
Baft **60** b
Bafwabalinga **42** u
Baga **41** j
Bagadzha **52** i
Bagamoyo **45** k
Bagan **62** e
Bagata **44** i
Bagazán **24** m
Bagdarin **52** s
Bagé **29** n
Baghdad **39** e
Baghin **60** a
Baghlan **60** j
Bago **67** h
Bagua **24** i
Baguinéda **40** i
Baguio **67** g
Bagzane **37** s
Bahamas **21** b
Bahariya, Oasis **38** g
Baharu **66** s
Bahawalpur **60** l
Baheh **39** e
Bahia Negra **28** k
Bahir Dar **43** b
Bahla **39** m
Bahrain **39** i
Bahrain, Gulf of **39** i
Bahr al Arab **42** g
Bahr al Ghazel **42** i
Bahr al Milh **39** d
Bahu Kalat **60** q
Baia Mare **35** f
Baio **26** l
Baïbokoum **41** y
Baicheng **58** e
Baidarata Bay **50** m
Baidoba **43** t
Baie Comeau **17** c
Baie-St-Paul **17** a
Baihe **63** a
Bailundo **44** u
Baima **62** g
Baingoin **61** k
Baiquan **58** f
Baird Mountains **11** a
Bairin Youqi **57** y
Bairin Zouqi **57** x
Bairnsdale **78** p
Bais **67** h
Bait al Faqih **43** f
Baixingt **57** t
Baiyanghe **56** t
Baiyuda **38** v
Bajan **57** e
Bajan Baraat **57** d
Bajanchongor **57** a
Bajan-Dalaj **57** p
Bajanleg **57** n
Bajan-Öndör **56** z
Bajan Ovoo **57** h
Bajan-Uul **56** l
Bajan-Uul (Uldz Gol) **57** h
Bajeador, Cape **63** u
Bajo Caracoles **30** v
Bajram Ali **55** r
Bajsint **57** u
Bakchar **51** s
Baker (California) **14** x
Baker (Montana) **15** e
Baker (Oregon) **14** j
Baker, Lake **12** z
Baker Lake **12** z
Bakersfield **14** w
Bakhta **51** v
Bakhta (River) **51** w
Bakhtaran **39** f
Bakhynay **52** j
Bako **40** h
Bakrakondre **26** g
Baku **54** x
Balabac **67** q
Balabac Strait **67** q
Bala, Cerros de **25** p
Baladek **53** n
Balagannakh **53** c
Balagannakh **53** k
Balagannoye **53** u
Balagansk **57** e
Balaghat **61** r
Balaipungut **66** s
Balaka **46** i
Balakovo **54** j
Bala Murghab **60** f
Balancán **19** u
Balāngīr **61** t
Balashikha **54** d
Balashov **54** g
Balasore **61** v
Balaton, Lake **35** c
Balbi **71** u
Balboa **22** j
Bald Cape **76** e
Balearic Islands **34** t
Baley **57** j
Balgazya **56** l
Balgo **73** n
Bali **69** p
Bali Sea **69** q
Balikesir **35** v
Balikpapan **69** e
Balkans **35** g
Balkhab **60** i
Balkhash **55** l
Balkhash, Lake **55** k
Balladonia **76** j
Ballarat **77** w
Ballé **40** h
Ballenas Bay **18** f
Balleny Islands **86** w
Balombo **44** t
Balonne **78** d
Balotra **60** z
Balrampur **61** f
Balranald **77** j
Balsas (Brazil) **27** n
Balsas (Mexico) **19** n
Balsas (Peru) **24** i
Balla **25** n
Bălti **35** i
Baltic Sea **33** t
Baltîm **38** i
Baltimore **16** y
Baluchistan **60** e
Bam (Iran) **60** c
Bam (Russia) **52** w
Bama **41** k
Bamaga **73** k
Bamako **40** h
Bamba (Mali) **40** m
Bamba (Senegal) **40** e
Bambama **44** f
Bambamarca **24** i
Bambari **42** p
Bamberg **33** q
Bambesa **42** t
Bambesi **42** m
Bambili **42** t
Bamboi **40** m
Bamboo Creek **72** v
Bambui **29** g
Bambuyka **52** t
Bamenda **41** u
Bamian **60** i
Bamingui **42** b
Bampur **60** d
Bampur (River) **60** d
Bamyili **73** d
Banados del Atuel **30** k
Banados del Izozog **28** i
Banagi **45** h
Banalia **42** t
Banamba **40** i
Banana **74** s
Banas **61** o
Banas, Ras **38** y
Ban Ban **65** m
Banbar **62** c
Bancroft **16** k
Banda (India) **61** r
Banda (Dem. Rep. Congo) **44** j
Banda Aceh **65** s
Banda Sea **69** x
Bandama **40** w
Bandar Abbas **60** a
Bandar Anzeli **54** x
Bandar-e-Khomeini **39** h
Bandar-e-Lengeh **39** l
Bandar-e-Margam **39** j
Bandar-e-Rig **39** h
Bandarlampung (Telukbetung) **68** v
Bandar Seri Begawan **69** b
Bandar Sri Aman **68** m
Bandeirante **26** x
Bandera **28** v
Banderas Bay **18** x
Bandiagara **40** k
Bandirma **35** i
Bandundu **44** i
Bandung **68** x
Baneh **54** v
Banes **21** p
Banff **14** l
Banfora **40** k
Bangalore **64** f
Banganapalle **64** g
Bangassou **42** r
Bangeta **71** n
Banggai Islands **69** i
Banggi **67** q
Bangka **68** w
Bangkalan **69** n
Bangkok (Krung Thep) **66** e
Bangladesh **61**
Bangor **17** c
Bangoran **42** c
Bangued **67** f
Bangui **42** n
Bangweulu, Lake **45** q
Bani **40** j
Bani Sar **39** p
Bania **41** y
Baniara **74** f
Banja Luka **35** b
Banjarmasin **69** o
Banjul **40** a
Banket **46** r
Banks Island (Canada) **12** d
Banks Islands (Micronesia) **75** e
Banks Peninsula **79** u
Banks Strait **78** p
Bankura **61** v
Ban Me Thout **66** k
Ban Na San **66** q
Bannu **60** k
Ban Pong **66** d
Ban Pu Kroy **66** j
Banswara **61** n
Ban Thateng **66** i
Banvalot **61** i
Banya Fort **43** n
Banyo **41** v
Banyuwangi **69** p
Banzare Coast **87** d
Banzart (Bizerte) **34** y
Baoji **62** l
Bao Loc **66** j
Baoshan **62** s
Baouna **40** l
Bap **60** m
Baqên **62** c
Ba'qubah **39** e
Bara (Nigeria) **41** h
Bara (Sudan) **42** j
Barabinsk **56** b
Baracoa **21** q
Baragoi **43** o
Barahona **21** s
Baralaba **74** r
Baram **69** b
Baran **61** o
Baranga **42** u
Baranoa **22** m
Baranof Island **11** x
Baranovichi **33** x
Barat Daya Islands **69** y
Barbacena **29** h
Barbados **23** k
Barbas, Cap **36** n
Barbuda **21** z
Barcaldine **74** p
Barcelona (Spain) **34** h
Barcelona (Venezuela) **23** 9
Barcelos **23** u
Bareoo **74** o
Bardai **37** x
Barddhaman **61** v
Bardera **43** s
Bardeskan **60** b
Bareilly **61** d
Barentsburg **92** o
Barents Island **92** p
Barents Sea **50** d
Barga **61** e
Barguzin **52** q
Bari **35** c
Barinas **23** c
Baring, Cape **12** e
Baringo Lodge **43** n
Baripada **61** v
Baris **38** u
Barisal **61** x
Barisan Mountains **68** s
Barito **69** c
Barka **38** z
Barkal **61** z
Barkam **62** i
Barkly East **49** p
Barkly Tableland **73** s
Barkly West **49** n
Barkol Hu **56** w
Barkol Kazak **56** w
Barlee, Lake **76** g
Barlow **11** j
Barmer **60** y
Barnaul **56** e
Barnes Ice Cap **13** i
Baro **41** e
Baroda (Vadodara) **60** z
Barquisimeto **23** c
Barra **27** p
Barraca da Boca **26** j
Barra de Navidad **18** x
Barra do Corda **27** o
Barra do So Manuel **25** k
Barranca **24** j
Barrancabermeja **23** a
Barrancas (Argentina) **30** j
Barrancas (Venezuela) **23** h
Barranquilla **22** m
Barre **17** n
Barreira do Campo **26** y
Barreiras **27** o
Barreiros **27** w
Barretos **29** e
Barrhead **14** l
Barrie **16** w
Barrier Range **77** h
Barrington Tops **78** f
Barrow **11** c
Barrow, Cape **11** c
Barrow Creek **73** r
Barrow Island **72** r
Barrow Strait **13** a
Barsa Kel'mes **55** c
Barsi **61** o
Barstow **14** x
Bartica **23** y
Bartle Frere **74** c
Bartlesville **15** v
Barumun **68** e
Baruun Turuun **56** k
Baruun-Urt **57** i
Barwani **61** n
Barysh **54** i
Basac **66** h
Basankusu **42** p
Bascuñan, Cabo **28** o
Bashi Channel **63** u
Basilan **67** t
Basilan (Island) **67** u
Bas-Kouilou **44** d
Basle (Basel) **34** k
Basoko **42** r
Basotho **49** b
Basra **39** g
Bassari **41** a
Bassas da India **46** y
Bassein **65** f
Basse Santa Su **40** d
Basseterre **21** y
Bassett **15** t
Bassikounou **40** i
Bassila **41** b
Bass Strait **77** x
Bastak **39** k
Bastia **34** k
Bata **41** u

Batagay 53 a
Batakan 69 o
Batama 42 u
Batamay 52 m
Batang 62 f
Batangafo 42 n
Batangas 67 g
Batan Islands 63 v
Batatais 29 f
Bataysk 54 e
Bat Cengel 57 b
Batchelor 73 b
Batha 42 b
Bathurst (Australia) 78 e
Bathurst (Canada) 17 e
Bathurst, Cape 12 a
Bathurst Inlet 12 i
Bathurst Island (Canada) 12 l
Bathurst Island (Australia) 73 b
Bati 43 d
Batista, Serra da 27 r
Batna 34 w
Batomga 53 p
Baton Rouge 20 e
Batouri 41 x
Batovi 29 b
Batsireet 57 g
Battambang 66 f
Batterbee Range 82 x
Batticaloa 64 w
Battle Creek 16 t
Battle Mountain 14 x
Batui 69 h
Batumi 54 s
Baturino 56 f
Baubau 69 v
Bauchi 41 h
Baunt 52 s
Baures 25 s
Baurú 29 d
Bavispe 18 i
Bawean 69 n
Bawîti 38 g
Bawku 40 m
Ba Xian 62 l
Baxoi 62 e
Bay Islands 20 v
Bayamo 21 o
Bayamón 21 v
Bayanga 41 z
Bayan Har Shan 62 f
Bayan Mod 57 p
Bayan Obo 57 s
Bayawan 67 h
Bay City 16 u
Bayir 38 l
Baykal, Lake 52 p
Baykalovo 51 n
Baykal'skiy Mountains 52 q
Baykal'skoye 52 q
Baykit 51 y
Baykonyr 55 g
Baymak 55 b
Bayneu 54 m
Bayombong 67 g
Bayonne 34 f
Bayóvar 24 g
Baytown 19 e
Baza 36 k
Bazaruto 46 v
Bazman 60 d
Beagle Bay 72 k
Beagle Gulf 73 b
Beardmore Glacier 86 c
Beardstown 16 q
Bear Island 92 P
Bearskin Lake 15 l
Beatrice 15 v
Beatty 14 x
Beaufort 69 d
Beaufort Sea 11/89
Beaufort West 48 z
Beaumont 19 e
Beaver (River, Canada) 15 b
Beaver (River, USA) 15 t
Beaver Creek 11 i
Beavermouth 14 j
Beawar 61 n
Béboura 41 z
Beckley 16 v
Bedajós 26 b
Bedajós, Lago 26 b
Bedele 43 a
Bedford 16 s
Bedopa 51 y
Bedourie 73 u
Beed 61 o
Beer Sheba 38 k
Beethoven Peninsula 82 x
Befale 42 p
Befandriana Av 47 e
Befori 42 r
Beger 56 z
Begichev 52 f
Behbahan 39 i
Behring Point 21 o
Behsud 60 i
Bei'an 58 g
Beihai (Pakhoi) 62 z
Beijing 57 w
Beira 46 v
Beirut 38 k
Bei Shan 56 w
Beitbridge 49 e
Beitun 56 g
Bejaia 34 v
Beja (Portugal) 34 o
Beja (Tunisia) 34 x
Béka 41 w
Bekabad 55 v
Bekdas 54 y
Békéscsaba 35 d
Bela 60 v
Bela-Bela 46 q
Bel'agas 56 e
Belaja Šhola 56 d
Belarus 33 x
Bela Vista 49 t
Belaya Tserkov 33 z
Beleher Channel 13 a
Beleher Islands 13 s
Bélel 41 x
Belém 26 m
Belem de So Francisco 27 t
Belen (USA) 15 p
Belén (Argentina) 28 r
Belen'kaya 52 w
Beles 43 a
Belesc Cogani 43 r
Belet Huen 43 t
Beleutty 55 g
Belev 54 e
Beleya 43 a
Belfast 32 v
Belfield 15 e
Belgaum 64 d
Belgium 32-33/34
Belgorod 35 j
Belgrade 35 d
Beli 41 v
Belitung 68 y
Belize (City) 19 w
Belize (State) 19/20
Bel'kachi 53 n
Bel'kovskiy 53 b
Bell 49 p
Bellary 64 f
Bella Unión 28 y
Bella Vista (Argentina) 28 x
Bella Vista (Brazil) 28 m
Belle Isle, Strait of 17 j
Belleville 16 x
Bellingham 14 h
Bellingshausen 82 r
Bellingshausen Sea 80 l
Bello 22 m
Bell Peninsula 13 r
Bellyk 56 i
Belmopan 19 w
Belogorsk 58 h
Belo Horizonte 29 i
Beloit 16 r
Belo Monte 26 k
Belomorsk 50 n
Beloretsk 55 b
Belterra 26 h
Belyuen 73 b
Belyy Yar 51 t
Bembe 44 g
Bembéréké 41 c
Bemidji 15 j
Bena Dibele 44 l
Benadir 43 t
Bena-Tshadi 44 l
Bend 14 h
Benda Range 77 g
Bender Beila 43 l
Bendigo 77 x
Bene 46 g
Benevento 35 a
Bengal, Bay of 64 m
Bengamisa 42 t
Benghazi 35 q
Bengkolan Bay 66 x
Bengkulu 68 s
Benguela 44 r
Ben Guerdane 34 y
Benguerir 36 g
Beni 42 v
Beni Abbès 36 k
Beni Cunif 36 l
Beni Mazâr 38 h
Beni Mellal 36 i
Benin 41 b
Benin City 41 r
Beni (River) 25 q
Beni Saf 34 r
Beni Suef 38 i
Benito Juárez 31 b
Beni Walid 35 n
Benjamin Acéval 28 y
Benjamin Constant 25 b
Benmara 73 9
Bennetta 53 h
Benoni 49 q
Bentinck Island 73 i
Bentiu 42 j
Bentong 66 s
Benue 41 f
Beograd (Belgrade) 35 d
Berber 38 x
Berbera 43 h
Berberati 41 y
Berdale 43 w
Berdyansk 54 c
Berea 16 t
Bereda 43 l
Berekum 40 y
Berenice 38 x
Berens River 15 i
Berezniki 50 w
Berezovka 52 t
Berezovka (River) 53 k
Berezovo 50 z
Berezovo (Koryak Mountains) 10 e
Berezovskaya 52 u
Berga, Cape 51 m
Bergamo 34 l
Bergen 33 a
Berhala Strait 68 h
Berhampore 61 w
Berhampur 61 t
Beringovskiy 10 h
Bering Sea 10
Bering Strait 10 k
Berkakit 52 x
Berkner Island 83 s
Berlin (Germany) 33 r
Berlin (USA) 17 a
Bermejillo 18 l
Bermejo 28 g
Bermejo (River) 28 h
Bermejo (River) 28 r
Berne 34 j
Bernier Bay 13 b
Bernina 34 l
Bernstorff's Isfjörd 90 w
Beroubouay 41 c
Berrechio 36 h
Berriane 37 a
Bertes 53 f
Bertoua 41 w
Berwick upon Tweed 32 x
Besançon 34 j
Besar 69 q
Besham Qila 60 l
Besikama 69 x
Besima 60 h
Bessemer 20 h
Bestamak 56 b
Bestobe 55 k
Betafo 47 q
Bétaré-Oya 41 x
Betehor 43 c
Bethal 49 r
Bethanie 48 v
Bethany 15 w
Bethel 11 a
Bethlehem 49 p
Betioky 47 o
Betoota 77 h
Bet-Pak-Dala 55 i
Betroka 47 p
Belsiboka 47 q
Bettiah 61 h
Betul 61 p
Betwa 61 q
Beverley Springs 72 l
Beyla 40 h
Beysehir, Lake 35 x
Bhadrachalam 64 i
Bhadrakh 61 v
Bhadra Reservoir 64 e
Bhadrathalam 61 r
Bhadravati 64 e
Bhagalpur 61 w
Bhaktapur 61 i
Bhamo 62 r
Bhandara 61 q
Bhanjanager 61 t
Bharatpur 61 b
Bharuch 60 z
Bhatinda 61 a
Bhavnagar 60 y
Bhawanipatna 61 s
Bheri 61 f
Bhilwara 61 n
Bhima 64 e
Bhojpur 61 i
Bhopal 61 p
Bhor 60 z
Bhubaneswar 61 u
Bhuj 60 x
Bhutan 61 k
Bia 62 v
Biak 70 t
Bialystok 33 v
Biankbuma 40 v
Biaza 51 r
Bickle Knob 16 w
Bida 41 e
Bidar 61 p
Bidokht 60 c
Bielsko-Biala 35 d
Bien Hoa 66 j
Bienville, Lake 17 a
Big Bell 76 e
Big Delta 11 g
Biggar 15 b
Big Island 90 e
Bignona 40 b
Big Sandy 15 a
Big Sioux 15 v
Big Spring 19 a
Big Trout Lake 16 d
Bihác 35 b
Bihar 61 u
Bija 56 g
Bijapur 64 e
Bijar 39 g
Bijiang 62 s
Bijie 62 k
Bijsk 56 f
Bikaner 60 m
Bikin 58 l
Bikin (River) 58 m
Bikoro 44 i
Bilaspur 61 s
Bilbao 34 d
Bilecik 35 w
Bili 42 t
Bili (River) 42 s
Bilir 53 q
Biliran 67 i
Billabalon 76 d
Billabong 77 y
Billiluna 73 n
Billings 15 b
Bilma 37 v
Bilma, Grand Erg de 37 v
Biloela 74 s
Biloku 23 x
Biloxi 20 g
Biltine 42 d
Bimbo 42 n
Binalbagan 67 h
Binatang 69 a
Bindura 49 f
Binga 46 p
Bing Bong 73 f
Binghampton 16 y
Bing l 54 r
Binhai 63 g
Bintuhan 68 t
Bintulu 69 b
Binzert 37 f
Bio Bio 30 h
Bioko Island 41 t
Bir al Hasa 38 y
Birao 42 e
Biratnagar 61 j
Birdsville 77 g
Bir el Khzaim 36 t
Bir Enzarah 36 p
Bireuen 68 b
Bir Gandus 36 n
Birganj 61 h
Birikčul 56 i
Birilyussy 51 w
Birjand 60 d
Birmingham (UK) 32 x
Birmingham (USA) 19 j
Birnin Gwari 41 e
Bernin-Kebbi 41 d
Birobidzhan 58 k
Birou 40 f
Bir Oum Greine 36 r
Bir Shari 39 d
Biru 62 b
Bir Zar 34 y
Biscay, Bay of 34 d
Biscoe Island 82 t
Bishkek 55 x
Bishop 14 w
Biskra 34 w
Bismarck 15 g
Bismarck Archipelago 71 p
Bismarck, Cape 91 u
Bismarck Range 70 z
Bismarck Sea 71
Bison 15 f
Bissagos Islands 40 a
Bissau 40 b
Bitam 41 v
Bitkine 42 b
Bitola 35 e
Bitterfontein 48 w
Bitterroot Mountains 14 l
Biu 41 j
Biwa, Lake 58 z
Bjorne Peninsula 13 c
Bla 40 j
Blackall 74 p
Blackburn, Mount 11 h
Black Hills 15 r
Blackpool (UK) 32 w
Blackpool (USA) 14 i
Black Sea 35
Blackstone Camp 76 m
Black Volta 40 l
Blackwood 76 d
Blagoevgrad 35 f
Blagoveshchensk 58 h
Blagovsčenka 56 c
Blair Athol 74 q
Blanca, Bahía 31 a
Blanca, Laguna 30 v
Blanche, Lake 72 x
Blanc, Mont 34 j
Blanco, R. 25 t
Blanding 15 o
Blanquilla 23 g
Blantyre 46 h
Blenheim 79 u
Blida 34 u
Blind River 16 h
Blitta 41 a
Bloemfontein 49 o
Bloemhof 49 o
Bloemhof Reservoir 49 o
Bloomington (Illinois) 16 r
Bloomington (Indiana) 16 s
Bluefield 16 v
Bluefields 22 g
Blue Mountains 14 k
Blue Nile 42 m
Bluenose Lake 12 e
Blue Ridge 16 u
Blue River 14 i
Bluff Knoll 76 f

Blumenau **29** r
Blumut **66** t
Blythe **18** e
Blytheville **16** q
Bo **40** e
Boac **67** g
Boa Esperanca Reservoir **27** o
Boa Esperança **29** g
Boa Fé **24** m
Boatasyn **53** s
Boa Vista **23** w
Bobo-Dioulasso **40** k
Bobruysk **33** y
Boca del Pao **23** g
Boca do Acará **25** g
Boca do Acre **25** q
Boca do Curequeté **25** r
Boca do Jari **26** j
Boca do Moaco **25** c
Boca do Mutúm **25** c
Boca Grande **23** j
Bocaranga **41** y
Boco Mavaca **23** t
Boconó **23** c
Bodala **42** q
Bodaybo **52** s
Bodélé **42** a
Bodö **33** e
Boende **44** k
Boffa **40** c
Bogangolo **42** o
Boganida **51** l
Bogantungan **74** q
Bogda Feng **56** u
Bogdanovich **50** y
Bognes **33** f
Bogo **67** i
Bogor **68** w
Bogorodskoye **53** r
Bogotá **23** n
Bogotol **51** v
Bogra **61** x
Boguchany **51** y
Bogué **40** d
Bo Hai, Gulf of **58** o
Bohemia Downs **72** z
Bohodou **40** g
Bohol **67** i
Boise **14** x
Boise City **15** s
Bojador **36** o
Bojador, Cap **36** o
Bojeador, Cape **67** f
Bojnurd **55** o
Boké **40** d
Boknfjörd **33** a
Boko **44** g
Bokol Mayo **43** r
Boktor **58** m
Bolaños **18** y
Bole (China) **56** q
Bole (Ghana) **40** l
Bolgatanga **40** m
Bolgrad **35** i
Bolia **44** i
Bolvar (Peru) **24** i
Bolivar (USA) **15** x
Bolivia **25/26/28**
Bollon **74** q
Bolodzhak **53** n
Bologna **34** m
Bolognesi **24** y
Bolomba **42** o
Bolon **58** m
Bolonchén de Rejón **19** u
Bol’senarymskoje **56** f
Bolsevik **55** o
Bol’shaya Balakhnya **52** b
Bol’shaya Glushitsa **54** k
Bol’shaya Kheta **51** g
Bol’sherech’ye **51** q
Bol’shezemel’skaya Tundra **50** j
Bol’shiye Khatymy **52** x
Bol’shoy Lyakhovskiy **53** e
Bol’shoy Onguren **57** e
Bol’shoy Patom **52** t
Bol’shoy Shantar **95** j
Bol’shoy Yenisey **56** m
Bol’shoye Vlasévo **53** r
Bol’shoye Zimovye **53** g
Bolson de Mapimi **18** l
Bolu **35** k
Boluntay **62** b
Bolzano **34** l
Boma **44** e
Bomassa **41** z
Bombale **41** z
Bombay **60** z
Bomberai **70** r
Bömbögör **57** a
Bom Despacho **29** g
Bomi **62** d
Bomi Hills **40** s
Bomili **42** u
Bom Jardim **25** r
Bom Jesus da Lapa **27** q
Bom Jesus de Itabapoana **29** k
Bonaire **23** e
Bonaparte Archipelago **72** l
Bon, Cap **34** z
Bonda **44** e
Bondo **42** s
Bondoukou **40** k
Bone, Gulf of **69** t
Bonete **28** q
Bong **40** t
Bonga **43** n
Bongadanga **42** q
Bongor **41** l
Bongos, Massif des **42** d
Bonifacio **34** k
Bonifacio, Str. of **34** k
Bonin Islands **235** j
Bonito **28** m
Bonn **33** o
Bonner Ferry **14** k
Bonny **41** s
Böön Cagaan Nuur **57** n
Boonderoo **73** y
Boothby, Cape **85** a
Boothia, Gulf of **13** b
Boolhia Peninsula **13** a
Booué **44** e
Booylgoo Springs **76** g
Boqueiro **27** p
Boquilla Reservoir **18** k
Bor **42** x
Bora **42** g
Borah Peak **14** m
Borås **33** e
Borazjah **39** i
Borba **25** j
Borborema, Planalto da **27** u
Bordeaux **34** e
Borden Island **12** 9
Borden Peninsula **13** e
Bordj-Bou-Arreridj **34** v
Bordj Bourguiba **34** y
Bordj Flye Sainte Marie **36** k
Bordj Messouda **37** e
Bordj-Moktar **36** z
Bordj Omar Driss **37** e
Borisoglebsk **54** l
Borisov **33** y
Borja **24** j
Borkou **37** y
Borlänge **33** e
Borneo **69**
Bornholm **33** s
Borodino **56** l
Borohoro Shan **56** p
Borolgustakh **52** i
Borovskoye **55** e
Borroloola **73** f
Borshchovochnyy Mountains **52** u
Boru **53** e
Borujerd **39** 9
Bor-Yuryakh **52** h
Borzya **57** j
Bosaso **43** k
Bose **62** y
Bosing **63** f
Bosnia-Herzegovina **35** c
Bosobolo **42** p
Bosporus **35** i
Bossangoa **41** z
Bossembélé **41** z
Bosso **41** j
Bostan **56** r
Bosten Hu **56** t
Boston **17** o
Bothnia, Gulf of **33** h
Botoşani **35** h
Botswana **44/46/48-49**
Bouaflé **40** w
Bou Akba **36** g
Bouaké **40** w
Bouanga **44** g
Bouar **41** y
Bou Arfa **36** k
Bouca **42** o
Boufarik **37** b
Bougainville, Cape **72** l
Bougainville (Island) **71** t
Bougouni **40** i
Bougtob **34** t
Bou Izakarene **36** e
Boukombé **41** b
Boulder **15** q
Boulder City **14** y
Boulia **73** v
Boulsa **40** m
Boumango **44** e
Boumba **41** x
Boûmdeid **40** e
Boundary Peak **14** w
Boundiali **40** i
Boundji **44** g
Boungo **42** q
Bountiful **14** z
Bounty Islands **79** y
Bourail **75** p
Bourem **40** m
Bouressa **37** n
Bourke **77** k
Bournemouth **32** x
Bou Saada **34** u
Bousso **41** m
Boutilimit **40** c
Bouvet Island **234** z
Bova Marina **37** j
Bowen **74** r
Bowling Green **16** s
Bowman **15** e
Bo Xian **63** d
Boxing **58** o
Boyabo **42** o
Boyarka **51** l
Boyoma Falls **42** t
Bozeman **15** a
Bozoum **41** y
Brac **35** b
Braco Menor do Araguaia **26** y
Bradenton **20** k
Braeburn **11** k
Braga **34** b
Bragança **27** a
Brahmaputra **61** y
Brainerd **15** j
Brak **37** i
Brali **55** f
Branco **25** h
Brandberg **48** g
Brandon (Canada) **15** g
Brandon (USA) **20** l
Brandvlei **48** y
Branson **15** x
Brasilandia **29** g
Brasileia **25** p
Brasilia **29** f
Brasilia Legal **26** g
Brasov **35** h
Brass **41** r
Bratislava **35** c
Bratsk **57** a
Bratskoye Reservoir **57** b
Brattleboro **17** n
Braunschweig (Brunswick) **33** p
Brava **43** t
Bravo del Norte **18** l
Bray **49** a
Brazil **23-29**
Brazilian Plateau **29** h
Brazos **19** d
Brazzaville **44** f
Breadalbane **73** u
Breda **33** n
Bredy **55** c
Breeza Plains **74** b
Breidhi Fjord **32** a
Bremen **33** p
Bremerhaven **33** p
Bremerton **14** g
Brescia **34** l
Brest (France) **34** c
Brest (Belarus) **33** w
Bretaña **24** l
Breves **26** l
Brewarrina **78** c
Brewster, Cape **91** q
Brézina **34** t
Bria **42** q
Bridgeport **16** z
Bridgetown **23** k
Bridgewater **17** r
Brig **34** k
Brighton **32** y
Brikama **40** a
Brindisi **35** c
Brinkley **15** y
Brisbane **78** g
Bristol **32** x
Bristol Bay **11** o
Bristol Channel **32** w
British Columbia **11/14**
Britstown **48** z
Brive **34** g
Brno **33** s
Broadus **15** d
Brockville **16** l
Brod **35** c
Brodeur Peninsula **13** c
Broken Hill **77** h
Brokopondo **26** h
Brookhaven **20** f
Brookings **15** i
Brooks **14** m
Brooks Range **11** c
Broome **72** w
Broome (Mount) **72** l
Broughton Island **13** l
Browning **14** m
Brownsville **19** d
Brownwood **19** c
Brumado **29** j
Brunei Darussalem **69** c
Brunei Bay **69** o
Brunette Downs **73** s
Brunswick (Georgia) **20** m
Brunswick (Germany) **33** p
Brunswick (New Hampshire) **17** o
Brunswick Peninsula **30** u
Brunt Ice Shelf **83** g
Brussels **33** n
Bruxelles (Brussels) **33** n
Bruzual **23** d
Bryan **19** d
Bryanka **51** w
Bryansk **54** a
Bubiyan **39** g
Bubye **49** f
Buçaco **44** w
Bucaramanga **23** a
Buchanan **40** s
Buchanan, Lake **74** p
Buchan Gulf **13** h
Buchans **17** j
Buchara **55** r
Bucharest **35** h
Buckingham **16** l
Bucklin **15** t
București (Bucharest) **35** h
Budapest **35** c
Budaun **61** d
Budd Coast **87** e
Budjala **42** p
Buea **41** t
Buena Esperanza **30** l
Buenaventura (Mexico) **18** j
Buenaventura (Colombia) **22** x
Buenos Aires (Argentina) **31** c
Buenos Aires (Colombia) **23** n
Buenos Aires, Lago **30** u
Buerát el Hsun **35** o
Buffalo (New York) **16** x
Buffalo (South Dakota) **15** e
Buffalo (Wyoming) **15** c
Buffalo Narrows **12** v
Bug **33** v
Bugarikta **51** v
Bugat **56** x
Bugorkan **52** c
Bugsuk **67** q
Bugul’ma **54** l
Bugun **55** d
Bugunda **52** s
Buguruslan **54** l
Bu Hasa **39** w
Bui **50** q
Buitepos **48** k
Bujant **56** m
Bujant-Ovoo **57** q
Bujba **56** k
Bujedžul’ **56** j
Bujumbura **45** d
Bukačača **57** j
Bukavu **45** c
Bukittingti **68** e
Bukoba **45** f
Bula **70** p
Bulag Sum **57** w
Bulan **67** i
Bulanavo **54** m
Bulawayo **46** p
Buldana **61** n
Bulgan **57** p
Bulgan (High Altai) **56** w
Bulgan (Selenga) **57** b
Bulgar **57** i
Bulgaria **35**
Bulguntay **56** t
Buller **78** o
Bulloo **77** j
Bulloo Downs (W. Australia) **72** v
Bulloo Downs (Queensland) **74** n
Bulo Burti **43** u
Bultfontein **49** o
Bulun **53** s
Bulun (River) **53** k
Bulungu **44** j
Bulukol **55** y
Bumba **42** q
Buna **43** q
Bunbury **76** d
Bundaberg **74** t
Bu Ndjem **37** j
Bungo Channel **58** x
Bunguran (Natuna) **66** x
Buni **41** i
Bunia **42** w
Bunkeya **45** o
Buntok **69** c
Buol **69** h
Bourkhaya Bay **52** m
Bourkhaya, Cape **53** a
Bur **52** p
Bur Acaba **43** s
Buraidah **39** d
Buramo **43** g
Buran **56** f
Burandzha **53** o
Burang **61** e
Bur Anod **43** i
Burao **43** h
Burdekin **74** p
Burden, Mount **12** p
Bure **43** b
Bureya **58** i
Bureya (River) **58** j
Burgakhchan **53** f
Burgas **35** i
Burgersdorp **49** o
Burgon **56** k
Burgos **34** d
Bürhanpur **61** o
Burias **67** h
Buri Ram **66** f
Burke Island **81** a
Burketown **73** v
Burkhala **53** h
Burkina Faso **40-41**
Burlington (Colorado) **15** s
Burlington (Iowa) **15** y
Burlington (Vermont) **16** z
Burma **61/62/65/66**
Burney **14** u
Burnie **77** x
Burni-Nkonni **41** d
Burns **14** w
Burns Junction **14** w
Burqin **56** g
Burra **77** g
Burro, Sierra **19** a
Burrundie **73** c
Bursa **35** j
Buru **69** y
Burundi **45** d
Burwash Landing **11** i
Burylbaytal **55** x
Busanga **44** l

Busanga (Congo River) 45 n
Bushehr 39 i
Bushenyi 42 w
Businga 42 p
Busira 44 j
Busselton 76 c
Busseri 42 u
Busu-Djanoa 42 q
Buta 42 s
Butembo 42 v
Butiá 29 p
Butte, Mount 14 m
Butterworth 66 r
Butuan 67 j
Butung 69 v
Buturlinovka 54 f
Butwal 61 g
Buwatah 39 n
Buyr-Nur 58 b
Buzuluk 54 l
Byam Martin Channel 89 p
Bychez 52 b
Bydgoszcz 33 t
Bylot Island 13 f
Byrd 81 g
Byro 76 d
Byrranga Mountains 51 k
Byssa 52 z
Bystryy 51 p
Bysyttakh 53 a
Bytantay 53 a
Byuchennyakh 53 f
Byyangnyr 53 9

C

Caala 44 t
Cáatingas 27 r
Caazapá 28 z
Caballococha 25 a
Cabanatuan 67 g
Cabezas 28 h
Cabimas 23 c
Cabinda 44 d
Cabonga Reservoir 16 l
Cabool 16 p
Caboolture 74 u
Cabo Pantoja 24 k
Cabora Bassa Reservoir 46 f
Caborca 18 f
Cabot Strait 17 h
Cabrobó 27 s
Caçador 29 p
Cáeeres (Brazil) 25 x
Cáeeres (Spain) 34 o
Cache Creek 14 h
Cachi 28 s
Cachi, Mount 28 s
Cachimbo 26 u
Cachimbo, Serra do 25 z
Cachingues 44 u
Cachoeira 27 s
Cachoeira do Arari 26 l
Cachoeira do Sul 29 o
Cachoeiro de Itapemirim 29 k
Cacolo 44 w
Caconda 44 t
Cacula 44 s
Cadan 56 j
Cadelga 77 h
Cadillac 16 t
Cadiz (Philippines) 67 i
Cádiz (Spain) 34 o
Cádiz, Gulf of 34 o
Čadobec 51 z
Caen 34 f
Caetité 27 q
Cafayate 28 s
Cagaan Bogd Uul 56 z
Cagaan Gol 56 h
Cagaan Nuur 56 h
Cagaan-Olom 56 l
Cagaan-Uul 56 m
Cagayan de Oro 67 w
Cagayan Sulu 67 r
Cágliari 34 y
Caguán 24 l
Caguas 21 w
Čagyl 55 n
Caia 46 v
Caiabis, Serra dos 25 y
Caiaponia 29 c
Caicara de Orinoco 23 f
Caicó 27 u
Caima 26 i
Caird Coast 83 f
Cairns 74 c
Cairo (Al Qahirah) (Egypt) 38 i
Cairo (Illinois) 16 r
Caiundo 44 u
Cajabamba 24 i
Cajamarca 24 i
Cajatamboo 24 w
Cajázeiras 27 t
Čajek 55 x
Cajueiro 26 v
Calabar 41 s
Calabozo 23 e
Calacoto 25 o
Calais 32 z
Calalaste, Cord. de 28 r
Calama (Brazil) 25 g
Calama (Peru) 28 d
Calamar (Colombia) 23 n
Calamar (Venezuela) 22 m
Calamian Group 67 f
Calanaque 25 l
Calang 65 s
Calapan 67 g
Calatayud 34 f
Calbayog 67 j
Calca 25 n
Calçoene 26 k
Calcutta 61 w
Caldera 28 o
Caldwell 14 x
Caledon 48 x
Caleta el Cobre 28 o
Caleta Olivia 30 x
Calexico 18 d
Calgary 14 l
Cali 22 y
Calicut 64 d
Calie Corar 43 t
Caliente 14 y
California 14 v
California, Gulf of 18 f
Calkini 20 s
Callao 24 w
Caloocan 67 g
Calpulalpan 19 p
Caltanissetta 35 n
Calulo 44 t
Calvert Hills 73 g
Calverton 17 n
Calvinia 48 x
Camabatela 44 g
Camaçari 27 t
Camacho 18 m
Camacupa 44 u
Camagüey 21 n
Camaná 24 z
Camaná (River) 24 z
Camaqu 29 p
Camaqu (River) 29 o
Camarones 30 y
Camaxilo 44 j
Camballin 72 y
Cambell River 14 f
Cambay, Gulf of (Gulf of Khambhat) 60 y
Cambodia 66
Cambridge (Delaware) 16 y
Cambridge (Ohio) 16 v
Cambridge (UK) 32 y
Cambridge Bay 12 j
Camburinga 73 f
Camden 20 m
Cameroon 41
Cameroon, Mount 41 t
Cametá 26 l
Camiguin 67 j
Camiri 28 h
Camisea 24 z
Camocim 27 e
Camooweal 73 u
Camopi 26 j
Campana 30 s
Campbell Island 235 y
Campbell, Mount 11 j
Campbelltown 17 d
Campeche 19 u
Campeche, Bay of 19 s
Campiègne 34 h
Campina Grande 27 v
Campinas 29 f
Campina Verde 29 e
Camp Alegre 26 x
Campobasso 35 a
Campo de Diauarum 26 w
Campo dos Parecis 25 w
Campo Grande 29 a
Campo Largo 29 q
Campo Major 27 q
Campo Mouro 29 b
Campos 29 k
Campos Belos 27 n
Campos (Region) 27 o
Campos Sales 27 s
Cam Ranh 66 k
Canada 11-17/88-90
Cañada de Gómez 28 v
Canadian 15 t
Canadian (River) 15 r
Canairiktok 17 g
Canakkale 35 u
Canal do Norte (Brazil) 26 k
Canal do Sul 26 l
Canamaria 25 o
Cananea 18 h
Canary Islands 36 a
Canastra, Serra de 29 g
Canatlán 18 l
Canaveral, Cape 20 m
Canavierias 29 l
Canberra 78 d
Cancun 19 x
Candeias 27 t
Canelones 31 d
Cangchou 57 x
Cangkuang, Cape 68 v
Canguaretama 27 w
Canguçu 29 o
Caniapiscau 17 b
Caniapiscau Reservoir 17 b
Canicado 49 g
Canindé 27 t
Canindé (Juruá) 25 a
Canindé (River) 27 r
Canipole 18 g
Cannanore 64 d
Cannanore Islands 64 b
Canoas 29 q
Canobie 73 w
Canon City 15 q
Cantabrian Mountains 34 c
Cantaura 23 g
Canterbury Bight 79 t
Canterbury Plains 79 s
Can-Tho 66 i
Canto do Buriti 27 q
Canton (USA) 16 v
Canton (China) 63 p
Cantwell 11 f
Canudos 27 s
Canum 25 i
Canum (River) 25 j
Canyon 15 a
Canyushikino 54 k
Cao Bang 62 x
Caojian 62 s
Capanaparo 23 e
Capanema 27 a
Capo Bonito 29 f
Cap-de-la-Madeleine 17 a
Cap-Haïtien 21 r
Cape Coast 40 z
Cape Cara 20 l
Cape Cuvier 72 q
Cape Dorset 13 t
Cape Flattery 74 c
Cape Girardeau 16 q
Cape May 16 z
Cape Parry 12 c
Cape Town 48 w
Cape Verde 234 k
Cape York Peninsula 73 k
Capella 74 r
Capelongo 44 t
Capim 26 m
Capim Verde 29 a
Capivara, Serra de 29 g
Caponga 27 t
Caprivi Strip 44 x
Caquetá 25 a
Carabaya, Cord. de 25 n
Caracarai 23 v
Caracas 23 f
Caracol 27 p
Carajás 26 y
Carajás, Serra dos 26 x
Carandotta 73 u
Caráquez, Baia de 22 v
Caratianas 25 t
Caratinga 29 j
Carauari 25 d
Caravelas 29 l
Caravelí 24 y
Caraz 24 w
Carazinho 29 o
Carbó 18 g
Carcaraña 28 v
Carcar Mountains 43 j
Carcassonne 34 g
Carcross 11 k
Cardara 55 s
Cardarinskoje Reservoir 55 t
Cardenas 19 o
Cardiel, Lago 30 u
Cardill 32 w
Cardona 31 d
Čardžou 55 r
Carey, Lake 76 i
Carhue 30 m
Cariamanga 24 h
Caribbean Sea 20-21
Caribou 12 z
Caribou Mountains 12 s
Caripito 23 h
Carleton, Mount 17 d
Carlisle 32 w
Carlsbad 18 l
Carmacks 11 j
Carmelo 31 s
Carmem 25 z
Carmen 22 m
Carmen (Island) 18 h
Carmen Alto 28 d
Carmen de Patagones 30 m
Carmen, Sierra 18-19
Carmila 74 r
Carnarvon (Australia) 72 q
Carnarvon (South Africa) 48 y
Carnarvon Range (Western Australia) 72 v
Carnarvon Range (Queensland) 74 r
Carnegie 76 i
Carnegie, Lake 76 h
Carney Island 81 p
Carnicobar 65 q
Carnot 41 y
Carolina (Brazil) 27 n
Carolina (Puerto Rico, USA) 21 w
Carolina (South Africa) 49 r
Caroline Islands 70 m
Caroni 23 u
Carora 23 c
Carpathians 35 f
Carpentaria, Gulf of 73 h
Carrington 15 g
Carrizal Bajo 28 o
Čarsk 56 c
Carson City 14 v
Cartagena (Spain) 34 s
Cartagena (Venezuela) 22 m
Cartago 22 y
Cartier Island 72 k
Cartwright 17 j
Caruarú 27 v
Carúpano 23 h
Carysskoje 56 e
Casablanca 36 g
Casa Grande 18 f
Casamance 40 b
Casanare 23 o
Casa Nova 27 r
Casa Verde 29 b
Cascade 14 k
Cascade Range 14 u
Cascavel 29 n
Caserta 35 a
Casey 87 f
Casey Bay 84 l
Casigna 23 b
Casino 78 g
Casiquiare 23 s
Casma 24 v
Casper 15 p
Caspian Depression 54 j
Caspian Sea 54 x
Cassai 44 x
Cassamba 44 w
Cassiar 11 z
Cassilandia 29 e
Cassinga 44 t
Castanhal (Brazil) 26 m
Castanhal (Brazil) 26 p
Castelli 28 w
Castellón 34 s
Castilla 28 o
Castries 23 i
Castro (Brazil) 29 q
Castro (Chile) 30 g
Castrovirreyna 24 x
Cat 21 p
Catalo 29 f
Catamarca 28 s
Catandica 46 u
Catanduanes 67 j
Catania 35 o
Catanzaro 35 o
Cataract, 1st 38 w
Cataract, 2nd 38 u
Cataract, 3rd 38 u
Cataract, 4th 38 v
Cataract, 5th 38 x
Cataract, 6th 38 w
Catarman 67 j
Catatumbo 23 b
Catbalogan 67 j
Catende 27 v
Catengue 44 s
Catete 44 s
Catió 40 c
Catisimiña 23 u
Cat Lake 16 c
Catoche, Cape 20 v
Cato Island 74 w
Catriel 30 k
Catrimani 26 c
Catrimani (River) 26 b
Catskill Mountains 16 z
Catua 28 e
Catuane 49 t
Catyrk'ol, Lake 55 x
Čatyrtăs 55 y
Cauca 22 m
Caucasia 22 m
Caucasus 54 t
Caungula 44 i
Caura 26 a
Cauvery 64 g
Cavally 40 u
Cavanayen 23 v
Caviana 26 l
Cavinas 25 q
Caxias 25 a
Caxias do Sul 29 q
Caxito 44 f
Cayambe 22 x
Cayenne 26 j
Cayman Brac 20 z
Cayman Islands 20 z
Cayos Miskitos 22 h
Cazombo 44 z
Ceará-Mirim 27 v
Cebu 67 i
Cebu (Island) 67 i
Cecerleg 57 a
Cedar City 14 z
Cedar Lake 15 g
Cedar Rapids 15 y
Cedros 18 d
Ceduna 77 c
Ceilândia 29 e
Čekisler 54 z
Cekunda 53 n
Celaya 19 n
Celébes (Sulawesi) 69 t
Celébes Sea 69 g
Celeken 54 z
Cel'us 56 g
Čemolgan 55 y
Cencher Mandal 57 f
Cenderawasih 70 q
Cenderawasih, Gulf of 70 s
Central African Republic 41-42
Central Butte 15 c

Central City 16 s
Central Cordillera (Bolivia) 28 f
Central Cordillera (Colombia) 22 z
Central Cordillera (Peru) 24 j
Centralia 14 g
Central Makran Range 60 g
Central Plateau 29 f
Central Siberian Uplands 52 b
Central Ural Mountains 50 x
Cephalonia 35 q
Ceram 69 z
Ceram Sea 69 z
Čeremchovo 57 b
Čerepovec 50 o
Ceres (Argentina) 28 v
Ceres (Brazil) 29 d
Čerkessk 54 s
Černogorsk 56 i
Cernysevsk 52 t
Cerritos 19 n
Cerro de Pasco 24 w
Cerro Pinacate 18 e
Cerros Colorados Reservoir 30 j
Cerro Sombrero 30 w
Cervantes Island 76 c
Česa Bay 50 f
Cesar 23 a
České Budějovice 35 a
Cess 40 t
Ceuta 34 p
Cevennes 34 i
Ceyhan 54 p
Chabahar 60 r
Chacabuco 31 b
Chachani 25 n
Chachapoyas 24 i
Chacharramendi 30 k
Chaco Austral 28 v
Chaco Boreal 28 j
Chaco Central 28 i
Chad 37/38/41/42
Chad, Lake 41 k
Chadron 15 r
Chagai 60 h
Chagai Hills 60 g
Chagda 52 z
Chaghcharan 60 g
Chagos Archipelago 235 r
Chaibāsa 61 v
Chai Si 66 d
Chaim 52 q
Chaiten 30 h
Chaiyaphum 66 f
Chajari 28 y
Chakwal 60 m
Chala 24 y
Chalaxung 62 f
Chalbi Desert 43 o
Chalchaqui 28 w
Chalhuanca 24 z
Chalinze 45 k
Chalkidike 35 f
Chalkis 35 s
Challapata 28 f
Chalons-sur-Saône 34 i
Chama 45 s
Chamah 65 x
Chaman 60 i
Chamba (India) 61 b
Chamba (Tanzania) 45 w
Chambal 61 o
Chamberlain 15 t
Chamberlin, Mount 11 h
Chambeshi 45 r
Chamcha 52 f
Chamical 28 s
Chamo, Lake 43 o
Champaign 16 r
Champassak 66 i
Champlain, Lake 17 a
Champotón 20 s
Chañaral 28 o
Chan Bogd 57 r
Chancay 24 w
Chanch 57 a
Chanchiang 63 n
Chandigarh 61 c
Chandler 17 e
Chandpur 61 y
Chandrapur 61 q
Chang 66 f
Changada 51 m
Changara 46 u
Changchih 63 b
Changchow 63 r
Changchun 58 f
Changhua 63 t
Changji 56 t
Chang Jiang (Yangtze) 63 a
Changkiakow 57 w
Changma 56 z
Chang Mai 62 s
Changpai Shan 58 h
Changsha 63 c
Changshu 63 h
Changteh 63 b
Changwu 62 l
Chani 28 f
Channel Cape 76 f
Channel Islands (UK) 34 d
Channel Islands (USA) 18 b
Channel-Port-aux-Basques 17 i
Chantaburi 66 f
Chany, Lake 56 b
Chao'an 63 r
Chao Phraya 66 e
Chaoyang 63 r
Chapa 52 z
Chapada das Mangabeiras 27 n
Chapada Diamantina 27 r
Chapada do Araripe 27 s
Chapado dos Gauchos 29 c
Chapadinha 27 d
Chapais 16 m
Chapala, Lake 18 y
Chaparral 22 z
Chapayevo 54 k
Chapayevsk 54 k
Chapceranga 57 g
Chapeco 29 o
Chapleau 16 h
Chaplino 54 c
Chara 52 u
Chara (River) 52 u
Charaa 57 d
Charagua 28 h
Charaña 25 o
Charata 28 v
Charcas 19 n
Charcot Island 82 w
Chari 41 l
Charikar 60 j
Charleroi 33 n
Charleston (South Carolina) 20 m
Charleston (Virginia) 16 v
Charleville 74 p
Charlotte 20 m
Charlottesville 16 x
Charlottetown 17 f
Char Nuur 56 k
Charouïne 36 m
Charters Towers 74 p
Chartres 32 z
Char Us Nuur 56 j
Chasaat 57 b
Chascomús 31 c
Chasel'ka 51 r
Chasovnya-Vchurskaya 53 n
Chăsuri 54 t
Châteauroux 34 h
Chatgal 57 a
Chatrapur 61 q
Chatsworth (Australia) 73 v
Chatsworth (Zimbabwe) 49 e
Chattahoochee 20 j
Chattanooga 20 j
Chauk 62 p
Chaumont 33 n
Chau Phu 66 h
Chavantina 29 c
Chaves 26 l
Chavirga 57 j
Chaya 52 r
Chay-Povarnaya 53 d
Chebka du Mzab 34 u
Cheboksary 54 i
Chechaouen 34 q
Chech Erg 36 w
Cheduba 65 e
Cheepie 74 o
Cheetham, Cape 86 v
Cheffadene 37 t
Chegdomyn 58 k
Chegga 36 u
Chegutu 49 e
Cheju 58 t
Chekalio 54 c
Chekunda 58 k
Chekurovka 52 l
Chelago 43 q
Chelforó 30 k
Cheliff 34 t
Chelkar 55 c
Chellala 34 t
Chelyabinsk 55 d
Chembe 45 q
Chemdal'sk 52 o
Chemnitz 33 q
Chemull 14 u
Chenab 60 l
Chenachane 36 w
Chen Barag Qi 58 c
Chengchow 63 c
Chengdu (Chengtu) 62 j
Chengshan, Cape 58 r
Chengteh 57 x
Chengtu 62 j
Chennai (Madras) 64 h
Chenxi 63 a
Chen Xian 63 p
Chepes 28 r
Cherbourg 32 x
Cherchell 37 a
Cherdyn 50 w
Cherendey 52 u
Cherepanovo 56 e
Cherkashina 52 q
Cherkassy 35 k
Cherlak 55 l
Cherlen Gol 57 h
Chernaya 51 h
Chernaya (Pechora Sea) 50 j
Chernigov 33 v
Chernikovsk 55 a
Chernovtsy 35 g
Chernyshevskiy 52 e
Chernyy Mys 53 l
Chernyy-Yar 54 h
Cherpiya 51 n
Cherskiy 10 a
Cherskogo Mountains 53 d
Chertovo-Ulovo 53 p
Chesapeake Bay 16 y
Chester 15 a
Chesterfield Islands 74 y
Chesterfield Inlet 13 o
Chesterton Range 74 q
Chetumal 20 u
Chetwynd 12 p
Cheulik 53 c
Chevak 10 m
Chew Bahir, Lake 43 o
Cheyenne 15 r
Cheyenne (River) 15 r
Chhatarpur 61 q
Chhatrapur 61 u
Chhindwara 61 p
Chiai 63 u
Chiang Mai 66 d
Chianje 44 s
Chiapa de Corzo 19 t
Chiba 59 o
Chibagalakh 53 d
Chibagalakh (River) 53 d
Chibaranda 44 x
Chibembe 45 r
Chibia 44 s
Chibougamau 16 m
Chicago 16 s
Chicala 44 w
Chicama 24 i
Chicapa 44 k
Chichaoua 36 g
Chichas, Cordillera de 28 f
Chichón 19 t
Chickasha 15 u
Chiclayo 24 h
Chico (River) 30 w
Chico (USA) 14 u
Chicoutimi-Jonquiére 17 a
Chicualacuala 49 f
Chidley, Cape 13 y
Chiefland 20 k
Chiengi 45 d
Chifeng 57 x
Chifwefwe 45 q
Chignik 11 p
Chigorodó 22 m
Chigubo 49 h
Chigul'bach 53 p
Chihuahua 18 j
Chik Ballapur 64 f
Chikhacheva 53 g
Chikmagalur 64 d
Chikwa 46 g
Chilapa 19 o
Chilas 60 m
Childers 74 t
Chile 28/30
Chile Chico 30 u
Chilecito 28 r
Chililabombwe 46 c
Chillagoe 74 b
Chillán 30 h
Chilli 55 t
Chillicothe (Iowa) 15 x
Chillicothe (Ohio) 16 u
Chilliwack 14 h
Chiloe 30 f
Chilok 57 g
Chilok (River) 57 e
Chilonga 46 f
Chilpancingo 19 o
Chilumba 46 h
Chilung 63 v
Chilwa, Lake 45 v
Chimala 45 g
Chimanimani 49 g
Chimbay 55 p
Chimborazo 24 h
Chimbote 24 v
Chimoio 49 g
China (Mexico) 19 b
China (State) 52/55-58/61-63
Chinandega 22 e
Chincha Alta 24 w
Chincha Islands 24 w
Chincheros 24 y
Chinchilla 74 t
Chinchow 58 d
Chincolco 28 o
Chinde 49 j
Chindu 62 d
Chindwin 62 p
Chingola 46 c
Chinguetti 36 q
Chinju 58 u
Chinko 42 s
Chinle 15 o
Chinnamp 58 s
Chinsali 46 f
Chinsi 58 c
Chintalnar 61 r
Chíos 35 t
Chipata 46 f
Chipinge 49 g
Chiplun 60 z
Chippewa Falls 16 c
Chiquían 24 w
Chiquilá 19 x
Chiquita, Laguna 28 u
Chiredzi 49 f
Chirgis Nuur 56 j
Chirinda 51 m
Chiriqui, Gulf of 22 h
Chirripó 22 h
Chisasa 45 n
Chisasibi 16 j
Chisig-Ond r 57 b
Chisinău 35 i
Chiskovo 51 x
Chisoso 45 s
Chistoozernoye 55 l
Chistopol 54 k
Chita 57 h
Chitembo 44 u
Chitina 11 h
Chitradurga 64 f
Chitrāl 60 l
Chittagong 61 y
Chittoor 64 h
Chitunde 46 g
Chiume 44 x
Chiuta 46 g
Chivay 25 n
Chivhu 49 f
Chivilcoy 31 b
Chodzeili 55 p
Choele Choel 30 l
Choiseul 71 v
Choke Mountains 43 b
Chokudakh 53 h
Choma 46 p
Chon Buri 65 k
Ch'ŏngjin 58 i
Ch'ŏngju 58 u
Chongquing (Chungking) 62 l
Chŏnju 58 u
Chonos, Archipelago 30 s
Chorinsk 57 f
Chorog 55 w
Chorracho 27 t
Chorregon 74 n
Chōshi 59 p
Chos Malal 30 i
Choteau 14 m
Cholla, Cerro de 19 n
Chott Djerid 34 x
Chott ech Chergui 36 m
Chott Melrhir 34 w
Chotts ech Chergui 34 t
Choüm 36 q
Choushuidan 56 z
Chovd 57 o
Chovd Gold 56 i
Chövsgöl 57 s
Chövsgöl Nuur 57 a
Christchurch 79 t
Christensen Mountains 84 l
Christiana 49 o
Christian X Land 91 q
Christmas Creek 72 y
Christmas Island 68 u
Chubut 30 l
Chuchow 63 c
Chudinovo 55 e
Chudovo 33 m
Chugwater 15 q
Chuhoi 63 p
Chukchi Sea 10
Chukehagirskoye, Lake 53 p
Chuken 58 m
Chukochye 53 l
Chukot Peninsula 10 j
Chulucanas 24 h
Chulugli 52 t
Chuluman 25 q
Chulym 56 d
Chumakovo 56 d
Chumar 61 d
Chumbicha 28 s
Chumikan 53 o
Chum Phae 65 k
Chumphon 65 i
Chuna 51 y
Ch'unch'on 58 u
Ch'ungju 58 u
Chungking 62 l
Chunguj Gol 56 k
Chunhuhub 20 t
Chunskiyo 57 a
Chunya 45 g
Chuquibamba 24 z
Chuquicamata 28 d
Chur 34 l
Churapcha 53 a
Churchill 13 n
Churchill, Cape 13 n
Churchill Lake 12 v
Churchill Peak 12 o
Churchill (River, Quebec) 17 f
Churchill (River, Saskatchewan) 12 w
Churu 61 a
Churunguara 23 c
Chushakyl' 55 b
Chushul 61 c
Chutag 57 b
Chuūronjang 58 i
Chuuk Islands 71 e
Chu Xian 63 f
Chuxiong 62 x
Chuya 52 r
Chuya (River) 52 r
Chuyengo 52 b
Chyul'be 52 z
Cibitoke 45 d
Ciego de Avila 21 n
Ciénaga 23 a
Cienfuegos 20 z

Cilacap **68** y
Čilik **56** o
Cimarron **15** t
Čimkent **55** v
Cincinatti **16** t
Cintra, G. de **36** n
Cipó **27** t
Cirčik **55** u
Circle **11** h
Cirebon **68** y
Cisneros **22** m
Citlaltépetl **19** p
Citrusdal **48** x
Ciudad Acuña **19** a
Ciudad Altamirano **19** n
Ciudad Bolivar **23** h
Ciudad Camargo **18** l
Ciudad del Carmen **19** t
Ciudad Guayana **23** i
Cuidad Guzman **18** y
Ciudad Ixtepec **19** r
Ciudad Juárez **18** j
Ciudad Madero **19** p
Ciudad Mante **19** o
Ciudad Mier **19** b
Ciudad Obregón **18** h
Ciudad Ojeda **32** b
Ciudad Real **34** q
Ciudad Valles **19** o
Ciudad Victoria **19** p
Civitavecchia **34** m
Clanwilliam **48** x
Claraville **73** w
Clarence, Cape **13** b
Clarence Island **82** f
Clarión **18** r
Clarksburg **16** w
Clarksdale **20** f
Clarksville (Arkansas) **16** o
Clarksville (Kentucky) **16** s
Clavering Island **91** s
Clayton **15** r
Clear, Cape **32** t
Clearfield **16** x
Cleopatra Needle **67** e
Clermont **74** q
Clermont-Ferrand **34** h
Cleveland (North Carolina) **16** t
Cleveland (Ohio) **16** v
Clifton **78** f
Clifton Hills **77** g
Clingmans Dome **16** u
Clinton (Missouri) **15** x
Clinton (Oklahoma) **15** u
Clinton (South Carolina) **20** l
Clinton Colden Lake **12** v
Clinton Creek **11** i
Clipperton **234** e
Clocolan **49** p
Cloncurry **73** w
Cloquet **15** k
Clorinda **28** y
Cloud Peak **15** c
Clovis **18** m
Cluj Napoca **35** f
Clutha **79** r
Clyde **13** j
Coari **25** g
Coari (River) **25** f
Coast Range **14** t
Coast Mountains **14** g
Coats Island **13** r
Coats Land **83** g
Coatzacoalcos **20** q
Cobar **78** b
Cobija **25** p
Coca **22** y
Cocalzinho **27** b
Cochabamba **28** f
Cochin **64** e
Cochrane (Canada) **16** i
Cochrane (Chile) **30** t
Cockburn **77** h
Cockburn (Mount) **73** o
Coco **22** g
Cocobeach **41** t
Cocoparra Range **78** b
Cocos Islands **65** e
Cocula **18** y
Codajás **25** h
Cod, Cape **17** o
Codera, Cabo **23** l
Codó **27** p
Codrington **21** y
Čodro **56** h
Cody **15** b
Coen **73** k
Coeur d'Alene **14** k
Coffee Bay **49** f
Cogt-Ovoo **57** q
Coiba **22** i
Coihaique **30** u
Coimbatore **64** e
Coimbra **34** o
Coipasa, Lago de **28** e
Coixtlahuaca **19** q
Cojbalsan **57** i
Cojedes **23** e
Colatina **29** k
Colbeck, Cape **81** x
Cold Bay **11** n
Coles Bay **78** q
Colesberg **49** n
Colhué Huapi, Lago **30** w
Colima **18** y
Colima, Nevada de **18** y
Colinas **27** o
Collahuasi **28** d
College Park **20** j
Collie **76** d
Collier Bay **72** k
Collingwood **79** t
Collinsville **74** q
Colnett **18** d
Cologne **33** o
Colomb Bechar **36** k
Colombia **22-25**
Colombo **64** u
Colón (Cuba) **20** z
Colón (Panama) **22** j
Colorado **15** q
Colorado (River, Texas) **19** d
Colorado (River, Arizona) **14** y
Colorado (River, Argentina) **30** m
Colorado Plateau **14** z
Colorado Springs **15** q
Colorados **28** q
Columbia (Missouri) **15** x
Columbia (South Carolina) **20** l
Columbia (Tennessee) **16** s
Columbia (River) **14** j
Columbia Plateau **14** w
Columbus (Georgia) **19** l
Columbus (Mississippi) **19** j
Columbus (Nebraska) **15** u
Columbus (Ohio) **16** u
Colville **11** e
Colville Lake **12** b
Comai **61** m
Comalcalco **20** q
Comarapa **28** g
Combarbala **28** o
Comilla **61** y
Comitán **19** t
Commandante Ferraz **82** r
Committee Bay **13** c
Communism, Mount **55** v
Como (China) **61** h
Como (Italy) **34** k
Comodoro Rivadavia **30** w
Comores Islands **47** a
Comorin, Cape **64** s
Comoros **47** b
Compiègne **32** z
Compostela **18** x
Cona **61** m
Conakry **40** d
Conceiço do Araguaia **26** y
Concepción (Argentina) **28** s
Concepción (Bolivia) **28** i
Concepción (Chile) **30** g
Concepción (Paraguay) **28** l
Concepción del Oro **19** a
Concepción del Uruguay **28** x
Concepción, Laguna **25** v
Conchi **28** d
Conchos **18** k
Concord **17** n
Concordia (Argentina) **28** x
Concórdia (Brazil) **25** e
Concordia (Peru) **24** k
Concordia (USA) **15** u
Conde **27** u
Condobolin **78** b
Congo **41-42/44**
Congo, Democratic Republic of **42/44-45**
Congo (River) **44-45**
Conjuboy **73** z
Connecticut (River) **17** n
Connecticut (State) **17** n
Connemara **73** w
Conrad **14** m
Conselheiro Lafaiete **29** i
Constance, Lake **33** p
Constanța **35** i
Constantine **34** w
Constitución **30** g
Consuelo Peak **74** q
Contact **14** y
Contamaná **24** l
Contas **27** r
Contendas do Sincorá **27** r
Contwoyto Lake **12** h
Conway **20** e
Coober Pedy **77** d
Cook **77** a
Cook, Mount **79** r
Cook Islands **234** o
Cook Strait **79** u
Cooktown **74** c
Coolah **78** e
Coolgardie **76** g
Cooma **78** q
Coonabarabran **78** d
Coonamble **78** c
Coonapoor **64** d
Coongoola **77** k
Cooper **77** h
Cooperation Sea **85** e
Coorabulka **73** v
Coorong, The **77** t
Coorow **76** d
Coos Bay **14** t
Cootamundra **78** d
Cooya Pooya **72** s
Copenhagen **33** p
Copiapo **28** p
Copiapo, Mount **28** p
Copiapo (River) **28** p
Copper Center **11** h
Copper Harbor **16** e
Coppermine **12** f
Copperton **48** z
Coqên **61** h
Coquimbo **28** o
Coracora **24** z
Coral Harbour **13** r
Coral Sea **74** f
Corcovado, Gulf of **30** g
Cordillera Central **67** g
Cordillo Downs **77** h
Córdoba (Argentina) **28** u
Córdoba (Mexico) **19** q
Córdoba (Spain) **36** j
Cordoba, Sierra de **28** t
Cordova **11** g
Corfield **73** x
Corfu **35** q
Corfu (Island) **35** q
Coricudgy **78** e
Corigliano **35** o
Corinda **73** h
Corinda (Lake Galilee) **73** y
Corinth **20** g
Corinth, Gulf of **35** s
Corinto **29** h
Coriolei **43** t
Cork **32** u
Corleone **37** i
Corner Brook **17** i
Corno, Monte **35** a
Cornwall **16** l
Cornwall Island **12** m
Cornwallis Island **12** m
Coro **23** d
Coroatá **27** p
Corocoro **25** p
Coroico **25** q
Coromandel Coast **64** h
Coronation Gulf **12** g
Coronation Island **82** f
Coronel Murta **29** j
Coronel Oviedo **28** z
Coronel Pringles **31** a
Coronel Suárez **31** a
Coropuna **24** z
Corowa **78** b
Corpus Christi **19** d
Corque **28** e
Corrente **27** o
Corrente (River) **27** p
Correntina **27** o
Corrientes **28** y
Corrientes, Cabo **22** x
Corrientes (River, Argentina) **28** x
Corrientes (River, Peru) **24** k
Corsica **34** k
Cortina **34** m
Corubal **40** c
Corum **54** o
Corumba **28** l
Corumbá (River) **29** e
Corunna **34** a
Cosalá **18** k
Coscomatepec **19** q
Cosenza **35** o
Cosmonaut Sea **84** j
Costa Blanca **34** s
Costa Brava **34** h
Costa del Sol **34** q
Costa Marques **25** s
Costa Rica **22** g
Cotabato **67** v
Cotagaita **28** f
Cotahuasi **24** z
Cote d'Ivoire **40**
Cotonou **41** o
Cotopaxi **22** x
Cottbus **33** r
Cottonwood **14** z
Coulman Island **86** s
Coulomb Point **72** j
Council Bluffs **15** v
Courantyne **23** y
Coventry **32** x
Cowan, Lake **76** h
Cowargarzê **62** e
Cowarie **77** f
Cowell **77** e
Cowra **78** d
Cox's Bzr **61** y
Cozumel **20** v
Cradock **49** o
Craig **15** p
Craiova **35** f
Cranbrook **14** k
Crateús **27** s
Cravo Norté **23** p
Cree Lake **12** v
Creporizinho **25** l
Cres **35** a
Crescent City **14** t
Creston **15** w
Crestview **20** h
Creswell Downs **73** f
Crete (Island) **35** s
Crete, Sea of **35** t
Criciúma **29** r
Crimea **35** l
Cristalandia **26** y
Cristalina **29** f
Cristino Castro **27** p
Cristóbal Colón **23** a
Croker, Cape **73** d
Crooked Island **21** q
Crookston **15** i
Cross **41** s
Crowl **77** k
Crown Prince Christian Land **91** i
Crown Prince Olal Coast **84** j
Crown Princess Martha Coast **83** h
Croatia **35** b
Croydon **73** w
Crozet Island **235** p
Cruz Alta **29** n
Cruz das Almas **27** s
Cruz del Eje **28** t
Cruzeiro do Oeste **29** b
Cruzeiro do Sul **24** m
Crystal River **20** k
Ču **55** x
Ču (River) **55** i
Cuamba **45** v
Cuando **44** y
Cuangar **44** w
Cuango **44** v
Cuanza **44** t
Cuatrociénegas **18** m
Cuauhtémoc **18** j
Cuba **20-21**
Cubango **44** v
Cuchi **44** u
Cucuí **23** r
Cucúi **25** d
Cúcuta **23** a
Cuddalore **64** h
Cuddapah **64** g
Cue **76** e
Cuemba **44** v
Cuenca **24** h
Cuernavaca **19** o
Cuiabá **25** y
Cuiabá (River) **25** y
Cuito **44** w
Cuito Cuanavale **44** w
Cuiuni **25** f
Cuixi **63** e
Culbertson **15** d
Culiacán **18** j
Cullman **20** h
Culpeper **16** x
Culuene **26** w
Čuluut **57** a
Culver, Cape **76** j
Čulym **56** h
Cumaná **23** g
Cumberland **16** x
Cumberland, Lake **16** t
Cumberland Peninsula **13** k
Cumberland Plateau **16** t
Cumberland (River) **16** t
Cumberland Sound **13** k
Cunani **26** k
Cuñaré **23** n
Cundza **56** p
Cunene **48** g
Čunja **51** y
Cunnamulla **77** l
Cupica **22** l
Curaçao **23** d
Curacautin **30** h
Curacó **30** l
Curaray **24** j
Curbur **76** d
Curico **30** i
Curimatá **27** p
Curitiba **29** r
Curitibanos **29** q
Currais Novos **17** v
Currant **14** y
Currie **77** w
Curtin Springs **73** p
Curtis Island **79** m
Curua **26** u
Curua-Uná **26** i
Curupira, Serra **25** f
Curuzú Cuatiá **28** x
Curvelo **29** h
Cutral-Có **30** j
Cuttack **61** v
Cuvelai **44** u
Cuya **28** c
Cuyuni **23** j
Cuzco **25** n
Cyaring Hu **62** e
Cyclades **35** t
Cyprus **35/38**
Cyrenaica **38** b
Czech Republic **33-35**
Czestochowa **33** t

D

Da **62** w
Dabar Tichitt **36** s
Dabola **40** f
Dabou **40** x
Dabsan Hu **62** c
Dachechang **57** o
Dachstein **35** a
Dadanawa **23** x
Daet **67** h
Dagana **40** c
Dagandely **55** m
Dagupan **67** g
Dahab **38** k
Dahar Oualata **36** t

Daheba 62 f
Dahlak, Islands 43 d
Dahna 39 f
Dāhod 61 n
Dahongliutan 61 d
Dai Hai 57 u
Dailehk 61 f
Dainkog 62 e
Daintree 74 c
Dairen (Dalian) 58 q
Dairût 38 h
Dairy Creek 76 d
Dajarra 73 v
Dakar 40 a
Dakhla 36 n
Dakhla, Oasis 38 t
Dakshin Gangotri 84 b
Dal 33 f
Dala 44 x
Dalaba 40 e
Dalandzadgad 57 q
Dalandžargalan 57 s
Dalaoba 56 r
Da Lat 66 k
Dalbandin 60 g
Dalby 78 f
Dalgonally 73 w
Dalgoye 52 l
Dalhart 15 s
Dalhousie, Cape 11 m
Dalian 58 q
Dallas 19 d
Dal'nerechensk 58 l
Daloa 40 v
Dalou Shan 62 k
Dalrymple 74 r
Dal'stroy 53 c
Dalton 20 j
Dalton, Cape 91 q
Dalou 62 t
Daly 73 c
Daly River 73 b
Daly Waters 73 d
Damakar 41 j
Daman 60 z
Damanhûr 38 h
Damar 70 n
Damara 42 o
Damascus 38 l
Damaturu 41 i
Damavand 39 j
Damavand (Mount) 39 j
Damba 44 g
Dambulla 64 u
Dam Gamad 42 i
Damghan 54 z
Dammam 39 i
Damon 61 q
Dampier 72 s
Dampier Downs 72 w
Dampier Land 72 j
Dampier Strait 70 p
Dam Qu 62 b
Damxung 61 l
Dana 10 a
Danakil 43 e
Danané 40 u
Da Nang 66 k
Danba 62 h
Danbury 16 z
Dan, Cape 90 y
Dandelehura 61 d
Danells Fjord 90 t
Dang 56 y
Dangla 43 b
Dango 42 f
Daniel 15 n
Daniëlskuil 48 z
Dankhar 61 e
Dankowa, Pik 55 z
Danmark Fjord 91 i
Dannebrog Island 90 x
Dansville 16 x
Danube, Mouths of the 35 j
Danube 33 r
Danville (Illinois) 16 s
Danville (Virginia) 16 w
Danzig (Gdansk) 33 t
Danzig, Gulf of 33 t
Daora 36 d
Dao Timni 37 v
Dapango 41 a
Dapchi 41 i
Dapitan 67 v
Da Qaidam 56 y
Daqing 58 f
Daquan 56 y
Dar'a 38 l
Daran 39 h
Darasun 52 s
Daraut-Kurgan 55 w
Darband 60 d
Darbhanga 61 v
Darchan 57 d
Dar-el-Beida (Casablanca) 36 g
Dar-es-Salaam 45 k
Dargaville 79 h
Darién, Gulf of 22 l
Daringskiy 55 j
Darj 37 g
Darjeeling 61 j
Darlag 62 g
Darling 77 i
Darling Range 76 d
Darnah 35 s
Darnley, Cape 85 c
Dartmouth 17 f
Daru 70 y
Darvaza 55 o
Darvel Bay 67 r
Darwin 73 c
Darwin (Island) 22 n
Darwin River 73 b
Daryacheh-ye-Bakhtegan 39 l
Daryacheh-ye-Namak 39 i
Daryacheh-ye-Tashk 39 k
Darzin 60 c
Dashizhai 58 d
Dasht 55 n
Dasht (River) 60 r
Dasht-e-Kavir 39 k
Dasht-e-Lut 39 m
Dashuiqiao 62 f
Datong 57 v
Datu, Cape 68 l
Datuk Bay 68 m
Daugav'pils 33 x
Daulat Yar 60 h
Daule 22 w
Daund 61 n
Dauphin 15 g
Davangere 64 e
Davao 67 w
Davao Gulf 67 w
Davenport 16 q
Davenport Downs 73 v
Davenport Range 73 r
David 22 h
Davidson 15 d
Davidson Mount 73 p
Davies 76 m
Davis 85 q
Davis Sea 85 t
Davis Strait 13 m
Davlekanovo 54 m
Dawa 43 p
Dawson (Canada) 11 i
Dawson (USA) 19 m
Dawson Creek 12 q
Dawson Landing 12 t
Dawu 62 h
Daxian 62 l
Dayangshu 58 f
Dayong 63 a
Dayr az Zawr 39 a
Dayton 16 t
Daytona Beach 20 m
Dazhu 62 l
De Aar 49 n
Deacon, Cape 82 z
Deadhorse 11 f
Dead Sea 38 l
Deakin 76 m
Deán Funes 28 t
Dease Lake 11 z
Dease Strait 12 i
Death Valley 14 x
Debba 38 u
Debdou 36 k
Debin 53 i
Débo Lake 40 k
Deborah, Lake 76 f
Debre Birhan 43 d
Debrecen 33 u
Debre-Markos 43 b
Debre Tabor 43 c
Debre Zeit 43 c
Decatur (Alabama) 20 h
Decatur (Illinois) 16 r
Deccan 61 o
Dechang 62 i
Decorah 16 p
Dédougou 40 l
Dedza 46 h
Deep Well 73 r
Deer 15 a
Dêgê 62 f
Degeh Bur 43 g
Degtyarsk Sysert' 55 c
Dehak 60 d
Dehbid 39 k
Dehloran 39 g
Dehra Dun 61 c
Dehui 58 f
Deihuk 60 b
Dej 35 g
Dejen 43 c
De Juh 62 c
De Kalb 16 r
Dekar 48 l
De Kastruskoye 53 r
Dekemehare 43 c
Dekese 44 l
Dékoa 42 o
Delamere 73 c
Delano 14 w
Delareyville 49 n
Delaware 16 y
Delgado, Cape 45 y
Delger 56 m
Delgerdhaan 57 c
Delgo 38 u
Delhi 61 b
Delicias 18 l
Déline (Fort Franklin) 12 c
Delingha 56 z
Dell 14 m
Dellys 34 v
Deloge, Monts 16 k
Delphi 35 s
Del Rio 19 a
Delta 14 z
Delta Downs 73 j
Delta Junction 11 g
Demba 44 l
Dembia 42 s
Dembidolo 42 m
Deming 18 j
Demini 23 u
Demnate 36 h
Dempo 68 t
Demyanka 51 o
Dem'yanovka 55 f
Demyanskoye 51 n
Denau 55 t
Denežkin Kamen 50 x
Dengkou 57 q
Dengqên 62 d
Denham 76 b
Deniliquin 77 x
Denison 15 w
Denizli 35 w
Denman Glacier 85 s
Denmark 33
Denmark Strait 32 a
Denpasar 69 p
D'Entrecasteaux Islands 71 r
D'Entrecastleaux, Cape 76 c
Denver 15 q
Deogarh 61 u
Depósito 23 w
Deputatskiy 53 d
Dêgên 62 f
Dera Bugti 60 k
Dera Ghazi Khan 60 k
Dera Ismail Khan 60 k
Derbent 54 w
Derby (Australia) 72 k
Derby (UK) 32 x
Dergachi 54 j
Derm 48 j
Dêrong 62 g
De Rose Hill 73 r
Derudeb 38 y
Derwent 77 y
Derzhavinsk 55 g
Desaquadero (River, Argentina) 28 r
Desaquadero (River, Bolivia) 28 d
Descalvados 25 x
Dese 43 c
Deseado 30 x
Desertas 36 a
Des Moines 16 o
Des Moines (River) 15 w
Desna 54 a
Desolación 30 t
Dessau 33 q
Desterrada Pérez 19 v
Dete 46 p
Detroit 16 u
Det Udom 66 h
Deva 35 f
Devils Lake 15 h
Devon Island 13 c
Devonport 77 x
Dez 39 h
Dezful 39 g
Dezhgan 39 l
Dhahran 39 h
Dhaka 61 y
Dhamar 34 g
Dhanbad 61 v
Dhandhuka 60 z
Dharmapuri 64 g
Dharmsala 61 b
Dharwar 64 d
Dhaulāgiri 61 g
Dhofar 39 x
Dhuburi 61 y
Dhule 61 n
Diaca 45 x
Dialafara 40 f
Dialakoto 40 d
Diamaa 34 w
Diamante 28 w
Diamante (River) 28 r
Diamantina 26 q
Diamantina (Minas Gerais) 29 i
Diamantina (River) 73 w
Diamantina Lakes 73 w
Diamantino 25 y
Diamou 40 e
Dian-Chi 62 v
Dianópolis 27 n
Diapaga 41 b
Dibba 60 n
Dibella 41 j
Dibrūgarh 62 c
Dickinson 15 e
Didesa 43 a
Didiéni 40 h
Didwana 61 a
Diego de Almargro 28 p
Diéma 40 g
Digne 34 j
Digos 67 w
Digul 70 w
Dijon 34 i
Dikhil 43 e
Dikimdya 52 w
Dikson 51 f
Dikwa 41 j
Dila 43 p
Dilaram 60 g
Dili 69 x
Di Linh 66 k
Dilling 42 j
Dillingham 11 p
Dillon 14 m
Dilolo 44 y
Dimas 18 w
Dimbelenge 44 m
Dimbokro 40 w
Dinagat 67 j
Dinajpur 61 w
Dinar 39 i
Dinaric Alps 35 b
Dinder 42 m
Dingbian 57 r
Dinggyê 61 j
Dinguiraye 40 f
Dingxi 62 j
Ding Xian 57 v
Dinsor 43 s
Diourbel 40 b
Dipolog 67 u
Dirâng 62 b
Dire Dawa 43 e
Dirkou 37 v
Dirranbandi 78 c
Disappointment, Lake 72 x
Discovery Bay 77 u
Discovery, Cape 91 g
Disko Bay 90 k
Diu 60 y
Divinópolis 29 h
Divisoes, Serra das 29 d
Divnoye 54 g
Divriği 54 q
Dixon Entrance 14 b
Diyala 39 e
Diyarbakir 54 r
Dja 41 x
Djado 37 u
Djado Plateau 37 u
Djamaa 37 c
Djanet 37 s
Djebel Amour 37 a
Djebel Bani 36 g
Djebel Serkout 37 q
Djebel Ouarkziz 36 f
Djebel Tahat Telertheba 37 q
Djelfa 34 u
Djenné 40 j
Djerba 37 g
Djermaya 41 l
Djibo 40 m
Djibouti (City) 43 f
Djibouti (State) 43 f
Djidjelli 37 c
Djolu 42 r
Djougou 41 b
Djúpivogur 32 f
Dmitrov 50 p
Dneprodzerzhinsk 35 l
Dnepropetrovsk 35 l
Dnieper 35 l
Dniester 35 g
Doab-i Mikhe Zarin 60 i
Doba 41 m
Dobryanka 50 w
Doce 29 j
Docker River 73 o
Doda Betta 64 f
Dodge City 15 t
Dodoma 45 h
Dogai Coring 61 j
Dogondoutchi 41 c
Do'gyaling 61 j
Doha 39 w
Doi Inthanon 62 s
Dois Irmos, Serra do 27 r
Dolak Island 70 u
Dolgany 51 j
Dolo 43 s
Dolomites 34 m
Dolon 56 b
Dolores 31 c
Dolphin and Union Strait 12 f
Domba 62 d
Dombarovskiy 55 c
Dombås 33 b
Dome Fuji 84 f
Domeyko 28 o
Dominica 23
Dominican Republic 21
Don 54 e
Donald 77 w
Dondo (Angola) 44 s
Dondo (Mozambique) 46 v
Dondo Bay 69 g
Dondra, Head 64 v
Donegal Bay 32 u
Donets 54 d
Donetsk 54 d
Dong 57 n
Dongbatu 56 y
Dongfang 62 z
Dongfanghong 58 k
Donggala 69 f
Dong Hoi 66 j
Dong Jian 63 q
Dongjingcheng 58 l
Dongkalang 69 g
Dongle 57 n
Dongliao 58 e
Donglük 56 u
Dongnan Ouiling 63 o
Dongola 38 u

Dongou . . . 42 n
Dongqiao . . . 61 l
Dongsheng . . . 57 t
Dong-Ujimqin Oi . . . 57 x
Dongzhen . . . 57 p
Donors Hill . . . 73 v
Doomadgee . . . 73 u
Dora Lake . . . 72 w
Dorbod . . . 58 f
Dordabis . . . 48 j
Dordogne . . . 34 g
Dori . . . 40 m
Doro . . . 40 m
Döröö Nuur . . . 56 k
Dorowa . . . 49 f
Dorrego . . . 31 a
Dortmund . . . 33 o
Doruma . . . 42 u
Dos de Mayo . . . 24 k
Doshi . . . 60 j
Dosso . . . 41 c
Dothan . . . 19 l
Douala . . . 41 u
Doualayel . . . 41 w
Douaouir . . . 36 x
Douentza . . . 40 l
Douglas (UK) . . . 32 w
Douglas (Arizona) . . . 18 h
Douglas (Wyoming) . . . 15 q
Douglas Range . . . 82 w
Dourada, Serra . . . 26 z
Dourados . . . 29 a
Douro . . . 34 b
Dove Bay . . . 91 t
Dover . . . 32 z
Dover, Strait of . . . 34 g
Dover Foxcroft . . . 17 b
Dovsk . . . 33 z
Dowlatabad . . . 60 b
Dracena . . . 29 c
Drakensberg . . . 49 q
Drake Passage . . . 82 c
Drammen . . . 33 c
Drava . . . 35 c
Dreikikir . . . 70 y
Dresden . . . 33 r
Dresvyanka . . . 50 j
Drosh . . . 60 l
Drovyanaya . . . 51 b
Dr. Pedro P. Pen . . . 28 i
Drumheller . . . 14 m
Drummondville . . . 17 a
Druzhina . . . 53 g
Dryden . . . 15 k
Drygalski Island . . . 85 t
Drysdale . . . 72 m
Dschang . . . 41 u
Dua . . . 42 q
Duaringa . . . 74 r
Duba . . . 38 l
Dubai . . . 39 y
Dubawnt Lake . . . 12 x
Dubbo . . . 78 d
Dublin (Ireland) . . . 32 v
Dublin (USA) . . . 20 k
Dubna . . . 50 p
Dubois . . . 14 m
Dubovskoye . . . 54 g
Dubrovnik . . . 35 c
Dubuque . . . 16 q
Duchess . . . 73 v
Duero . . . 34 d
Dudinka . . . 51 h
Dudypta . . . 51 k
Duékoué . . . 40 v
Duero . . . 34 e
Dugiuma . . . 43 s
Dukhan . . . 39 v
Duki . . . 53 o
Dukou . . . 62 u
Dulan . . . 62 e
Dulce . . . 28 v
Dugla-Kyuyel' . . . 52 r
Dulia . . . 42 s
Duluth . . . 15 k
Dumaguete . . . 67 i
Dumai . . . 68 f
Dumka . . . 61 w
Dumont d'Urville Sea . . . 87 o
Dumont d'Urville . . . 87 o
Dumyat . . . 35 x
Duna . . . 35 c
Dunărea . . . 35 g
Dunay . . . 52 k
Dunbar . . . 73 k
Dundalk . . . 32 v
Dundas . . . 91 b
Dundas Harbour . . . 13 e
Dundas Peninsula . . . 12 g
Dundas Strait . . . 73 c
Dundee (South Africa) . . . 49 r
Dundee (UK) . . . 32 x
Dunedin . . . 79 s
Dungas . . . 41 g
Dungu . . . 42 v
Dungun . . . 66 t
Dungunab . . . 38 y
Dunhua . . . 58 h
Dunhuang . . . 56 x
Dunkwa . . . 40 z
Duqm . . . 60 o
Durack Range . . . 72 m
Durango (Mexico) . . . 18 x
Durango (USA) . . . 15 p
Duranzo . . . 28 z
Durban . . . 49 s
Durham . . . 16 x
Duri . . . 68 f
Durma . . . 39 r
Durrës . . . 35 d
Durrie . . . 77 g
Dusak . . . 55 p
Dusa Mareb . . . 43 v
Dushan . . . 62 y
Dushanbe . . . 55 u
Düsseldorf . . . 33 o
Dusti . . . 55 u
Dutch Harbor . . . 10 z
Dutlwe . . . 49 a
Dvina . . . 33/50
Dvina Bay . . . 50 p
Dwangwa . . . 45 t
Dwarka . . . 60 w
Dwarsberg . . . 49 c
Dyalinya . . . 53 b
Dyer, Cape . . . 13 m
Dyersburg . . . 16 q
Dzachuj . . . 56 y
Dzag . . . 56 m
Džambul . . . 55 v
Dzaoudzi . . . 47 c
Dzavchan Gol . . . 56 k
Dzereg . . . 56 j
Dzerzhinsk . . . 50 r
Dzerzhinsky . . . 54 f
Dzhaksy . . . 55 g
Dzhalinda . . . 52 x
Dzhambul . . . 55 e
Dzhamm . . . 53 a
Dzhankoy . . . 35 m
Dzhara . . . 52 d
Dzhardzhan . . . 52 k
Dzhelinde . . . 52 f
Dzhelon . . . 52 m
Dzhezkazgan . . . 55 h
Dzhungdzhur Mountains . . . 53 p
Dzhunkun . . . 52 e
Dzhusaly . . . 55 f
Džida . . . 57 d
Džida (River) . . . 57 d
Dzilam de Bravo . . . 19 v
Dzioua . . . 37 c
Džirgatal . . . 55 v
Džizak . . . 55 u
Dznankoy . . . 54 b
Dznetygara . . . 55 d

E

Eagle Plain . . . 11 k
Eagle Summit . . . 11 g
East Cape . . . 79 x
East China Sea . . . 63 k
Easter Group . . . 76 b
Easter Island . . . 234 r
Eastern Cordillera (Bolivia) . . 25 s
Eastern Cordillera (Colombia) . . . 22-23
Eastern Cordillera (Peru) . . . 24 k
Eastern Ghats . . . 61 s
Eastern Sierra Madre . . . 19 o
East Falkland . . . 31 p
East London . . . 49 q
Eastmain . . . 16 k
Eastmere . . . 74 p
East Millinocket . . . 17 c
East Sea . . . 58 v
Easter Siberian Sea . . . 53 i
East Turkestan . . . 56 o
Eau Claire . . . 16 c
Eau Claire, Lac à l' . . . 13 u
Eauripik Atoll . . . 70 l
Ebe . . . 52 j
Ebebiyin . . . 41 u
Ebinur Hu . . . 56 q
Ebola . . . 42 q
Ebolowa . . . 41 u
Ebro . . . 34 f
Echo Bay (Port Radium) . . . 12 e
Ecija . . . 34 p
Eclipse Sound . . . 88 y
Ecuador . . . 22/24
Ed Damazin . . . 42 m
Ede . . . 41 q
Edéa . . . 41 u
Eden . . . 15 o
Edessa . . . 35 e
Edge Island . . . 92 p
Edinburgh . . . 32 x
Edirne . . . 35 h
Edith Ronne Ice Shelf . . . 83 o
Edith Ronne Land . . . 83 s
Edjeleh . . . 37 f
Edmonton . . . 14 m
Edmundston . . . 17 c
Edsel Ford Range . . . 81 u
Edson . . . 14 k
Eduardo Castex . . . 30 l
Edward, Lake . . . 45 c
Edward River . . . 73 j
Edwards Plateau . . . 19 a
Edward VII Peninsula . . . 81 x
Edzo . . . 12 r
Efate . . . 75 e
Effingham . . . 16 r
Egadi . . . 34 z
Egersund . . . 33 a
Egijn Gol . . . 57 b
Egmont . . . 79 u
Eğridir, Lake . . . 35 w
Egvekinot . . . 10 h
Egypt . . . 35/38
Eidfjord . . . 33 b
Eight Degree Channel . . . 64 o
Eighty Mile Beach . . . 72 v
Eil . . . 43 k
Eilerts-de-Haan-Gebergte . . . 26 g
Einasleigh . . . 73 y
Eindhoven . . . 33 n
Eirunepé . . . 25 b
Eje Volcanico . . . 19 o
Ejido Insurgentes . . . 18 f
Ejin Qi . . . 57 n
Ejutla . . . 19 q
Ekalaka . . . 15 d
Ekibastuz . . . 55 l
Ekimchan . . . 53 n
Ekiti . . . 41 q
Ekoli . . . 44 m
Ekonda . . . 52 b
El Arco . . . 18 f
El Aaiun . . . 36 d
El Aricha . . . 36 k
Elat . . . 38 k
Elato Atoll . . . 71 a
Elaziğ . . . 54 q
Elba . . . 34 l
El Banco . . . 23 a
El Barcat . . . 37 t
El Baúl . . . 23 e
El Bayadh . . . 34 s
Elbe . . . 33 q
Elbert, Mount . . . 15 p
Elblag . . . 33 u
El Bordo . . . 22 y
Elbrus . . . 54 s
El Bur . . . 43 v
Elburz Mountains . . . 54 x
El Cajon . . . 18 c
El Calafate . . . 30 u
El Callao . . . 23 i
El Canton . . . 23 c
El Carmen . . . 25 t
El Carrizalillo . . . 18 k
El Cerro . . . 25 u
El Chilicote . . . 18 l
El Chinero . . . 18 e
El Dab . . . 43 i
El Dere . . . 34 v
El'dikan . . . 53 b
El Djem . . . 34 y
El Djouf . . . 36 t
El Dorado (Arkansas) . . . 19 g
Eldorado (Brazil) . . . 29 a
Eldorado (Uruguay) . . . 29 n
Eldorado (Canada) . . . 12 c
El Dorado (Kansas) . . . 15 v
El Dorado (Mexico) . . . 18 w
El Dorado (Venezuela) . . . 23 i
Eldoret . . . 43 n
El Eglab . . . 36 v
Elektrostal . . . 54 d
Elemi Triangle . . . 42 z
El Encanto . . . 23 o
Elephant Island . . . 82 e
Eleuthera . . . 21 c
El Fasher . . . 42 g
El Fuerte . . . 18 i
El Fugha . . . 37 k
El Gal . . . 43 l
El Gezira . . . 42 l
Elgin (UK) . . . 32 k
Elgin (USA) . . . 16 r
El Goléa . . . 37 a
Elgon . . . 42 z
El Hamurre . . . 43 x
El Hank . . . 36 t
El Hasheisa . . . 42 m
Eliseu Martins . . . 27 p
Elista . . . 54 h
Elizabeth . . . 16 z
Elizabeth City . . . 16 y
Elizabeth Reef . . . 78 k
Elizabethtown . . . 16 t
El Jadida . . . 36 f
El Jebelein . . . 42 l
Elkedra . . . 73 s
El Kere . . . 43 s
El Khenachich . . . 36 v
Elkins . . . 16 w
Elko . . . 14 x
Ellef Ringnes Island . . . 12 k
Ellensburg . . . 14 i
Ellesmere Island . . . 91 f
Elliot (S. Africa) . . . 49 q
Elliot Bay . . . 77 x
Elliot, Mount (Queensland) . . 74 q
Elliott (Australia) . . . 73 e
Elliott, Mount (W. Australia) . 72 z
Ellsworth Land . . . 81 b
Ellsworth Mountains . . . 81 d
El Maia . . . 36 x
El Maitén . . . 30 i
El Mango . . . 23 s
El Mansour . . . 36 l
El Meghaier . . . 37 c
Elmira . . . 16 y
El Mreïti . . . 36 t
El Obeid . . . 42 j
El Oso . . . 23 t
El Oued . . . 34 w
El Paso . . . 18 k
El Porvenir . . . 18 k
El Real . . . 22 k
El Salado . . . 30 w
El Salto . . . 18 x
El Salvador (State) . . . 22
El Salvador (Chile) . . . 28 p
El Sauz . . . 18 k
Elsey . . . 73 e
El Sherana . . . 73 d
El Socorro . . . 18 f
El Sombrero . . . 23 e
El Sueco . . . 18 k
Eltanin Bay . . . 80 m
El Tigre . . . 23 h
Elton . . . 54 i
El Triunfo . . . 18 u
El Tule . . . 19 a
El Uach . . . 43 r
Eluru . . . 64 i
Elvira . . . 24 m
El Wak . . . 43 q
Ely . . . 14 y
Emalahleni . . . 49 r
Emba . . . 55 b
Emba (River) . . . 55 a
Embarcación . . . 28 g
Embu . . . 43 o
Emden . . . 33 o
Emerald . . . 74 r
Emi Koussi . . . 37 y
Emmen . . . 33 n
Emmet . . . 74 o
Emory Peak . . . 18 m
Empalme . . . 18 h
Emporia (Kansas) . . . 15 v
Emporia (Virginia) . . . 16 x
Empredrado . . . 28 y
Emu . . . 58 h
Emu (River) . . . 77 v
Encarnación . . . 28 z
Encounter Bay . . . 77 s
Endako . . . 14 f
Ende . . . 69 u
Enderby Land . . . 84 l
Endicott Mountains . . . 11 d
Endybal . . . 52 m
Eneabba . . . 76 d
Engaño, Cape . . . 67 h
Engel's . . . 54 i
Engenheiro Dolabela . . . 29 i
Engerdyakh . . . 52 h
Enggano . . . 68 s
England . . . 32 x
English Channel . . . 32 x
English River . . . 16 d
Engoordina . . . 77 d
Enid . . . 15 u
Enkan . . . 53 q
Enna . . . 37 i
Ennadai Lake . . . 12 x
Ennedi . . . 38 p
Ennis . . . 15 a
Enschede . . . 33 o
Ensenada . . . 18 c
Enshi . . . 62 m
Entebbe . . . 42 y
Enterprise . . . 12 r
Entroncamento . . . 27 n
Enugu . . . 41 s
Envira . . . 25 b
Envira (River) . . . 25 n
Enyélé . . . 42 n
Equatorial Guinea . . . 41
Erba . . . 38 y
Erdene . . . 57 t
Erdene Büren . . . 56 i
Erdene Cagaan . . . 57 w
Erdenedalaj . . . 57 p
Erdenet . . . 57 c
Erdi . . . 38 o
Erebus . . . 86 r
Ereğli . . . 35 y
Ereğli (Black Sea) . . . 35 j
Erenhot . . . 57 u
Erexim . . . 29 o
Erfoud . . . 36 i
Erfurt . . . 33 p
Ergedzhey . . . 52 h
Erg el Ahmar . . . 36 w
Erger Raoui . . . 36 k
Erguig . . . 42 a
Ergun . . . 52 v
Ergun Yuoqi . . . 52 v
Erie . . . 16 w
Erie, Lake . . . 16 v
Erigavo . . . 43 j
Erimbet . . . 55 r
Erimo . . . 59 d
Eritrea . . . 43 d
Erldunda . . . 73 q
Ermelo . . . 49 r
Er Mreyer . . . 36 t
Ernabella . . . 77 b
Ernakulam . . . 64 e
Erode . . . 64 f
Eromanga . . . 73 x
Erozionnyy . . . 53 h
Er Rahad . . . 42 j
Errego . . . 46 j
Erromango . . . 75 s
Ershiyizhan . . . 52 x
Ertai . . . 56 u
Ertix He . . . 56 f
Erzincan . . . 54 r
Erzurum . . . 54 s
Esbjerg . . . 33 o
Escalante . . . 15 n

Escanaba 16 f
Escondido 18 c
Escondido (River) 22 g
Escoumins 17 b
Escuinapa 18 x
Escuintla 22 c
Ese 42 u
Eséka 41 u
Eshowe 49 s
Esk 74 t
Eskimo Point 13 n
Eskişehir 35 w
Eslamada-e-Gharb 39 f
Esmeralda 73 x
Esmeraldas 22 w
Esmeraldas (River) 22 w
Espanola 16 i
Española 22 p
Esperance 76 h
Esperanza (Antarctica) 82 g
Esperanza (Argentina) 28 v
Esperanza (Argentina) 30 v
Esperanza (Mexico) 18 h
Esperanza (Peru) 25 n
Espinal 22 z
Espinnaco, Serra do 29 i
Espiritu Santo 75 d
Espiritu Santo Bay 20 u
Espiritu Santo (Island) 18 h
Espita 19 w
Espoo 33 j
Espungabera 49 g
Esquel 30 h
Esquina 28 x
Essaouira 36 f
Essen 33 o
Essendon, Mount 72 v
Essequibo 23 x
Esso 53 y
Estación Babicora 18 j
Estacion Trincheras 18 g
Estancia 27 u
Estcourt 49 r
Estelí 22 e
Estevan 15 e
Estherville 15 w
Estmain 16 l
Estonia 33 j
Etah 61 d
Etap Bajanmönch 57 t
Etawah 61 d
Ethiopia 42-43
Ethiopian Highlands 43 b
Etna 35 n
Etosha Pan 48 h
Etykan 53 g
Etzatlán 18 y
Eubboea 35 s
Eucla 76 m
Eucumbene, Lake 77 z
Eufaula 20 i
Eugene 14 g
Eulo 74 o
Euphrates 39 e
Eureka (California) 14 t
Eureka (Canada) 13 c
Eureka (Nevada) 14 x
Eureka (Utah) 14 z
Eurinilla 77 h
Europe Island 46 z
Evans Lake 16 j
Evans Strait 13 r
Evanston 15 n
Evansville 16 s
Everard, Lake 77 c
Everard Park 77 b
Everest 61 i
Everett 14 h
Everglades 20 m
Evinayong 41 u
Evje 33 b
Evora 34 o
Èvreux 32 y
Ewirgol 56 t
Ewo 44 g
Exeter 32 w
Exeter Sound 13 m
Exmouth 72 q
Exmouth Gulf 72 r
Eyakit-Tërdë 52 f
Eyik 52 h
Eyre 76 k
Eyre, Lake 77 e
Eyre Peninsula 77 d
Ezequil Ramos
Mexia Reservoir 30 j
Ez Zemoul el Akbar 37 d

F

Fachi 37 t
Facundo 30 v
Fada 38 o
Fada-Ngourma 41 a
Faeroe Islands 32 i
Fafen 43 t
Faggo 41 h
Fagnano, Lago 30 w
Faguibine Lake 40 k
Faidzabad 55 v
Fairbanks 11 g
Fairmont (Minnesota) 15 w
Fairmont (W. Virginia) 16 w
Fairweather, Mount 11 x
Faisalabad 60 m
Faizabad 61 f
Fakfak 70 q
Falcon Reservoir 19 b
Falkland Islands/
Islas Malvinas 31 p
Fallon 14 w
Falster 33 q
Famagusta 35 y
Famatina, Sierra de 28 q
Fana 40 i
Fandriana 47 r
Fangshan 63 t
Fang Xian 63 a
Fangzheng 58 i
Fan si Pan 62 v
Faradje 42 w
Farafangana 47 r
Farafra 38 l
Farafra, Oasis 38 g
Farah 60 e
Farahrod 60 f
Farah Rud 60 f
Faranah 40 f
Farasen, Island 39 o
Faraulep Atoll 70 m
Farewell, Cape 90 t
Farewell Island 80 l
Fargo 15 i
Faribault 15 k
Faridpur 61 x
Farim 40 c
Farkovo 51 h
Farmington 15 o
Faro (Brazil) 26 g
Faro (Canada) 11 l
Faro (Portugal) 34 o
Farrukhabad 61 d
Farsi 60 f
Fartak, Ras 43 m
Fasa 39 k
Faskari 41 e
Fataki 42 x
Fatala 40 d
Fatehpur 61 r
Fátima 26 y
Fatsa 54 p
Fatunda 44 i
Fatzeh 54 b
Fauresmith 49 o
Fauro 71 u
Faxa Bay 32 a
Faya 37 z
Fayetteville (Arkansas) 15 w
Fayetteville (Carolina) 16 x
Fayetteville (Tennessee) 16 s
Fayu 71 d
Fazilka 60 m
Fdérik 36 q
Federated States of Micronesia 71 c
Fedorovskiy 53 d
Fedulki 50 z
Feijó 25 o
Feira de Santana 27 s
Felipe Carrillo Puerto 20 u
Fengfeng 63 c
Fengje 62 m
Fengning 57 w
Fengqing 62 s
Feodosiya 54 p
Ferdows 60 b
Ferfer 43 u
Fergana 55 w
Fergus Falls 15 i
Fériana 34 x
Ferkéssédougou 40 k
Ferlo 40 c
Fernandina 22 o
Fernandina Beach 20 m
Fernando de Noronha 27 l
Fernandópolis 29 d
Ferrara 34 m
Ferreira Gomes 26 k
Ferreñafe 24 h
Fès 36 i
Feshi 44 i
Festus 16 q
Fethiye 35 w
Fetisovo 54 z
Fezzan 37 u
Fiambalá 28 r
Fianarantsoa 47 p
Fiche 43 c
Ficksburg 49 q
Figuig 36 l
Fiji 75 l
Fik 43 e
Filadelfia 28 k
Filchner Ice Shelf 83 t
Filingué 41 c
Filtu 43 r
Fimbul Ice Shelf 83 j
Finch Hatton 74 r
Findlay 16 u
Fingoè 45 r
Finike 35 w
Finisterre, Cape 34 a
Finke 77 d
Finke (River) 77 d
Finland 33
Finland, Gulf of 33 i
Firat 54 q
Firenze (Florence) 34 m
Firozpur 61 a
Firuzabad 39 j
Fish 48 x
Fisher Strait 13 q
Fishguard 32 w
Fiske, Cape 83 n
Fitzcarrald 24 z
Fitzroy (Queensland) 74 s
Fitzroy (Western Australia) 72 x
Fitzroy Crossing 72 z
Flagstaff 15 n
Flathead Lake 14 m
Flattery, Cape 14 f
Flensburg 33 p
Flinders 73 w
Flinders Bay 76 c
Flinders Island 78 q
Flinders Range 77 g
Flin Flon 15 f
Flint 16 u
Flint (River) 20 j
Floraville 73 v
Florence (Italy) 34 m
Florence (South Carolina) 20 m
Florence (Tennessee) 20 h
Florencia 22 z
Florentino Ameghino
Reservoir 30 y
Flores 20 s
Flores (Island) 72 j
Flores Sea 69 t
Floriano 27 p
Florianópolis 29 s
Florida (Uruguay) 31 d
Florida (USA) 20
Florida Bay 20 m
Florida, Straits of 20 m
Flower's Cove 17 j
Fly 70 x
Foggia 35 b
Foley 13 g
Fomich 52 c
Fond du Lac 16 r
Fonseca, Gulf of 22 d
Fontanillas 25 w
Fonte Boa 25 e
Fontur 32 e
Foochow 63 t
Forbes 78 c
Forécariah 40 d
Forel, Mt 90 z
Formosa (Argentina) 28 x
Formosa (Brazil) 29 f
Formosa do Rio Preto 27 o
Formosa, Serra 25 z
Formoso 29 g
Forrest River 73 a
Forsayth 73 x
Forsyth 15 c
Fort Albany 16 i
Fortaleza (Bolivia) 25 q
Fortaleza (Brazil) 27 g
Fortaleza (Brazil) 25 s
Fort Beaufort 49 o
Fort Black 12 v
Fort Chipewyan 12 t
Fort Collins 15 q
Fort-de-France 23 i
Fort Dodge 15 w
Forte Carumbo 44 k
Fortescue 72 u
Forte Veneza 26 v
Fort-Foureau 41 k
Fort Frances 15 k
Fort Franklin 12 c
Fort Gardel 37 r
Fort Good Hope 12 a
Fort Hope 16 e
Fortin Garrapatal 28 i
Fortin General Diaz 28 j
Fortin Gral. Eugenio Garay 28 i
Fortin Hernandarias 28 i
Fortin Juan de Zalazar 28 k
Fortin Madrejón 28 k
Fortin Max Parades 28 k
Fortin Ravelo 28 j
Fort Lauderdale 20 m
Fort Laird 12 p
Fort Macleod 14 m
Fort McMurray 12 t
Fort McPherson 11 k
Fort Morgan 15 r
Fort Myers 20 l
Fort Nelson
(British Columbia) 12 p
Fort Nelson (Manitoba) 13 n
Fort Norman 12 o
Fort Peck Reservoir 15 d
Fort Pierce 20 m
Fort Portal 42 w
Fort Providence 12 r
Fort Qarain 38 d
Fort Resolution 12 t
For Rupert 16 k
Fort Sandeman 60 j
Fort Scott 15 w
Fort-Sevčenko 54 x
Fort Severn 13 p
Fort Simpson 12 p
Fort Smith (Canada) 12 u
Fort Smith (USA) 15 w
Fort St. John 12 q
Fort Stockton 18 m
Fort Walton Beach 20 i
Fort Wayne 16 t
Fort William 32 v
Fort Worth 19 c
Fort Yukon 11 g
Foshan 63 o
Fossil Bluff 82 x
Foumban 41 v
Foum El Aassane 36 g
Fourdou 40 d
Fouta Djalon 40 d
Foveaux Strait 79 q
Fowler's Bay 77 b
Foxe Basin 13 g
Foxe Channel 13 s
Foxe Peninsula 13 t
Fox Harbour 17 j
Fox Islands 10 z
Foz do Cunen 48 e
Foz do Igua (Iguaçú) 29 n
Foz do Mamoriá 25 d
Foz do Pauini 25 d
Franca 29 f
France 32-33/34
Franceville 44 f
Francisco Beltro 29 o
Francistown 49 c
Frankfort 49 q
Frankfurt 33 r
Frankfurt (am Main) 33 p
Frankland 76 d
Frankland Range 78 o
Franklin Mountains 12 o
Franklin Strait 12 m
Franz Josef Land 50 h
Fraserburg 48 y
Fraserdale 16 i
Fraser (Newfoundland) 13 y
Fraser
(River, British Columbia) 14 h
Fraser Island 74 u
Fraser Range 76 i
Frazier Downs 72 w
Frederick 15 h
Frederick E. Hyde Fjord 91 j
Fredericksburg 16 x
Fredericton 17 d
Frederikshavn 33 c
Fredonia 16 w
Fredrikstad 33 d
Freeport 16 q
Freetown 40 d
Fregon 77 a
Freiburg 33 o
Fremantle 76 d
French Guiana 26 i
French Polynesia 234 p
Fresco 26 x
Fresnillo 18 z
Fresno 14 v
Frewena 73 r
Fria 40 d
Frias 28 s
Frio, Cabo (Brazil) 29 j
Frio, Cape (Namibia) 48 e
Frisian Islands 33 n
Frobisher Bay (Iqaluit) 13 w
Frobisher Lake 12 v
Frome, Lake 77 g
Frontera 19 t
Front Range 15 q
Frutal 29 e
Frylinckspan 48 z
Fuding 63 h
Fuerte 18 i
Fuerte Olimpo 28 k
Fuerteventura 36 c
Fujairah 39 y
Fujin 58 k
Fujiyama 59 n
Fuka 38 f
Fukui 58 z
Fukuoka 58 v
Fukushima 59 o
Fukuyama 58 x
Funchal 36 a
Fundy, Bay of 17 e
Fünen 33 p
Funhalouro 49 h
Funing 62 x
Funtua 41 f
Furmanovka 55 w
Furmanovo 54 k
Furnas Reservoir 29 g
Furneaux Group 78 p
Fury and Hecla Strait 13 e
Fushun 58 f
Fusin 58 d
Fu Xian 58 q
Fuyang 63 d
Fuyu 58 f
Fuyuan 58 l
Fuyun 56 h
Fuzhou (Foochow) 63 e

G

Ga 40 l
Gabela 44 t
Gabès 37 f
Gabès, Gulf of 34 y
Gabon 41/44
Gabon (River) 44 c
Gaborone 46 o
Gadag 64 e
Gadsden 19 l
Gaferut 71 a
Gafoua 41 k

Gafsa **37** e
Gagarin **54** b
Gagnoa **40** w
Gagnon **17** c
Gagyu Hu **63** g
Gainesville (Florida) **20** k
Gainesville (Georgia) **20** j
Gairdner, Lake **77** d
Gairo **45** i
Gai Xian **58** e
Gaktsynka **53** p
Gala **61** k
Galanino **51** x
Galapagos Islands **22** q
Galati **35** h
Galbraith **73** j
Galcaio **43** v
Galdhöpiggen **33** b
Galena **11** c
Galesburg **16** q
Galilee, Lake **73** z
Galimyy **53** k
Galiwinku **73** f
Galle **64** u
Gallegos **30** v
Gallinas, Cabo **23** b
Gällivare **33** h
Gallup **15** o
Galt **57** a
Galveston **19** e
Galway **32** u
Galway Downs **77** i
Gama **29** f
Gambaga **40** m
Gambela **42** m
Gambell **10** k
Gambia **40**
Gambo **42** q
Gamboma **44** g
Gamboola **73** l
Gambos **44** s
Ganda **44** t
Gandadiwata **69** s
Gandajika **44** m
Gandak **61** g
Gander **17** k
Gandi **42** f
Gandia **34** s
Gandsya **54** v
Gandu **27** s
Gangan **30** j
Ganganagar **60** m
Gangapur **61** b
Gangaw **62** p
Gangca **57** n
Ganges **61** t
Ganges Delta **61** x
Ganhezi **56** u
Gan Jiang **63** d
Gannett Peak **15** o
Ganta **40** u
Gao **41** a
Gao'an **63** d
Gaotai **57** n
Gaoua **40** k
Gaoual **40** d
Garalo **40** i
Garanhuns **27** v
Garapu **26** w
Garba Tula **43** p
Garberville **14** t
Garda, Lake **34** m
Garden City **15** t
Gardez **60** j
Gardo **43** j
Garhchiroli **61** q
Gariep Reservoir **49** o
Garies **48** v
Garimpo Cumaru **26** x
Garissa **43** q
Garland **19** d
Garmabe **42** w
Garmenka **57** b
Garmsar **39** j
Garoe **43** j
Garonne **34** g
Garoua **41** j
Garoua Boulaï **41** x
Garry Lake **12** l
Garwa **61** t
Gary **16** s
Garyarsa **61** d
Gar Zangbo **61** e
Garzê **62** g
Garzón **22** z
Gascoyne **72** s
Gascoyne Junction **72** r
Gas Hu **56** v
Gashua **41** i
Gaspar Strait **68** w
Gaspé **17** e
Gastre **30** j
Gatton **74** t
Gauhati **61** z
Gavbandi **39** k
Gavio **27** s
Gävle **33** g
Gawan **59** b
Gawler **77** g
Gawler Ranges **77** d
Gaxun Nur **57** n
Gaya (India) **61** u
Gaya (Niger) **41** c
Gayndah **78** f
Gaza **38** k
Gaziantep **54** q
Gbadolite **42** p
Gbarnga **40** t
Gdansk **33** t
Gdynia **33** t
Gebe **67** z
Gedaref **43** a
Geelong **77** w
Geelvink Channel **76** c
Gē'gyai **61** e
Geidarn **41** i
Geita **45** f
Gejiu (Kokiu) **62** v
Gela **35** n
Gelib **43** s
Gêlo **25** y
Gernena **42** p
Gemsa **38** j
Gemsbok **48** x
General Acha **30** l
General Alvear **30** k
General Belgrano **83** s
General Belgrano Plateau **83** o
General Bernardo O'Higgins **82** s
General Carrera, Lago **30** t
General Conesa **30** l
General Güemes **28** t
General Juan Madariaga **31** d
General Pico **30** m
General Roca **30** k
General San Martin **82** v
General Santos **67** w
General Viamonte **31** a
General Villegas **30** m
Geneva **34** i
Geneva, Lake **34** j
Genève (Geneva) **34** i
Genoa **34** k
Genoa, Gulf of **34** k
Genova (Genoa) **34** k
Genovesa **22** p
Geographical Society Island **91** r
George **48** z
George IV Sound **82** z
George, Lake (USA) **20** l
George, Lake (Uganda) **45** d
George Land **50** h
Georgetown (Australia) **73** x
Georgetown
(Cayman Islands, UK) **20** y
Georgetown (Gambia) **40** c
Georgetown (Guyana) **23** l
George Town
(Pinang, Malaysia) **65** w
Georgetown (USA) **21** a
George V Land **86** y
Georgia **20** k
Georgia (State) **54** t
Georgian Bay **16** i
Georgia, Strait of **14** g
Georgiievka **56** d
Georgina **73** v
Georg von Neumayer **83** g
Gera **33** q
Geral de Goiás, Serra **29** f
Geral do Parana, Serra **29** f
Geraldton (Australia) **76** c
Geraldton (Canada) **16** f
Gerasmikova **51** p
Gerede **35** k
Gerlachovsky **35** d
Germania Land **91** t
Germany **33**
Germiston **49** q
Gerona **34** h
Gêrzê **61** g
Gettysburg **15** g
Getz Ice Shelf **81** q
Geureudong **65** u
Gewane **43** d
Geyhan **35** z
Ghadames **37** f
Ghadir **43** g
Ghaem-Shahr **54** z
Ghaghara **61** t
Ghana **40-41**
Ghandi Sagar Reservoir **61** n
Ghanzi **48** m
Ghardaia **34** u
Ghariyan **34** z
Ghat **37** s
Ghazaouet **36** k
Ghaziabad **61** c
Ghazipur **61** t
Ghazni **60** j
Ghelinsor **43** v
Ghent **34** h
Ghion **43** b
Ghorian **60** e
Ghubeish **42** h
Ghugri **61** v
Giahel **43** k
Giamama **43** s
Gibb River **72** m
Gibraltar **34** p
Gibraltar, Str. of **34** p
Gibson Desert **72** y
Gidgealpa **77** g
Gidgee **76** g
Gidole **43** o
Gifford **88** x
Gifu **58** z
Giganta, Siera de la **18** g
Gijón **34** c
Gila Desert **18** f
Gilbert River **74** a
Gilbués **27** o
Giles
Meteorological Station **73** n
Gilf Kebir Plateau **38** r
Gilgil **43** n
Gilgit **61** a
Gillam **13** n
Gillette **15** d
Gimbi **43** a
Gingin **76** d
Gingoog **67** w
Giohar **43** u
Giresun **54** q
Giri **42** n
Girishk **60** g
Gironde **34** f
Gisborne **79** x
Gisenyi **45** d
Gitarama **45** d
Giza **38** h
Gizhduvan **55** r
Gizhiga **53** m
Gjoa Haven **12** m
Gjumri **54** t
Glace Bay **17** h
Gladmar **15** d
Gladstone **74** t
Glasgow (Kentucky) **16** t
Glasgow (Montana) **15** c
Glasgow (UK) **32** w
Glazov **50** v
Glenayle **76** h
Glendive **15** d
Glenelg **77** u
Glen Falls **16** z
Glengyle **77** g
Glen Helen **73** q
Glen Innes **78** f
Glenmorgan **78** d
Glennallen **11** g
Glenns Ferry **14** y
Glenormiston **73** u
Glenroy **72** m
Glenwood Springs **15** p
Globe **18** h
Glubinnoye **58** l
Goa **64** d
Goageb **48** v
Goalpara **61** y
Goba **43** q
Gobabis **48** k
Gobernador Gregores **30** v
Gobi **56-57**
Gochas **48** j
Godavari **64** i
Gode **43** s
Goderich **16** v
Godhavn **90** k
Godhra **60** z
Godinlave **43** v
Gods Lake **15** j
Godthåb (Nuuk) **90** u
Goiána **27** w
Goianésia **29** e
Goiânia **29** d
Goiás **29** d
Goiatuba **29** d
Gokwe **46** q
Gol (Norway) **33** b
Gol (Russia) **52** t
Golan Heights **38** l
Gol'chikha **51** g
Gold Coast (Australia) **78** g
Gold Coast (Ghana) **40** z
Golden **14** k
Goldfield **14** x
Goldsboro **16** x
Goldworthy **72** u
Golmud **62** c
Golyshmanovo **55** g
Goma **45** c
Gombari **42** w
Gombe **41** i
Gomel **33** y
Gomera **36** a
Gómez Palacio **18** l
Gonaïves **21** r
Gonam **52** z
Gonam (River) **52** z
Gonbad-e-Kavus **54** z
Gonder **43** b
Gongbo'gyamda **62** b
Gonggar **61** l
Gonghe **62** g
Gongola **41** h
Gongpoquan **56** z
Gongshan **62** f
Gongxi **63** e
Gonzales **19** d
Good Hope, Cape of **48** w
Gooding **14** y
Goodnews Bay **11** n
Goondiwindl **78** e
Gooty **64** f
Gorakhpur **61** f
Gordon **77** x
Gordon Downs **73** n
Gore (Chad) **41** m
Gore (Ethiopia) **43** a
Gore (New Zealand) **79** r
Goreis **48** h
Gorel'de **55** q
Gorelova **53** m
Gorgan **54** z
Gorgona **22** x
Gorgora **43** b
Gor'kiy (Nizhiy Novgorod) **50** r
Gorlovka **54** d
Gornak **56** d
Gorno-Chuyskiy **52** r
Gornyy **58** m
Gornyy Kazymsk **51** a
Goroka **70** z
Gorongosa **46** u
Gorontalo **69** i
Gorotiré **26** x
Gorshechnoye **54** d
Gorzow-Wikp **33** s
Gosford **78** f
Göteborg **33** d
Gotland **33** g
Göttingen **33** p
Goudoumaria **41** i
Gouin Reservoir **16** m
Goukou **58** d
Goulburn **78** e
Goulmina **36** i
Goumeré **40** x
Goundam **40** k
Goura **41** g
Gourara **36** m
Gouré **41** h
Gouro **37** z
Gov'Chonin **56** w
Governador Valadares **29** j
Govorovo **52** l
Goya **28** x
Goyder's Lagoon **77** g
Goyoum **41** x
Goz Beïda **42** d
Gozo **35** n
Graaff-Reinet **49** n
Grabo **40** v
Gracias á Dios, Cape **22** h
Gradaús, Serra **26** x
Grafton (Australia) **78** g
Grafton (USA) **15** h
Graham Bell Island **50** m
Graham Island **13** b
Graham Land **82** u
Grahamstown **49** o
Grain Coast **40** t
Grajaú **27** n
Grajaú (River) **27** o
Granada (Nicaragua) **22** f
Granada (Spain) **34** q
Gran Canaria **36** b
Gran Chaco **28** h
Grand Bahama Island **21** a
Grand Bank **17** j
Grand-Bassam **40** x
Grand Caicos **21** s
Grand Canal **57** w
Grand Canyon **14** z
Grand Cayman **20** z
Grand Centre **15** a
Grande (Brazil) **29** h
Grande (Nicaragua) **22** g
Grande (Venezuela) **23** i
Grande, Bahia **30** w
Grande Cache **14** j
Grande Comore **47** b
Grande de Gurapa **26** k
Grande de Santiago, Rio **18** y
Grande Prairie **14** j
Grand Erg de Bilma **37** u
Grande, Rio (River, Bolivia) **28** g
Grande, Rio
(River, Campos, Brazil) **27** p
Grande, Rio (River, Brazil) **29** h
Grande, Rio (River, USA) **19** b
Grand Forks **15** i
Grand Island **15** u
Grand Junction **15** o
Grand-Lahou **40** w
Grand Marais **16** d
Grand Mére **17** a
Grand Pantanal **28** m
Grand Rapids (Minnesota) **15** k
Grand Rapids (Michigan) **16** t
Grand Santi **26** i
Grangeville **14** k
Granite Downs **77** c
Granite Peak (Australia) **76** h
Granite Peak (USA) **15** a
Gran Laguna Salada **30** x
Gran Sabana **23** v
Grant Island **81** r
Grants **15** p
Grants Pass **14** t
Granville **32** x
Gravatá **27** v
Gray, Cape **87** o
Grayling **16** g
Graz **35** b
Great Abaco Island **21** b
Great Anyuy **10** a
Great Australian Bight **76** k
Great Bahama Bank **21** n
Great Barrier Reef **74** b
Great Barrier Island **79** i
Great Basin **14** x
Great Bear Lake **12** d
Great Bitter Lake **38** j
Great Britain and
Northern Ireland **32**
Great Dividing Range **78**
Great Eastern Erg **34** w
Great Exuma Island **21** p

Great Falls 15 a
Great Fish 49 o
Great Inagua Island 21 q
Great Karasberg 48 w
Great Karoo 48 z
Great Khingan Mountains 58 d
Great Plain 63 c
Great Salt Lake 14 z
Great Sandy Desert 72 x
Great Slave Lake 12 t
Great Smoky Mountains 16 u
Great Tiraz 48 u
Great Victoria Desert 76 l
Great Wall 57 u
Great Western Erg 36/37
Great Whale River 16 l
Greater Antilles 20/21
Greece 35
Greeley 15 q
Greely Fjord 91 e
Green Bay 16 e
Green Head 76 c
Greenland 90-91
Greenland Sea 91 w
Greenough 76 d
Green River 15 n
Green River (River) 15 o
Greensboro 16 w
Greensburg 16 w
Greenvale 74 o
Greenville (Alabama) 20 i
Greenville (Liberia) 40 t
Greenville (Mississippi) 20 f
Greenville (North Carolina) 16 x
Greenville (South Carolina) 20 k
Gregory Downs 73 u
Gregory, Lake 72 z
Gregory Range 74 n
Greifswald 33 r
Grenada 23 j
Grenoble 34 j
Grey Bull 15 b
Greymouth 79 s
Grey Range 77 i
Greytown 49 s
Gribingui 42 b
Griekwastad 48 z
Griffith 77 k
Grinell Peninsula 13 a
Grise Fjord 13 e
Groblersbrug 46 p
Groblersdal 46 q
Groblershoop 48 y
Grodno 33 w
Grong 33 d
Groningen (Netherlands) 33 n
Groningen (Suriname) 26 g
Groot 49 n
Groote Eylandt 73 g
Grootfontein 48 j
Groot Winterhoek 48 x
Grosseto 34 m
Gross Glockner 34 m
Groznyi 54 u
Grudziadz 33 t
Grünau 48 w
Grungedal 33 b
Gryazi 54 e
Guaçu 29 a
Guadalajara (Mexico) 18 z
Guadalajara (Spain) 34 e
Guadalcanal 71 x
Guadalquivir 34 p
Guadalupe 18 z
Guadalupe (Island) 18 c
Guadalupe Peak 18 l
Guadalupe Victoria 18 y
Guadalupe-y-Calvo 18 j
Guadeloupe 21
Guadiana 34 o
Guadix 36 k
Guafo, Gulf of 30 f
Guaicurus 28 l
Guainia 23 q
Guaira 29 n
Guajará Mirim 25 r
Guajira Peninsula 23 b
Gualeguay 28 x
Gualeguay (River) 28 x
Gualeguaychu 28 x
Guallatiri 28 c
Guam 235 j
Guamini 31 a
Guanahacabibes Peninsula 20 w
Guanaja 20 v
Guanajuato 19 n
Guanambi 29 i
Guanare 23 d
Guandacol 28 q
Guangnan 62 x
Guangyuan 62 k
Guangzhou (Canton) 63 p
Guanipa 23 h
Guantiánamo 21 p
Guan Xian 62 i
Guapi 22 x
Guaporé 25 s
Guaqui 25 o
Guarabira 27 v
Guarai 26 z
Guarapuava 29 p
Guaratinguetá 29 g
Guarda 34 b
Guardafui, Cape 43 l
Guàrico 23 e
Guarulhos 29 f
Guasave 18 i
Guásimas 18 h
Guatemala 19/22
Guatemala (City) 22 c
Guaviare 23 q
Guayabero 23 n
Guayaquil 24 g
Guayaquil, Gulf of 24 g
Guaymas 18 g
Guazapares 18 j
Guba 43 a
Gubakha 50 w
Gudur 64 h
Gudvangen 33 a
Guéckedou 40 g
Guedi 42 b
Guelb er Richat 36 r
Guélengdeng 41 l
Guelma 37 d
Guelta Zemmur 36 q
Guémar 37 d
Güera 36 n
Guerara 34 v
Guercif 36 j
Guernsey 34 e
Guga 53 p
Gugu 43 d
Guguang 69 d
Guia 36 b
Guiana Highlands 23 s
Guidjiba 41 k
Guiglo 40 u
Guijá 46 t
Guimaras 67 h
Guinan 62 g
Guinea 40
Guinea Bissau 40
Guinea, Gulf of 41 p
Güines 20 y
Guir 36 k
Güiria 23 i
Guiyang (Kweiyang) 62 x
Gujarkhan 60 m
Gujranwala 60 m
Gujrat 60 m
Gulbarga 61 p
Gul'ča 55 x
Gulf of Mexico 20
Gulfport 20 g
Gulian 52 v
Gulistan 55 u
Gulkana 11 h
Gulu 42 x
Gulya 52 v
Gumare 48 l
Gumbiro 45 v
Gumti 61 s
Gümüşhane 54 q
Guna 61 p
Guna (Mount) 43 c
Gungu 44 j
Gunnbjrn Fjeld 91 n
Gunnedah 78 e
Gunpowder 73 v
Gunter 61 r
Guntur 64 h
Guoju 63 i
Gurdaspur 61 a
Gurgei 42 f
Gurguéia 27 p
Gurha 60 y
Guri, Lago de 23 i
Guro 46 u
Gurskoye 59 a
Gurué 46 j
Gurun 54 q
Gurupi 26 y
Gurupi (River) 27 a
Gurupizinho 26 m
Guru Sikhar 60 z
Gurvan Ozagal 57 i
Gusau 41 e
Gushan 58 r
Gusinoozersk 57 e
Gutara 56 m
Guthrie 15 u
Guyana 23/25/26
Guyang 57 t
Guymon 15 s
Guyuan 62 k
Gvardeyets 53 g
Gwa 62 p
Gwai 49 c
Gwai River 49 c
Gwalior 61 p
Gwanda 49 d
Gweru 49 e
Gwydir 78 d
Gyaring Co 61 j
Gyda 51 e
Gydan Bay 51 d
Gydan Peninsula 51 d
Gyirong 61 h
Gyldenlves Fjord 90 w
Gympie 78 g
Gynym 52 z
Győr 35 c

H

Haarlem 48 z
Haast 79 q
Haast Bluff 73 p
Hab 60 v
Habaswein 43 q
Hab Chauki 60 v
Habshan 39 x
Hachioji 59 n
Ha Coi 62 y
Hadarba, Ras 38 z
Haddad 42 c
Hadejia 41 h
Hadejia (River) 41 g
Hadera 38 k
Hadilik 56 s
Hadley Bay 12 i
Hadramaut 43 j
Haeju 58 s
Haena 18 p
Hafun, Ras 43 l
Hagadera 43 q
Hagere Hiywot 43 c
Hagerstown 16 x
Ha Giang 62 x
Hague, The 33 n
Hagunia 36 d
Haifa 38 k
Haikow 63 n
Ha'il 39 b
Hailar 58 c
Hailar (River) 58 b
Hainan 63 n
Hainan Strait 63 n
Haines 11 x
Haines Junction 11 k
Haiphong 62 y
Haiti 21
Haiya 38 y
Haiyan 62 h
Hajiabad 39 l
Hajjah 43 f
Hakkâri 54 t
Hakodate 59 c
Halaban 39 q
Halaib 38 z
Halawa 18 r
Halcon 67 g
Hale 76 e
Hal Flood Range 81 t
Halifax 17 s
Halil Rud 60 b
Hall Beach 13 e
Halle 33 q
Hallett 86 s
Halley Bay 83 f
Hall Island 50 j
Hall Islands 71 e
Hall Land 91 f
Hall Peninsula 13 x
Halls Creek 73 n
Halmahera 69 m
Halmstad 33 r
Ham 41 l
Hama 38 m
Hamada du Guir 36 j
Hamada du Tinrhert 37 d
Hamada el Haricha 36 w
Hamada el Homra 37 g
Hamada of Dra 36 h
Hamadan 39 h
Hamada Safia 36 v
Hamadat Tingharat 37 g
Hamamatsu 59 n
Hamar 33 d
Hamarro Hadad 43 f
Hambantota 64 v
Hamburg 33 p
Hämeenlinna 33 i
Hamelin Pool 76 c
Hamersley Range 72 s
Hamhung 58 u
Hami 56 x
Hamilton (Alaska) 11 a
Hamilton (Australia) 77 v
Hamilton (Canada) 16 w
Hamilton (Idaho) 14 l
Hamilton (New Zealand) 79 v
Hamilton Downs 73 q
Hamilton Inlet 17 i
Hammami 36 t
Hammam Lif 37 g
Hammerfest 33 j
Hammond 20 f
Hamoud 40 e
Hampton (Oregon) 14 v
Hampton (Virginia) 16 y
Hampton Tableland 76 l
Hamrat al Shaikh 42 h
Hamun-e Jaz Murian 60 c
Hancheng 63 a
Handeni 45 j
Handgai 57 l
Hangayn Mountains 57 a
Hangchow 63 h
Hanggin Houqi 57 q
Hangzhou (Hangchow) 63 h
Hankö 33 i
Hann 72 l
Hanna 14 m
Hannah Bay 16 j
Hannibal 16 p
Hannover 33 p
Hanoi 62 x
Hanover (Hannover) 33 p
Hanover (Island) 30 s
Hanover (South Africa) 49 n
Hanover (USA) 17 n
Han Shui 63 b
Hantan 57 v
Hanzhong 62 k
Haouach 38 n
Happy Valley-Goose Bay 17 g
Hapur 61 c
Harad 39 u
Harare 46 s
Haraza 42 j
Haraze Mangueigne 42 c
Harbin 58 g
Harda 61 p
Hardangerfjörd 33 a
Hardec Králove 33 s
Hardin 15 c
Harding 49 r
Hardy 16 p
Harer 43 e
Hargeisa 43 g
Hargele 43 s
Har Hu 56 z
Hari 68 h
Hari Rud 60 f
Harlingen 19 c
Harlowtown 15 a
Harper 40 u
Harqin 58 c
Harricana 16 k
Harrington Harbour 17 h
Harrisburg 16 y
Harrismith 49 q
Hartford 17 n
Harts Range 73 s
Harudj el Asued 37 l
Hasa 39 h
Hasalbag 56 n
Hassan 64 e
Hassel Sound 12 l
Hassi bel Guebbour 37 d
Hassi Inifel 37 b
Hassi Maroket 38 a
Hassi Messaoud 34 w
Hassi Touareg 37 c
Hastings (New Zealand) 79 w
Hastings (USA) 15 u
Hatches Creek 73 s
Ha Tinh 62 x
Hat Lek 66 g
Hatteras, Cape 16 y
Hattiesburg 20 f
Hat Yai 66 q
Hauberg 83 o
Haugesund 33 a
Hauraki Gulf 79 v
Haute Kotto 42 r
Hauts Plateaux 36 m
Havana 20 x
Haverhill 17 o
Havre 15 a
Havre-St-Pierre 17 f
Hawaii 18 r
Hawaiian Islands 18 o
Hawera 79 u
Hawi 18 r
Hawke Bay 79 w
Hawr al Hammar 39 f
Hawthorne 14 w
Hay 77 j
Hay River 12 s
Hay River (River) 12 r
Hayes 13 a
Hayes Halv 91 c
Haynan 43 j
Hays (USA) 15 t
Hays (Yemen) 43 f
Hazard 16 u
Hazaribagh 61 u
Hazelton 14 e
Head of Bight 77 a
Healesville 78 o
Heard Island 235 r
Hearst 16 h
Hearst Island 82 x
Hebel 74 q
Heber City 15 n
Hebrides 32 h
Hecate Strait 14 c
Hechi 62 y
Hefei 63 f
Hefeng 63 a
Heihe 58 g
Heilbron 62 p
Helena, Mount 14 m
Helen Reef 70 d
Heligoland Bight 33 o
Hell-Ville 47 e
Helmand 60 f
Helmeringhausen 48 v
Helsingborg 33 r
Helsingfors (Helsinki) 33 j
Helsingör 33 q
Helsinki 33 j
Helwân 38 i
Henbury 73 q
Hendersonville 16 u
Hendijan 39 h
Hengchun 63 u
Hengduan Shan 62 e
Hengyang 63 b
Henrietta Maria, Cape 16 i
Henryetta 15 v
Henry Kater Peninsula 13 j
Henties Bay 48 g
Henzada 65 f
Heppner 14 i
Herat 60 f

Herbert Vale . . . . 73 u
Heredia . . . . 22 g
Hereford . . . . 15 s
Hermosillo . . . . 18 g
Herrick . . . . 78 p
Herschel . . . . 11 j
Hervey Bay . . . . 74 u
Hexigten Qi . . . . 57 x
Heze . . . . 63 d
Hibbing . . . . 15 k
Hibbs, Point . . . . 77 x
Hickory . . . . 16 v
Hidalgo del Parral . . . . 18 k
Hidden Valley . . . . 73 d
High Altai . . . . 56 v
High Atlas . . . . 36 h
High Knob . . . . 16 u
Highlands (Scotland) . . . . 32 w
High Level . . . . 12 s
High Prairie . . . . 12 r
High River . . . . 14 l
Hiiumaa . . . . 33 i
Hilla . . . . 39 e
Hillsboro . . . . 16 u
Hilo . . . . 18 s
Hima' . . . . 16 r
Himalaya 61
Himatnagar . . . . 60 z
Himeji . . . . 58 y
Himeville . . . . 49 r
Hindmarsh, Lake . . . . 77 u
Hindu Kush . . . . 60 j
Hinlopen Strait . . . . 92 o
Hirafok . . . . 37 q
Hirakud Reservoir . . . . 61 t
Hiroshima . . . . 58 x
Hisar . . . . 61 a
Hoa Binh . . . . 62 x
Hoare Bay . . . . 13 l
Hobart (Tasmania) . . . . 77 y
Hobart (USA) . . . . 15 u
Hobbs . . . . 18 m
Hochfeld . . . . 48 j
Hochwan . . . . 62 k
Ho Chi Minh (Saigon) . . . . 66 j
Hodal . . . . 61 b
Hodda . . . . 43 k
Hoddur . . . . 43 s
Hodeida . . . . 43 f
Hoeryŏng . . . . 58 i
Hohoi . . . . 63 c
Hoh Xii Shan . . . . 61 l
Hoima . . . . 42 x
Hoit Taria . . . . 56 y
Hokang . . . . 58 i
Hokitika . . . . 79 s
Hokkaidō . . . . 59 d
Hola . . . . 43 q
Holbrook . . . . 15 n
Hold wilh Hope Peninsula . . . 91 s
Holguín . . . . 21 p
Hollick Kenyon Plateau . . . . . 81 d
Holman Island . . . . 12 e
Holstebro . . . . 33 o
Holt Sum . . . . 57 x
Holy Cross . . . . 11 b
Holyhead . . . . 32 w
Hombori . . . . 40 l
Home Bay . . . . 13 k
Homodji . . . . 41 k
Homs . . . . 38 m
Honda . . . . 22 z
Honda Bay . . . . 67 e
Hondo . . . . 20 t
Hondo Reservoir . . . . 28 t
Hondo (River) . . . . 19 w
Honduras 19-20/22
Honduras, Gulf of . . . . 20 v
Hönefoss . . . . 33 c
Hong Ha . . . . 62 w
Hong Kong . . . . 63 q
Hongliuyuan . . . . 56 y
Hongyuan . . . . 62 i
Hongze . . . . 63 g
Honiara . . . . 71 x
Honolulu . . . . 18 p
Honshū . . . . 59 n
Hood Cape . . . . 76 g
Hooker, Cape . . . . 86 t
Hooker Creek . . . . 73 p
Hooker Island . . . . 50 i
Hooper Bay . . . . 10 m
Hopa . . . . 54 s
Hope . . . . 14 h
Hopedale . . . . 17 h
Hopes Advance, Cape . . . . . . 13 w
Hopetown . . . . 49 n
Hopin . . . . 62 q
Hopkins, Lake . . . . 73 n
Hormuz, Strait of . . . . 39 m
Horn, Cape (Chile) . . . . 30 x
Horn, Cape (Iceland) . . . . 32 b
Horn, Cape (Norway) . . . . 91 p
Hornavan . . . . 33 g
Horqin Youyi Qianqi . . . . 58 d
Horqin Zuoyi Zhongqi . . . . . . 58 e
Horsham . . . . 77 v
Hoseinabad . . . . 39 l
Hoshab . . . . 60 t
Hoshangabad . . . . 61 p
Hoshiarpur . . . . 61 b
Hospet . . . . 64 e
Hossororo . . . . 23 k
Hoste . . . . 30 w
Hotan . . . . 56 p
Hotan (River) . . . . 56 p
Hotazel . . . . 48 z
Hot Springs . . . . 15 r
Hottah Lake . . . . 12 e
Houailu . . . . 75 p
Houck . . . . 15 o
Houghton . . . . 16 e
Houma (China) . . . . 63 a
Houma (USA) . . . . 19 h
Houndé . . . . 40 l
Houston . . . . 19 d
Houtman Abrolhos . . . . 76 b
Houxia . . . . 56 t
Howar . . . . 38 q
Howe, Cape . . . . 78 r
Howship . . . . 73 d
Howship (Mounl) . . . . 73 d
Hoyo . . . . 42 w
Hradec Králové . . . . 35 b
Hsinchu . . . . 63 u
Hsuchang . . . . 63 c
Huachinango . . . . 20 n
Huacho . . . . 24 v
Huacrachuco . . . . 24 w
Huadian . . . . 58 g
Huahaizi . . . . 56 x
Hau Hin . . . . 65 j
Huai . . . . 63 d
Huai' an . . . . 57 v
Huaibei . . . . 63 e
Huaiji . . . . 63 o
Huainan (Hwainan) . . . . 63 e
Huairen . . . . 57 u
Huaiyin . . . . 63 g
Huajuapan de León . . . . 19 p
Hualien . . . . 63 v
Huallaga . . . . 24 k
Huamanrazo . . . . 24 y
Huancabamba . . . . 24 h
Huancané . . . . 25 o
Huancapi . . . . 24 y
Huancavelica . . . . 24 x
Huancayo . . . . 24 y
Huanchaca, Serra de . . . . 25 v
Huang . . . . 63 a
Huangchuan . . . . 63 d
Huan Jiang . . . . 62 l
Huanren . . . . 58 f
Huanta . . . . 24 y
Huánuco . . . . 24 x
Huapanapan . . . . 19 p
Huaping . . . . 62 u
Huara . . . . 28 c
Huaral . . . . 24 w
Huaraz . . . . 24 v
Huarmey . . . . 24 v
Huascaran . . . . 24 w
Huasco . . . . 28 o
Huasco (River) . . . . 28 p
Huashixia . . . . 62 f
Huatabampo . . . . 18 h
Huauchinango . . . . 19 p
Hudiksvall . . . . 33 g
Hudson . . . . 16 z
Hudson, Cape . . . . 86 y
Hudson Bay . . . . 13 q
Hudson Bay (Saskatchewan) 15 f
Hudson Land . . . . 91 s
Hudson Strait . . . . 13 v
Hue . . . . 66 j
Huehuetenango . . . . 19 u
Huelva . . . . 34 o
Huesca . . . . 34 f
Huéscar . . . . 36 j
Hufuf . . . . 39 u
Hughenden . . . . 73 y
Hughes . . . . 11 d
Hugli . . . . 61 w
Huhehot . . . . 57 u
Hüchŏn . . . . 58 t
Huila, Nev. de . . . . 22 y
Huili . . . . 62 u
Huinan . . . . 58 f
Huinca Renancó . . . . 30 m
Huixtla . . . . 19 t
Huize . . . . 62 v
Huizhou . . . . 63 q
Hulaifah . . . . 39 a
Hulin . . . . 58 k
Hull . . . . 16 l
Hulls Cove . . . . 17 p
Huluun Nuur . . . . 57 j
Huma . . . . 58 g
Humahuaca . . . . 28 f
Humaitá . . . . 26 o
Humansdorp . . . . 49 n
Humboldt . . . . 14 w
Humera . . . . 43 b
Hume Reservoir . . . . 78 c
Humphreys Peak . . . . 15 n
Hun . . . . 37 j
Húna Bay . . . . 32 b
Hunchun . . . . 58 i
Hungary 33/35
Hungchui . . . . 62 y
Hungtze, Lake . . . . 63 f
Hunter Island . . . . 77 w
Huntington . . . . 16 u
Hunt, Mount . . . . 12 n
Huntsville (Canada) . . . . 16 j
Huntsville (Texas) . . . . 19 e
Huntsville (Tennessee) . . . . . 19 k
Hunyani . . . . 46 e
Huolongmen . . . . 58 g
Huo Xian . . . . 63 b
Hurghada . . . . 38 k
Huron . . . . 15 h
Huron, Lake . . . . 16 h
Hutanopan . . . . 68 d
Hutehinson . . . . 15 u
Huth . . . . 43 g
Huxley, Mount . . . . 72 z
Huyani . . . . 49 e
Hval Sound . . . . 91 c
Hvar . . . . 35 b
Hwainan . . . . 63 e
Hwang (Huang He) . . . . 63 f
Hwange . . . . 46 o
Hwangju . . . . 58 t
Hwangshih . . . . 63 d
Hyderabad (India) . . . . 61 q
Hyderabad (Pakistan) . . . . 60 w
Hyesanjin . . . . 58 h
Hyndman Peak . . . . 14 y

I

Iaco . . . . 25 p
Iași . . . . 35 h
Iba . . . . 67 f
Ibadan . . . . 41 q
Ibagué . . . . 22 z
Ibarra . . . . 24 i
Ibb . . . . 43 g
Ibba . . . . 42 v
Ibembo . . . . 42 s
Iberia . . . . 25 o
Ibi . . . . 41 h
Ibiapaba, Serra da . . . . 27 r
Ibicui . . . . 28 z
Ibitunane . . . . 27 q
Ibiza . . . . 34 t
Ibologero . . . . 45 g
Ibotirama . . . . 27 q
Ibra . . . . 42 f
Ibri . . . . 39 z
Ica . . . . 24 x
Içá . . . . 25 b
Ica (River) . . . . 24 x
Içana . . . . 25 d
Iceland 32
Içel (Mersin) . . . . 35 y
Icha . . . . 53 x
Ichang . . . . 63 a
Ichera . . . . 52 q
Ichilo . . . . 25 s
Ichinskaya Sopka . . . . 53 y
Ich'on . . . . 58 t
Icó . . . . 27 t
Icy, Cape . . . . 11 a
Idaho . . . . 14 k
Idaho Falls . . . . 14 z
Idaye . . . . 41 i
'Idd al Ghanam . . . . 42 e
Ideles . . . . 37 p
Iderijn Gol . . . . 57 a
Idfu . . . . 38 w
Idhan Awbari . . . . 37 g
Idhan Murzuq . . . . 37 u
Idi . . . . 65 u
Idiofa . . . . 44 j
Idlib . . . . 38 l
Idri . . . . 37 i
Idutywa . . . . 49 q
Idyum . . . . 52 z
Idžim . . . . 56 j
Ifakara . . . . 45 i
Ifalik Atoll . . . . 70 l
Ife . . . . 41 q
Ilérouane . . . . 37 r
Iffley . . . . 73 w
Igaluit . . . . 90 f
Iganga . . . . 42 y
Igarka . . . . 51 h
Igbetti . . . . 41 c
Ighil . . . . 36 h
Igli . . . . 36 k
Igluligaarjuk
(Chesterfield Inlet) . . . . 13 o
Igombe . . . . 45 f
Igra . . . . 50 v
Iguaçú . . . . 29 o
Iguala . . . . 19 o
Iguatu . . . . 27 s
Iguidi Erg . . . . 36 i
Iharaña . . . . 47 f
Ihosy . . . . 47 p
Ija . . . . 57 a
Ijebu Ode . . . . 41 q
Ika . . . . 52 p
Ikanbujmal . . . . 56 u
Ikela . . . . 42 r
Ikelenge . . . . 44 z
Ikerasassuaq . . . . 90 t
Ikerre . . . . 41 q
Ikom . . . . 41 t
Il'a . . . . 57 h
Ilagan . . . . 67 h
Ilam . . . . 39 f
Ilan . . . . 63 v
Ilaro . . . . 41 p
Ilbenge . . . . 52 k
Ilebo . . . . 44 k
Île de France . . . . 91 u
Ilek . . . . 54 l
Ilesha . . . . 41 g
Ilha do Bananal . . . . 26 y
Ilhéus . . . . 29 l
Ili . . . . 56 p
Iliamna Lake . . . . 11 q
Iligan . . . . 67 v
Ilimniir . . . . 53 i
Ilimpeya . . . . 52 b
Ilimsk . . . . 57 c
Ilir . . . . 57 a
Ilirney . . . . 10 c
Illampu . . . . 28 d
Illapel . . . . 28 o
Illéla . . . . 41 d
Illela . . . . 41 d
Illimani . . . . 25 q
Illinois . . . . 16 q
Illizi . . . . 37 r
Illmani . . . . 28 e
Ilmer, Lake . . . . 33 m
Ilo . . . . 25 n
Iloilo . . . . 67 h
Ilorin . . . . 41 d
Il'pyrskiy . . . . 10 a
Imataca, Serrania de . . . . 23 i
Imatra . . . . 33 l
Imbros . . . . 35 t
im Chapyeva . . . . 53 i
im Gastello . . . . 53 g
Imi . . . . 43 s
Imilili . . . . 36 o
Imperatriz . . . . 27 n
Imperia . . . . 34 k
Imperial . . . . 15 s
Impfondo . . . . 42 n
Imphāl . . . . 62 p
In Abalene . . . . 37 n
In Afeleh . . . . 37 s
In Alay . . . . 36 x
Inambari . . . . 25 o
In Amenas . . . . 37 e
In Amguel . . . . 37 p
Iñapari . . . . 25 o
Inari, Lake . . . . 33 k
In Azaoua . . . . 37 r
In Azaraf . . . . 37 n
Inca de Oro . . . . 28 p
Ince, Cape . . . . 35 m
Inch'ŏn . . . . 58 t
Independence . . . . 15 w
Independence Fjord . . . . 91 i
Inderborskiy . . . . 54 k
India 60-61/64
Indiana (Australia) . . . . 73 s
Indiana (USA) . . . . 16 s
Indianapolis . . . . 16 s
Indian Head . . . . 15 e
Indian Ocean 23 1
Indiara . . . . 29 d
Indigirka . . . . 53 g
Indonesia 68-70
Indore . . . . 61 n
Indragiri . . . . 68 g
Indravati . . . . 61 r
Indus 60
Indus Delta . . . . 60 v
In Ebeggi . . . . 37 q
In Ekker . . . . 37 p
Inezgane . . . . 36 f
Infiernillo Reservoir . . . . 18 z
Inga . . . . 44 f
Ingende . . . . 44 j
Ingeniero Guillermo
Nueva Juárez . . . . 28 i
Ingeniero Jacobacci . . . . 30 j
Ingham . . . . 74 c
Ingili . . . . 53 o
Inglefield Land . . . . 91 d
Inglewood . . . . 78 e
Ingoda . . . . 52 t
Ingolf Fjord . . . . 91 j
Ingraj Bazar . . . . 61 w
Ingrid Christensen Coast . . . . 85 n
In Guezzam . . . . 37 p
Inhambane . . . . 46 v
Inhaminga . . . . 46 v
Inhassoro . . . . 46 v
Ining . . . . 56 q
Inirida . . . . 23 q
Inja . . . . 56 g
Injaptuk . . . . 57 f
Injune . . . . 74 r
Inkerman . . . . 73 j
Inkisi-Kisantu . . . . 44 g
Inland Sea . . . . 58 x
Inn . . . . 33 q
Innamincka . . . . 77 h
Inner Mongolia 57
Innisfail . . . . 74 c
Innsbruck . . . . 34 m
Inongo . . . . 44 i
Inoni . . . . 44 g
In Salah . . . . 37 a
Inscription, Cape . . . . 76 a
Insein . . . . 65 g
Inta . . . . 50 k
In Talak . . . . 41 c
Intelewa . . . . 26 h
International Dateline . . . . . . 86 q
Internalional Falls . . . . 15 k
Inthanon . . . . 66 c
Inukjuak . . . . 13 t
Inuvik . . . . 11 k
Invercargill . . . . 79 q
Inverell . . . . 78 e
Inverness . . . . 32 j
Inverway . . . . 73 b
Investigator Strait . . . . 77 r
Investigator Group . . . . 77 c

Inya 53 s
Inya (River) 53 s
Inyanga 46 t
Inza 54 h
Iolotan 55 r
Iona 48 f
Ionian Sea 35 p
Iowa 16 o
Iowa City 16 p
Iparneri 29 f
Ipatinga 29 j
Ipiales 24 j
Ipiau 29 l
Ipin 62 j
Ipixuna 25 g
Ipoh 66 r
Iporanga 29 r
Ipswich (Australia) 74 u
Ipswich (UK) 32 z
Ipu 27 r
Iqaluit (Frobisher Bay) 13 w
Iqe 56 x
Iquique 28 c
Iquitos 24 m
Iracema 25 u
Irai 29 o
Iráklion 35 t
Iramaia 27 r
Iran 39/54-55/60
Iranshahr 60 e
Irapuato 19 n
Iraq 39/54
Irbid 38 l
Irebu 44 h
Irece 27 r
Ireland 32
Irgiz 55 d
Irharharene 37 d
Irigui 40 j
Iringa 45 h
Iriomote 63 w
Iriri 26 v
Irish Sea 32 w
Irkestam 55 x
Irkutsk 52 o
Irondro 47 q
Iron Knob 77 e
Iron Mountain 16 e
Iron Range 73 l
Ironwood 16 d
Irrawaddy 62-65
Irrawaddy, Mouths of the 65 f
Irta 50 t
Irtysh 51 n
Irving 19 d
Iryupinsk 54 f
Isabal, Lago de 20 s
Isabela 22 o
Isabella 22 g
Isabella, Cordillera 22 f
Isačenko 51 h
Isa Fjord 92 o
Isaljördur 32 a
Isalo Mountains 47 p
Isangi 42 s
Isangila Falls 44 f
Isayevskiy 51 l
Isbil 43 g
Iseyin 41 c
Isfahan 39 j
Ishasha River 45 d
Isherton 23 x
Ishim 51 n
Ishim (River) 51 o
Ishinomaki 59 p
Isil'kul 55 j
Isinga 57 g
Isiolo 43 o
Isiro 42 v
Isisford 73 y
Iskenderun 35 z
Iskine 54 l
Islamabad 60 m
Island Falls 12 x
Island Lagoon 77 e
Island Lake 51 j
Islands, Bay of 79 h
Islay 32 v
Ismâilîya 38 j
Isna 38 v
Isoka 45 s
Israel 35/38
Israelite Bay 76 j
Issyk-Kul 55 y
Issyk-Kul, Lake 56 n
Istanbul 35 i
Istlán 18 y
Itabaiana 27 u
Itaberaba 27 r
Itaboca 26 o
Itabuna 29 l
Ilacaiúnas 26 y
Itacajá 27 n
Itacoatiara 25 j
Itaituba 25 l
Itajai 29 r
Italy 34-35
Itamaraju 29 l
Itamalaré 27 a
Itanagar 62 b
Itaobim 29 j
Itapetinga 29 k
Itapetininga 29 e
Itapeva 29 e
Itapicuru 27 t
Itapicuru
(Serra das Alpercatas) 27 p
Itapicuru Mirim 27 c
Itapicuru, Serra do 27 o
Itapipoca 27 f
Itaporanga 27 t
Ithaca 16 y
Itigi 45 g
Itinga 26 z
Itiquira 25 y
Itonamas 25 s
Ittoqqortoormiit 91 r
Itui 25 a
Ituiutaba 29 d
Itumbiara 29 d
Itungi 45 t
Ituri 42 t
Iturup 59 g
Ituxi 25 r
Iuaretê 25 b
Iultin 10 h
Ivaipor 29 p
Ivalo 33 k
Ivangrad 35 d
Ivanhoe 78 a
Ivano Frankovsk 35 g
Ivanovo 50 q
Ivdel' 50 y
Ivinhema 29 b
Ivittuut 90 t
Ivohibe 47 q
Ivory Coast 40 v
Ivujivik 13 t
Iwaki 59 p
Iwo 41 p
Ixiamas 25 p
Ixopo 49 r
Ixtlán 18 y
Izadkhast 49 j
Izamal 19 v
Izhevsk 50 v
Izhma 50 v
Izimbiri 42 r
Izki 60 p
Izmir 35 u
Izmit 35 j
Izúcar de Matamoros 19 p
Izvestkovyy 58 j
Izyum 54 c

J

Jabal adh Dhanna 39 w
Jabal al Qamar 39 w
Jabal al Qara 39 x
Jabal at Tinef 39 a
Jabal Nuqay 37 y
Jabalpur 61 q
Jabal Tuwaya 39 s
Jablanica 35 c
Jaboato 27 w
Jabung, Cape 68 i
Jaburu 25 f
Jaca 34 f
Jaçara 29 c
Jacaré 27 r
Jacareacanga 25 k
Jacaretinga 25 v
Jaciara 25 z
Jaciparaná 25 s
Jackson (Alabama) 19 j
Jackson (Michigan) 16 t
Jackson (Mississippi) 19 i
Jackson (Tennessee) 16 r
Jackson (Wyoming) 15 n
Jackson Island 50 j
Jacksonville (Florida) 20 m
Jacksonville (Illinois) 16 q
Jacksonville (North Carolina) 21 a
Jacmel 21 r
Jacobabad 60 i
Jacobina 27 r
Jacona 18 z
Jacques Cartier, Mount 17 d
Jacui 29 o
Jacundá 26 y
Jacundá (River) 26 l
Jadu 34 z
Jaen 24 h
Jaén 34 q
Jafene 36 u
Jaffa (Yafo) 38 k
Jaffna 64 h
Jafura 39 v
Jagdalpur 61 s
Jaggang 61 d
Jaghbub, Oasis 38 d
Jaghin 60 b
Jagtial 61 q
Jaguaro 29 o
Jaguaribe 27 t
Jagüe 28 q
Jahra 39 g
Jahrom 39 k
Jaicós 27 r
Jailolo 69 l
Jaipur 61 b
Jaisalmer 60 k
Jakarta 68 w
Jakes Corner 11 l
Jakobstad (Pietarsaari) 33 i
Jalaid Qi 58 e
Jalalabad 60 k
Jalapa 19 q
Jalaula 39 e
Jalda 54 o
Jalgaon 61 n
Jalingo 41 i
Jalna 61 o
Jalpaiguri 61 k
Jalu 38 b
Jalu Oasis 38 b
Jamaica 21
Jamalpur 61 y
Jamanxim 25 l
Jamari 25 t
Jamarovka 57 g
Jambi 68 h
Jambogan 67 q
James (River, S. Dakota) 15 h
James (River, Virginia) 16 x
James Bay 16 k
James Bay (Bay) 16 i
Jameson Land 91 q
James Ross Island 82 t
Jamestown (North Dakota) 15 h
Jamestown (Pennsylvania) 16 w
Jammu 61 a
Jammu and Kashmir 61 a
Jamnagar 60 x
Jamshedpur 61 u
Jamuna 61 x
Jamysevo 56 b
Janaúba 29 i
Janaucú 26 l
Jane Peak 79 q
Janesville 16 r
Jangijul' 55 u
Jánina 35 q
Janjira 60 z
Jan Mayen 32 h
Janos 18 j
Januária 29 h
Janzur 37 h
Japan 58-59
Japan, Sea of 58 y
Japurá 25 c
Jaramillo 30 x
Jardim 28 m
Jari 26 j
Jarma 37 u
Jaroslavl 50 p
Jartai 57 q
Jarú 25 t
Jarud Qi 58 d
Jaruu 56 m
Jashpurnagar 61 t
Jask 60 o
Jason Peninsula 82 u
Jasper (Canada) 14 j
Jasper (Indiana) 16 s
Jasper (Texas) 19 f
Jatai 29 c
Jati 60 w
Jatobá 26 w
Jatobal 26 y
Jatuarana 25 i
Jaú 25 g
Jauaperi 26 c
Jauja 24 x
Jaumave 19 o
Jauperi 25 h
Java 68 x
Java Sea 69 z
Jawand 60 g
Jawhar 60 z
Jaya (Mount) 70 t
Jayapura 70 w
Jaypur 61 s
Jebba 14 d
Jebel Ali 39 y
Jebel Igdet 36 g
Jebel Marra 42 f
Jeberos 24 j
Jebri 60 h
Jeddah 39 n
Jefferson City 16 p
Jeffreys Bay 49 n
Jekabpils 33 w
Jeleninskij 52 s
Jember 69 o
Jenda 45 t
Jens Munk Island 13 f
Jequié 27 s
Jequitinhonha 29 k
Jerbent 55 p
Jeremoabo 27 t
Jerez 18 z
Jerez de la Frontera 34 p
Jerramungup 76 p
Jersey 32 x
Jerusalem 35 y
Jessore 61 x
Jesup 20 l
Jesús Carranza 20 p
Jevlach 54 v
Jhang-Maghiana 60 l
Jhansi 61 q
Jialing Jiang 62 k
Ji'an 63 p
Jianchuan 62 t
Jiangcheng 62 u
Jiangjumiao 56 u
Jiangjuntai 56 z
Jiangmen (Kongmoon) 63 o
Jiangquanzi 56 z
Jianping 57 y
Jianshui 62 v
Jianyang 62 j
Jiaohe 58 h
Jiao Xian 58 p
Jiaozuo 63 c
Jiayin 58 i
Jiayuguan 56 z
Jibóia 25 b
Jieyang 63 q
Jigzhi 62 h
Jihlava 33 s
Jijiga 43 f
Jili Hu 56 g
Jilin (Kirin) 58 g
Jima 43 b
Jiménez 19 p
Jiménez (Mexican Plateau) 18 l
Jinan (Tsinan) 63 e
Jinchang 57 o
Jinchuan 62 h
Jingdong 62 u
Jinghong 62 t
Jingning 62 k
Jing Xian 62 z
Jingyu 58 g
Jingyuan 62 j
Jinhe (Borohoro Shan) 56 r
Jinhe 58 d
Jining 63 e
Jinja 42 z
Jinka 43 n
Jinping 62 v
Jinsha 62 k
Jinsha Jiang 62 t
Jiparaná 25 u
Jipijapa 24 g
Jirin Gol 57 x
Jishou 62 m
Jiuding Shan 62 j
Jiuquan 56 z
Jiwani 60 r
Jizan 43 f
Joaçaba 29 p
Joo Monlevade 29 i
Joo Pessoa 27 w
Joaquín V. González 28 u
Jodhpur 60 z
Joensuu 33 l
Jofane 49 h
Johannesburg 49 p
John Day 14 j
John Day (River) 14 i
Johnsons Crossing 11 l
Johnston, Lakes 76 g
Johnstown 16 w
Johor 68 h
Johor Baharu 68 h
Joinville 29 r
Joinville Island 82 g
Jojoca 26 v
Jojutla 19 o
Joliet 16 r
Jolo 67 t
Jolo Group 67 t
Jomda 62 e
Jonesboro 16 q
Jones Sound 13 d
Jönköping 33 e
Joplin 15 w
Jordan (USA) 15 c
Jordan (Stale) 38
Jorhāt 62 p
Jorullo 19 n
Jos 41 g
José Bonifácio 25 v
José de San Martin 30 v
Joseph Bonaparte Gulf 73 a
Joseph, Lake 17 e
José Rodrigues 26 x
Jryupinsk 54 f
Juan Aldama 18 z
Juan de Fuca, Strait 14 g
Juan de Nova (Island) 47 a
Juan Fernández Islands 30 c
Juanjuí 24 j
Juan Sola 28 h
Juarez, Sierra de 18 d
Juàzeiro 27 s
Juàzeiro do Norte 27 s
Juazohn 40 u
Juba 42 x
Jubany 82 s
Juba (River) 43 s
Jubbah 39 b
Júcar 34 s
Juchatengo 19 q
Juchipila 18 z
Juchilán 20 p
Juiz de Fora 29 i
Juli 25 o
Juliaca 25 o
Julia Creek 73 w
Julian 62 j
Juliana Top 23 y
Jullundur 61 a
Junagadh 60 x
Junan 63 g
Junction City 15 v
Jundah 77 i
Jundiai 29 g
Juneau 11 y
Junee 78 c
Jungarian Gobi 56 w
Junggar Pendi 56 t
Junín 31 a
Junin de los Andes 30 h
Juntuliang 57 s

Jura 34 j
Juruá 25 d
Juruena 25 w
Juruena (River) 25 w
Justo Daract 28 s
Jutai 25 c
Jutaí (River) 25 c
Juticalpa 22 e
Jutland 33 p
Juyom 39 k
Jwaneng 49 a
Jyväskylä 33 k

K

K2 55 z
Kaabong 42 z
Kabaena 69 u
Kabakly 55 q
Kabala 40 f
Kabale 45 d
Kabalebo Reservoir 23 z
Kabalo 45 b
Kabanjahe 68 c
Kabba 41 r
Kabinda 45 a
Kabo 42 a
Kabompo 44 z
Kabompo (River) 44 z
Kabondo Dianda 45 n
Kabongo 45 a
Kaboudia, Cape 34 z
Kabul 60 j
Kabul (River) 60 l
Kabului 18 r
Kabunga 45 c
Kabwe 45 p
Kabyrga 55 e
Kachalola 45 r
Kachikattsy 52 m
Kachiry 56 a
K'achta 57 e
Kadali 52 s
Kadan 65 h
Kadina 77 f
Kadiri 64 g
Kadnoye 54 d
Kadoma 49 e
Kadugli 42 j
Kaduna 41 f
Kaduna (River) 41 e
Kadzherom 50 w
Kadzhi 51 s
Kaédi 40 d
Kaesŏng 58 t
Kafanchan 41 f
Kafue 46 d
Kafue, Lake 49 c
Kaga 55 b
Kaga Bandoro 42 o
Kagan 55 s
Kagizman 54 t
Kagora 41 g
Kagoschima 58 v
Kahal Tabelbala 36 k
Kahama 45 f
Kahan 60 j
Kahemba 44 j
Kahnuj 60 b
Kahoolawe 18 q
Kahului 18 r
Kaiama 41 d
Kaidu 56 s
Kaifeng 63 d
Kai Islands 70 q
Kaikoura 79 u
Kaili 62 y
Kailua 18 r
Kailu He 58 d
Kaimana 70 r
Kainji Reservoir 41 d
Kairouan 34 y
Kaiser Franz Josephs Fjord 91 r
Kaitaia 79 g
Kaiyuan 58 f
Kaiyuan 62 v
Kajaani 33 k
Kajabbi 73 v
Kajiado 45 i
Kajnar 56 a
Kajo-Kaji 42 x
Kajrakkumskoje Reservoir 55 v
Kaka 43 c
Kakamas 48 y
Kakamega 42 z
Kakenge 44 l
Kakhovskoye Reservoir 54 c
Kakinada 64 j
Kaktovik 11 h
Kakuma 42 z
Kalaallit Nunaat (Greenland) 90-91
Kalabagh 60 l
Kalachinsk 55 l
Kalach-na-Donu 54 g
Kalahari 48 m
Kalakan 52 u
Kalamai 35 r
Kalamare 49 b
Kalamazoo 16 t
Kalamba 42 o
Kalandula 44 u
Kalao 69 u
Kalaqi 58 d
Kalar 52 u
Kalasin 66 g
Kalat 60 i
Kalbarri 76 b
Kalekol 43 n
Kalemie 45 d
Kalemyo 62 p
Kalevala 33 m
Kalgoorlie 76 g
Kalima 45 b
Kalimantan 69
Kaliningrad 33 u
Kalininsk 54 g
Kaliro 42 y
Kalispell 14 l
Kalisz 33 t
Kalkfeld 48 h
Kalkfontein 48 l
Kalkrand 48 i
Kalli 76 e
Kalmar 33 s
Kalni 61 y
Kaloko 45 a
Kalole 45 c
Kalomo 45 n
Kalpa 61 c
Kaltag 11 b
Kaluga 54 c
Kalumburu 72 m
Kalyazin 50 p
Kama 50 v
Kamanga 52 n
Kamanjab 48 g
Kamba 41 c
Kambia 40 e
Kambove 45 n
Kamchatka 10/53
Kamchatsk 10 n
Kamen 51 k
Kamenets-Podol'skiy 33 x
Kamenka 54 g
Kamenka (Angara) 51 x
Kamenka (Don) 54 e
Kamen-na-Obi 56 c
Kamensk 57 e
Kamensk-Shakhinskiy 54 e
Kamensk-Ural'skiy 50 y
Kamet 61 d
Kamiji 44 m
Kamileroi 73 v
Kamina 45 a
Kamkaly 55 v
Kamloops 14 i
Kamo 51 y
Kamoa Mts 23/26
Kampala 42 x
Kampar 66 r
Kampar (River) 66 s
Kampene 45 b
Kamskoje Reservoir 50 v
Kamuli 42 y
Kamyshin 54 h
Kan 56 k
Kanaaupscow 16 l
Kanab 14 z
Kananda 52 b
Kananga 44 l
Kanash 54 i
Kanawolo 40 j
Kanazawa 58 z
Kanbe 66 a
Kanbetlet 62 p
Kanchanaburi 65 i
Kanchipuram 64 h
Kanchow 63 p
Kand 60 i
Kandahar 60 h
Kandalaksa 50 a
Kandalaksa Gulf 50 a
Kandangan 69 p
Kandi 41 c
Kandrian 71 p
Kandy 64 v
Kandychan 53 v
Kane 16 x
Kánea 35 t
Kane Basin 91 e
Kanem 41 l
Kaneohe 18 q
Kang 48 m
Kangalassay 52 m
Kangan (Australia) 72 u
Kangan (Iran) 39 j
Kangaroo Island 77 r
Kangavar 39 f
Kangchenjunga 61 j
Kangding 62 h
Kangean 69 p
Kangerlussuaq 91 n
Kangiqsualujjuaq 13 x
Kangiqsujuaq 13 v
Kangmar 61 k
Kangmar (Zhari Namco) 61 h
Kangnŭng 58 v
Kangro 61 h
Kangto 62 b
Kangxiwar 61 d
Kani 40 i
Kaniama 44 m
Kanin Nos, Cape 50 d
Kanin Peninsula 50 e
Kankakee 16 r
Kankan 40 g
Kanker 61 s
Kankossa 40 f
Kankroli 60 z
Kannapolis 16 w
Kannauj 61 e
Kano 41 g
Kanona 45 r
Kanpur 61 d
Kansas 15 u
Kansas City 16 o
Kansk 56 l
Kantchari 41 a
Kantemirovka 54 e
Kanuku Mts 23/26
Kanye 49 a
Kanyu 49 a
Kaohsiung 63 t
Kaolack 40 b
Kaoma 45 n
Kaoyui Lake 58 o
Kapalala 45 q
Kapanga 44 m
Kapatu 46 e
Kapčagaj 55 y
Kapchagayskoye Reservoir 56 o
Kapedo 43 n
Kapia 44 j
Kapingamarangi Atoll 71 g
Kapiri Mposhi 46 d
Kapka 42 d
Kapoeta 42 z
Kapona 45 d
Kapsabet 43 n
Kapsan 58 h
Kapuas 69 a
Kapuskasing 16 h
Kapustin-Yar 54 i
Karaaul 56 b
Kara-Balta 55 x
Karabas 55 m
Kara-Bogaz-Gol 54 z
Karabük 35 k
Karabutak 55 e
Karachi 60 v
Kara Dag 54 q
Karaganda 55 k
Karagayly 55 l
Karaginskiy Island 10 o
Karaj 39 i
Karakax 61 c
Karakoram Range 61 a
Karakore 43 d
Karakul, Lake 55 x
Kara Kum 55 o
Karakumskiy Canal 55 r
Karalon 52 t
Karalundi 76 f
Karaman 35 y
Karamay 56 r
Karamea Bight 79 s
Karamysheva 51 z
Karana 39 v
Karanambo 23 x
Karand 39 e
Karang 40 b
Karaozek 55 o
Karasaj 55 z
Karasay 56 r
Karasayo 61 g
Karasburg 48 w
Kara Sea 50 l
Karashoky 55 l
Karasino 51 h
Karasuk 56 b
Karatal 56 n
Kara Tau 55 v
Karaton 54 m
Karaul 51 g
Karaulkeldy 55 a
Karayulgun 56 p
Karazhingil 55 j
Karbala 39 d
Karbi 65 v
Karchev 54 b
Karet 36 s
Kargasok 51 s
Kargil 61 b
Kargo 51 l
Kargopol 50 p
Kari 41 h
Kariba 46 q
Kariba Reservoir 46 q
Karima 38 v
Karimata 68 l
Karimata Strait 68 k
Karimganj 61 z
Karimnagar 61 q
Karisimbi 45 d
Karkheh 39 f
Karkinitskiy Bay 35 k
Karla-Aleksandra 50 k
Karlik Shan 56 x
Karlovac 35 b
Karlsbad 33 r
Karlskrona 33 s
Karlsruhe 33 p
Karlstad 33 d
Karma 42 c
Karmé 41 l
Karnak 38 j
Karnal 61 b
Karnali 61 f
Karnaprayag 61 d
Karoi 49 e
Karonga 46 h
Karora 38 z
Karoy 55 l
Kárpathos 35 u
Karrats Fjord 90 m
Karridale 76 c
Kars 54 t
Karshi 54 y
Karsi 55 t
Karskiye Vorota, Proliv 50 j
Kartaly 55 d
Karumba 73 i
Karun 39 h
Karunjie 72 m
Karwar 64 c
Karymskoje 57 i
Kasai 44 j
Kasaji 44 z
Kasala 43 a
Kasama 45 r
Kasambule 45 b
Kasanga 45 e
Kasar, Ras 39 n
Kasba Lake 12 x
Kasembe 45 p
Kasempa 45 n
Kasenga 45 p
Kasese 42 w
Kashan 39 j
Kashgar 55 y
Kashgert 50 z
Kashing 63 h
Kashiwazaki 59 n
Kashkanteniz 55 j
Kashmar 60 c
Kashmir 61 a
Kasindi 45 d
Kaskalen 55 y
Kasli 55 c
Kasongo 45 b
Kasongo-Lunda 44 h
Kass 42 f
Kassel 33 p
Kasserine 34 y
Kastamonu 35 l
Kasulu 45 e
Kasungu 45 t
Kasur 61 a
Kata 52 n
Kataba 49 a
Katamobagu 67 w
Katanda 44 m
Katanning 76 e
Katchall 65 r
Katesh 45 h
Katete 45 s
Katgi 61 r
Katherina 38 j
Katherine 73 c
Kathiawar 60 x
Kathmandu 61 h
Kati 40 h
Katima Mulilo 49 a
Katiola 40 j
Katompi 45 a
Katoomba 78 d
Katowice 33 t
Katsina 41 f
Kattakurgan 55 s
Kattegat 33 q
Katun' 56 f
Katyl'ga 51 q
Kauai 18 p
Kaula 18 o
Kaunas 33 v
Kaura Namoda 41 f
Kavalerovo 58 l
Kavali 64 h
Kaval'kan 53 p
Kaválla 35 f
Kavendou 40 e
Kavieng 71 r
Kavkaz 54 p
Kavumu 42 v
Kaw 26 j
Kawambwa 45 q
Kawasaki 59 n
Kawatop 26 h
Kawio 69 k
Kawio Islands 67 w
Kawlin 62 q
Kaya 40 m
Kayan 69 d
Kayanza 45 d
Kaycee 15 p
Kayes 40 e
Kayonza 45 e
Kaysatskoye 54 i
Kayseri 35 z
Kayunga 42 y
Kazabegi 54 u
Kazach 54 u
Kazachinskoye 57 f
Kazach'ye 53 c
Kazakdarya 55 p
Kazakstan 55 e
Kazak Uplands 55 k
Kazalinsk 55 d
Kazan' 54 j
Kazandžik 55 n
Kazerun 39 j
Kazhim 50 u
Kazumba 44 l
Kazungula 49 a
Kazym 51 o
Kazymskaya 51 n
Kazyr 56 k
Kearney 15 t
Keban Lake 54 q

Kebili **37** f
Kebkabiya **42** f
Kebnekaise **33** g
Kebri Dehar **43** t
Kediri **69** n
Kédougou **40** e
Kedrovyy **51** n
Kedva **50** v
Keele Peak **11** m
Keeroongooloo **77** i
Keetmanshoop **48** w
Keflavík **32** a
Kegen **56** p
Keibeb **48** j
Keimoes **48** y
Kéita ou Doka **42** b
Kekuk **53** y
Kelʹ **52** k
Kelafo **43** t
Kelang **68** f
Kelantan **66** s
Kelem **43** n
Kélibia **34** z
Kéllé **44** g
Kellerberrin **76** e
Kellog **51** u
Kelo **41** l
Kelowna **14** i
Keluang **68** h
Kembolcha **43** d
Kemerovo **56** f
Kemi **33** j
Kemi (River) **33** k
Kemijärvi **33** k
Kemkara **53** q
Kemmerer **15** n
Kemp Land **85** a
Kempsey **78** g
Ken **61** r
Kenadsa **36** j
Kenai **11** e
Kenai Peninsula **11** e
Kenamuke Swamp **42** z
Kencha **53** q
Kendari **69** v
Kendégué **42** b
Kenema **40** s
Kenge **44** h
Kenhardt **48** y
Kenitra **36** h
Kenmare **15** f
Kenmaur **49** d
Kennedy **62** p
Kenner **19** h
Kennya **53** c
Keno Hill **11** k
Kenora **15** j
Kenosha **16** r
Kentau **55** u
Kent Group **78** p
Kenton **16** u
Kentucky **16** s
Kenya **43/45**
Kenya (Mount) **43** o
Kenzou **41** x
Keokuk **16** p
Kerchʹ **54** c
Kerekyano **52** k
Kerema **71** n
Keren **43** b
Kerguelen **235** r
Kericho **43** n
Keirči **55** s
Kerinci **68** f
Keriske **53** a
Keriya **61** f
Kerkenna Islands **34** z
Kerma **38** u
Kermadec Islands **79** l
Kerman **60** b
Kérouané **40** h
Kerrville **19** b
Kerzaz **36** l
Kesagami Lake **16** j
Kesen **59** p
Keshvar **39** g
Ket **51** s
Keta, Lake **51** i
Ketapang **68** l
Ketchikan **11** y
Key Afer **43** n
Keyano **17** a
Keyi **56** q
Key Largo **20** m
Key West **20** y
Kez **50** v
Khabalakh **52** r
Khabarovsk **58** m
Khabary **56** b
Khairpur **60** j
Khakhar **53** o
Khakhea **48** m
Khalesavoy **51** r
Khalʹmer-Yu **50** m
Khamasin **39** r
Khambhat **60** y
Khamis Mushait **39** q
Kham-Keut **66** h
Khammam **64** h
Khampa **52** j
Khamra **52** s
Khandwa **61** o
Khandyga **53** b
Khangokurt **50** z
Khanh Hung **66** i
Khani **52** v
Khanka, Lake **58** k
Khantau **55** x
Khantayka **51** h
Khantayskoye, Lake **51** i
Khantayskoye Reservoir **51** h
Khanty-Mansiysk **51** n
Khanyangda **53** r
Khara-Aldan **53** a
Kharabalio **54** i
Kharagpur **61** v
Kharalakh **52** k
Kharan **60** h
Khara-Tala **53** j
Khara-Tas **53** b
Khargon **61** o
Kharkov **54** c
Kharstan **53** d
Khartoum **42** k
Khartoum North **42** l
Khasalakh **52** k
Khash **60** e
Khashm al Girba **43** a
Khash Rud **60** f
Khastakh **53** e
Khatanga **52** a
Khatanga (River) **52** a
Khatyngnakh **53** n
Khatyngnakh (Chereskogo Mountains) **53** g
Khatyrka **10** f
Khay **39** o
Khayber **39** a
Khemarat **66** h
Khemis Miliana **34** t
Khemisset **36** h
Khenchela **34** w
Kherson **35** k
Kheta **51** m
Kheta (River) **51** m
Khiaw **65** k
Khmelʹnitskiy **35** h
Khobolchan **53** d
Khochot **52** j
Khokiley **51** f
Kholmsk **59** c
Khomokashevo **52** p
Khomotovka **54** b
Khoms **35** n
Khong **66** i
Khongo **53** i
Khongsey **53** i
Khon Kaen **65** l
Khonu **53** f
Khoppuruo **52** v
Khor (Iran) **39** l
Khor (Russia) **58** l
Khordogoy **52** g
Khorgo **52** f
Khorixas **48** g
Khormuj **39** i
Khoronkhu **52** m
Khoronnokh **52** j
Khorramabad **39** g
Khorramshahr **39** g
Khouribga **36** h
Khoy **54** u
Khroma **53** f
Khrom-Tau **55** b
Khuff **39** q
Khujand **55** u
Khulais **39** n
Khulm **60** i
Khulna **61** x
Khurais **39** t
Khurasan **55** o
Khurd **60** h
Khusf **60** c
Khuzdar **60** i
Khuzdzhakh **53** f
Khyber Pass **60** l
Kiambi **45** c
Kiamusze **58** j
Kibangou **44** e
Kiberg, Cape **33** m
Kibiti **45** k
Kibombo **45** a
Kibondo **45** e
Kibre Mengist **43** q
Kibuye **45** d
Kichi Kichi **37** x
Kidal **37** n
Kidira **40** e
Kiel **33** p
Kiel Bay **33** p
Kielce **33** u
Kieta **71** u
Kiev **35** j
Kiev Reservoir **33** y
Kiewietskuil **48** z
Kiffa **40** e
Kigali **45** d
Kigilyakh **53** d
Kigisa **13** k
Kigoma **45** d
Kiik **55** j
Kiikkaškan **55** m
Kii Strait **58** y
Kijabe **43** n
Kikagati **45** e
Kikao **49** a
Kikiakki **51** s
Kikki **60** s
Kikondja **45** a
Kikori **70** z
Kikwit **44** i
Kilembe **44** k
Kilibo **41** o
Kilifi **45** l
Kilimanjaro **45** j
Kilindoni **45** k
Kilkoris **42** z
Killarney **32** u
Killeen **19** c
Kilʹmez **50** u
Kilosa **45** i
Kimball, Mount **11** h
Kimberley **49** n
Kimberley Downs **72** l
Kimberley Plateau **72** m
Kimchaek **58** i
Kimpese **44** f
Kimvula **44** g
Kinabulu **67** p
Kindersley **15** b
Kindia **40** d
Kindu **45** a
Kineshma **54** f
Kinesma **50** r
Kingaroy **74** t
King Christian IX Land **90** z
King Christian X Land **91** r
King City **14** u
King Frederick VI Coast **90** v
King Frederick VIII Land **90** n
King George Island **82** e
King Island **77** w
King Leopold and Queen Astrid Coast **85** r
King Leopold Range **72** l
Kingman **14** z
Kingombe **45** b
Kingoonya **77** d
King Oscars Fjord **91** r
King Peninsula **81** a
Kingri **60** k
King Salmon **11** p
King Sound **72** k
Kingsport **16** u
Kingston (Canada) **16** y
Kingston (Jamaica) **21** o
Kingston (USA) **16** z
Kingston upon Hull **32** y
Kingstown **23** j
Kingsville **19** c
Kingtehchen **63** f
King William Land **90** t
King Wilhelm II Land **85** s
King William Island **12** m
King Williams Town **49** p
Kinhwa **63** g
Kinkala **44** g
Kinoosao **12** x
Kinshasa **44** g
Kintampo **40** m
Kintus **51** o
Kinyeti **42** y
Kipaka **45** b
Kipembawe **45** g
Kipushi **45** o
Kirakira **75** a
Kirbey **52** b
Kirbey (Arga-Sala) **52** d
Kirenga **52** q
Kirensk **52** q
Kirgiz Steppe **55** a
Kiri **44** j
Kiribati **235** l
Kirikkale **35** y
Kirim **42** i
Kirin **58** g
Kirinyaga (Kenya) **43** o
Kirit **43** h
Kiritimati **234** b
Kiriwina or Trobriand Island **71** r
Kiriyestyakh **52** l
Kirkenes **33** l
Kirkland Lake **16** j
Kirkpatrick, Mt **86** c
Kirksville **15** x
Kirkuk **39** e
Kirkwood **49** o
Kirov **50** t
Kirovograd **35** k
Kirovsk **50** a
Kirovskiy **58** l
Kirrikkale **54** o
Kirs **50** u
Kirsanov **54** g
Kirthar Range **60** i
Kiruna **33** h
Kirundu **45** a
Kisangani **42** s
Kisarawe **45** k
Kiseleva **53** h
Kiselʹovsk **56** g
Kishtwar **61** a
Kisi **58** j
Kisigu **45** g
Kisii **45** h
Kisima **43** o
Kisimaio **43** s
Kiska **10** t
Kislovodsk **54** s
Kissidougou **40** g
Kisumu **45** g
Kita **40** g
Kita-Kjūshū **58** v
Kitale **43** n
Kitami **59** d
Kit Carson **15** r
Kitchener-Waterloo **16** w
Kitgum **42** y
Kithira **38** c
Kitimat **14** d
Kitui **45** j
Kitunda **45** f
Kitutu **45** c
Kitwe **45** p
Kiukiang **63** e
Kivitoo **13** l
Kivu, Lake **45** c
Kizema **50** r
Kizilirmak **35** l
Kiziltepe **54** r
Kizʹlar **54** v
Kizyl-Arvat **55** n
Kizyl Jilga **61** d
Klabat **67** w
Klaeng **65** k
Klagenfurt **35** a
Klamath Falls **14** u
Klar **33** e
Klerksdorp **49** e
Kliapeda **33** u
Klin **57** k
Klin (Moscow) **50** o
Klintsy **54** a
Knob, Cape **76** f
Knox Coast **87** f
Knoxville (Iowa) **15** x
Knoxville (Tennessee) **16** u
Knud Rasmussens Land **91** d
Knysna **48** z
Kobbo **43** d
Kobdo **56** i
Kobe **58** y
København (Copenhagen) **33** q
Kobenni **40** g
Koblenz **33** o
Kobuk **11** c
Kobya **52** l
Kočečum **51** m
Koch Bihr **60** k
Kochechum **52** b
Kochi **58** y
Kochikha **51** k
Kochubey **54** v
Kodarma **61** v
Kodiak **11** r
Kodiak (Island) **11** r
Koës **48** x
Koforidua **40** z
Kogon **40** c
Koh-e-Baba **60** i
Kohima **52** o
Koidern **11** j
Koidu **40** f
Kokand **55** v
Kokaral **55** c
Kokchetav **55** i
Kokiu **62** v
Kok-Jangak **55** x
Kokkola **33** i
Kokoma **16** s
Kokonau **70** t
Kokpekty **56** d
Koksoak **13** w
Kokstad **49** r
Kokuora **53** f
Kokwo **43** n
Kola **50** c
Kolaka **69** u
Kolar **64** g
Kolari **33** j
Kolarovgrad **35** h
Kolbio **43** r
Kolda **40** c
Kolding **33** p
Kole **42** t
Kolesovo **53** i
Kogujev **50** g
Kolhapur **64** c
Kolikʹyegan **51** q
Kolin **33** s
Kolkata (Calcutta) **61** w
Kolmya **53** j
Köln (Cologne) **33** o
Kolokani **40** h
Kolomna **54** e
Kolomyya **35** g
Kolono **69** u
Kolpashevo **51** t
Kolpino **33** m
Koltovkinda **52** u
Kolva **50** w
Kolva (Bolʹshezemelʹskaya Tundra) **50** j
Kolwezi **46** b
Kolyma **53** j
Kolmyskiy Mountains **53** j
Kolymskiy Plain **53** j
Komadugu **41** j
Komanda **42** w
Komandor Islands **10** p
Komatipoort **49** g
Komelek **53** a
Komoé **40** x
Kom Ombo **38** w
Komotini **35** g
Kompong Cham **66** i
Kompong Chhnang **66** h
Kompong Som **66** g
Komsberg **48** y
Komsomolets **55** e

Komsomolets (Island) **51** k
Komsomol'sk **51** u
Komsomol'skaya **85** n
Komsomol'skiy **50** z
Komsomol'sk-na-Amure **53** p
Konda **51** n
Kondakovo **53** i
Kondoa Irangi **45** i
Kondon **58** m
Kong **40** k
Kongbo **42** q
Kongmoon **63** o
Kongor **48** m
Kongor **42** x
Kongsvinger **33** d
Kongwa **45** i
Koniakari **40** f
König Wilhelm Land **91** s
Konkouré **40** d
Konongo **40** z
Konosa **50** p
Konotop **54** a
Konqi **56** t
Konso **43** o
Kontagora **41** e
Kontum **66** j
Konya **35** y
Konzaboy **53** l
Kopet Dag **55** o
Kopiago **70** y
Korča **35** e
Korčula **35** b
Korda **51** y
Kordofan **42** j
Korea, North **58**
Korea, South **58/63**
Korea Bay **58** r
Korea Strait **58** v
Korennoye **52** c
Koreyver **50** k
Korf **10** b
Korhogo **40** j
Kor-i-Hisar **60** h
Korinth **35** s
Kōriyama **59** o
Korkodon **53** j
Korkodon (River) **53** k
Korla **56** s
Korliki **51** s
Koro **40** i
Korobovskiy **55** l
Korogwe **45** j
Koronadal **67** w
Koror **70** f
Koro Sea **75** m
Korosten **35** i
Koro Toro **42** b
Korsakov **59** d
Korti **38** v
Korti Linchang **56** g
Koryaken Mountains **10** d
Kos-Agač **56** h
Kosciuszko, Mount **77** z
Kosha **38** u
Košice **35** d
Kosovo **35** de
Kosrae **71** m
Kossou Reservoir **40** v
Kosti **42** l
Kostino **51** u
Kostroma **50** q
Kosyu **50** x
Kószalin **33** s
Kota **61** o
Kota Baharu **66** s
Kotabaru **69** q
Kotabumi **68** u
Kota-i Ashro **60** j
Kota Kinabalu **69** c
Kotamobagu **69** j
Kotel'nich **50** t
Kotel'nikovo **54** g
Kotel'nyy **53** b
Kotel'nyy (Island) **53** d
Kotido **42** z
Kotikova **58** l
Kotka **52** k
Kotlas **50** s
Koto **59** b
Kotri **60** v
Kotto **42** q
Kotturu **64** e
Kotui **51** l
Kotuykan **52** b
Kotzebue **11** a
Kotzebue Sound **10** m
Kouango **42** p
Koudougou **40** l
Kougouleu **41** u
Kouilou **44** d
Koukdjuak **13** i
Koulamoutou **44** d
Koulikoro **40** i
Koulou **41** c
Koumpentoum **40** c
Koumra **41** m
Kounala **40** j
Koundara **40** d
Koundougou **40** k
Kourou **26** j
Kouroussa **40** g
Koutouba **40** k
Kouyou **44** g
Kova **51** z
Kovel' **33** w
Kovrizhka **10** a
Kovroy **54** f
Kowares **48** g
Koxlax **56** p
Koyma Reservoir **64** c
Koyuk **11** a
Kozan **35** z
Kozáni **35** r
Kozhevnikovo **52** e
Kpalimé **41** n
Kpandu **41** n
Kpessi **41** a
Kra, Isthmus of **66** d
Krakatau **68** v
Krakow **33** t
Kramatorsk **54** c
Kramfors **33** f
Krasino **50** i
Krasnaya Yaranga **10** j
Krasnoarmeyesk **55** i
Krasnoarmeyskiy **10** e
Krasnobrodskij **56** g
Krasnodar **54** r
Krasnojarsk **56** j
Krasnokamsk **50** v
Krasnosel'kup **51** s
Krasnoslobodsk **54** h
Krasnotur'insk **50** x
Krasnoufimsk **50** w
Krasnousol'skiy **55** a
Krasnoyarskiy **55** b
Krasnoyarskoye Reservoir **56** i
Krasnyye-Baki **50** s
Krasnyy-Yar **54** i
Kratie **66** i
Kreb en Naga **36** u
Kremenchug **35** l
Kremenchugskoye Reservoir **35** l
Kresty **51** i
Krichev **33** z
Krishna **64** g
Krishnagiri **64** f
Kristianstad **33** r
Kristiansund **33** b
Krivoy Rog **35** k
Krk **35** a
Kronok **53** z
Kronprins Frederiks Bjaerge **90** z
Kroonstad **49** P
Kropotkin **54** f
Kropotkin
(Patomskoye Plateau) **52** t
Krotovka **54** l
Kruglikovo **58** m
Kruis, Cape **48** f
Krung Thep (Bangkok) **66** e
Ksabi **36** l
Ksabi Talsinnt **36** j
Ksar el Boukhari **37** a
Ksar el Kebir **34** o
Ksar es Souk **36** j
Ksar Torchane **36** q
Kuala Belait **69** b
Kuala Kubu Baharu **68** l
Kuala Lipis **66** s
Kuala Lumpur **66** r
Kuala Terengganu **66** t
Kuancheng **57** y
Kuandang **69** i
Kuandian **58** f
Kuango **44** h
Kuantan **66** t
Kuba **54** w
Kuban' **54** d
Kubor **70** z
Kuching **68** m
Kudu-Kyuyel' **52** v
Kudymkar **50** v
Kufrah Oasis **38** p
Kugluktuk (Coppermine) **12** f
Kugulta **54** g
Kuhak **60** f
Kuh Lab, Ras **60** q
Kuito **44** u
Kukan **58** k
Kul **39** l
Kul'ab **55** u
Kulagino **54** k
Kulai **68** h
Kulakshi **55** a
Kulaly **54** x
Kulanoy **55** b
Kular **53** a
Kule **48** k
Kulgera **77** b
Kulinda **52** o
Kul'sary **54** m
Kultay **54** y
Kulu **61** c
Kulunda **56** b
Kulundinskaya Step **56** b
Kulundinskoje, Lake **56** b
Kulynigol **51** s
Kuma **54** t
Kumamoto **58** w
Kumanovo **35** e
Kumarina **76** f
Kumasi **40** z
Kumertau **55** a
Kumkuduk **56** w
Kunar **60** l
Kunašir **59** f
Kuná-Urgenč **55** o
Kunduz **60** j
Kungrad **55** o
Kungu **42** o
Kungur **50** w
Kungur (Mount) **55** y
Kunlun Shan **61** h
Kunming **62** v
Kunsan **58** t
Kuntuk **53** t
Kununurra **73** a
Kuogastakh **53** a
Kuopio **33** k
Kuoyka **52** i
Kupang **69** v
Kupino **56** b
Kuqa **56** q
Kura **54** t
Kuragino **56** j
Kuraj **56** g
Kurčum **56** e
K rdžali **35** g
Kureyka **51** j
Kureyskaya **52** r
Kurgal'dzhino **55** i
Kurgan **55** f
Kurgan-T'ube **55** u
Kuria Muria Islands **60** n
Kuri Bay **72** i
Kuril Islands **59** g
Kurleja **57** l
Kurmuk **42** m
Kurnool **64** f
Kursk **54** c
Kurtalan **54** s
Kurtamysh **55** e
Kuruman **48** z
Kurundi **73** s
Kurunegala **64** u
Kurun Uryak **53** o
Kur'ya **50** w
Kusagan-Olokh **53** a
Kushiro **59** e
Kushmurun **55** f
Kuska **60** e
Kuskokwim **11** b
Kuskokwim Mountains **11** c
Kustanay **55** e
Kut **66** f
Kutacane **68** d
Kütahya **35** v
Kutaisi **54** t
Kutch, Gulf of **60** w
Kutch, Rann of **60** x
Kutno **33** t
Kutu **44** i
Kuujjuaq **13** w
Kuujjuarapik **16** k
Kuusamo **33** i
Kuwait **39** g
Kuwait (State) **39**
Kuybysehevskoye Reservoir **54** j
Kuyygan **55** x
Kuzmov'ka **51** w
Kuznetsk **54** i
Kvigtind **33** e
Kwale **45** k
Kwanga **45** a
Kwangju **58** t
Kwango **44** h
Kweilin **63** n
Kweiyang **62** x
Kwe Kwe **46** r
Kwethluk **11** a
Kwigillingok **11** a
Kwiha **43** d
Kwikila **74** d
Kwilu **44** i
Kwoko **70** q
Kyaikto **65** h
Kyaka **45** e
Kyanjojo **42** x
Kyaukme **62** r
Kyaukpyu **62** o
Kyauktaw **62** o
Kycham-Kyuyel' **53** g
Kyrgyzstan **55** x
Kynuna **73** x
Kyoga, Lake **42** y
Kyōto **58** z
Kyrbana **53** j
Kyren **57** b
Kyrgyday **52** j
Kyrta **50** w
Kystalyam **52** j
Kysyl-Suluo **53** c
Kysyl-Yllyk **52** f
Kytyl-Zhura **52** k
Kyuekh-Bulung **52** f
Kyushe **55** n
Kyūshū **58** w
Kyusyur **52** l
Kyyiv (Kiev) **33** z
Kyyjärvi **33** j
Kyzyk **54** y
Kyzyl **56** k
Kyzyl-Chaja **56** h
Kyzyl-Chem **56** l
Kyzyl-Dzhar **55** i
Kyzyl Kum **55** r
Kyzyl-Orda **55** s
Kyzyluy **55** f
Kzyltu **55** j

L

La Almarcha **34** r
La Asunción **23** h
La Banda **28** t
La Barca **18** z
Labaz, Lake **51** l
Labé **40** e
Labe (Elbe) **33** r
Laboulaye **30** m
Labrador City **17** c
Labrador Sea **13** z
Lábrea **25** e
Labuha **67** y
Labuk Bay **67** q
Labytnangi **50** m
La Calera **28** o
La Campana **18** m
La Carbonera **19** c
La Carlota **28** u
La Ceiba **19** x
Lacepede Bay **77** t
Lachlan **78** a
La Chorrera **24** m
Lac la Martre **12** r
La Concepción **23** b
Laconia **17** n
La Crosse **16** q
Ladoga, Lake **33** m
Lady Frere **49** p
Lady Grey **49** p
Ladysmith (South Africa) **49** r
Ladysmith (USA) **16** d
Lae **71** o
Laem Ngop **66** g
La Escalera **23** w
La Esmeralda **23** t
La Esperanza **25** u
Lafayette (Indiana) **16** s
Lafayette (Lousiana) **19** g
Lafia **41** g
Laghouat **34** u
Lagos (Nigeria) **41** p
Lagos (Portugal) **36** g
La Grande **14** j
La Grande Rivière **16** m
Lagrange **72** w
Laguna de Bay **67** g
Lagunas **28** c
Lagunillas **28** g
Lahad Datu **69** e
Lahat **68** u
Lahej **43** g
La Higuera **28** o
Lahore **61** a
Lahti **33** j
Lai **41** m
Laibin **62** z
Lai Chau **62** v
Laichou Bay **58** p
Laingsburg **48** y
Laisamis **43** p
Laiyang **58** q
Laiyuan **57** v
Lajeado **29** o
Lajes **29** q
Lajes (Cáatingas) **27** v
Lajinha **25** k
La Julia **30** v
La Junta **15** r
La Junta **18** j
Lake Charles **19** g
Lake City (Florida) **20** l
Lake City (South Carolina) **20** m
Lake Harbour **13** w
Lake Havasu Clty **18** e
Lake Jackson **19** d
Lakeland **20** l
Lake Nash **73** u
Lake Ozark **15** x
Lake Providence **20** e
Lake River **16** h
Lake Seul **15** k
Lakes Toshka **38** v
Lakeview **14** v
Lakhpat **60** w
Lakki **60** l
Lakota **40** w
Lakselv **33** j
Lakshadweep **64** b
Lalara **44** d
Laleh Zar **60** b
Lalin **58** g
Lalitpur **61** p
Lalyo **42** x
Lama-Kara **41** a
La Mancha **18** m
Lama,Lake **51** i
Lamar **15** s
La Martre, Lake **12** r
Lambaréné **44** c
Lambasa **75** m
Lambayeque **24** g
Lambert Glacier **84** y
Lambert Land **91** i
La Merced **24** x
Lamia **35** s
La Misión **18** c
Lamon Bay **67** h
La Montaña **24/25**
Lamotrek Atoll **71** a
Lampalla **24** y
Lampang **66** d
Lampedusa **34** z
Lanai **18** q
Lanbi **65** h
Lancang **62** t
Lancang Jiang **62** t

Lancaster 14 w
Lancaster (California) 18 c
Lancaster (Pennsylvania) 16 y
Lancaster Sound 13 d
Lanchow 62 j
Lândana 44 d
Landeck 34 l
Lander 15 o
Landes 34 f
Landless Corner 45 p
Land's End 32 v
La'nga Co 61 e
Langano, Lake 43 c
Langdon 15 h
Langhko 62 r
Langkawi 66 q
Langöyo 33 e
Langsa 66 p
Lang Son 62 x
La Noria 25 v
Lansing 16 t
Lanus 31 c
Lanxi 58 g
Lanzarote 36 c
Lanzhou (Lanchow) 62 j
Laoag 67 f
Laoang 67 j
La Oroya 24 x
Laos 62/65-66
La Palma 22 k
La Paragua 23 h
La Parilla 18 y
La Paz (Argentina) 28 r
La Paz (Bolivia) 25 p
La Paz (Honduras) 22 d
La Paz (Mexico) 18 u
La Paz (Paraná, Argentina) 28 x
La Paz Bay 18 h
La Pedrera 25 b
La Pérouse Strait 59 c
La Piedad 18 z
La Plata 31 c
Lappeenranta 33 k
Lappland 33 j
Laptev Sea 52 h
Laptev Strait 53 e
La Puntilla 24 g
La Quiaca 28 f
Lar 39 k
Larache 36 h
La Radisson 16 k
Laramie 15 q
Laranjal 25 k
Larba 52 w
Laredo 19 b
Largo 20 k
La Rioja 28 r
Lárisa 35 s
Larkana 60 i
Larnaca 35 y
La Rochelle 34 e
La-Roche-sur-Yon 34 e
La Romana 21 u
La Ronge, Lake 15 d
La Rons'yer 50 l
Laropi 42 x
Larrimah 73 e
Larsen Ice Shelf 82 v
Larvik 33 c
Lar'yak 51 s
Las Anod 43 i
Las Bocas 18 h
Las Cruces (Mexico) 18 u
Las Cruces (USA) 18 j
Las Davé 43 j
Las Entimias 18 m
La Serena 28 o
Las Flores 31 b
Las Heras 30 w
Lashio 62 r
Lashkargah 60 g
Lasho Joâyin 60 e
Las Koreh 43 j
Las Lajas 30 i
Las Lajitas 28 t
Las Lomas 24 h
Las Lomitas 28 w
Las Palmas 36 b
La Spezia 34 l
Las Piedras 25 n
Las Plumas 30 x
Lastoursville 41 w
Las Trincheras 23 g
Las Vegas (Arizona) 14 y
Las Vegas (New Mexico) 15 q
Latacunga 24 h
Latady Island 82 x
Latady Range 82 z
La Tagua 24 k
Latakia 35 z
Latina 34 m
La Tuque 17 a
Latur 61 o
Latvia 33 w
Launceston 77 y
La Unión 24 w
Laura 73 l
La Urbana 23 f
Laurel 19 i
Laurie Island 82 g
Lausanne 34 j
Laut 69 q
Lavapié, Punta 30 g
La Vega 21 s
La Venturosa 23 q
La Vibora 18 m
La Victoria 23 f
Lavumisa 49 s
Lawa 62 r
Lawdar 43 h
Lawn Hill 73 u
Lawra 40 l
Lawrenceburg 16 s
Lawton 19 c
Layla 39 s
Lazarev Sea 83 j
Learmonth 72 q
Leavenworth 15 w
Lebanon (State) 38
Lebanon (USA) 16 p
Lebap 55 q
Lebú 30 g
Lecce 35 c
Leduc 14 m
Leeds 32 x
Leesburg 20 l
Leeton 78 b
Leeuwarden 33 n
Leeward Islands 21 x
Legazpi 67 i
Leghorn (Livorno) 34 l
Legnica 33 s
Leh 61 c
Le Havre 34 f
Leiah 60 k
Leicester 32 y
Leigh Creek 77 f
Leipzig 33 q
Leiria 34 n
Leiyang 63 o
Le Kef 34 x
Lekki 41 p
Le Mans 34 g
Lema Shilindi 43 s
Lemmon 15 f
Lemsid 36 p
Lena 52 l
Leng Hu 56 w
Lenina 55 w
Leningradskaya 86 w
Leninsk 55 e
Len'ki 56 c
Lensk 52 s
Léo 40 l
Leoben 35 b
Leonardville 48 j
León (Mexico) 19 n
León (Nicaragua) 22 e
León (Spain) 34 c
Leonora 76 h
Lephalale 49 c
Lepsy 56 b
Le Puy 34 h
Lerdo 18 l
Lerida 73 y
Lérida (Colombia) 25 b
Lérida (Spain) 34 g
Lerwick 32 k
Lesbos 35 t
Les Cayes 21 q
Leschenault, Cape 76 c
Leskovac 35 e
Lesnaya 53 z
Lesosibirsk 51 w
Lesotho 49 p
Lesozavodsk 58 l
Lesser Antilles 23 e
Lesser Slave Lake 12 s
Lesser Sunda Islands 69 r
Lesser Zab 39 d
Leszno 33 s
Lethbridge 14 m
Lethem 23 x
Leticia 25 b
Leti Islands 69 z
Letlhakane 49 b
Letlhakeng 49 a
Letsok-Aw 66 c
Leuser 65 u
Levelland 18 m
Levéque, Cape 72 j
Levkas 35 q
Lewis 32 i
Lewis Hill 17 i
Lewiston (Idaho) 14 k
Lewiston (Maine) 17 o
Lewiston (Montana) 15 b
Lexington 16 t
Lexington Park 16 y
Leyte 67 j
Lhari 62 b
Lhasa 61 l
Lhazê 61 j
Lhazhong 61 i
Lhokseumawe 65 u
Lhoksukon 65 u
Lhorong 62 d
Lhozhag 61 l
Lhünzê 62 b
Liadong Wan 57 z
Liangcheng 63 g
Liangpran 69 c
Lianjiang 63 n
Lian Xian 63 o
Lianyuan 63 b
Liao 58 e
Liaodun 56 w
Liaotong Bay 58 d
Liaoyang 58 e
Liaoyuan 58 f
Liard 12 o
Libby 14 k
Libenge 42 o
Liberal 15 t
Liberdade 25 e
Liberia (State) 40
Liberia (Costa Rica) 22 f
Libertador General San Martin 28 g
Liboi 43 q
Libreville 41 t
Libya 37/38
Libyan Desert 38 d
Libyan Plateau 38 e
Licancabur 28 e
Licata 37 i
Lichinga 46 i
Lichtenburg 46 o
Lichuan 62 m
Lida 33 w
Lidköping 33 e
Liebig, Mount 73 p
Liechtenstein 34
Liège 34 i
Lienyunkang 63 g
Liepāja 33 u
Ligurian Sea 34 k
Lihue 18 p
Lijiang 62 g
Likasi 45 o
Likati 42 r
Likoto 45 a
Likouala 41 x
Likouala-aux-Herbes 41 z
Lille 32 z
Lillehammer 33 d
Lilongwe 45 t
Lima (Paraguay) 28 m
Lima (Peru) 24 w
Lima (USA) 16 t
Limassol 35 y
Limay 30 j
Limbe (Cameroon) 41 t
Limbe (Malawi) 46 i
Limerick 32 u
Lime Village 11 d
Limmen Bight 73 f
Limnos 35 t
Limoeiro 27 v
Limoges 34 g
Limon 15 r
Limón 22 h
Limpopo 49 g
Linah 39 d
Linares 19 b
Lincang 62 t
Lincoln (Argentina) 31 a
Lincoln (USA) 15 v
Lincoln Sea 91 h
Linda Downs 73 u
Linde 52 k
Linden 23 y
Lindi 45 x
Lindi (River) 42 t
Lindo 28 l
Line Islands 234 a
Linganamakki Reservoir 64 d
Lingayen 67 f
Lingayen Gulf 67 f
Lingbao 63 a
Lingga Islands 68 i
Lingomo 42 q
Lingshan 62 z
Lingsugur 64 f
Linguère 40 c
Linhai 58 f
Linhares 29 l
Linhe 57 r
Linkiang 58 g
Linköping 33 f
Linkou 58 i
Linosa 34 z
Linqing 63 d
Lins 29 d
Linsia 62 i
Lintao 62 i
Linton 15 g
Linxi 57 x
Linz 35 a
Linze 57 n
Lion, Gulf of 34 i
Liouesso 41 y
Lipa 67 g
Lipari Islands 35 n
Lipetsk 54 e
Lipez, Cord, de 28 e
Lira 42 y
Lisala 42 p
Lisboa (Lisbon) 34 n
Lisbon 34 n
Lisburne, Cape 10 m
Lishui 63 g
Lismore 78 g
Listowel Downs 77 k
Listv'anka 57 c
Litang 62 z
Litang 62 g
Litchfield 16 q
Lithgow 78 e
Lithuania 33 v
Litke 53 q
Litke (Novaya Zemlya) 50 j
Litovko 58 m
Little Andaman 65 d
Little Anyuy 10 a
Little Cayman 20 z
Little Current 16 i
Little Diomede 10 l
Little Falls 15 j
Little Karoo 48 y
Little Khinggan Mountains 58 h
Little Rock 15 x
Little Salmon 11 l
Little Scarcies 40 e
Little Swartberge 48 y
Litunde 45 v
Liuba 62 l
Liuchou 62 z
Liuhe 58 g
Liupo 45 y
Liushuquan 56 w
Live Oak 20 k
Liverpool 32 w
Liverpool Land 91 r
Livingston 15 a
Livingstone 46 o
Livingstone Mountains 45 u
Livingstonia 46 h
Livingston Island 82 s
Livny 54 d
Livorno 34 l
Liwa Oasis 39 x
Li Xian 63 a
Li Yuba 42 u
Lizarda 27 n
Ljamin 51 o
Ljubljana 35 a
Ljungan 33 f
Ljusnan 33 f
Llanos 23 b
Llanos de Chiquitos 28 i
Llanos de los
Caballos Mestenos 18 l
Llanos de Mojos 25 s
Llanquihue, Lago 30 g
Llata 24 w
Llera 19 o
Lloydminster 15 a
Llullaillaco 28 q
Loa 28 c
Lobatse 46 o
Lobaye 42 n
Loberia 31 b
Lobito 44 r
Lobos 31 b
Lobos, Cape 18 f
Lobos Island 24 g
Lobva 50 x
Lockhart River 73 l
Loddon 77 w
Lodja 44 m
Lodwar 43 n
Łódź 33 t
Loeriesfontein 48 x
Loffa 40 s
Lofoten Islands 33 d
Log 54 g
Logan 15 n
Logan, Mount 11 i
Logone 41 l
Logroño 34 e
Loikaw 65 h
Loi-lem 62 r
Loire 34 f
Loiyangalani 43 o
Loja 24 h
Lokandu 45 a
Lokichar 43 n
Lokichokio 42 z
Lokitaung 43 n
Lokoja 41 r
Lokomo 41 x
Lokori 43 n
Lokshak 52 z
Loks Land 90 e
Lolland 33 q
Lolworth 74 p
Lomami 45 a
Loma Mts 40 f
Lombarda, Serra 26 j
Lombardina 72 j
Lombok 69 q
Lomé 41 o
Lomela 44 m
Lomela (River) 44 k
Lomza 33 u
Londonderry 32 v
Londonderry, Cape 72 m
London (Canada) 16 v
London (UK) 32 y
Londrina 29 d
Longa 44 w
Long Beach 18 b
Longchuan 63 q
Longhua 57 x
Long Island (Bahamas) 21 p
Long Island (USA) 17 n
Longjiang 58 e
Longlac 16 f
Long Range 17 i
Longreach 73 y
Long Strait 10 g
Longview (Texas) 19 e
Longview (Washington) 14 g
Longxi 62 j
Long Xuyen 66 h
Longyan 63 r
Longyearbyen 92 o
Longzhen 58 g
Lonkin 62 q
Lönsdal 33 e

Lontra 26 x
Lookout Mountain 17 f
Loongana 76 l
Lop Buri 65 j
Lopcha 52 w
Lopez 67 h
Lop Nur 56 v
Lopori 42 q
Loptyuga 50 t
Lora 60 h
Loralai 60 j
Lorca 34 r
Lord Howe Island 78 k
Lordsburg 18 i
Lorenço 26 j
Loreto (Bolivia) 25 s
Loreto (Mexico) 18 g
Lorica 22 m
Lorient 34 d
Loruk 45 i
Lorukumu 43 n
Los Alamos 15 p
Los Angeles (Chile) 30 h
Los Angeles (Mexico) 18 f
Los Angeles (USA) 18 f
Losap Atoll 71 f
Los Blancos 28 i
Los Burros 18 g
Los Menucos 30 j
Los Mochis 18 i
Los Vientos 28 p
Lot 34 g
Lotta 33 l
Louangphrabang 62 v
Loubomo 44 e
Louchuan 62 b
Louga 40 b
Lougheed Island 12 j
Louisiade Archipelago 74 g
Louisiana 20 d
Louisville 16 t
Louis XIV, Point 16 j
Lovat 33 m
Lovelock 14 w
Løvenørn, Kap 90 x
Lowa 45 b
Lower California 18 f
Lower Egypt 38 h
Lower Hutt 79 v
Lower Red Lake 15 j
Loxton 48 z
Loyalty Islands 75 r
Loyang 63 b
Loyoro 42 z
Luachimo 44 k
Lua Dekere 42 p
Luambo 45 o
Luan 57 w
Luanda 44 e
Luang 66 q
Luang Prabang 65 l
Luang, Thale 66 r
Luangua 45 r
Luangwa 45 s
Luanhaizi 62 b
Luanshya 45 p
Luapula 45 p
Luau 44 y
Luba 41 t
Lubang 67 f
Lubango 44 s
Lubao 45 a
Lubbock 19 a
Lübeck 33 p
Lübeck Bay 33 p
Lubefu 44 m
Lubero 45 c
Lubilanji 44 m
Lublin 33 v
Lubny 35 l
Lubuklinggau 68 t
Lubuksikaping 68 e
Lubumbashi 45 p
Lubungu 45 n
Lubutu 45 b
Lucala 44 t
Lucapa 44 k
Lucas 25 z
Lucena (Philippines) 67 g
Lucena (Spain) 36 j
Lucerne 34 k
Luchegorsk 58 l
Luchou 62 k
Lüchun 62 v
Lucira 44 r
Luckhoff 49 n
Lucknow 61 e
Lucusse 44 x
Lucy Creek 73 t
Lüda (Dalian) 58 q
Luddington 16 s
Lüderitz 48 t
Ludhiana 61 b
Ludlow 18 d
Luebo 44 k
Luena (Angola) 44 w
Luena (River) 44 y
Luena (Congo, Dem. Rep. of) 45 n
Luengué 44 x
Lüeyang 62 k
Lufeng 63 q
Lufkin 19 e
Luga 33 l
Lugansk 54 e
Lugenda 45 v
Lugh Ganana 43 s
Lugovatka 51 u
Lugovoi 55 w
Lugu 61 g
Luhombero 45 w
Luiana 44 y
Luidzhi 50 i
Luiza 44 l
Luján 31 b
Lujlang 63 e
Lukacheck 53 n
Lukenie 44 k
Lukolela 44 i
Lukuga 45 b
Lukula 44 e
Lukulu 44 z
Lukumburu 45 u
Lukunskiy 52 b
Lule 33 h
Luleá 33 i
Luleburgaz 35 i
Lulong 57 y
Lulonga 42 o
Lulonga (River) 42 o
Lumbala 44 y
Lumbala N'Guimbo 44 x
Lumberton 21 a
Lumbo 45 y
Lumding 62 o
Lurni 70 x
Lumsden 79 q
Lumu 69 f
Lün 57 d
Lundazi 45 s
Lunga 45 o
Lunggar 61 g
Lungue-Bungu 44 g
Luni 60 y
Luobei 58 j
Luobuzhuang 56 t
Luocuan 62 m
Luoding 63 n
Luoguhe 57 m
Luohe 63 c
Luoshan 63 c
Luozi 44 f
Lupar 68 m
Luputa 44 m
Luqu 62 h
Luremo 44 i
Lúrio 45 w
Lusaka 46 c
Lusarnbo 44 m
Lushoto 45 k
Lusk 15 q
Lutembo 44 x
Luton 32 y
Lutselk'e (Snowdrift) 12 u
Lutsk 35 g
Lützow Holm Bight 84 h
Luvua 45 b
Luvuei 44 x
Luwingu 45 r
Luxembourg (City) 33 n
Luxembourg (State) 33-34
Luxi 62 s
Luxor 38 j
Luzi 44 x
Luzon 67 f
Luzon Strait 63 u
L'viv 35 g
L'vovka 51 q
Lycksele 33 g
Lydenburg 49 e
Lynchburg 16 w
Lyndhurst 74 o
Lyndon 72 r
Lynn Lake 12 x
Lynwood 74 n
Lyon (Lyons) 34 i
Lyon Inlet 13 e
Lyons (France) 34 i
Lyons (River) 72 r
Lyubinskiy 55 j

M

Ma'an 38 l
Maas (Netherlands) 33 n
Maas (Somalia) 43 v
Maasin 67 j
Maatsuyker Island 77 x
Maba 42 d
Mabote 49 h
Mabrouk 36 y
Macaé 29 j
McAlester 15 v
McAllen 19 c
MacAlpine Lake 12 k
McArthur 73 f
Macane 49 g
Macao 63 p
Macapá 26 k
Macas 24 i
Macau 27 u
Macauley Island 79 m
McClintock 13 n
McClintock Channel 12 k
McClintock Island 50 j
McClure Strait 12 d
McComb 20 f
McCook 15 t
McDermitt 14 w
Macdonald, Lake 73 n
Macdonell Ranges 73 q
Macedonia (F. Y. R.) 35 e
Maceió 27 v
Macenta 40 g
McGrath 11 d
Mach 60 i
Machachi 24 h
Machačkala 54 w
Machaila 49 g
Machakos 45 j
Machala 24 g
Macharetí 28 h
Machettie, Lake 73 v
Macheng 63 d
Machilipatnam 64 i
Machiques 23 b
Machuan 61 g
Macia 49 g
Macina 40 j
Mack 15 o
Mackay 74 r
Mackay, Lake 73 o
Mackenzie 12 p
Mackenzie (River) 11-12
Mackenzie Bay 11 j
Mackenzie King Island 12 g
Mackenzie Mountains 12 o
Mackinaw City 16 g
McKinlay 73 w
McKinley, Mount 11 e
Maclear 49 q
McLennan 12 s
McLeod Lake 14 g
McLeod, Lake 72 q
Macrnillan 11 l
McMurdo 86 r
Macomia 45 y
Mâcon 34 i
Macon (Georgia) 20 k
Macon (Missouri) 16 p
Macquarie 78 d
Macquarie Harbour 77 w
Macquarie Island 235 x
MacRobertson Land 85 b
Macuarichic 18 j
Macuje 24 m
Macusani 28 c
Macuspana 19 t
Madagascar 47
Mada'in Salih 38 m
Madama 37 u
Madang 71 n
Madaoua 41 e
Madeira (Island) 36 a
Madeira (River) 25 j
Madenijet 56 b
Madera 18 j
Madidi 25 p
Madikeri 64 e
Madimba 44 g
Madingou 44 f
Madison 16 r
Madiun 69 n
Mado Gashi 43 q
Madoi 62 f
Madrakah 39 o
Madrakah, Ras 60 p
Madras (India) 64 h
Madras (USA) 14 h
Madre de Dios 25 q
Madre, Laguna 19 d
Madre, Sierra 67 g
Madre del Sur, Sierra 19 p
Madre Occidental, Sierra 18 k
Madre Oriental, Sierra 19 o
Madrid 34 d
Madula 42 t
Madura 69 o
Madura Strait 69 o
Madurai 64 f
Madziwadzido 49 d
Maebashi 59 o
Mae Sot 66 d
Maevatanana 47 q
Mafeteng 49 p
Mafia Island 45 l
Mafra 29 q
Mafraq 38 l
Magadan 53 v
Magadi 45 i
Magdagachi 52 y
Magdalen Islands 17 g
Magdalena (Argentina) 31 c
Magdalena (Bolivia) 25 s
Magdalena (Island) 30 t
Magdalena (Mexico) 18 h
Magdalena (River) 22 z
Magdalena (USA) 18 j
Magdeburg 33 q
Magelang 68 z
Magellan Straits 30 v
Maggieville 73 j
Maghama 40 e
Magnetic North Pole 89 r
Magnitogorsk 55 c
Magrur 42 g
Maguarinho, Cabo 26 m
Magude 49 g
Magwe 62 q
Mahabad 54 v
Mahabo 47 p
Mahaddei Uen 43 u
Mahagi 42 x
Mahajan 60 m
Mahajanga 47 d
Mahakam 69 d
Mahalapye 49 c
Mahalevona 47 f
Mahallat 39 h
Mahânadi 61 s
Mahanoro 47 r
Maha Sarakham 65 l
Mahavavy 47 p
Mahbubnagar 64 f
Mahd adh Dhahab 39 o
Mahdia 37 g
Mahenge 45 i
Mahesana 60 y
Maicao 23 b
Maiduguri 41 k
Maigudo 43 o
Maihar 61 r
Maiko 42 t
Maimana 60 h
Maimeča 51 m
Main 34 l
Mai-Ndombe, Lake 44 i
Maine 17 b
Mainé-Soroa 41 i
Mainling 62 b
Mainoru 73 e
Maiodao Islands 57 z
Maiquetía 23 e
Mait 43 i
Maitencillo 28 o
Maitri 83 m
Maizhokunggar 61 m
Majene 69 s
Maji 43 n
Makjadikgadi Pans 46 n
Makalu 61 i
Makambako 45 h
Makanza 42 o
Makarikha 50 j
Makarov 59 d
Makarova 51 g
Makassar Strait 69 e
Makat 54 m
Makaw 62 r
Makeni 40 e
Makeyevka 54 d
Makgadikgadi Pans 49 b
Makhado 49 e
Makinsk 55 i
Makkovik 17 h
Makokou 41 w
Makongolosi 45 g
Makopong 48 m
Makoua 44 g
Makrah 43 l
Makri 61 s
Maksimicha 57 f
Maksimovka 59 a
Makteir 36 r
Maku 54 u
Makurdi 41 t
Makushino 55 g
Makuyuni 45 i
Mâl 40 d
Mal. Lyakhovskiy 53 e
Malabar Coast 64 d
Malabo 41 t
Malacca 68 f
Malacca, Strait of 68 d
Malad City 14 z
Málaga (Colombia) 23 b
Málaga (Spain) 34 q
Malagarasi 45 e
Malaita 71 y
Malakal 42 k
Malang 69 n
Malanga 45 v
Malange 44 u
Mälar, Lake 33 f
Malargue 30 i
Malaripó 26 h
Malaspina 30 x
Malatya 54 q
Malawi 45-46/49
Malawi, Lake 46 i
Malaya (Malaysia) 66 s
Malaya (Russia) 53 i
Malaya Baranikha 10 b
Malayan Peninsula 66 r
Malayan Sea 67 o
Malaybalay Bislig 67 w
Malayer 39 h
Malaysia 65-69
Malčin 56 j
Maldives 64
Maldonado 31 e
Male 64 p
Maléa, Cape 35 s
Malegaon 61 n
Malekula 75 d
M. Aléman Reservoir 19 q
Malemba Nkulu 45 b
Malème-Hodar 40 c
Malhada 29 i
Malheur Lake 14 w
Mali 36-37/40-41
Malik 69 i
Malindi 45 l
Malka 53 x
Malkachan 53 x
Mallapunyah 73 f
Mallorca 34 u
Malmesbury 48 w
Malmö 33 r
Malmyzh 50 u

Maloca 26 h
Maloca Velha 26 h
Malonga 44 z
Målöy 33 a
Malozemelskaja Tundra 50 h
Malpelo 22 v
Malta (State) 35/37
Malta (USA) 15 b
Maltahöhe 48 i
Maluwe 40 l
Malykay 52 g
Malyutka 52 z
Malyy Lyakhovskiy 53 e
Mama 52 s
Mama (Cherskogo Mountains) 53 f
Mamadysh 54 k
Mamanguape 27 w
Mamantel 19 u
Mamazair 55 x
Mambasa 42 v
Mamberamo 70 v
Mamfe 41 l
Mamoré 25 r
Mamoriá 25 d
Mamou 40 e
Mampikony 47 e
Mampong 40 z
Man 40 u
Man, Isle of 32 w
Mana (French Guiana) 26 i
Mana (Hawaiian Islands) 18 o
Mana (River) 56 j
Mana Elena 28 c
Manacapuru 25 h
Manado 69 j
Managua 22 f
Managua, Lake 22 e
Manakara 47 q
Manama 39 i
Manambolo 47 o
Manamo 23 i
Mananara 47 f
Mananara (River) 47 q
Mananjary 47 r
Manankoro 40 i
Manar 43 g
Manas 56 t
Manas Hu 56 s
Manaus 25 j
Manchan 56 i
Manchester (UK) 32 x
Manchester (USA) 17 n
Manchouli 57 j
Manchuria 58 l
Máncora 24 g
Mand 39 j
Mandabe 47 p
Mandach 57 r
Mandal 57 e
Mandal (Hangayn Mountains) 57 a
Mandala 70 w
Mandalay 62 q
Mandalgov' 57 r
Mandal-Ovoo 57 p
Mandasor 61 n
Mandaue 67 i
Mandera 43 r
Mandimba 46 i
Mandla 61 r
Mandoto 47 p
Mandritsara 47 e
Mandurah 76 d
Mandvi 60 w
Mandya 64 e
Mangalmé 42 e
Mangalore 64 d
Manglares, Cabo 24 h
Mangnai 56 w
Mangochi 46 i
Mangoky 47 o
Mangole 69 k
Mangueni Plateau 37 u
Mangui 52 w
Manhattan 15 v
Manhuaçu 29 j
Mani 40 l
Maniamba 46 i
Manicoré 25 i
Manicouagan 17 c
Manicouagan Reservoir 17 c
Maniitsoq 90 v
Manila 67 l
Manila Bay 67 f
Manily 10 b
Maningrida 73 e
Manipa, Strait of 69 y
Manisa 35 v
Manitoba 15 g
Manitoba, Lake 15 h
Manitoulin Island 16 i
Manitouwadge 16 g
Manitowoc 16 s
Maniwaki 16 l
Manizales 22 z
Manjacaze 49 h
Manjimup 76 d
Manjra 61 p
Manjuri 25 e
Man Kabat 62 e
Mankato 15 j
Manlaj 57 r
Manley Hot Springs 11 f
Mannar 64 u
Mannar, Gulf of 64 t
Mannheim 33 p
Manning 12 r
Manoa 25 r
Manokwari 70 r
Manombo 47 n
Manonga 45 g
Manono 45 b
Mano River 40 s
Manpojin 58 g
Mansa 45 q
Mansel Island 13 e
Mansield (Ohio) 16 u
Mansfield (Pennsylvania) 16 y
Mansi 62 q
Manso 29 b
Mansôa 40 b
Manta 24 g
Mantalingajan 67 r
Mantaro 24 x
Mantecal 23 d
Mantena 29 k
Mantiqueira, Serra da 29 h
Manú 25 n
Manuel 19 p
Manuel Urbano 25 o
Manuelzinho 25 m
Manyami 45 k
Manyapa 69 d
Manych Gudilo Lake 54 g
Manyoni 45 g
Manyuan 57 o
Manzanillo 21 o
Manzini 49 s
Manzurka 57 d
Mao (Chad) 41 l
Mao (Dominican Republic) 21 s
Maoke (Mountains) 70 t
Mapaga 69 f
Mapai 49 g
Mapam Yumco 61 f
Mapati 44 f
Mapia Islands 70 f
Mapinhane 49 i
Maple Creek 15 a
Mappi 70 v
Mapuera 25 k
Maputo 49 t
Maqên 62 g
Maqên Gangri 62 f
Maquela do Zombo 44 g
Maquinchao 30 j
Mar, Serra do 29 r
Mara 45 g
Mara 25 e
Marabá 26 y
Maracá 26 l
Maracaçumé 27 b
Maracaibo 23 b
Maracaibo, Lake 23 b
Maracaju 29 a
Maracaju, Serra de 29 a
Maracay 23 e
Maradah 37 l
Maradi 41 f
Maragheh 54 v
Marajó 26 l
Marajó Bay 26 m
Maralal 43 n
Marambio 82 t
Marand 54 v
Maranguape 27 t
Maranoa 78 c
Marañón 24 k
Marape 25 y
Maraş 54 p
Marathon 16 f
Maravatio 19 o
Maraveh-Tappeh 55 n
Marawah 39 w
Marawi 67 v
Marble Bar 72 u
Marcelino 25 d
Marchena 36 i
Marchena (Island) 22 p
Marcy, Mount 16 z
Mardan 60 m
Mar del Plata 31 c
Mareeba 74 b
Margai Caka 61 i
Margaret River 72 z
Margarita 23 g
Margaritovo 58 m
Margherita Peak 42 w
Margilan 55 w
Marhoum 34 s
Maria Cleofas 18 x
Maria Madre 18 w
Marea Magdalena 18 w
Mariana Islands, Northern 235 j
Marianna 20 j
Maribor 35 a
Maridi 42 w
Maridi (River) 42 w
Marie Byrd Land 81 d
Marihamn 33 h
Marienlal 48 j
Marietta (Georgia) 20 j
Marietta (Ohio) 16 v
Mariinsk 56 g
Mariinskoye 53 q
Marilia 29 d
Marinduque 67 g
Marinette 16 f
Maringa (Dem. Rep. Congo) 42 q
Maringá (Brazil) 29 b
Marioka 59 p
Marion 16 v
Marion Downs 73 v
Maripa 23 g
Mariscal Esligarribia 28 j
Maritime Alps 34 j
Mariupol 54 d
Markakol', Lake 56 f
Markam 62 f
Markapur 64 g
Marken 49 d
Markha 52 g
Markit 56 o
Markoka 52 f
Markovo 10 d
Marlborough (Australia) 74 s
Marlborough (Guyana) 23 k
Marmara, Sea of 35 i
Marmarica 35 s
Marneuli 54 u
Maro 42 b
Maroantsetra 47 e
Maroochydore 74 u
Maroua 41 k
Marovoay 47 c
Marowijne 26 h
Marqua 73 t
Marquesas Islands 234 p
Marquette 16 f
Marrakech 36 g
Marree 77 f
Marresale 51 a
Marrupa 45 w
Marsa Alam 38 x
Marsabit 43 p
Marsala 34 z
Marsa Matruh 35 u
Marseilles 34 i
Marsella 24 k
Marshall (Arkansas) 15 x
Marshall (Minnesota) 15 i
Marshall (Missouri) 15 x
Marshall (Texas) 20 c
Marshall Islands 235 l
Marshal'skiy 53 d
Marshfield 16 d
Martaban, Gulf of 65 g
Martinez 31 b
Martinez de la Torre 19 q
Martinique 23 j
Martinique Passage 23 i
Martinsville 16 w
Martuk 55 a
Marungu Mountains 45 d
Mary 55 q
Maryborough (Queensland) 74 u
Maryborough (Victoria) 77 v
Marydale 48 y
Maryevka 54 j
Mar'yevka 55 h
Mary Kathleen 73 v
Maryland 16 y
Maryvale 73 y
Masai Steppe 45 i
Masaka 42 x
Masalembo 69 p
Masamba 69 t
Masan 58 u
Masasi 45 x
Masawa 43 d
Masbate 67 i
Masbate (Island) 67 i
Mascara 36 m
Maseru 49 p
Mashhad 55 p
Masia-Mbio 44 h
Masi-Manimba 44 i
Masindi 42 x
Masinloc 67 f
Masirah 60 p
Masisea 24 y
Masisi 45 c
Masjed-e-Soleiman 39 h
Maskana 38 m
Mason City 15 x
Massachusetts 17 n
Massaguet 41 l
Massakori 41 l
Massangena 49 g
Massangulo 45 u
Massenya 41 l
Massif Central 34 h
Massinga 49 i
Massingir 49 g
Mastabah 39 n
Masakh 52 j
Masteksay 54 k
Masterton 79 v
Mastuj 60 m
Masturah 39 n
Masuda 58 w
Masuguru 45 x
Masvingo 49 f
Matadi 44 f
Matagalpa 22 f
Matagami 16 k
Mataj 56 o
Matam 40 d
Matamoros 19 d
Matamoros (Mexican Plateau) 18 m
Matane 17 d
Matankari 41 c
Matanzas 20 y
Matapan, Cape 35 r
Mataram 72 f
Mataranka 73 d
Mategua 25 t
Matehuala 19 n
Mateur 37 f
Mathura 61 b
Matías Romero 20 p
Mato Grosso 25 w
Mato Grosso Plateau 26/29
Mato, Sierra de 23 t
Matruh 38 f
Matsue 58 x
Matsumoto 59 n
Matsusaka 58 z
Matsuyama 58 x
Matucana 24 w
Matun 60 k
Matundu 42 s
Maturín 23 h
Maturuca 23 w
Matyl'ka 51 t
Mau, Cape 66 u
Maúa 45 w
Maués 25 k
Maués (River) 25 k
Maui 18 r
Maule 30 h
Maulvi Bazar 61 y
Maumere 69 v
Maun 48 m
Mauna Kea 18 r
Mauna Loa 18 r
Maunaloa 18 q
Maurice, Lake 77 a
Mauritania 36/40
Mauritius 47 y
Mavinga 44 x
Mawson 85 b
Maxcanú 19 u
Maxwelton 73 x
May, Cape 16 z
Maya 68 l
Maya (River) 53 n
Maya (River) 53 p
Mayaguana Island 21 r
Mayagüez 21 v
Mayamba 44 h
Mayamey 55 n
Maychew 43 d
Mayer 18 g
Maykain 55 l
Maykop 54 r
Maymakan 53 o
Maynopil'gyn 10 g
Mayo 11 k
Mayo (River) 24 j
Mayoko 44 e
Mayon 67 i
Mayor-Krest 53 f
Mayotte 47 c
May Pen 21 n
Maysville 16 u
Mayumba 44 c
Mayupa 23 v
Mayuram 64 h
Mayya 52 m
Mazabuka 49 c
Mazalet 41 g
Mazán 24 m
Mazanillo 18 y
Mazar 56 n
Mazar-i Sharif 60 i
Mazaruni 23 w
Mazatenango 22 a
Mazatlán 18 w
Mazo Cruz 25 o
Mazunga 49 e
Mazunzut 62 e
Mbabane 49 s
Mbaïki 42 n
Mbaké 40 c
Mbala 46 f
Mbale 42 z
Mbalmayo 41 v
Mbandaka 44 i
M'Banza-Congo 44 g
Mbanza-Ngungu 44 g
Mbarara 42 x
Mbari 42 r
Mbé 41 k
Mbesuma 45 s
Mbeya 45 f
Mbini 41 t
Mbini (River) 41 u
Mbombela 49 s
Mbomou 42 s
Mbour 40 a
Mbout 40 e
Mbrés 42 p
Mbuji Mayi 44 m
Mbulu 45 h
Mbuyuni 45 i
Mcherrah 36 i
Mchinji 46 g
Mdrak 66 k
Mead Lake 14 y
Meadow Lake 15 b
Meadville 16 w
Mearim 27 n
Mecejana 27 g
Mecca 39 n
Méehéria 34 s
Medan 68 d
Medea 37 a

Medellín **22** z
Médenine **34** y
Mederdra **40** c
Medford **14** t
Medi **42** x
Medicine Hat **15** a
Medina **39** n
Mediterranean Sea **34-35**
Medje **42** u
Mednogorsk **55** a
Mednyy, Cape **51** j
Medvežjegors **50** o
Meekatharra **72** u
Meerut **61** e
Mega **43** p
Megalo **43** r
Meghna **61** y
Meharry, Mount **72** u
Mehdia **36** h
Mehran **39** f
Mehriz **39** k
Meiganga **41** x
Meighen Island **12** l
Meigu **62** i
Meiktla **62** q
Meishan **62** i
Mei Xian **63** r
Mejicana **28** q
Mejillones **28** b
Mékambo **41** w
Mekele **43** c
Meknès **36** i
Mekong **66** g
Mekong Della **66** i
Mekongga **69** u
Melambes **35** t
Melanesia **74** i
Melbourne (Australia) **77** x
Melbourne (USA) **20** m
Meletsk **51** v
Melfi **42** a
Melfort **15** d
Melilla **34** r
Melitopol **35** m
Melo **29** n
Melville **15** e
Melville Bay **91** b
Melville Island (Canada) **12** g
Melville Island (Auslralia) **73** e
Melville Lake **17** h
Melville Land **91** i
Melville Peninsula **13** d
Memmingen **33** p
Memphis (Egypt) **38** h
Memphis (USA) **16** q
Ménaka **41** b
Menawashei **42** g
Mendebo Mountains **43** q
Menderes **35** v
Mendi (Ethiopia) **42** m
Mendi (Papua New Guinea) **70** y
Mendocino, Cape **14** s
Mendoza **28** q
Mengoub **36** k
Menkere **52** j
Menominee **16** f
Menongue **44** v
Menorca **34** v
Menyapa **67** p
Menyuan **62** h
Menza **57** f
Menzies **76** g
Meoqui **18** k
Meraia **36** u
Merak **68** w
Merauke **70** w
Merca **43** t
Merced **14** v
Mercedario **28** p
Mercedes (Argentina) **30** l
Mercedes (Buenos Aires, Argentina) **31** b
Mercedes (Corrientes, Argentina) **28** y
Mercedes (Uruguay) **31** c
Mergenevo **54** l
Mergol **56** x
Mergui **65** i
Mergui Archipelago **65** h
Mérida (Mexico) **19** v
Mérida (Spain) **34** p
Mérida (Venezuela) **23** b
Mérida, Cordillera de **23** b
Meridian **20** g
Merir **70** d
Meriruma **26** h
Merowe **38** v
Merrill **16** d
Mersa Metruh **38** f
Mersa Teklay **39** n
Mersín **35** y
Mértola **36** g
Mertz Glacire **87** n
Meru **45** j
Meru (Mount) **45** i
Mesa **18** g
Mesa del Nayar **18** y
Mesa de Yambi **25** a
Meseta del Cerro Jáua **23** t
Meseta de Somuncurá **30** k
Mesfinto **43** b
Messaad **34** u
Messalo **46** j
Messina (Italy) **35** n
Messoyakha **51** e
Meta **23** r
Meta Incognita Peninsula **13** w
Metán **28** t
Metema **43** a
Metlaoui **37** e
Metlili Chaamba **34** u
Metoro **45** y
Metsiamonong **49** a
Metu **43** a
Metz **33** n
Meulaboh **65** t
Mexiana **26** l
Mexicali **18** d
Mexican Plateau **18-19**
Mexican Water **15** o
Mexico **18-19/20**
Mexico City **19** o
Meymeh **39** i
Mezcalapa **20** q
Meždurečensk **56** g
Mezen' **50** e
Mezen' Gulf **50** e
Mezen' (River) **50** t
Mezquital **18** y
M'guiden **37** a
Mhangura **49** e
Mhluzi **49** n
M. Hidalgo Reservoir **18** i
Mhow **61** n
Miahuatlán **19** q
Miami (Florida) **20** m
Miami (Oklahoma) **15** w
Mianeh **54** w
Miang **65** j
Mianwali **60** l
Mianyang **62** j
Mianyang **63** c
Miaodao Qundao **58** q
Miarinarivo **47** r
Michigan **16** t
Michigan, Lake **16** s
Michurinsk **54** f
Micronesia **71** a
Midar **36** j
Middle Andaman **65** e
Middle Atlas **36** i
Middleboro **15** j
Middlesbrough **32** y
Middleton (Australia) **73** w
Middleton (Canada) **17** e
Middleton Reef **78** k
Midelt **36** i
Midi **39** p
Midland (Canada) **16** w
Midland (Michigan) **16** u
Midland (Texas) **18** m
Midway Island **235** m
Mieso **43** d
Miguel Auza **18** l
Mikhaylova **54** i
Mikhaylova (Taimyr Peninsula) **51** g
Mikhaylovka **54** g
Mikumi **45** j
Mikun **50** l
Milan **34** k
Milang **62** u
Milano (Milan) **34** k
Mildura **77** i
Mile **62** v
Miles **78** e
Miles City **15** d
Milestone **15** d
Mileura **76** e
Milford **14** z
Milford Sound **79** p
Milgun **76** f
Milingimbi **73** e
Milk **15** b
Mil'kovo **53** y
Milk River **14** m
Mille, Lake **15** k
Millerovo **54** e
Millicent **77** u
Millstream **72** t
Millungera **73** x
Milne Land **91** p
Milos **35** t
Milparinka **77** i
Milwaukee **16** r
Mimongo **44** d
Mina **56** k
Minab **39** m
Minas **31** e
Mina Saud **39** g
Minas Gerais **29** h
Minalitlán **19** s
Minch **32** i
Mindanao **67** x
Minden **19** f
Mindoro **67** f
Mindoro Strait **67** f
Mineiros **29** b
Minfeng **56** q
Minga **45** p
Mingecaurskoje Reservoir **54** v
Mingenew **76** d
Mingoyo **45** x
Mingshui **56** y
Minhe **62** i
Minicoy **64** p
Min Jiang **63** r
Min Jiang (Wuyi Shan) **62** i
Minle **57** n
Minna **41** f
Minneapolis-St. Paul **16** b
Minnedosa **15** g
Minnesota (State) **15** j
Minnesota (River) **15-16**
Minnie Creek **72** s
Miño **34** a
Minot **15** f
Minqing **63** s
Minsk **33** x
Minto Atoll **71** h
Minto Inlet **12** e
Minto, Lake **13** u
Minusinsk **56** j
Min Xian **62** j
Minya Konka **62** h
Min'yar **55** b
Mirabad **60** e
Miracema do Norte **26** z
Miraflores **24** m
Miramar **31** c
Miran **56** u
Miranda **28** m
Miranda Downs **73** k
Mirbat **39** y
Mirbat, Ras **39** y
Miri **69** b
Miriam Vale **74** t
Mirim, Lagoa **29** o
Mirjaveh **60** e
Mirny **85** s
Mirnyy **52** f
Mirpur Khas **60** w
Mirsale **43** w
Mirzapur **61** t
Miscou, Point **17** e
Misgar **61** a
Misheguk Mountain **11** a
Misión San Fernando **18** e
Miskolc **35** d
Misool **70** o
Missinaibi **16** h
Mississippi (Stale) **20** f
Mississippi (River) **16** q
Mississippi Delta **20** g
Missoula **14** l
Missour **36** j
Missouri (River) **15** w
Missouri (Stale) **15** x
Mistassini **17** a
Mistassini, Lake **16** m
Misurata **35** o
Mitatib **43** a
Mitchell (Australia) **78** e
Mitchell (USA) **15** u
Mitchell (River) **73-77**
Mitchell River **73** j
Mitla **19** q
Mito **59** o
Mitrovica **35** d
Mitú **25** b
Mitumbar Mountains **45** b
Mitwaba **45** b
Mitzic **41** v
Miyako **63** x
Miyazaki **58** w
Miyun **57** x
Mizan Teferi **43** n
Mizda **34** z
Mjösa, Lake **33** c
Mkomazi **45** j
Mkurnbura **46** f
Mkuze **49** t
Mmabatho **49** n
Moab **15** o
Moa (Island) **73** k
Moa (River) **40** s
Moanda **44** e
Mobe **45** d
Mobaye **42** p
Mobayi-Mbongo **42** q
Mobeka **42** o
Mobile **20** h
Mobridge **15** g
Mocajuba **26** m
Moçambique **45** y
Mocimboa da Praia **45** y
Mocoa **24** j
Moctezuma **18** h
Moctezuma (River) **19** o
Mocuba **45** v
Modena **34** l
Modesto **14** u
Modimolle **49** d
Modjamboli **42** q
Moe **77** x
Moengo **26** h
Moenkopi **15** n
Mogadishu **43** u
Mogdy **58** l
Mogilev **33** z
Mogilev Podol'skiy **35** i
Mogoča **52** u
Mogochin **51** s
Mogojto **52** r
Mogok **62** r
Moharnrnadia **34** s
Moheli **47** a
Mohoro **45** k
Moincêr **61** e
Mointy **55** k
Mo-i-Rana **33** e
Mojave **14** w
Mojo **28** f
Mojones **28** r
Moju **26** m
Mokambo **45** p
Mokil Atoll **71** l
Mokolo **41** k
Mokopane **49** e
Mokowe **45** l
Mokp'o **58** t
Molde **33** b
Moldova **35** i
Molepolole **49** b
Molgary **56** b
Moline **16** q
Mollendo **24** z
Molócuè **46** k
Molodetschno **33** x
Molodezhnaya **84** l
Molodo **52** j
Molokai **18** q
Molopo **49** n
Moloundou **41** y
Moluccas **69** l
Molucca Sea **69** k
Moma **46** y
Mombasa **45** k
Momboyo **44** j
Mompós **23** a
Monaco **34** k
Monapo **46** l
Mona-Quimbundo **44** x
Mončegorsk **50** a
Mönch Chajrchan **56** j
Monclova **19** a
Moncton **17** e
Mondo **41** l
Mondy **57** a
Monga **42** r
Mongalla **42** y
Monghyr **61** v
Möng Küng **62** r
Mongo **42** b
Mongola **42** p
Mongolia **56-57**
Mongolo **52** d
Möng Ton **62** s
Mongu **44** y
Monguno **41** j
Möng Yai **62** r
Monkira **77** h
Monkoto **44** k
Mono **71** u
Mon'obut **52** z
Monroe **19** g
Monrovia **40** s
Montalvo **24** j
Montana **15** a
Mont aux Sources **49** q
Monte Alegre **26** i
Monte Alto **29** e
Monte Dourado **26** j
Montego Bay **21** n
Monemorelos **19** b
Montenegro **35** d
Montepuez **45** x
Monte Quemado **28** u
Monterey **14** u
Montería **22** l
Montero **25** t
Monteros **28** s
Monterrey **19** b
Montes Claros **29** i
Montevideo (Uruguay) **31** d
Montevideo (USA) **15** i
Montgomery **20** i
Monticello **15** o
Mont-Joli **17** c
Mont Laurier **16** m
Montmagny **17** b
Monto **74** s
Montpelier **15** n
Montpellier **34** h
Montréal **16** m
Montrose **15** o
Monts des Ouled Nail **37** b
Montserrat **21** y
Montsinery **26** j
Monywa **62** p
Moomou **42** s
Moonie **78** e
Moora **76** d
Moore, Lake **76** e
Moorhead **15** i
Moose Jaw **15** d
Moose Lake **15** g
Moosonee **16** i
Mopeia **46** v
Mopti **40** k
Moquegua **25** n
Moradabad **61** d
Morafenobe **47** p
Moran **15** n
Morava (River, Slovakia) **35** b
Morava (River, Serbia) **35** e
Moray Firth **32** j
Mordaga **57** l
Morden **15** h
Mordovo **54** f
Moree **78** e
Morehead **16** u
Morehead City **21** b
Morelia **19** n
Morella **74** o
Morelos **19** s
Moreton Island **78** g
Morgan **77** g
Morichal **23** p
Morin Dawa **58** f

Morley River 11 m
Mornington 73 h
Mornington (Island) 73 h
Moro 60 j
Morobe 71 o
Morocco 36
Morogoro 45 j
Moro Gulf 67 u
Morokovo 10 e
Morokweng 49 n
Moroleón 19 n
Morombe 47 n
Morón 21 n
Mörön 57 a
Morona 24 j
Morondava 47 o
Morondo 40 i
Moroni 47 a
Moron Us He 61 m
Morotai 69 m
Moroto 42 z
Morozovsk 54 g
Morrinhos 29 d
Morris 16 r
Morro Grande 26 h
Morshansk 54 f
Morteros 28 v
Mortes 26 x
Mortlock Islands 71 g
Morven 74 q
Morvi 60 y
Moryakovskiy Zaton 51 s
Moscow (Russia) 54 c
Moscow (USA) 14 k
Mosegda 52 u
Moses Lake 14 i
Mosetse 49 b
Moshi 45 j
Mösjöen 33 d
Moskva (Moscow) 54 c
Mosomane 49 b
Mosquitos Coast 22 g
Mosquitos, Gulf of 22 i
Mossaka 44 h
Mossel Bay 48 z
Mossendjo 44 e
Mossman 74 b
Mossoró 27 u
Mostaganem 34 s
Mosting, Cape 90 w
Mosul 54 t
Mota 43 b
Motagua 22 c
Motihari 61 h
Motorchuna 52 i
Motril 36 j
Motul 20 t
Motykleyka 53 u
Moudjéria 40 e
Mouila 44 d
Moul 41 j
Mould Bay 12 e
Moulins 34 h
Moulmein 65 h
Moulouya 36 j
Moundou 41 l
Mount Coolon 74 q
Mount Dalrymple 74 r
Mount Darwin 49 f
Mount Douglas 74 q
Mount Dutton 77 d
Mount Elliot 74 q
Mount Fletcher 49 q
Mount Florence 72 t
Mount Frere 49 r
Mount Gambier 77 u
Mount Hagen 70 z
Mount House 72 m
Mount Isa 73 v
Mount Magnet 76 f
Mount Minnie 72 s
Mount Morgan 74 s
Mount Seabrook 76 e
Mount Sturgeon 74 n
Mount Surprise 74 b
Mount Vernon (Australia) 72 t
Mount Vernon (Illinois) 16 r
Mount Vernon (Washington) 14 h
Mount Wedge 73 q
Mountain Village 11 a
Moura 25 h
Mourdi Depression 38 o
Moussoro 41 m
Moutong 69 h
Moutouroua 41 k
Moven 77 l
Mowming 63 y
Moxico 44 x
Moyale 43 p
Moyeni 49 q
Moyero 52 b
Moyobamba 24 j
Moyu 61 d
Mozambique 45-46/49
Mozambique Channel 46 z
Mozhaysk 54 e
Mozyr 33 y
Mpaathutlwa Pan 48 l
M'Pala 44 f
Mpanda 45 e
Mpata 44 j
Mpessoba 40 i
Mpika 45 s
Mporokoso 45 r
Mpulungu 46 e
Mpwapwa 45 i
Msaken 37 g
Msata 45 j
Mtito Andei 45 j
Mtubatuba 49 t
Mtwara 45 y
Muanda 44 d
Muang Chiang Rai 65 i
Muang Khoa 62 u
Muar 68 g
Muarabadak 69 e
Muarabungo 68 g
Mubarek 55 s
Mubende 42 x
Mubi 41 j
Mucajá 25 k
Muchinga Mountains 45 r
Muconda 44 x
Mucubela 46 x
Mucuim 25 l
Mucuri 29 k
Mucusso 48 l
Muddy Gap 15 p
Mudgee 78 e
Mudginbarry 73 d
Mueda 45 x
Muerto Meer 20 p
Mufulira 45 p
Muğla 35 v
Mugrab 55 r
Muhairiqah 39 r
Muhammad Qol 38 z
Muhammad, Ras 38 k
Mühling Hoffman Mountains 83 j
Muinak 55 p
Muji 56 o
Mukachevo 33 v
Mukana 45 o
Mukden (Shenyang) 58 c
Mukuku 45 q
Mukumbura 49 f
Mulanje 46 i
Mulata 26 i
Mulga Park 77 a
Mulgrave 17 g
Mulhouse 34 j
Muligort 50 z
Muling 58 j
Mull 32 v
Mullaittivu 64 v
Mullewa 76 d
Mulongo 45 b
Multan 60 l
Mulu 67 o
Mulundo 44 t
Mulym'ya 50 z
Mumbai (Bombay) 60 z
Mumbué 44 v
Mumbwa 45 o
Mumra 54 j
Muna (Island) 69 v
Muna (River) 52 h
München (Munich) 34 l
Muncie 16 t
Mundo Novo 27 s
Mundri 42 x
Mundubbera 74 s
Mundybas 56 g
Mungindi 78 d
Munich 34 l
Münster 33 o
Muntok 68 i
Munugudzhak 53 l
Mununga 45 q
Muonio 33 j
Muqdisho (Mogadishu) 43 u
Muraé 29 j
Murallón 30 t
Murashi 50 t
Murat Nehri 54 r
Muratus Mountains 69 p
Murchison 76 d
Murchisson 41 g
Murcia 34 r
Murdochville 17 e
Mures 35 e
Murfreesboro 16 s
Murgab (River, Afghanistan) 60 h
Murgab (River, Tajikistan) 55 w
Murgenella 73 d
Murgon 74 t
Murgoo 76 d
Muriege 44 x
Murillo Atoll 71 f
Murmansk 50 a
Murom 54 f
Muroran 59 c
Murray 77 g
Murray Bridge 77 g
Murray Lake 70 x
Murrumbidgee 77 j
Murukta 52 a
Mururoa 234 p
Murwillumbah 78 g
Murzuq 37 v
Musa Ali 43 f
Musala 35 f
Musandam, Ras 60 n
Muscat 60 p
Musgrave Ranges 77 a
Musina (Zimbabwe) 49 e
Muskegon 16 s
Muskogee 15 w
Muslimbagh 60 j
Musmar 38 y
Musoma 45 g
Mussende 44 u
Mussera 44 f
Mustahil 43 t
Mustang 61 g
Musters, Lago 30 w
Muswellbrook 78 f
Mût 38 t
Mutankiang 58 h
Mutare 49 g
Mutoko 49 g
Mutomo 45 k
Mutoray 51 z
Muttaburra 74 o
Mutuali 45 w
Muxima 44 s
Muzaffarabad 60 m
Muzaffarnagar 61 c
Muzaffarpur 61 u
Muztag 61 j
Muztag (Yurungkax He) 61 e
Muz Tau 56 f
Mvolo 42 w
Mvomero 45 j
Mvurwi 46 s
Mwanza 45 f
Mwatate 45 k
Mweka 44 k
Mwene Ditu 44 m
Mwenezi 49 e
Mwenga 45 c
Mweru, Lake 45 c
Mwingi 45 j
Mwinilunga 44 z
Myajlar 60 x
Myakit 53 i
Myanaung 66 a
Myanmar (Burma) 62/65/66
Myingyan 62 q
Myitkyina 62 r
Mymensingh 61 x
Mynaral 55 x
Mynbulak 55 q
Myroodah 72 x
Myrtle Beach 21 a
Mys 53 m
Mysore 64 f
Mys Shmidta 10 h
Mys Zelaniya 51 a
My Tho 66 i
Mytisci 54 d
Myylybulak 55 l
Mzimba 45 t
Mzuzu 45 t

N

Naalehu 18 r
Naba 62 q
Nababeep 48 v
Nabberu, Lake 76 g
Naberezhnyje Celny 54 l
Nabk 38 l
Nabk Abu Qasr 39 a
Nabukjuak 13 g
Nacala 45 y
Nacaome 22 e
Nacaroa 45 y
Nachičevan 54 u
Nachingwea 45 w
Nachodka 58 k
Nacogdoches 19 e
Nadiad 60 z
Nador 36 j
Nadym 51 p
Nafud 39 h
Naga 67 i
Nagano 59 n
Nagarjuna Reservoir 64 g
Nagasaki 58 v
Nagaur 60 m
Nagercoil 64 s
Nagornyy 52 x
Nagoya 58 z
Nagpur 61 p
Nagqu 61 l
Nagykanizsa 35 c
Naha 63 z
Nahanni Butte 12 p
Nahuel Huapi, Lago 30 h
Naiband 60 b
Naij Tal 62 c
Naiman Qi 58 d
Nain (Canada) 13 z
Nain (Iran) 39 k
Nairobi 45 j
Naivasha 45 i
Najaf 39 d
Najafabad 39 i
Najd 39 c
Najin 58 j
Najran 39 q
Nakamura 58 x
Nakasongola 42 x
Nakfa 38 z
Nakhl 38 k
Nakhodka 51 e
Nakhon Pathom 65 j
Nakhon Ratchasima 65 k
Nakhon Sawan 65 j
Nakhon Si Thammarat 65 w
Nakina 16 f
Naknek 11 p
Nakonde 45 s
Nakuru 45 i
Nal'čik 54 t
Nalgonda 61 q
Nalut 34 z
Nalychevo 53 y
Nama 28 d
Namaacha 49 t
Namacurra 46 w
Namakzar-e Shadad 60 c
Namanga 45 i
Namangan 55 v
Namanyere 45 e
Namapa 46 l
Nambour 78 g
Nambuangongo 44 f
Nam Can 66 u
Nam Co 61 k
Namco 61 l
Nam Dinh 62 x
Nametil 46 l
Namialo 45 x
Namib 48 g
Namibe 48 e
Namibia 44/48
Namies 48 x
Namjagbarwa Feng 62 c
Nam Kaub 48 k
Namlea 69 y
Namoi 78 d
Namolok 71 f
Namonuito Atoll 71 c
Nampala 40 j
Nampula 46 k
Namutoni 44 u
Namwala 45 n
Nan 65 j
Nanaimo 14 g
Nanbu 62 k
Nancha 58 i
Nanchang 63 e
Nanchung 62 k
Nancy 34 i
Nanda Devi 61 d
Nanded 61 p
Nandi 75 k
Nanfeng 63 e
Nanga Eboko 41 w
Nanga Parbat 61 a
Nanga Pinoh 69 a
Nanga Sokan 68 m
Nangdoi 62 h
Nangomba 45 w
Nangqên 62 e
Nangsin Sum 57 r
Nangtud 67 h
Nangurukuru 45 k
Nang Xian 62 b
Nan Hulsan Hu 62 d
Nanjing (Nanking) 63 g
Nanking 63 g
Nanning 62 z
Nanortilak 90 s
Nanpang Jiang 62 w
Nanping 63 s
Nansen Land 91 i
Nansen Sound 89 s
Nanshan Islands 67 a
Nansio 45 g
Nantes 34 f
Nantulo 45 x
Nantung 63 h
Nanuque 29 k
Nanxiong 63 p
Nanyang 63 b
Nanyuki 43 o
Nanzhang 63 a
Nanzhao 63 b
Nao, Cabo de la 34 s
Napalkovo 51 c
Nape 62 w
Napier 79 w
Naples (Florida) 20 l
Naples (Italy) 35 a
Napo 24 k
Napoleon 16 t
Napoli (Naples) 35 a
Nara 40 i
Naracoorte 77 u
Narasun 57 h
Narat 56 r
Narathiwat 65 x
Narbonne 34 h
Nares Strait 91 d
Naribis 48 h
Narijnteel 57 n
Narmada 60 z
Narodnaja 50 k
Narodnaya 50 x
Narok 45 h
Narrabri 78 e
Narrandera 78 b
Narrogin 76 e
Narromine 78 c
Narsarsuaq 90 t
Narsimhapur 61 q
Narsinghgarh 61 p
Narubis 48 w
Narva 33 k
Narvik 33 f
Nar'yan Mar 50 i
Naryn 56 m
Naryn (Naryn Taragay) 55 y
Narynkol 56 p
Naryn Taragay 56 n
Nashville 16 s

Näsi, Lake 33 j
Nasik 60 z
Nasir 42 l
Naskaupi 17 g
Nassau 21 b
Nasser, Lake 38 v
Nata 49 b
Natal 27 w
Natanz 39 i
Natashquan 17 g
Natashquan (River) 17 g
Natchitoches 20 d
Nathan River 73 f
Natitingou 41 b
Natividade 26 z
Natron, Lake 45 i
Natuna 66 x
Natuna Islands 66 x
Naulila 44 t
Nauru 235 y
Naurzum 55 f
Nauta 24 l
Nauzad 60 g
Navadwip 61 w
Navarin, Cape 10 h
Navio, Serra do 26 j
Navirai 29 a
Navlya 54 b
Navoi 55 s
Navojoa 18 h
Navolato 18 i
Navrongo 40 m
Navsari 60 z
Náxos 35 t
Nayakhan 53 l
Naykhin 58 m
Naypidaw (Pyinmana) 62 q
Nazaré 27 t
Nazarovka 54 f
Nazarovo 56 i
Nazas 18 l
Nazca 24 y
Nazimovo 51 w
Nazina 51 r
Nazret 43 d
Nazwa 60 o
Nazym 51 n
Nazyyayevsk 55 i
Nchelenge 45 q
N'Dalatando 44 t
Ndareda 45 h
Ndélé 42 c
Ndendé 44 d
N'Djamena 41 l
Ndji 42 e
Ndjolé (Cameroon) 41 v
Ndjolé (Gabon) 41 v
Ndola 46 d
Neagh, Lough 32 v
Neale, Lake 73 o
Near Islands 10 r
Nebit-Dag 54 z
Nebo 74 r
Nebraska 15 s
Nebraska City 15 v
Nechako Reservoir 14 e
Nechi 22 m
Necochea 31 c
Necoclí 22 l
Nêdong 61 m
Neds Creek 76 f
Neepawa 15 g
Neftekamsk 54 m
Negage 44 g
Negele 43 q
Negoiu 35 g
Negola 44 t
Negotka 51 s
Negro Reservoir 28 z
Negro (River, Argentina) 30 m
Negro (River, Brazil) 25 e
Negro (River, Uruguay) 29 n
Negros 67 h
Nehbandan 60 d
Nehe 58 f
Nehri 54 s
Neijiang (Neikiang) 62 k
Neiriz 39 k
Neishabur 55 o
Neiva 22 z
Nekempt 43 b
Neleger 52 m
Nel'gese 53 b
Nelidovo 50 n
Nel'kan 53 p
Nellore 64 h
Nelson (Canada) 14 k
Nelson (New Zealand) 79 t
Nelson (River) 13 n
Nelson Forks 12 p
Nelson Strait 30 s
Néma 40 i
Nembudziya 46 r
Nemen 33 v
Nemuro 59 e
Nemuy 53 p
Nenana 11 f
Nenjiang 58 f
Nen Jiang (River) 58 f
Neosho 15 w
Nepa 52 q
Nepal 61
Nepalganj 61 f
Nepa (River) 52 p
Nephi 14 z
Nepoko 42 v
Nera 53 f
Nerča 52 t
Nerchinskiy Zavod 52 u
Nerpich'ye, Lake 53 m
Nerrima 72 x
Neryungri 52 w
Ness, Loch 32 j
Neter 53 r
Netherlands 33
Netherlands Antilles 21 x/23 e
Netta 37 e
Nettilling Lake 13 j
Neubrandenburg 33 r
Neuquén 30 k
Neuquén (River) 30 i
Neuruppin 33 r
Neuva Alejandría 24 l
Nevada (Missouri) 15 w
Nevada (State) 14 w
Nevada, Mount 28 q
Nevada, Sierra 36 k
Nevado, Sierra del 30 j
Nevanka 56 m
Nevel' 33 y
New 23 y
Newala 45 x
New Amsterdam 23 y
Newark (New York) 16 z
Newark (Ohio) 16 u
New Bedford 17 o
New Bern 16 y
New Braunfels 19 c
New Britain 71 q
New Brunswick 17 d
Newburgh 16 z
New Caledonia 75 o
Newcastle (Australia) 78 f
Newcastle (Canada) 17 d
Newcastle (South Africa) 49 r
Newcastle (USA) 15 q
Newcastle upon Tyne 32 x
Newcastle Waters 73 d
New Crown 77 d
New Delhi 61 b
Newfoundland 17 jk
Newfoundland and Labrador 17 gh
New Georgia 71 w
New Georgia Islands 71 v
New Glasgow 17 f
New Guinea 70
New Hampshire 17 n
New Haven 17 n
New Hebrides 75 f
New Iberia 19 g
New Ireland 71 s
New Jersey 16 z
New Kandla 60 x
New Liskeard 16 j
Newman 72 v
Newman, Mount 72 u
New Meadows 14 k
New Mexico 15 p
New Norfolk 77 y
New Orleans 20 f
New Plymouth 79 u
Newport (Arkansas) 16 p
Newport (Tennessee) 16 u
Newport News 16 y
New Providence Island 21 o
New Richmond 17 d
New Siberian Islands 53 d
New South Wales 78 c
Newton 15 x
New Ulm 15 j
New York (City) 16 z
New York (Stale) 16 y
New Zealand 79
Neyto, Lake 51 a
Nezhin 33 z
Ngabang 68 m
Ngamring 61 i
Ngangala 42 y
Ngangla Ringco 61 g
Ngaoundéré 41 x
Ngatik Atoll 71 j
Ngauruhoe 79 v
Ngidinga 44 g
N'Giva 48 h
Ngo 44 g
Ngoring 62 e
Ngoura 41 m
Ngouri 41 l
Ngourti 41 j
Ngudu 45 g
Nguigmi 41 j
Ngukurr 73 f
Ngulu Atoll 70 h
Nguru 41 h
Nhamunda 25 k
Nha Trang 66 l
Nhill 77 u
Nhulunbuy 73 g
Niafounké 40 k
Niagara Falls 16 w
Niak 61 h
Niamey 41 b
Niangara 42 u
Nia Nia 42 u
Nias 68 c
Nicaragua 22
Nicaragua, Lake 22 f
Nice 34 j
Nichatka, Lake 52 u
Nicholls Town 21 a
Nicholson 73 n
Nička 55 r
Nicobar 65 q
Nicobar, Great 65 r
Nicobar, Little 65 r
Nicosia 35 y
Nicoya, Gulf of 22 g
Nicoya Peninsula 22 f
Nidym 51 z
Niefang 41 u
Niemba 45 e
Niénokoué 40 v
Nieuw Nickerie 23 z
Nieuwoudtville 48 x
Niğde 35 z
Niger 37/41
Niger (River) 41 e
Nigeria 41
Nihau 18 o
Niigata 59 o
Nikolayev 35 k
Nikolayevsk 54 h
Nikolayevsk-na-Amure 53 r
Nikol'sk 50 s
Nikol'skiy 55 g
Nikopol 35 l
Nikshahr 60 q
Nilang 61 d
Nile 38 u
Nile Delta 35 w
Nilka 56 q
Niman 53 n
Nîmes 34 h
Nimule 42 y
Ninbing 73 a
Nindigully 74 r
Nine Degree Channel 64 p
Ninety Mile Beach (Australia) 77 z
Ninety Mile Beach (New Zealand) 79 g
Ning'an 58 i
Ningde 63 t
Ningdu 63 q
Ningjin Shan 62 d
Ningming 62 y
Ningpo 63 h
Ningshan 62 m
Ningyuan 63 o
Ninnis Glacier 86 z
Niobrara 15 s
Nioki 44 h
Niokolo Koba 40 d
Nioro du Sahel 40 g
Nipigon 16 e
Nipigon Lake 16 d
Niquero 21 n
Nis 35 e
Nishinoyima 58 y
Nitchequon 17 b
Niterói 29 i
Niubiziliang 56 v
Nizamabad 64 f
Nizhinevartovsk 51 q
Nizhneangarsk 57 f
Nizhneimbatskoye 51 v
Nizhniy Novgorod 50 r
Nizhniy Tagil 50 x
Nizhniy Yenangsk 50 s
Niznaya Tunguska 52 a
Nižneudinsk 56 m
Niž Taimyra 51 m
Njombe 45 u
Nkambe 41 u
Nkawkaw 40 z
Nkayi 49 d
Nkhata Bay 45 u
Nkhotakota 45 t
Nkolabona 41 v
Nkongsamba 41 u
Noatak 11 a
Noatak River 11 a
Nochixtlán 20 o
Nockatunga 77 i
Nogales 18 h
Noginsk 54 d
Noginskiy 51 v
Nohar 61 a
Nok Kundi 60 f
Nokou 41 l
Nola 41 y
Nolinsk 50 u
Nom 56 x
Nome 10 m
Nomgon 57 q
Nomhon 62 e
Nonacho Lake 12 v
Nonburg 50 h
Nong Khai 65 l
Noonamah 73 c
Noorvik 11 b
Noranda 16 j
Nordenshel'da Arch 51 k
Nordostrundingen 91 k
Nordvik 52 e
Nordvik, Cape 52 f
Norfolk (Nebraska) 15 u
Norfolk (Virginia) 16 y
Norfolk Island 79 d
Noril'sk 51 h
Normanby (Island) 74 g
Normanton 73 j
Norman Wells 12 b
Norovtin 57 h
Norrköping 33 f
Norrtälje 33 g
Norseman 76 h
Norte, Cabo 26 l
Norte, Puntá 30 m
Norte, Puntá (Samboronbón Bay) 31 d
Norte, Serra do 25 w
Northam 76 e
Northampton 76 c
North Andaman 65 e
North Anyuskiy Mountains 10 b
North Battleford 15 b
North Bay 16 j
North Baykal Plateau 52 r
North, Cape (Canada) 17 h
North Cape (New Zealand) 79 g
North Cape (Norway) 33 k
North Caribou Lake 16 c
North Carolina 16 v
North Channel (UK) 32 v
North Dakota 15 f
North Dvina 50 r
North East Land 92 e
Northern Ireland 32 v
Northern Mariana Islands 235 j
Northern Sporades 35 t
Northern Territory 73 q
Northern Ural Mountains 50 x
Northern Uvaly 50 s
North Fork 14 x
North Horr 43 o
North Island 79 u
North Lincoln Land 13 f
North Natuna 68 k
North Pagai 68 r
North Platte 15 t
North Pole 92 a
North Saskatchewan 15 a
North Sea 32 z
North Sosva 50 y
Northumberland Strait 17 f
Northway Junction 11 h
North West Cape 72 q
North-West River 17 g
Northwest Territories 11-12
Norton 15 t
Norton Sound 11 a
Norvegia, Cape 83 g
Norway 33
Norway House 15 h
Norwegian Bay 13 b
Norwegian Sea 33
Norwich 32 z
Noshul' 50 t
Nosratabad 60 c
Nossob 48 y
Nosy-Bé 47 e
Nosy Boraha 47 s
Noto 37 j
Nôtre Dame Bay 17 k
Nottingham 32 y
Nottingham Island 13 s
Nouâdhibou 36 n
Nouâdhibou, Cape 36 n
Nouakchott 36 n
Nouamrhar 36 o
Nouméa 75 q
Nouna 40 k
Nova Alvorada 29 a
Novabad 55 v
Nova Crixas 29 d
Nova Gaia 44 u
Nova Goleg 46 u
Nova Olinda do Norte 25 j
Novara 34 k
Nova Russas 27 r
Nova Scotia 17 f
Novay 51 m
Novaya Kakhova 54 b
Novaya Kakhovka 35 l
Novaya Sibir' 53 h
Novaya Zemlya 50 j
Novgorod 33 l
Novillo Reservoir 18 h
Novisad 35 d
Novo Aripuan 25 i
Novocherkassk 54 e
Novodolinka 55 k
Novo Friburgo 29 j
Novograd Volynskiy 33 x
Novo Hamburgo 29 q
Novoilinovka 59 a
Novoishimskiy 55 h
Novokiyevskiy Uval 58 i
Novokuzneck 56 g
Novolazarevskaya 83 m
Novomoskovsk (Russia) 54 e
Novomoskovsk (Ukraine) 54 c
Novorossiyka 52 z
Novorossiysk 54 q
Novoryonye 52 b
Novoshakhtinsk 54 e
Novosibirsk 56 e
Novo Uzensk 54 j
Novo-Vyatsk 50 u
Novoyerundinskiy 51 x
Novyy 52 c
Novyy (Vilyuy) 52 f
Novyy Port 51 b
Nowgong 62 o
Nowra 78 e
Noyabr'sk 51 q
Nsah 44 g
Nsanje 46 v
Ntadembele 44 h

Ntam 41 w
Ntoroko 42 w
Nuanetsi 49 g
Nuatja 41 n
Nuba Mountains 42 j
Nubian Desert 38 x
Nueltin Lake 12 y
Nueva Rosita 19 a
Nueve de Julio 31 a
Nuevitas 21 o
Nuevo 20 r
Nuevo Casas Grandes 18 j
Nuevo Gulf 30 m
Nuevo Laredo 19 a
Nu Jiang 62 e
Nukiki 71 w
Nullagine 72 v
Nullarbor 77 a
Nullarbor Plain 76 l
Numan 41 i
Numkaub 48 k
Numto 51 p
Nunap Isua (Cape Farewell) 90 t
Nunatak 13 k
Nunavut 12-13
Nungnain Sun 57 y
Nungo 45 w
Nunivak 10 m
Nunligran 10 j
Nuoro 34 k
Nura 55 i
Nurabad 39 i
Nuremberg 34 m
Nuri 18 i
Nürnberg (Nuremberg) 34 m
Nur Turu 56 w
Nuru 55 k
Nusaybin 54 s
Nushki 60 h
Nutak 13 z
Nutwood Downs 73 e
Nuuk 90 u
Nuussuaq Halv. 90 l
Nuweiba 38 k
Nuyts Archipelago 77 c
Nuyts, Cape 76 d
Nyahururu 43 o
Nyainqentangla Shan 62 c
Nyainrong 61 m
Nyakanazi 45 e
Nyaksimvol 50 y
Nyala 42 f
Nyalam 61 i
Nyamapanda 49 g
Nyamboyoto 51 e
Nyanga 49 g
Nyanyadzi 49 f
Nyasa, Lake (Lake Malawi) 45 u
Nyazepetrovsk 55 b
Nyda 51 c
Nyeboe Land 91 g
Nyêmo 61 l
Nyima 61 j
Nyimba 45 r
Nyingchi 62 c
Nyíregyháza 35 e
Nyiri Desert 45 j
Nyiru 43 n
Nyngan 78 c
Nyong 41 u
Nysh 53 r
Nyunzu 45 c
Nyurba 52 h
Nyurchan 53 v
Nyuya 52 t
Nyuya (River) 52 r
Nyvrovo 53 r
Nzega 45 f
Nzérékoré 40 u
N'Zeto 44 e

O

Oahe, Lake 15 g
Oahu 18 p
Oajazeiras 27 t
Oakes 15 h
Oakland 14 u
Oaklands 78 o
Oakley 15 t
Oak Ridge 16 t
Oala Panja 55 w
Oamaru 79 s
Oates Land 86 w
Oatlands 77 y
Oaxaca 20 o
Ob' (River) 56 e
Ob', Gulf of 51 c
Obbia 43 w
Oberá 29 n
Obi 69 l
Obidos 25 m
Obihiro 59 d
Oblukovino 53 x
Obninsk 54 c
Obo 42 u
Obo Liang 56 w
Obombo 48 g
Oboyan' 54 c
Obrique 36 i
Obsči Syrt 54 l
Obuasi 40 z
Ocala 20 l
Ocaña 23 a
Ocean Falls 14 d
Oceanside 18 c
Ocha 53 r
Ocoña 24 z
Ocoña (River) 24 z
Ocotlán 18 z
Oda (Ghana) 40 z
Oda (Mountain) 38 y
Odemira 36 f
Odense 33 p
Oder 33 r
Odessa (USA) 18 m
Odessa (Ukraine) 35 k
Odienné 40 i
Odintsovo 54 c
Oeiras 27 q
Oenpelli 73 d
Ofira 38 k
Ogaden 43 t
Ogbomosho 41 c
Ogden 14 z
Ogeechee 20 l
Ogilvie Mountains 11 i
Ogoamas 69 g
Ogoja 41 t
Ogoki 16 f
Ogooué 44 c
Ogoron 52 z
Ogun 41 p
Ogurčinskij 54 y
Ohanet 37 e
O' Higgins, Lago 30 t
Ohio (State) 16 u
Ohio (River) 16 u
Ohrid Lake 35 d
Oiapoque 26 k
Ōita 58 w
Ojgun Nuur 56 l
Ojinaga 18 l
Ojo Caliente 18 z
Ojo de Liebre, Lac 18 e
Ojos del Salado 28 q
Oka 54 f
Oka (Vostochny Mountains) 57 a
Okahandja 48 i
Okakarara 48 j
Okara 60 m
Okaukuejo 48 h
Okavango 48 m
Okavango Delta 48 m
Okayama 58 x
Okeechobee, Lake 20 l
Okene 41 r
Okha 53 s
Okhotsk 53 s
Okhotsk, Sea of 59 g
Okhotskiy-Perevoz 53 b
Oki 58 x
Okinawa 63 z
Oklahoma 15 t
Oklahoma City 15 u
Okondja 44 e
Okoyo 44 g
Oktyabr'isk 51 y
Oktyabrskiy 52 z
Oktyabr'skiy 54 m
Oktyabr'skiy (Kamchatka) 53 x
Oktyabr'skoy Revolutsii 51 k
Öland 33 s
Olavarría 31 a
Olbia 34 l
Old Crow 11 j
Oldéani 45 h
Olds 14 l
Olean 16 x
Olekma 52 v
Olekminsk 52 v
Olenëk 52 e
Olenëk (River) 52 g
Oleng-Sala 52 h
Oleni 51 d
Ölgij 56 j
Olinda 27 w
Oliva 28 u
Oliva, Cord. de 28 p
Olivares 28 p
Oliveira 29 h
Ollague 28 d
Ollita, Cord. de 28 p
Olmos 24 h
Oloitokitok 45 i
Ol'okama 52 v
Olomouc 33 s
Olongapo 67 f
Olov'annaja 57 i
Oloyskiy Mountains 10 a
Olsztyr 33 u
Olympia 14 g
Olympus 35 y
Olyutorskiy 10 d
Olyutorskiy, Cape 10 q
Ōma 59 c
Omaha 15 v
Omak 14 i
Oman 39/60
Oman, Gulf of 60 p
Omaruru 48 h
Omboué 44 b
Ombu 61 i
Omchali 54 z
Omdurman 42 k
Omeo 78 p
Omnya 53 n
Omo 43 n
Omolon 53 l
Omoloy 53 a
Omsk 55 k
Omsukchan 53 k
Onangué, Lake 41 u
Oncócua 48 f
Ondangwa 48 h
Ondo 41 q
Öndör 57 j
Öndör Chaan 57 g
Öndör Sum 57 u
Onega 50 p
Onega Bay 50 o
Onega, Lake 50 o
Onega (River) 50 p
Onegva Yar 51 q
O'Neill 15 u
Onekotan 59 k
Ongjin 58 s
Ongniud Qi 57 x
Ongole 64 h
Onitsha 41 s
Onkuchakh 52 f
Onon Gol 57 g
Onseepkans 48 w
Onslow 72 r
Ontario 14 k
Ontario (Province) 16 c
Ontario, Lake 16 x
Oobagooma 72 k
Oodnadatta 77 d
Ooldea 77 b
Oologah Lake 15 w
Ooratippra 73 t
Ootacamund 64 e
Opal 55 y
Opala 42 s
Opelika 20 i
Ophir 68 e
Opienge 42 u
Opochka 33 y
Opornyy 54 m
Oporto 34 a
Oppdal 33 c
Opuwa 48 f
Oradea 35 e
Orai 61 q
Oran (Algeria) 34 s
Oran (Russia) 52 t
Orange 78 d
Orange (River) 48 y
Orange, Cabo 26 k
Orangegebirge 26 h
Orbost 78 q
Orcadas 82 g
Orchontuul 57 d
Ord (Mount) 72 l
Ord (River) 73 a
Ord River 73 a
Ordos 57 s
Ordynskoye 56 d
Örebro 33 e
Orechovo-Zujevo 54 e
Oregon 14 h
Orel 54 c
Orel', Lake 53 q
Orellana 24 i
Orellana (Ucayali) 24 k
Orenburg 54 m
Orense 34 a
Orgejev 35 i
Orgon Tal 57 u
Orillia 16 j
Orinoco 23 h
Orinoco Delta 23 i
Oristano 34 x
Orlximinà 25 m
Orizaba 19 q
Orkney Islands 32 k
Orlando 20 m
Orléans 32 z
Orleans, Isle d' 17 b
Orlik 57 a
Orlinga 57 d
Ormara 60 t
Ormoc 67 j
Ornsköldsvik 33 g
Orocué 23 o
Orog Nuur 57 n
Orög Nuur 56 i
Oroluk Atoll 71 i
Oron (Russia) 57 h
Oronoque 23 y
Oroqen Zizhiqi 58 e
Oroquieta 67 v
Orotukan 53 i
Oroumieh 54 u
Orovada 14 w
Oroville (California) 14 u
Oroville (Washington) 14 i
Oroyek 53 j
Orsha 33 y
Orsk 55 b
Orto-Ayan 52 m
Orto-Kyuyel' 52 m
Ortonville 15 i
Orupembe 48 f
Oruro 28 e
Os 55 w
Ōsaka 58 z
Oscrovnoy 94 i
Oshakati 48 h
Oshawa 16 w
Oshkosh 16 r
Oshmarino 51 f
Oshogbo 41 d
Oshtoran 39 h
Oshwe 44 j
Osijek 35 c
Osinovo 51 v
Oskaloosa 15 x
Oskara, Cape 51 l
Oslo 33 d
Oslofjord 33 c
Osnabrück 33 o
Osório 29 q
Osorno 30 g
Ossa 77 x
Ossora 10 o
Ostashkov 50 n
Ostend 32 z
Östersund 33 e
Ostrava 35 c
Ostrov 33 l
Ostrovnoy 10 b
Ōsumi Channel 58 v
Osvaldo Cruz 29 d
Otaru 59 b
Otavalo 24 i
Otavi 48 i
O.T.Downs 73 e
Otish Mountains 17 b
Otjiwarongo 48 h
Otog Qi 57 r
Ot-Siyen 52 i
Ottawa (Canada) 16 m
Ottawa (USA) 15 w
Ottawa (River) 16 k
Ottawa Islands 13 s
Ottuk 55 y
Ottumwa 15 x
Oturkpo 41 t
Otuzco 24 i
Otway, Cape 77 v
Ouadane 36 r
Ouadda 42 e
Ouagadougou 40 m
Ouagat 42 d
Ouahigouya 40 l
Ouahila 36 i
Ouake 42 p
Oualata 36 u
Oualidia 36 f
Ouallam 41 b
Ouandago 42 o
Ouandja 42 d
Ouangolodougou 40 i
Ouaqui 26 i
Ouara 42 s
Ouarane 36 s
Ouargla 34 v
Oubangui 42 p
Oudeïka 36 y
Oudtshoorn 48 y
Oued Zem 36 h
Ouémé 41 b
Ouerrha 36 j
Ouesso 41 y
Ouezzane 36 i
Ouham 42 a
Ouidah 41 o
Oujda 34 r
Oujeft 36 q
Ouled Djellal 34 v
Oulou 42 d
Oulu 33 j
Oulu, Lake 33 k
Oum 36 g
Oum Hadjer 42 b
Ounianga Kebir 38 n
Ouricuri 27 s
Ourinhos 29 d
Ourique 36 g
Outat-el-Hadj 36 j
Outjo 48 h
Ouyen 77 v
Ouyu Bezze Denga 41 j
Ovalle 28 o
Ovamboland 48 i
Oviedo 34 c
Ovoot 56 k
Ovoot Chural 57 o
Övre Årdal 33 b
Owando 44 h
Owensboro 16 s
Owen Sound 16 i
Owen Stanley Range 71 o
Owo 41 r
Owyhee 14 x
Oxapampa 24 x
Oxford 32 x
Oyapock 26 j
Oyem 41 v
Oyero Arys 55 g
Oymyakon 53 e
Oyo 41 p
Oysurdakh 53 j
Oyun-Kyuyel' 53 c
Oyun-Yurege 53 a
Ožalilabad 54 w
Ozamiz 67 v
Ozark Plateau 15 x
Ozhogina (River) 53 x
Ozhogino, Lake 53 g
Ozhogino 53 h

P

Paamiut 90 t
Pabna 61 x

Pacáas Novos, Serra dos 25 s
Pacajá 26 k
Pacaraima, Serra 23 v
Pacasmayo 24 h
Pacaya 24 k
Pachino 37 j
Pachitea 24 y
Pachuca 19 o
Pacific Ocean 234-235
Pacoval 26 i
Padang 68 e
Padangpanjang 68 f
Padangsidempuan 68 d
Padauiri 25 f
Padre Bernardo 29 e
Padre Paraiso 29 k
Padua 34 m
Paducah 16 r
Padunskoje, More 33 m
Paektu-san 58 h
Pafuri 49 f
Pag 35 a
Paga-Conta 26 u
Pagadian 67 u
Pagalu 41 r
Page 15 n
Pahokee 20 m
Päijänne, Lake 33 j
Paita 24 g
Pajule 42 y
Pakanbaru 68 g
Pakaraima Mountains 26
Pakhoi 62 z
Pakistan 60
Pakkoku 62 p
Pak Sane 66 g
Pakse 66 i
Pala 41 k
Palana 53 z
Palangkaraya 69 b
Palanpur 60 z
Palapye 49 c
Palata 20 l
Palatka 53 v
Palau 70 e
Palawan 67 d
Paleleh 69 h
Palembang 68 u
Palenque 20 r
Palermo (Colombia) 24 m
Palermo (Italy) 35 n
Palghat 64 f
Palgrave 72 s
Palgrave, Mount 72 s
Pali 60 z
Palikir 71 k
Palk Strait 64 h
Palliser, Cape 79 v
Palma 34 u
Palmares 27 w
Palma Soriano 21 p
Palmeirais 27 p
Palmeira los Indios 27 v
Palmeiras 27 n
Palmer (Alaska) 11 f
Palmer (Graham Land) 82 t
Palmer Land 82 x
Palmerston North 79 v
Palmira 22 y
Palm Springs 18 d
Palmyra 38 m
Palmyra Island 234 a
Palmyras Point 61 w
Paloieh 42 l
Palomani 25 p
Palopo 69 t
Palo Santo 28 x
Palpa 24 x
Palparara 73 w
Palu 69 g
Pamangkat 68 k
Pambarra 49 h
Pamir 55 x
Pamlico Sound 16 y
Pampa de las Salinas 28 r
Pampa de los Guanacos 28 v
Pampas 24 y
Pampas (River) 24 y
Pamplona (Colombia) 23 b
Pamplona (Spain) 34 f
Pamyat 56 j
Pamzal 61 c
Panaji 64 c
Panamá (City) 22 j
Panama 22
Panamá Canal 21 j
Panamá, Gulf of 22 j
Panay 67 g
Panda 49 h
Pandamatenga 49 a
Pandharpur 61 n
Pandie Pandie 77 g
Panevėžhys 33 w
Panfilov 56 o
Pangi 45 b
Pangin 62 c
Pangi Range 61 a
Pangkalanbrandan 68 d
Pangkalanbuun 69 n
Pangkal Kalong 66 s
Pangkalpinang 68 j
Pangnirtung 13 l
Pangody 51 c
Pangong Tso 61 d
Panipat 61 b
Panjgur 60 g
Pankshin 41 h
Panna 61 r
Pannawonica 72 s
Panovo 51 o
Panova (Angara) 51 z
Pantelleria 34 z
Panter Downs 72 l
Pánuco 19 p
Pan Xian 62 w
Panyabungan 68 e
Paoting 57 v
Paotow 57 t
Papai 23 z
Papantla 19 q
Paphos 35 x
Paposo 28 o
Papua, Gulf of 70 z
Papua New Guinea 70-71/73/74
Pará 26 m
Parabel 51 r
Paraburdoo 72 t
Paracatu 29 f
Paracatu (River) 29 g
Paracel Islands 66 m
Paragominas 27 a
Paraguá (Brazil) 25 u
Paragua (Venezuela) 23 u
Paraguacu 27 s
Paraguai 25 x
Paraguaná Peninsula 23 d
Paraguari 28 z
Paraguay 28-29
Paraiba 27 v
Paraiba do Sul 29 j
Paraïso 29 b
Paraíso do Norte de Goiás 26 y
Parakou 41 c
Paramaribo 26 h
Paramirim 27 q
Paramušir 59 l
Paraná (Argentina) 28 w
Paraná (Brazil) 26 z
Paraná (River, Argentina) 28-29
Paraná (River, Brazil) 26
Paranaguá 29 s
Paranaiba 29 d
Paranam 26 h
Paranapanema 29 c
Paranapiacaba, Serra 29 r
Paranavai 29 e
Paraparaumu 79 u
Parapeti 28 i
Paratunka 53 y
Parbhani 61 o
Parbig 51 s
Parding 61 j
Pardo 29 b
Pardo (Brazilian Plateau) 29 l
Parepare 69 s
Parecis, Serra dos 25 u
Pariaguán 23 g
Paria, Gulf of 23 i
Parima, Serra 23 u
Pariñas, Cabo 24 f
Parintins 25 l
Paris (France) 34 h
Paris (USA) 19 e
Parkano 33 i
Parker Range 76 f
Parkersburg 16 v
Parkes 78 c
Parma 34 l
Parnaíba 27 e
Parnaíba (River) 27 d
Parnamirim 27 s
Parnu 33 j
Paromay 53 s
Paroo 77 k
Parras 18 m
Parry Islands 12 h
Parry Sound 16 j
Parsons 16 n
Partida, Roca 18 s
Paru 26 i
Paru de Oeste 25 l
Parvalipuram 61 t
Paryang 61 g
Parys 49 p
Pasadena 19 e
Pasangkayu 69 f
Pascagoula 20 g
Pasig 67 g
Pasni 60 t
Paso de los Indios 30 w
Paso de los Toros 28 z
Paso Limay 30 i
Paso Real 20 p
Passau 35 a
Passmore 77 g
Passo Fundo 29 o
Passos 29 g
Pastakh 53 i
Pastaza 24 j
Pasto 24 j
Pastos Bons 27 p
Patagonia 30 u
Patandar 60 g
Paterson 16 z
Pathankot 61 a
Patia 24 i
Patiala 61 b
Patience 26 i
Pativilca 24 v
Patkai Mountains 62 p
Patna 61 u
Patnos 54 t
Pato Branco 29 o
Patom 52 u
Patomskoye Plateau 52 s
Patos 27 u
Patos de Minas 29 f
Patos, Lagoa dos 29 p
Patquía 28 r
Pátrai (Patras) 35 r
Patras 35 r
Patrasuy 50 x
Patrocinio 29 f
Patta Island 43 r
Pattani 68 f
Pattullo, Mount 11 z
Patuca 22 g
Patzcuaro 19 n
Pau 34 f
Pau de Ferros 27 u
Pauini 25 c
Paulatuk 12 c
Paulistana 27 r
Paulo Alfonso 27 t
Pauri 61 d
Pausa 27 p
Pavlodar 55 m
Pavlogard 54 c
Pavlovsk 54 e
Pavlovsk 56 d
Pavon 23 o
Paxson 11 g
Payakumbuh 68 f
Pay-Khoy 50 l
Paynes Find 76 e
Paysandú 28 y
Payturma 51 k
Pay-yer 50 m
Peace River 12 s
Peace (River) 12 t
Peak Charles 76 g
Peak Downs 74 r
Peak Hill 76 f
Pearl City 18 p
Pearsall 19 b
Pearson Island 77 c
Peary Channel 89 r
Peary Land 91 i
Pebane 46 x
Pebas 25 a
Pečenga 33 m
Pechora 50 w
Pechora (River) 50 w
Pechora Sea 50 i
Pecked 77 m
Pecos 18 m
Pecos (River) 18 m
Pécs 35 c
Pedasi 22 j
Pedirka 77 d
Pedra Azul 29 k
Pedreiras 27 o
Pedro Alfonso 26 z
Pedro Gomes 25 z
Pedro Luro 30 m
Peel Bay 76 c
Peel Sound 12 m
Peg Ariak 70 r
Pegu 65 g
Pehuajó 31 a
Peipus, Lake (Lake Chud) 33 l
Peixe 26 z
Pekalongan 68 z
Pelagie Islands 34 z
Peleng 69 i
Peljesac 35 c
Pelly 11 l
Pelly Bay 13 b
Pelly (Bay) 13 b
Pelly Crossing 11 j
Pelotas 29 p
Pelsart Group 76 b
Pelvoux 34 j
Pematangsiantar 66 q
Pemba 45 y
Pemba Island 45 l
Pembroke 16 k
Penas Gulf 30 s
Pendembu 40 f
Pendleton 14 j
Penedo 27 v
Pene-Tungu 42 t
Penganga 61 p
Penge 45 a
Penglai 58 q
Pengpu 63 e
Penitente, Serra do 27 n
Penki 58 e
Penner 64 h
Pennsylvania 16 w
Penny Ice Cap 13 k
Penong 77 c
Penonomé 22 i
Pensacola 20 h
Pensacola Mountains 83 u
Penticton 14 i
Pentland 74 p
Pentland Firth 32 i
Penukonda 64 f
Penza 54 h
Penzance 34 c
Penzhino 10 c
Penzhinskaya, Gulf of 10 a
Penzhinskiy Mountains 10 b
Peoria 16 q
Perabumulih 68 v
Perak 66 r
Perambalur 64 g
Percival Lakes 72 y
Pereira 22 z
Pereshchepino 35 m
Pérez 20 t
Pergamino 31 a
Péribonca 17 a
Pericos 18 j
Perija, Sierra de 23 b
Periquitos 29 q
Perito Moreno 30 v
Perlas Archipelago 22 j
Perm′ 50 w
Perote 19 p
Perpignan 34 h
Perry 20 k
Perry Island 12 k
Perryton 15 t
Perryville 16 q
Persepolis 39 j
Perseverancia 25 t
Perth (Australia) 76 d
Perth (Canada) 16 l
Perth (UK) 32 w
Peru 24-25
Perugia 34 m
Pervomaysk 33 z
Pervoural′sk 50 x
Pesagi 68 u
Pescara 35 a
Peschanyy, Cape 92 j
Peshawar 60 l
Peski 55 g
Pesqueira 27 v
Petaling Jaya 68 f
Petatlán 19 n
Petauke 45 r
Peterborough (Australia) 77 g
Peterborough (Canada) 16 x
Peter Island 80 j
Petermann Range 76 m
Peter Pond Lake 12 u
Petersburg 16 x
Peters Mine 23 x
Petitsikapau Lake 17 d
Peto 19 v
Petoskey 16 g
Petra 38 k
Petrel 82 g
Petrolina 27 r
Petropavlovka 52 y
Petropavlovsk 55 i
Petropavlovsk-Kamchatskiy 53 y
Petrópolis 29 i
Petrovka 55 h
Petrovsk-Zabajkal′skij 57 f
Petrozavodsk 50 n
Petukhovo 55 g
Petushkova 10 e
Peureulak 65 u
Pevek 10 d
Phalaborwa 49 f
Phalsund 60 y
Phangan 65 j
Phatthalung 68 e
Phayao 65 j
Phenix City 20 i
Phetchabun 65 k
Phetchaburi 65 i
Phiafay 66 i
Philadelphia 16 z
Philippines 67
Philippine Sea 235 i
Philipsburg 21 y
Phitsanulok 65 j
Phnom Penh 66 h
Phoenix 18 g
Phoenix Island 235 z
Phongsali 62 u
Phrae 65 j
Phu Chong 66 i
Phuket 65 u
Phulabāni 61 t
Phu Quoc 66 g
Piacenza 34 l
Pianguan 57 t
Piauí 27 q
Piauí, Serra de 27 q
Pibor Post 42 z
Pico da Neblína 25 e
Pico de Aneto 34 g
Picos 27 r
Pico Truncado 30 w
Picton 79 u
Pic Toussidé 37 w
Picuún Leufú 30 i
Pidi 45 a
Pidurutalagala 64 v
Piedrabuena 30 w
Piedras Negras 19 a
Pielis, Lake 33 l
Pierce 14 k
Pierre 15 g
Pietarsaari (Jakobstad) 33 i
Pietermaritzburg 49 s
Piet Retief 49 r
Pietrosu 35 g
Pigu 40 m
Pigüe 31 a
Pijijiapan 19 t
Pik-Aborigen 53 i
Pikelot 71 b

Pikes Peak 15 q
Pikeville 16 u
Pik Grandioznyi 56 l
Pikhtovka 56 d
Pikounda 41 y
Pik Pobedy 56 p
Pilo Arcada 27 q
Pilar 28 y
Pilaya 28 g
Pilcomayo 28 g
Pilibhit 61 d
Pilsen (Plzeň) 35 a
Pim 51 o
Pimenta Bueno 26 q
Pimenteiras 25 v
Pinacate Desert 18 c
Pinang (Indonesia) 68 l
Pinang
(George Town, Malaysia) 65 w
Pinang (Island) 65 w
Pinar del Río 20 w
Pinawa 15 i
Pindobal 26 m
Pindus 35 r
Pine Bluff 19 g
Pine City 15 k
Pine Creek 73 c
Pinega 50 r
Pinega (River) 50 r
Pine Island Bay 81 b
Pingelap Atoll 71 l
Pingguo 62 y
Pingle 63 n
Pingliang 62 l
Pingnan 63 n
Pingsiang 63 c
Pingtingshan 63 b
Pingtung 63 u
Pingxiang 62 x
Pinheiro 27 b
Pini 65 v
Pinnaroo 77 u
Pinos 19 n
Pinos, Isla de 20 y
Pinotepa Nacional 19 p
Pinsk 33 w
Pinta 22 p
Pintados 28 c
Pinto 28 u
Pioner 51 j
Pionerskaya 85 r
Pionerskiy 50 y
Piorini 25 f
Piorini, Lago 25 g
Piotrków 33 u
Pipangikum Lake 15 k
Pipmuacan Reservoir 17 b
Piquiri 29 o
Piracuruca 27 d
Piraeus 35 s
Pirajiba 27 p
Piraju 29 e
Pirané 28 x
Piranhas 25 g
Pirapetinga 25 g
Pirapora 29 h
Pires do Rio 29 e
Piripiri 27 q
Piritu 23 g
Pirot 35 f
Pisa 34 l
Pisagua 28 c
Pisco 24 x
Pishin 60 r
Pissis 28 q
Pita 40 e
Pitcairn Island 234 q
Pite 33 h
Pitești 35 g
Pithara 76 e
Pittsburgh 16 w
Pittsfield 16 z
Piura 24 g
Piura (River) 24 g
Pixa 61 d
Pizhma 50 s
Pjasina 51 i
Pjasina, Lake 51 h
Pjatigorsk 54 t
P.K.le Roux Reservoir 49 n
Placa 29 b
Placetas 21 n
Plainview 19 a
Plamennyy 10 f
Plara-Açu 26 v
Plast 55 d
Platbakkies 48 w
Plate, River 31 d
Plato 22 m
Platte 15 u
Plauen 34 m
Plavsk 54 c
Playa Azul 18 z
Plaza Huincul 30 j
Pleiku 66 j
Plenty, Bay of 79 w
Plétipi, Lake 17 b
Pleven 35 g
Plibo 40 u
Ploiești 35 h
Plovdiv 35 g
Plumtree 49 c
Plymouth 32 w
Plymouth (Island) 21 z
Plzeň 35 a
Po 40 m
Po (River) 34 m
Pocatello 14 z
Poconé 25 y
Pocos de Caldas 29 g
Podgorica 35 c
Podkamennaya Tunguska 51 w
Podkamennaya Tunguska
(River) 51 x
Podolsk 54 d
Podporoze 50 o
Podukkottai 64 g
Podutesnaya 10 p
Pofadder 48 x
Pogo 40 j
Po Hai 57 y
Poinsett, Cape 87 g
Pointe-à-Pitre 21 z
Pointe-Noire 44 d
Point Hope 10 m
Poitiers 34 g
Pokacheva 51 p
Pokaran 60 l
Pokhara 61 g
Poko 42 u
Pokrovka 56 a
Pokrovsk 52 l
Pokrovsk
(Vostochny Mountains) 56 m
Poland 33
Poli 41 j
Poligus 51 x
Polillo Islands 67 h
Politovo 50 t
Polkan 51 w
Pol'kyko 51 m
Polokwane 49 d
Polotsk 33 y
Polovinka 53 n
Polovinka
(Patomskoye Plateau) 52 t
Polson 14 l
Poltava 35 m
Polui 51 a
Polunochnoye 50 y
Polur 64 h
Pol'yanovo 50 z
Pomeranian Bay 33 r
Pomona 48 u
Pon 33 h
Ponape 71 h
Ponca City 15 v
Ponce 21 v
Pondicherry 64 h
Pond Inlet 13 g
Pongo 42 u
Pongola 49 s
Ponta Branca 29 b
Ponta da Pedra 25 x
Ponta Grossa 29 q
Ponta Pora 28 m
Pontdrif 49 d
Ponterrada 34 b
Pontianak 68 l
Pontine Mountains 54 o
Ponton 76 i
Poopó, Lake 28 e
Popayán 22 Y
Popigay 52 d
Popigay (River) 52 e
Poplar Bluff 16 q
Popocatépetl 19 o
Popokabaka 44 h
Popomanaseu 71 y
Popondetta 71 p
Porangatu 26 y
Porbandar 60 w
Porcupine 11 i
Porog 50 w
Poronaysk 59 d
Porpoise Bay 87 q
Portachuelo 25 s
Portadown 32 v
Portage 16 q
Portage la Prairie 15 h
Port Albert 78 p
Portales 18 l
Port Alfred 49 p
Port Alsworth 11 d
Port Angeles 14 g
Port Arthur 78 p
Port Augusta 77 l
Port-au-Prince 21 r
Port-Bergé-Vaovao 47 d
Port Burwell 13 x
Port-Cartier 17 d
Port Charlotte 20 k
Port Davey 77 x
Port-de-Paix 21 q
Port Edward 49 r
Portel 26 k
Port Elgin 16 v
Port Elizabeth 49 o
Port-Gentil 41 t
Port Harcourt 41 r
Port Hardy 14 e
Port Hedland 72 t
Portimo 36 f
Port Keats 73 b
Portland (Australia) 77 u
Portland (Maine, USA) 17 o
Portland (Oregon, USA) 14 g
Port Lincoln 77 d
Port Louis 47 x
Port Macquarie 78 g
Port Menier 17 e
Port Moresby 71 n
Port Nolloth 48 v
Porto (Oporto) 34 a
Pôrto Alegre 29 p
Pôrto Alegre (Purus) 25 q
Pôrto Alegre
(Serra do Roncador) 26 x
Pôrto Alegre (Xingu) 26 w
Porto Alexandre (Tômbua) 48 d
Porto Amboim 44 r
Pôrto Artur 26 u
Pôrto Atlántico 25 y
Pôrto de Moz 26 j
Pôrto do Cajueiro 25 z
Pôrto dos Gauchos 25 x
Pôrto dos Meinacos 26 w
Pôrto Esperança 28 l
Pôrto Esperidio 25 w
Pôrto Franco 27 n
Port of Spain 23 i
Pôrto Grande 26 k
Pôrto Jofre 25 y
Pôrto Murtinho 28 k
Pôrto Naçional 26 z
Pôrto Novo 41 o
Pôrto Poet 26 i
Pôrto Santana 26 k
Pôrto So José 29 b
Pôrto Seguro 29 l
Pôrto Velho 25 s
Portoviejo 24 g
Port Philip Bay 77 x
Port Pirie 77 f
Port Radium 12 e
Port Safaga 38 k
Port Said 35 x
Port Shepstone 49 s
Portsmouth
(New Hampshire) 17 o
Portsmouth (Ohio) 16 u
Portsmouth (Virginia) 16 y
Port St. Johns 49 r
Port Sudan 38 z
Portugal 34
Porva 54 m
Porvenir (Bolivia) 25 p
Porvenir (Chile) 30 v
Posadas 28 z
Poso 69 g
Pospelicha 56 c
Posse 29 g
Postavy 33 x
Postmasburg 48 z
Posto Alto Manissaua 26 u
Posto Cocraimore 26 w
Posto Curuá 25 m
Potapovo 51 h
Potchefstroom 49 p
Poteau 20 c
Potenza 35 b
Poti (Brazil) 27 q
Poti (Georgia) 54 s
Potiskum 41 h
Potosí 28 f
Potrero del Llano 20 n
Potsdam 33 q
Pottuvil 64 w
Pouso Alegre 29 g
Povungnituk 13 t
Powder 15 d
Powell Basin 82 g
Powell, Lake 15 n
Powell River 14 l
Poyang, Lake 63 e
Poza Rica 19 q
Poznan 33 t
Pozo Colorado Monte 28 k
Pozo del Tigre 28 w
Prachin Buri 65 k
Prachuap Khiri Khan 65 j
Prague 35 a
Praha (Prague) 35 a
Prainha 26 i
Prainha Nova 25 i
Pratt 15 u
Precordillera 28 q
Predivinsk 51 x
Preparis 65 e
Prescott 18 f
Presidencia de la Plaza 28 w
Presidencia Roque
Sâenz Peña 28 w
Presidente Dutra 27 o
Presidente Hermes 25 u
Presidente Prudente 29 c
Presnogor'kovka 55 g
Presnovka 55 g
Prešov 35 e
Prespa Lake 35 e
Presque Isle 17 c
Prestea 40 y
Preto do Igapó-Acu 25 i
Priargunsk 57 k
Pribilof Islands 10 y
Price 15 n
Prieska 48 z
Priluki 35 k
Primavera 82 s
Primorsk 54 h
Primorsko-Akhtarsk 54 d
Prince Albert 15 d
Prince Albert Peninsula 12 f
Prince Albert Road 48 y
Prince Albert Sound 12 f
Prince Albert, Cape 12 b
Prince Charles Island 13 g
Prince Charles Mountains 84 m
Prince Edward Island
(Canada) 17 f
Prince Edward Island
(S. Africa) 235 o
Prince George 14 g
Prince Gustav Adolf Sea 12 i
Prince Harald Coast 84 g
Prince Karls Forland 91 z
Prince of Wales Strait 12 e
Prince of Wales Island (Canada) 12 l
Prince of Wales Island (USA) 11 y
Prince of Wales Island
(Australia) 73 j
Prince Patrick Island 12 d
Prince Regent Inlet 13 b
Prince Rupert 14 c
Princess Astrid Coast 83 l
Princess Charlotte Bay 73 l
Princess Elizabeth Land 85 q
Princess Ragnhild Land 84 c
Princess Ragnhild Coast 84 d
Princeton 16 r
Principe 41 s
Prins Christian Sund 90 t
Prinzapolca 22 h
Pripet 33 y
Pripet marshes 33 x
Pristina 35 d
Prizren 35 e
Probolinggo 69 o
Prochladnyl 54 t
Prof. van Bloomenstein Lake 26 h
Progreso 20 s
Prokopjevsk 56 f
Prome 65 f
Propriá 27 u
Proserpine 74 r
Providence 17 o
Providencia 25 n
Providencia (Island) 22 i
Provideniya 10 k
Provo 15 n
Prudhoe Bay 11 f
Prudhoe Land 91 c
Prut 35 h
Przemyśl 33 v
Przeval'sk 55 z
Pskov 33 l
Pskov, Lake 33 l
Pucallpa 24 y
Pucheng 63 f
Pudukkottai 64 g
Puebla 19 p
Pueblo 15 q
Puelches 30 l
Pu'er 62 t
Puerto Acosta 28 d
Puerto Aisén 30 t
Puerto Alegre 25 u
Puerto Angel 19 q
Puerto Armuelles 22 h
Puerto Asis 24 k
Puerto Ayacucho 23 r
Puerto Barrios 19 v
Puerto Berrio 23 a
Puerto Boyaca 23 n
Puerto Cabello 23 e
Puerto Cabezas 22 h
Puerto Carreño 23 r
Puerto Cisnes 30 t
Puerto Cortés (Honduras) 19 w
Puerto Cortés (Costa Rica) 22 g
Puerto Cumarebo 23 d
Puerto del Rosario 36 c
Puerto Deseado 30 y
Puerto Escondido 19 p
Puerto Heath 25 p
Puerto Huitoto 24 l
Puerto Inca 24 y
Puerto Juárez 19 x
Puerto La Cruz 23 f
Puerto Leguizamo 24 l
Puerto Lempira 22 g
Puerto Limón 23 n
Puertollano 34 q
Puerto Lobos 30 l
Puerto Lopez 23 o
Puerto Madryn 30 l
Puerto Maldonado 25 o
Puerto Marquez 25 r
Puerto Miraña 25 a
Puerto Montt 30 g
Puerto Morelos 19 x
Puerto Natales 30 t
Puerto Nuevo 23 q
Puerto Pardo 24 k
Puerto Peñasco 18 e
Puerto Piñasco 28 l
Puerto Plata 21 t
Puerto Portillo 24 z
Puerto Presidente Stroessner 29 n
Puerto Princesa 67 e
Puerto Rico (Bolivia) 25 q
Puerto Rico (Peru) 24 y
Puerto Rico (USA) 21 v
Puerto San Augustín 25 a
Puerto San Julián 30 w
Puerto Sastre 28 k
Puerto Suárez 28 l
Puerto Vallarta 18 y
Puerto Villamil 22 o
Puerto Villarroel 25 s

Puesta Varadero 24 z
Pugal 60 m
Pujehun 40 r
Pukaki, Lake 79 s
Pukch'ong 58 h
Pula 35 a
Pular 28 d
Pullman 14 k
Pulo Anna 70 d
Pulog 67 g
Pulu 61 e
Pumasillo 24 z
Puná 24 h
Pune 60 z
Punia 42 u
Puno 25 o
Punta Aguja 24 f
Punta Alta 31 a
Punta Arenas 30 u
Punta Chirambirá 22 x
Punta Delgada (Chile) 30 w
Punta Delgada (Argentina) 30 m
Punta Galera 22 v
Punta Manzanillo 22 k
Punta Norte 30 m
Punta Prieta 18 e
Puntarenas 22 f
Punto Fijo 23 c
Puquio 24 y
Pur 52 j
Pur (Siben'an Uvaly) 51 e
Pura 51 h
Puri 61 v
Puricama 23 u
Purnia 61 v
Purros 48 f
Pursat 66 g
Purtuniq 13 u
Puruliya 61 v
Purus 25 h
Purwokerto 68 y
Pusan 58 v
Pushchino 53 y
Pushkino 54 i
Pusur 61 x
Putao 62 e
Putian 63 s
Putorana Mountains 51 j
Putrajaya 68 f
Putre 28 c
Puttalam 64 u
Putumayo 24 l
Puzla 50 v
Pweto 45 c
Pyaozero, Lake 33 m
Pyapon 65 g
Pyasina Bay 51 f
Pyatigory 50 v
Pyinmana (Naypidaw) 62 q
Pymta 53 x
Pyongyang 58 s
Pyramid Lake 14 v
Pyramids 38 h
Pyrenees 34 l
Pyrgos 35 r
Pyu 62 r

Q

Qaanaaq (Thule) 91 c
Qaen 60 c
Qagan 57 k
Qagan Nur 57 v
Qageaka 61 f
Qahan Oulut 57 k
Qaisar 60 g
Qala Panja 60 l
Qal'at Bishah 39 p
Qalat-i Chilzai 60 h
Qamar, Bay 39 w
Qamdo 62 e
Qaminis 35 q
Qânâq (Thule) 91 c
Qapqal 56 p
Qaqortoq 90 t
Qara 38 e
Qarah Tarai 60 h
Qareh Su 39 h
Qarhan 56 y
Qarqan 56 t
Qarqi 56 s
Qaryat al 'Ulya 39 f
Qasr 38 g
Qatar 39 v
Qatrun 37 v
Qattara Depression 38 f
Qazvin 54 x
Qen 57 n
Qena 38 j
Qeqertarsuaq (Godhavn) 90 k
Qeqertarsuaq (Island) 90 l
Qeshm 60 a
Qeshm (Island) 60 a
Qeys 39 k
Qian Gorlos 58 f
Qianjiang 62 m
Qianning 62 g
Qiansuo 62 t
Qianwei 62 j
Qian Xian 62 l
Qianyang 63 a
Qiaowan 56 z
Qiemo 56 s
Qijiang 62 k
Qijiaojing 56 w
Qila Ladgasht 60 f
Qila Saifullah 60 j
Qilian Shan 56 y
Qing'an 58 h
Qingchuan 62 j
Qingdao (Tsingtao) 58 p
Qinghai Hu 57 n
Qingyang 62 l
Qingyuan 58 f
Qing Zang Gaoyuan 62 a
Qinhuangdao 57 y
Qinling Shan 62 l
Qinzhou 62 z
Qionglai 62 i
Qiqian 52 v
Qiqihar (Tsitsihar) 58 f
Qiquanhu 56 u
Qira 61 e
Qitai 56 u
Qitaihe 58 j
Qi Xian 63 c
Qlom 53 a
Qma 61 g
Qojur 54 w
Qom 39 i
Qomsheh 39 i
Qotbabad 60 a
Quala-I Nau 60 f
Quanzhou 63 n
Quarai 28 z
Quchan 55 o
Queanbeyan 78 d
Québec 17 a
Quebec (Province) 13 u
Queen Astrid and King Leopold Coast 85 r
Queen Charlotte Strait 14 e
Queen Charlotte Islands 14 b
Queen Elizabeth Islands 12 d
Queen Louise Land 91 s
Queen Mary Land 85 s
Queen Maud Gulf 12 k
Queen Maud Land 84 c
Queen Maud Range 81 k
Queensland 74 r
Queenstown (Australia) 77 x
Queenstown (New Zealand) 79 r
Queenstown (South Africa) 49 p
Queimada 26 k
Queimadas 27 s
Quélésseboudou 40 i
Quelimane 46 w
Quelpart Island (Cheju) 58 t
Querétaro 19 o
Querrha 34 p
Quesnel 14 h
Quetta 60 i
Quezaltenango 22 b
Quezon 67 g
Quibala 44 t
Quibaxe 44 g
Quibdo 22 y
Quilán, Cabo 30 f
Quillabamba 24 z
Quillacollo 28 f
Quillagua 28 c
Quilon 64 r
Quilpie 77 j
Quimbele 44 h
Quinee Mil 25 n
Quiney 16 p
Quines 28 s
Quinghai Hu 62 g
Quinhon 66 l
Quiñones 18 t
Quirinópolis 29 d
Quissico 49 i
Quitapa 44 v
Quiteve 44 t
Quitexe 44 g
Quito 24 h
Quixadá 27 s
Qujing 62 w
Qumar Heyan 62 b
Qumarlêb 62 d
Qunaitra 38 l
Qungtag 61 j
Quorn 77 f
Quraiyat 60 p
Qus 38 j
Qusair 38 k
Quseima 38 k
Quynh Luu 62 x
Quzhou 63 g

R

Raahe 33 j
Raas Caseyr (Cape Guardafui) 43 l
Raba 72 h
Rabableh 43 j
Rabak 42 l
Rabang 61 e
Rabat 36 h
Rabaul 71 r
Rabbit Lake 12 x
Rabyanah 38 o
Race, Cape 17 l
Rach Gia 66 h
Racine 16 r
Radhanpur 60 x
Radom 33 u
Rae 12 s
Rae Isthmus 13 c
Rae Lakes 12 r
Raeside, Lake 76 h
Rafaela 28 v
Rafaii 42 r
Rafha 39 d
Rafsanjan 39 l
Raga 42 g
Ragusa 35 n
Raha 69 v
Rahad 43 a
Rahad al Berdi 42 f
Rahimyar-Khan 60 l
Raichur 64 f
Rainier, Mount 14 h
Rainy Lake 15 k
Raipur 61 s
Raisut 39 x
Rajada 27 s
Rajahmundry 64 i
Rajang 69 a
Rajanpur 60 k
Rajasthan Canal 60 m
Rajgarh 61 o
Rajkot 60 x
Raj Nandgaon 61 r
Rakaposhi 61 a
Raka Zangbo 61 i
Raleigh 16 x
Rama 22 g
Ramah 13 y
Ramanathapuram 64 t
Ramdurg 64 d
Rameswaram 64 u
Ramhormoz 39 h
Ramlat al Wahiba 60 p
Ramlat Rabyana 37 x
Ramlu 43 e
Ramos 26 i
Rampur 61 d
Rampur (Rann of Kutch) 60 w
Rampura 61 n
Ramree 65 e
Ramsar 54 x
Ramshir 39 h
Ramu 43 r
Ramu (River) 70 z
Rancagua 30 i
Ranchi 61 u
Ranco, Lago de 30 h
Randa 43 f
Randers 33 p
Randolph 15 v
Rangoon 65 g
Rangpur 61 x
Raniganj 61 w
Rankin Inlet 13 n
Ranong 65 i
Rantauprapat 68 e
Rantekombola 69 s
Rantoul 16 r
Rao Go 66 h
Rapak 69 e
Rapel 30 h
Rapid City 15 e
Rapti 61 g
Rapulo 25 r
Raqqah 39 a
Ras al Khaimah 39 l
Ras-al-Ma 36 l
Ras Dashén 43 e
Ras Ghârib 38 j
Ras Halileh 39 i
Rasht 54 x
Ras Koh 60 h
Raskoh 60 g
Ras Nuh 60 s
Rason, Lake 76 j
Rassokha 53 h
Ras Tannurah 39 i
Ratangarh 61 a
Rat Islands 10 t
Ratlam 61 n
Ratnagiri 64 c
Raton 15 q
Ratta 51 t
Ratz, Mount 11 y
Raub 65 x
Raub Kuala Kubu 66 s
Rauch 31 b
Ravansar 39 e
Ravenna 34 m
Ravensthorpe 76 g
Ravor 60 b
Rawalpindi 60 m
Rawlinna 76 k
Rawlins 15 p
Rawson 30 l
Raya 69 a
Raychikhinsk 58 h
Rayones 19 a
Rayong 65 k
Razan 39 h
Razdolnoye 53 i
Razmak 60 k
Reading (USA) 16 y
Reading (UK) 32 x
Read Island 12 g
Reboly 33 m
Rebricha 56 d
Recherche Archipelago 76 i
Recife 27 w
Reconquista 28 w
Recovery Glacier 83 v
Recreio 25 w
Recreo 28 s
Red Bluff 14 u
Red Deer 14 m
Redding 14 u
Redenço 26 y
Redeyef 37 d
Red Lake 15 k
Red Pass 14 j
Red River (Canada) 15 i
Red River (USA) 20 d
Red Sea 38 y
Red Wing 15 k
Regensburg 34 m
Reggane 36 y
Reggio 35 o
Regina 15 d
Régina 26 j
Registan 60 g
Registro 29 s
Rehoboth 48 i
Reims 34 i
Reindeer Lake 12 x
Reitz 49 q
Reivilo 49 n
Reliance 12 v
Relizane 34 t
Remada 34 y
Remanso 27 q
Remarkable, Mount 73 a
Remeshk 60 c
Renascença 25 d
Rengat 68 g
Renk 42 l
Renland 91 q
Renmark 77 h
Rennell Island 74 m
Renner Springs 73 r
Rennes 34 e
Rennick Glacier 86 u
Reno 14 v
Repetek 55 r
Republican 15 u
Repulse Bay 13 c
Requena 24 l
Resistencia 28 x
Resolute 12 m
Resolution Island 13 y
Retkucha 10 d
Retreal 77 j
Réunion 47 w
Revelstoke 14 j
Revilla Gigedo Islands 18 t
Rewa 61 r
Rewari 61 b
Rey 22 k
Reyes 25 q
Reykjanes 32 a
Reykjavik 32 b
Reynosa 19 b
Rhallamane 36 s
Rhine 34 k
Rhinelander 16 e
Rhode Island 17 o
Rhodes 35 v
Rhodes (Island) 35 v
Rhodope 35 g
Rhône 34 i
Riau Islands 68 h
Ribas do Rio Pardo 29 a
Ribat 60 e
Ribauè 46 k
Ribeiro Preto 29 l
Riberalta 25 r
Riboiro Gonçalves 27 o
Rice Lake 15 l
Rich 36 j
Richards Bay 49 t
Richardson 11 g
Richfield 14 z
Richland 14 i
Richmond (Australia) 74 n
Richmond (Indiana) 16 t
Richmond (Virginia) 16 x
Richmond Range 79 t
Richwood 16 v
Riddle 14 x
Ridgecrest 14 w
Rif 36 l
Rifle 15 p
Rift Valley 42/45
Riga (Latvia) 33 w
Riga (Russia) 52 p
Riga, Gulf of 33 i
Riggins 14 k
Rigolet 17 i
Rig-Rig 41 k
Rihand Reservoir 61 s
Riiser Larsen Peninsula 84 h
Riiser Larsen Sea 84 d
Riiser Larsen Ice Shelf 83 g
Rijeka 35 a
Rime 42 b
Rimini 34 m
Rimouski 17 c
Rinčin Lchumbe 57 a
Ringwood 73 s
Riobamba 24 h
Rio Branco 25 p
Rio Claro 29 l
Río Colorado 30 m
Río Cuarto 30 m
Rio de Janeiro 29 i
Rio de Oro 36 n
Río Gallegos 30 w

Rio Grande (Argentina) 30 x
Rio Grande (Brazil) 29 p
Rio Grande (Mexico) 18 z
Riohacha 23 a
Rioja 24 j
Rio Largo 27 v
Río Mayo 30 v
Rio Mulatos 28 e
Riosucio 22 l
Río Tercero 28 t
Río Turbio 30 u
Río Verde (Mexico) 19 n
Rio Verde (Brazil) 29 c
Rio Verde de Mato Grosso 29 a
Riposto 37 j
Rissani 36 i
Rito 44 w
Ritscher Highland 83 i
Rivadavia 30 m
Rivas 22 l
Rivera 28 z
River Falls 15 k
Riversleigh 73 u
Riverton (Canada) 15 i
Riverton (USA) 15 o
Rivière du Loup 17 b
Rivungo 44 y
Riwoqê 62 d
Riyadh 39 s
Riyan 43 k
Rjazan 54 e
Roanoke 16 w
Roanoke (River) 16 y
Roanoke Rapids 16 x
Roatán 20 v
Robe 77 h
Roberval 17 a
Robeson Channel 91 g
Robinhood 73 y
Robinson, Cape 82 v
Robinson Crusoe 30 d
Robinson River
(Papua New Guinea) 74 e
Robinson River (Australia) 73 g
Roboré 28 k
Robson, Mount 14 j
Roçadas 44 s
Roca, General 30 k
Rocas, Atol das 27 w
Rocha 31 e
Rochelle 16 r
Rochester (Maine) 17 o
Rochester (Minnesota) 16 p
Rochester (New York) 16 x
Rockefeller Plateau 81 g
Rockford 16 r
Rockhampton 74 t
Rock Hill 20 l
Rockingham 20 m
Rock Springs 15 o
Rockstone 23 x
Rockwood 16 t
Rocky Harbour 17 i
Rocky Mount 16 x
Rocky Mountains 12/14
Rocky Mountain House 14 g
Roda Velha 27 o
Rodeo 18 l
Rodina 51 x
Rodino 56 c
Rodrigues 24 m
Roebuck Plains 72 x
Roes Welcome Sound 13 p
Rogachevka 54 e
Rogaguado, Lago 25 r
Rogagua, Lago 25 r
Rogers, Mount 16 v
Rohtak 61 b
Roi Et 66 g
Rokel 40 e
Rolla 16 p
Rolleston 74 q
Roma 74 r
Roma (Rome) 34 m
Romaine 17 f
Romania 33/35
Romano, Cape 20 l
Romanova 57 b
Romanovka 57 h
Rome 34 m
Rommani 36 h
Roncador, Serra do 26 w
Ronda (Pakistan) 61 a
Ronda (Spain) 36 i
Rondônia 25 u
Rondonôpolis 29 a
Rong'an 62 z
Rongjiang 62 z
Ronne Entrance 82 y
Roodepoort 49 p
Roosevelt Island 81 z
Roper 73 f
Roper Bar 73 e
Roques, Islas los 23 f
Roraima 26 d
Roraima, Mount 23 w
Rosa, Monte 34 k
Rosário 27 c
Rosario 28 f
Rosario (Paraná) 28 w
Rosario de Arriba 18 d
Rosario de la Frontera 28 t
Rosário do Sul 29 n
Rosario Oeste 25 y
Rosarito 18 c
Rosarito
(Sabastían Vizcaino Bay) 18 e
Rosa Zárate 24 h
Roscrea 32 u
Roseau (Dominica) 23 j
Roseburg 14 t
Roseires, Lake 42 m
Rose River 73 l
Rosetown 15 b
Rosetta 38 h
Rosiori 35 g
Roslavl 54 a
Ross Barrier 86 o
Ross Ice Shelf 81 l
Ross Island 86 r
Rosso 40 b
Ross River 11 m
Ross Sea 81 v
Rossville 73 m
Rostock 33 q
Rostov 54 e
Roswell 18 l
Rothera 82 v
Roti 69 v
Rotorua 79 w
Rotterdam 33 n
Rouen 34 g
Round 78 l
Roundup 15 b
Rourkela 61 v
Rovno 35 h
Rovnoye 54 i
Rowley 90 k
Rowley Island 13 g
Roxas 67 h
Rosborough Downs 73 u
Roydale Isle 16 e
Rtishchevo 54 h
Ruacana 48 g
Ruahine Range 79 v
Ruapehu 79 v
Rubačje 56 n
Rub'al Khali 39 t
Rubcovsk 56 d
Ruby 11 d
Rudbar 60 f
Rudnaya Pristan 58 m
Rudnya 54 g
Rudolf 50 k
Rufiji 45 j
Rufino 31 a
Rufunsa 46 e
Rugby 15 g
Rügen 33 r
Ruhengeri 45 d
Rui Barbosa 27 r
Ruijin 63 r
Ruki 44 i
Rukwa, Lake 45 l
Rumbek 42 w
Rum Cay 21 q
Rumford 17 b
Rumphi 45 t
Rundu 48 k
Rungu 42 v
Rungwa 45 g
Rungwa (River) 45 f
Rungwe 45 t
Ruoqiang 56 u
Rupat 68 f
Rupert 16 l
Rupisi 49 f
Rupununi 26 e
Rurópolis 26 u
Rusape 46 t
Ruse 35 h
Russas 27 t
Russell Range 76 i
Russia 10/33/50-59
Russkaya 81 s
Russkaya (River) 51 f
Russkaya Gavan 50 l
Russkiy 51 l
Rustavi 54 u
Rustenburg 49 p
Ruteng 69 t
Rutenga 46 s
Rutland 17 n
Rutshuru 42 v
Ruvuma 46 j
Ruzayevka 55 g
Rwanda 42/45
Rybinsk 50 p
Rybinsker Reservoir 50 p
Rycroft 12 r
Ryūkyu Islands 63 x
Rzeszów 35 e
Ržev 54 b

S

Saadatabad 39 k
Saarbrücken 34 j
Saaremaa 33 h
Sab'Abar 38 m
Sabah 69 d
Sabana de Cardona 23 s
Sabaya 28 d
Sabha 37 i
Sabie 49 e
Sabinas 19 a
Sabinas Hidalgo 19 a
Sabkhat Matti 39 w
Sable, Cape (Canada) 17 r
Sable, Cape (USA) 20 l
Sabrina Coasl 87 e
Sabun 51 s
Sabyndy 55 i
Sabzevar 55 n
Sabzevaran 60 c
Šachrisabz 55 t
Sachs Harbour 12 c
Saco Uen 43 s
Sacramento 14 u
Sacramento (River) 14 u
Sacramento, C. del 31 c
Sa'dah 43 g
Sado 59 n
Safaniyah 39 g
Safar 60 g
Safford 18 h
Safi 36 l
Saga (China) 61 h
Saga (USSR) 55 e
Šagan 56 b
Sagar 61 q
Sagastyr 52 l
Saginaw 16 u
Sagiz 54 m
Šagonar 56 k
Sagres 36 f
Sagua la Grande 20 z
Sagyr 52 a
Sahara 36/37
Saharan Atlas 36/37
Saharanpur 61 c
Sahel 40 i
Sahil al Jazir 60 o
Sahiwal 60 m
Sahuaripa 18 i
Sahuayo 18 z
Sai Buri 65 x
Saigon (Ho Chi Minh) 66 j
Saihan Toroi 57 n
Saikhoa Ght 62 d
Saimaa, Lake 33 k
Ste-Agathe-des-Monts 16 m
St. Alban's 17 j
Sto. André 29 g
Ste-Anne-des-Monts 17 d
St. Anthony 17 k
St.-Augustin 17 i
St.-Augustin (River) 17 i
St. Augustine 20 m
St. Brieuc 34 d
St. Catharines 16 w
St. Christopher and Nevis 21 x
St. Cloud 15 j
St. Croix 21 w
Saint-Denis 47 w
St. Eleanors 17 f
St. Elias Mountains 11 j
St. Étienne 34 h
Ste-Foy 17 a
St. George (Australia) 74 q
St. George (USA) 14 z
Saint George's 23 i
St.-Georges (Canada) 17 b
St. Georges (French Guiana) 26 j
St. George's Channel 32 v
St. Helena 234 z
St.-Hyacinthe 17 a
St. Jean, Lac 17 a
St.-Jérôme 16 m
Saint John 17 d
St. John's
(Antigua and Barbuda) 21 z
St. John's (Canada) 17 l
St. Joseph 15 w
St. Joseph, Lake 15 l
St. Kitts/Nevis 21 x
St. Laurent 26 i
St. Lawrence 74 r
St. Lawrence, Gulf of 17 g
Saint Lawrence Island 10 l
St. Lawrence (River) 17 c
St. Louis (Senegal) 40 b
St. Louis (USA) 16 q
Saint Lucia 23 j
St. Lucia, Lake 49 t
St. Lucia Passage 23 j
St. Malo, Gulf of 32 x
St.-Marc 21 r
Ste.-Marguerite 17 d
Sta. Maria de Nieva 24 i
St. Martin 21 y
St. Marys 78 q
Saint Matthew 10 k
St. Matthew's 66 c
St. Nazaire 34 d
St. Paul (Liberia) 40 t
St. Paul (River) 17 i
St. Paul Island 235 r
St. Paul-Minneapolis 16 b
St. Petersburg (Russia) 33 l
St. Petersburg (USA) 20 k
St. Pierre and Miquelon 17 j
St. Quentin 34 h
St. Roch Basin 12 m
St.-Siméon 17 b
St. Stephen 17 c
St. Thomas 16 v
Saint Vincent and
Grenadines 23 j
St. Vincent, Cape 34 n
St. Vincent Gulf 77 f
St. Vincent Passage 23 j
Saintes 34 f
Sajak 56 b
Sajama 25 p
Sajanogorsk 56 j
Sajnsand 57 s
Sak 48 y
Saka 43 q
Sakai 58 z
Sakakah 39 b
Sakakawea Reservoir 15 f
Sakami 16 m
Sakata 59 o
Sakhalin 53/59
Sakhandzha 52 l
Sakhon Nakhon 66 h
Sakir 60 i
Salaberry-De-Valleyfield 16 m
Salada Laguna 18 d
Saladillo 31 a
Salado 30 k
Salado (Bahia Samboronbón) 31 c
Salado (Chaco Austral) 28 u
Salaga 40 m
Salajar 69 t
Salal 42 a
Salalah 39 x
Salamanca (Chile) 28 o
Salamanca (Mexico) 19 n
Salamanca (Spain) 34 c
Salamat 42 b
Salar de Antofalle 28 q
Salar de Arizaro 28 r
Salar de Atacama 28 d
Salar de Pipanaco 28 s
Salar de Pocitos 28 r
Salar de Uyuni 28 d
Salar Punta Negra 28 p
Saldanha 48 v
Sale 77 y
Salé 36 h
Salekard 50 m
Salem (India) 64 f
Salem (USA) 14 g
Salerno 35 a
Salgueiro 27 t
Sali 36 z
Salima 45 u
Salina (Kansas) 15 u
Salina (Utah) 14 z
Salina Cruz 20 p
Salinas (Brazil) 25 w
Salinas (Ecuador) 24 g
Salinas (Mexico) 19 n
Salinas (USA) 14 u
Salinas Grande 28 t
Salinas Grandes 28 f
Salinópolis 27 a
Salisbury 16 z
Salisbury Island (Canada) 13 t
Salisbury Island (Russia) 50 i
Saljany 54 w
Salluit 13 u
Sallûm 35 t
Sal'um 50 k
Salmon 14 l
Salmon (River) 14 l
Salmon River Mountains 14 l
Salsacate 28 s
Sal'sk 54 f
Salta 28 t
Saltillo 19 a
Salt Lake City 14 z
Salt Lakes 76 g
Salto 28 y
Salton Sea 18 d
Salvador 27 t
Salvatierra 19 n
Salwa 39 v
Salween 65 h
Salzburg 35 a
Samagaltaj 56 l
Samar 67 j
Samara 54 k
Samarga 59 a
Samarinda 69 d
Samarkand 55 t
Samatiguila 40 i
Samaumá 25 j
Samba 45 a
Samba (Maringa) 44 l
Samba Caju 44 g
Sambalpur 61 u
Sambas 68 l
Sambava 47 f
Sambhar 61 a
Samborombón Bay 31 d
Same 45 j
Samfya 45 q
Samirah 39 c
Samoa 235 z
Sámos 35 u
Sampit 69 o
Samrong 66 g
Samsang 61 g
Samsun 35 m
Samui 65 j
San'a 43 g
San 40 j
Sanae 83 i
Sanaga 41 v
San Agustín de Valle Fértil 28 r
Sanandaj 39 f
San Andrés 22 i
San Andrés Tuxtla 19 r
San Angelo 19 b
San Antonio (Chile) 30 h
San Antonio (USA) 19 c

San Antonio de los Cobres 28 r
San Antonio Oeste 30 l
Sanaw 39 v
San Benedicto 18 u
San Bernardino 18 c
San Bernardo 30 i
San Blas 18 i
San Blas, Cape 20 i
San Borja 25 q
San Buenaventura 19 a
San Carlos (Colombia) 23 r
San Carlos (Luxon, Philippines) 67 f
San Carlos (Mexico) 18 g
San Carlos (Negros, Philippines) 67 i
San Carlos (Nicaragua) 22 g
San Carlos de Bolivar 31 a
San Carlos de Bariloche 30 i
San Carlos del Zulia 23 b
Sanch 57 b
Sanchakou 56 o
San Clemente 18 b
San Cristóbal 75 a
San Cristóbal (Argentina) 28 v
San Cristóbal (Columbia) 24 m
San Cristóbal (Ecuador) 22 q
San Cristóbal (Venezuela) 23 b
San Cristóbal las Casas 19 t
Sancti Spiritus 21 n
Sandakan 69 e
Sanderson 18 m
Sandia 25 o
San Diego 18 c
San Diego, Cabo 30 y
Sandoa 44 y
Sandoway 62 p
Sandpoint 14 k
Sandstone 76 f
Sandusky 16 u
Sandy Bay 76 i
Sandy Cape 77 w
Sandy Lake 15 k
San Estanislao 28 z
San Felipe (Chile) 28 o
San Felipe (Lower California, Mexico) 18 e
San Felipe (Mexico) 19 n
San Felipe (Venezuela) 23 d
San Fernando (Chile) 30 i
San Fernando (Philippines) 67 g
San Fernando (Venezuela) 23 j
San Fernando de Atabapo 23 r
San Fernando de Apure 23 e
San Francisco (Argentina) 28 u
San Francisco (USA) 14 t
San Francisco de Macoris 21 t
San Francisco de Rincón 18 z
San Francisco del Oro 18 k
San Gabriel 22 y
Sangan 60 g
Sangareddi 61 p
Sange 45 c
Sanggau 68 m
Sangha 41 z
Sanghar 60 w
Sangihe 69 k
Sangihe Islands 69 k
Sangijn Dalaj Nuur 56 m
Sangli 61 n
Sangmelima 41 w
Sangre de Cristo Range 15 q
Sangruma 62 f
Sangyyakhtakh 52 x
San Guillermo 18 k
San Ignacio (Apere, Bolivia) 25 r
San Ignacio (Argentina) 28 y
San Ignacio (Bolivia) 25 v
San Ignacio (Peru) 24 i
San Ignacio, Laguna 18 f
San Ildefonso Peninsula 67 h
San Javier 25 u
San Joaquin 25 s
San Jorge 22 m
San Jorge, Gulf of 30 x
San Jose (USA) 14 u
San José (Mexico) 18 h
San José (Costa Rica) 22 g
San Jose (Philippines) 67 g
San José (Venezuela) 23 r
San José de Amacuro 23 j
San Jose de Buenavista 67 g
San José de Chiquitos 25 v
San José de Jáchal 28 q
San José del Guaviare 23 o
San Jose de Mayo 31 d
San José de Ocune 23 p
San Juan (Argentina) 28 q
San Juán (Chile) 24 x
San Juan (Dominican Republic) 21 s
San Juan (Puerto Rico) 21 w
San Juan (Venezuela) 23 s
San Juan (River, Columbia) 22 y
San Juan (River, USA) 15 n
San Juan (River, Argentina) 28 r
San Juan (River, Costa Rica) 22 g
San Juan Bautista 28 z
San Juan del Norte 22 g
San Juan de los Lagos 18 z
San Juan do los Morros 23 e
San Juan Mountains 15 p
San Juanico 18 f
San Julián 30 w
San Justo 28 w
Sankuru 44 m
San Lázaro, Cape 18 f
San Lorenzo (Ecuador) 24 h
San Lorenzo (USA) 18 j
San Lorenzo (Mount) 30 u
San Lucas 18 u
San Lucas, Cape 18 u
San Luis 28 s
San Luis de la Paz 19 o
San Luis de Tapajós 25 l
San Luis, Lago de 25 s
San Luis Obispo 14 v
San Luis Potosi 19 n
San Luis-Rio Colorado 18 e
San Malo, Golfe de 34 e
San Marcos 19 o
San Marino 34 m
San Martin (Argentina) 28 q
San Martin (Bolivia) 25 t
San Martín de los Andes 30 i
San Martin (River) 25/28
San Matías 25 w
San Matias Gulf 30 m
Sanmenhsia 63 a
San Miguel (Bolivia) 25 t
San Miguel (El Salvador) 22 d
San Miguel del Monte 31 c
San Miguel de Tucumán 28 t
Sanming 63 s
San Nicolas (Argentina) 31 a
San Nicolas (USA) 18 a
Sanniquellie 40 u
Sanouyah 40 f
San Pablo (Bolivia) 28 e
San Pablo (River, Bolivia) 25 u
San Pablo (Philippines) 67 g
San Pédro 40 v
San Pedro (Argentina) 28 t
San Pedro (Paraguay) 28 y
San Pedro (Paraná, Argentina) 28 w
San Pedro de Atacama 28 d
San Pedro de las Colonias 18 m
San Pedro de Lloc 24 h
San Pedro Sula 19 w
San Rafael (Argentina) 30 j
San Rafael (Venezuela) 23 b
San Ramón de la Nueva Orán 28 g
San Salvador 22 d
San Salvador (Island, Bahamas) 21 q
San Salvador (Island, Ecuador) 22 p
San Salvador de Jujuy 28 s
Sansanné-Mango 41 a
San Sebastián 34 e
Santa 24 v
Santa Ana (Beni, Bolivia) 28 e
Santa Ana (Bolivia) 25 r
Santa Ana (Serra de Santiago, Bolivia) 28 k
Santa Ana (El Salvador) 22 c
Santa Ana (Mexico) 18 h
Santa Ana (USA) 18 c
Santa Barbara (Mexico) 18 k
Santa Barbara (USA) 14 v
Santa Barbara (Venezuela) 23 s
Santa Catalina (Chile) 28 p
Santa Catalina (Island, USA) 18 c
Santa Catalina (Island, Mexico) 18 h
Santa Clara (Colombia) 25 b
Santa Clara (Cuba) 20 z
Santa Clotilde 24 l
Santa Corazón 25 w
Santa Cruz (Bolivia) 28 h
Santa Cruz (Brazil) 24 m
Santa Cruz (Chile) 28 o
Santa Cruz (Peru) 24 h
Santa Cruz (Spain) 36 b
Santa Cruz (USA) 14 u
Santa Cruz (Island, California) 18 a
Santa Cruz (Island, Eduador) 22 p
Santa Cruz (River, Argentina) 30 v
Santa Cruz Island 75 d
Santa Elena, Cabo 22 f
Santa Elena de Uairen 23 v
Santa Fé (Argentina) 28 w
Santa Fe (USA) 15 q
Santa Filomena 27 o
Santa Helena 25 l
Santai 62 k
Santa Inés (Brazil) 27 b
Santa Inés (Chile) 30 t
Santa Isabel (Argentina) 30 k
Santa Isabél (Brazil) 25 v
Santa Isabel (Ecuador) 24 h
Santa Isabel (Peru) 24 l
Santa Isabel (Solomon Islands) 71 x
Santa Lucia 24 m
Santa Maria (Amazon, Brazil) 25 j
Santa María (Argentina) 28 r
Santa Maria (Brazil) 29 o
Santa Maria (Ecuador) 22 o
Santa Maria (Mexico) 18 k
Santa Maria (Peru) 24 l
Santa Maria (Venezuela) 23 e
Santa Maria da Vitória 27 p
Santa Maria di Leuca, Capo 35 p
Santa Marta 23 a
Santana 27 p
Santana do Araguaia 26 x
Santana do Ipanema 27 u
Santana do Livramento 29 n
Santander (Colombia) 22 y
Santander (Spain) 34 d
Santanghu 56 x
Santarém (Brazil) 25 m
Santarém (Portugal) 34 o
Santa Rosa (Argentina) 30 l
Santa Rosa (Brazil) 25 n
Santa Rosa (California) 14 t
Santa Rosa (Colombia) 23 p
Santa Rosa (Honduras) 22 d
Santa Rosa (New Mexico) 15 q
Santa Rosa (Island) 18 a
Santa Rosa del Conlara 28 s
Santa Rosalia 18 g
Santa Teresa 73 r
Santa Teresinha 26 x
Santiago (Argentina) 29 n
Santiago (Chile) 28 o
Santiago (Dominican Republic) 21 t
Santiago (Panama) 22 l
Santiago (River) (Mexico) 18 y
Santiago (River) (Peru) 24 i
Santiago de Chuco 24 v
Santiago de Compostela 34 b
Santiago de Cuba 21 p
Santiago del Estero 28 t
Santiago Ixcuintla 18 x
Santiago Papasquiaro 18 k
Santiago, Serra de 28 k
San Tiburcio 19 n
Santo Amaro 27 t
Santa Angelo 29 n
Santo Antônio 25 t
Santo Antônio (Negro) 25 i
Santo Antônio de Içá 25 d
Santo Antônio de Jesus 27 s
Santo Corazón 28 k
Santo Domingo 21 t
Santos 29 g
Santos Dumont 25 c
Santo Tomás 24 z
Santo Tomé 28 z
San Valentin 30 t
San Vincente 18 c
San Vicente de Cañete 24 w
San Yanaro 23 q
Sanyati 45 q
Sanyuan 62 m
Sanza Pombo 44 h
São Borja 28 z
São Carlos 29 e
São Félix 26 z
São Félix (Illha do Bananal) 26 x
São Félix do Xingu 26 x
São Francisco (Island) 29 s
São Francisco (River) 29 i
São Francisco, Serra 27 r
São Gabriel 29 n
São Hill 45 h
São Joo de Araguaia 26 z
São Joo do Piauí 27 q
São José 29 r
São José (Negro) 25 d
São José do Anauá 25 h
São José do Egito 27 u
São José do Rio Preto 29 d
São José do Ribamar 27 c
São José dos Campos 29 g
São José do Xingu 26 w
Saolat 69 m
São Lorenço 25 z
São Luis 27 b
São Luis de Montes Belos 29 c
São Marcos Bay 27 c
São Mateus 29 l
São Miguel de Gyamá 27 a
São Miguel do Araguaia 26 x
São Miguel dos Campos 27 v
São Paulo 29 g
São Paulo de Olivença 25 e
São Pedro do Piauí 27 q
São Raimundo das Mangabeiras 27 n
São Raimundo Nonato 27 q
São Romo 29 g
São Roque, Cabo 27 w
São Sebastião 26 v
São Sebastio (São Paulo) 29 h
São Tomé 41 s
São Tomé (Island) 41 r
São Tomé and Principe 41/44
Sapele 41 r
Saposoa 24 j
Sapporo 59 c
Sapri 35 o
Sapt Kosi 61 i
Saqqez 54 v
Saquenay 17 b
Sarajevo 35 c
Sarakhs 55 q
Saraktash 55 a
Saran 55 j
Sarangani Islands 67 w
Sarangarh 61 s
Saranpaul 50 y
Saransk 54 h
Sarapul 54 m
Sarapulskoye 58 l
Sarar 39 g
Sarasota 20 k
Saratov 54 i
Saravan 60 e
Sarawak 69 b
Saraya 40 e
Saraysu 55 g
Sarbaz 60 e
Sarbisheh 60 d
Sarbulak 56 h
Sardinia 34 l
Sarek 33 g
Sarempaka 69 d
Sargasso Sea 21 d
Sargento Lores 24 i
Sargodha 60 l
Sarh 42 b
Sari Bulak 56 n
Sari-i Pul 60 h
Sarikei 68 m
Sarina 74 r
Sarkand 56 p
Šarlauk 55 n
Sarmi 70 v
Sarmiento 30 w
Sarmiento Peninsula 30 w
Särna 33 e
Sarnia 16 v
Sarolangun 68 t
Sarychik 54 l
Saryg-Sop 56 l
Sarykamysskoje, Lake 55 n
Saryozek 55 z
Sásabe 18 g
Sasaram 61 t
Sasebo 58 v
Saser 61 c
Saskatchewan (Province) 12/15
Saskatchewan (River) 15 e
Saskatoon 15 c
Saskylakh 52 f
Sassandra 40 w
Sassandra (River) 40 v
Sassari 34 k
Sasykkol', Lake 56 c
Satadougou 40 f
Satara (India) 64 d
Satara (South Africa) 49 f
Satawal 71 b
Satawan Atoll 71 g
Satipo 24 y
Šatlyk 55 o
Sātpura Range 61 n
Satu Mare 33 v
Satun 65 w
Saucillo 18 l
Saudi Arabia 38-39
Saul 26 i
Sault Ste. Marie 16 g
Sauqira, Bay 39 z
Sauré 25 k
Saurimo 44 x
Sava 35 c
Savalou 41 o
Savannah Downs 73 x
Savannah (Georgia) 20 m
Savannah (Tennessee) 20 g
Savannakhet 66 h
Savanur 64 d
Savate 44 u
Saveh 39 h
Savé (Benin) 41 o
Save (Mozambique) 49 h
Sawla 40 l
Sawu 69 u
Sawu Sea 69 u
Say 41 b
Sayaboury 62 u
Sayan Mountains 56 i
Saydy 53 b
Sayhut 43 l
Sayram Hu 56 q
Sayula 18 y
Sayun 43 k
Say-Utes 54 z
Sayylyk 53 b
Sbaa 36 m
Scarborough 23 j
Šcerbakty 56 b
Schefferville 17 c
Schenectady 16 z
Schizuoka 59 n
Schmidta 51 i
Schreiber 16 f
Sehuangliao 58 e
Schwabenland 83 k
Schwangyashan 58 j
Schweizer-Reneke 49 o
Schwerin 33 q
Sciacca 34 z
Scoresby Land 91 q
Scoresby Sound 91 q
Scorpion Bay 76 m
Scotia Sea 82 d
Scotland 32 w
Scott Base 86 r
Scottsbluff 15 r
Scranton 16 y
Sea of Crete 38 e
Seabra 27 q
Searchlight 14 y
Seattle 14 h
Sebastian Vizcaino Bay 18 e
Sebuku Bay 69 e
Sedalia 16 o
Sedan 34 i
Seddon, Kap 91 b
Sededema 53 i
Sedova 50 i
Sefid 54 x
Sefrou 36 i

Segamat 65 y
Segera 45 k
Ségou 40 j
Segovia 34 d
Seguedine 37 v
Séguéla 40 i
Segura 34 r
Sehwan 60 i
Seine 34 g
Seke Banza 44 f
Sekoma 49 a
Sekondi-Takoradi 40 z
Selaru 70 p
Selatan, Cape 69 p
Selby 15 g
Selebi-Phikwe 49 d
Selenga 57 e
Selety-Tengiz Lake 55 j
Sélibabi 40 e
Šelichov 57 c
Selma 20 h
Semarang 68 z
Sembé 41 x
Semeru 69 n
Semipalatinsk 56 c
Seman 39 k
Semnu 37 j
Semonaicha 56 d
Sem-Tripa 26 v
Sena 45 u
Senador Pompeu 27 s
Sena Madureira 25 o
Senanga 44 z
Sencha 52 l
Sendai 59 p
Seneca, Lake 16 x
Senegal 40
Senegal (River) 40 c
Senekal 49 p
Seney 16 f
Senhor do Bonfim 27 s
Senja 33 f
Senjavin Group 71 k
Sennar 42 m
Senneterre 16 k
Seoni 61 q
Seoul 58 u
Sepik 70 y
Sepone 66 i
Sepopa 48 l
Sept-Iles 17 d
Serbia 35 d
Serebr'ansk 56 d
Sere Kunda 40 a
Seremban 68 f
Serengeti 45 h
Serenje 45 q
Sergach 54 h
Sergino 50 z
Sergiyev Posad 50 p
Sermata 70 o
Serowe 49 b
Serpent's Mouth 23 i
Serpuchovo 54 d
Sérrai 35 f
Serra Acarai 25 l
Serra Aguapei 25 w
Serra Curupira 25 f
Serra da Batista 27 r
Serra da Canastra 29 g
Serra da Capivara 29 g
Serra da Ibiapaba 27 r
Serra da Mantiqueira 29 h
Serra das Alpercatas 27 p
Serra das Araras 29 o
Serra das Divisoes 29 c
Serra da Tabatinga 27 p
Serra de Huanchaca 25 v
Serra de Maracaju 29 a
Serra de Pianí 27 q
Serra de Santiago 28 k
Serra do Apiaú 23 v
Serra do Cachimbo 25 z
Serra do Dois Irmos 27 r
Serra do Espinnaco 29 i
Serra do Itapicuru 27 o
Serra do Mar 29 r
Serra do Navio 26 j
Serra do Norte 25 w
Serra do Penitente 27 n
Serra do Roncador 26 w
Serra dos Apiacás 25 x
Serra dos Caiabis 25 y
Serra dos Carajás 26 x
Serra dos Pacáas Novos 25 s
Serra dos Parceis 25 u
Serra do Tiracambu 27 a
Serra do Tombador 25 x
Serra Dourada 26 z
Serra do Urucui 27 o
Serra do Valentim 27 p
Serra Formosa 25 z
Serra Geral de Goiás 27 n
Serra Geral do Parana 29 f
Serra Gradaús 26 x
Serra Lombarda 26 j
Serra Pacaraima 23 u
Serra Paranapiacaba 29 r
Serra Parima 23 u
Serra So Francisco 27 r
Serra Talhada 27 t
Serra Tapirapecó 23 t
Serra Tumucumaque 26 i
Serrania de Imataca 23 i
Serrinha 27 t
Sêrtar 62 g
Serule 49 c
Sêrxu 62 e
Sesayap 69 d
Sese Islands 42 y
Sesheke 44 z
Sesriem 48 h
Sessa 44 x
Sete Lagoas 29 i
Setif 34 v
Settat 36 h
Setúbal 34 n
Seul, Lac 16 c
Sevan, Lake 54 u
Sevastopol 35 k
Severn 16 e
Severnaja 51 i
Severnaya Zemlya 51 j
Severnoye 54 l
Severodvinsk 50 q
Severomuysk 52 u
Sevi 56 k
Sevilla 34 p
Seville (Sevilla) 34 p
Sewa 40 f
Seward 11 f
Seward Peninsula 10 m
Seychelles 235 p
Seym 54 b
Seymchan 53 j
Seymour 77 x
Sfax 34 y
Shache 55 z
Shackleton Ice Shelf 85 t
Shackleton Range 83 u
Shadrinsk 55 e
Shahdol 61 r
Shahgarh 60 k
Shahjahanpur 61 e
Shahjui 60 i
Shahr-e-Kord 39 i
Shajapur 61 o
Shakamku 48 k
Shakawe 48 l
Shakhty (Russia) 54 e
Shakhty (Kazakstan) 55 b
Shala, Lake 43 b
Shaliangzi 56 x
Shamattawa 13 o
Shamrock 15 t
Shandan 57 o
Shangdu 57 v
Shanghai 63 i
Shangjao 63 f
Shangkiu 63 d
Shangnan 63 a
Shangshui 63 d
Shang Xian 62 m
Shangyi 57 v
Shangzhi 58 h
Shannon (Island) 91 t
Shannon (River) 32 u
Shanshan 56 v
Shantar 53 p
Shantung 58 p
Shaohing 63 h
Shaokwan 63 p
Shaowu 63 f
Shaoyang 63 b
Shaqra 39 r
Sharbithat 39 y
Sharbithat, Ras 60 o
Sharjah 39 y
Shark Bay 76 b
Sharlyk 54 m
Sharqat 39 d
Shashe 46 q
Shashemene 43 p
Shashi 49 d
Shasi 63 b
Shasta, Mount 14 u
Shatsk 54 f
Shaunavon 15 b
Shawinigan 17 a
Shawnee 15 v
Shay Gap 72 v
Shazaoyuan 56 x
Shcherbakovo 53 m
Shehuch'ye 51 a
Sheberghan 60 h
Sheboygan 16 r
Sheffield 32 x
Shelayevo 56 m
Shelburne 17 r
Shelby 14 m
Shelekhova, Gulf of 53 x
Shenandoah 16 n
Shenber 55 f
Shendi 38 w
Shenmu 57 t
Sheno 43 d
Shepparton 77 x
Sherbro Island 40 q
Sherbrooke 17 a
Sherda 37 x
Sheridan 15 c
Sherman 19 d
Sherman Basin 12 m
Sherman Island 80 m
Shestakova 53 g
Shetland Islands 32 k
Shetpe 54 y
Shevli 53 n
Sheya 52 h
Shibeli 43 e
Shibing 62 l
Shiguaigou 57 t
Shihezi 56 s
Shihkiachwang 57 u
Shihtsuishan 57 q
Shikar Pur 60 i
Shikoku 58 x
Shilabo 43 t
Shiliguri 61 k
Shilkan 53 t
Shillong 62 n
Shimanovsk 52 y
Shimonoseki 58 v
Shinas 39 z
Shindand 60 e
Shingwedzi 49 f
Shintuya 25 n
Shinyanga 45 g
Shiping 62 u
Shiquanhe 61 d
Shir 39 k
Shirane-san 59 n
Shiraz 39 j
Shisanzhan 58 g
Shishinda 43 n
Shishmaref 10 l
Shivera 51 w
Shivpuri 61 o
Shiyan 63 a
Shkodër 35 d
Shokal'skogo Str. 51 m
Sholapur 61 o
Shornvukva 50 u
Shorobe 49 a
Shoshong 46 o
Shostka 54 a
Showak 43 a
Show Low 18 h
Shreveport 19 f
Shuangcheng 58 g
Shuangjiang 62 s
Shuangliao 58 e
Shubra al Kheima 38 h
Shuga 55 l
Shuga (Gulf of Ob') 51 b
Shule 56 x
Shumikha 55 e
Shumougou 57 l
Shunte 63 p
Shuo Xian 57 t
Shurgaz 60 d
Shush 39 g
Shwangliao 58 e
Shwebo 62 q
Shyok 61 c
Si 63 o
Siakwan 62 t
Sialkot 61 a
Sian 62 m
Siangfan 63 b
Siangtan 63 b
Siargao 67 k
Siauliai 33 v
Šibenik 35 b
Siberia 51 o
Siberian Uvaly 51 w
Siberut 68 e
Sibi 60 j
Sibirjakov 51 e
Sibiti 44 e
Sibiu 35 g
Sibolga 68 d
Sibu 69 a
Sibut 42 o
Sibuyan 67 h
Sicasica 28 e
Sichuan Pendi 62 j
Sicilian Channel 34 z
Sicily 35 n
Sicuani 25 n
Sidatun 58 l
Sîdi Barrâni 38 e
Sidi-bel-Abbes 34 s
Sidi Bennour 36 g
Sidi Ifni 36 e
Sidi Kacem 36 i
Sidley, Mount 81 f
Sidney (Montana) 15 e
Sidney (Nebraska) 15 r
Sidorovoko 51 f
Siedlce 33 v
Siegen-Kyugel 52 m
Siena 34 l
Sierra Blanca Peak 18 k
Sierra Burro 19 a
Sierra Carmen 18/19
Sierra Colorada 30 j
Sierra de Aguas Calientes 28 r
Sierra de Cordoba 28 t
Sierra de Famatina 28 q
Sierra de Juarez 18 d
Sierra de la Giganta 18 g
Sierra del Nevado 30 j
Sierra de Mato 23 t
Sierra de Perija 23 b
Sierra de Velasco 28 r
Sierra Leone 40
Sierra Madre 67 g
Sierra Morena 34 p
Sierra Nevada (Spain) 34 q
Sierra Nevada (USA) 14 v
Sierra Vista 18 h
Sigil 65 t
Signy 82 g
Siguiri 40 g
Sihabuhabu 65 v
Sihochac 19 u
Sikar 61 a
Sikasso 40 i
Sikhote-Alin 58 m
Sikonge 45 g
Siktyakh 52 k
Silao 19 n
Silay 67 h
Silchar 62 o
Silet 37 o
Silifke 35 y
Siligir 52 f
Siling Co 61 k
Siljan, Lake 33 e
Šilka 57 j
Silkwood 73 m
Sillanjnuay 28 d
Silver City 18 i
Simao 62 t
Simcoe, Lake 16 j
Simenga 52 d
Simeulue 68 b
Simferopol 54 o
Simikot 61 f
Simla 61 c
Simpson Desert 73 t
Simpson Peninsula 13 b
Simušir 59 i
Sinadogo 43 u
Sinai Peninsula 38 j
Sinawan 37 g
Sincelejo 22 m
Sinclair Milis 14 h
Sind 61 p
Sindi 42 f
Šinedžinst 57 n
Sines 34 n
Sinewit 71 r
Singa 42 m
Singapore 66 t
Singatoka 75 k
Sing Buri 65 j
Singida 45 h
Singkawang 68 l
Singkep 68 i
Singkilbaru 66 p
Singtai 63 c
Sining 57 o
Sinjai 69 t
Sinjar 54 s
Sinkat 38 y
Sinn 58 m
Sinnamary 26 j
Sinop 35 m
Sintang 68 m
Sinú 22 m
Sinugif 43 j
Sinŭiju 58 s
Sinyaya 52 k
Sinyuga 52 s
Sioux City 15 v
Sioux Falls 15 v
Sioux Lookout 16 c
Siple (Mountain) 81 b
Siple (Research Sta.) 81 q
Siple Island 81 q
Sipora 68 e
Siquijor 67 i
Sira 56 i
Siracha 65 j
Siracusa (Syracuse) 35 o
Siraj 60 b
Sir Edward Pellew Group 73 g
Siret 35 h
Sirha 61 i
Sir James McBrien, Mount 12 o
Sirjan 60 a
Sir Joseph Banks Group 77 e
Sirohi 60 z
Sironcha 64 h
Sirpur 64 g
Sirsa 61 a
Sirte 35 o
Sirte Desert 37 k
Sirte, Gulf of 35/38
Sirwa 43 h
Si Sa Ket 66 g
Sishen 48 z
Sisimiut 90 i
Sisophon 66 g
Sisseton 15 i
Sitapur 61 e
Sitka 11 x
Sittang 65 g
Sittwe 62 o
Siuri 61 w
Sivas 54 p
Siverek 54 r
Sivuch 53 v
Siwa 38 e
Siwa, Oasis 38 e
Sixtymile 11 i
Siziman 59 b
Skagen, Cape 33 c
Skagerrak 33 b
Skagway 11 x
Skardu 61 a
Skeena 14 d
Skellefte 33 g
Skellefteå 33 h
Skibotn 33 h
Skikda 34 w
Sklad 52 j
Skoer Fjord 91 u

Skopin 54 e
Skopje 35 e
Skovorodino 52 x
Skukuza 49 f
Skye 32 v
Slamet 68 y
Slana 11 h
Slave Coast 41 o
Slave Lake 14 l
Slave River 12 u
Slessor Glacier 83 u
Slidell 19 i
Sligo 32 u
Slippers Bay 48 w
Sliven 35 h
Slonim 33 w
Slovakia 35 c
Slovenia 35 a
Sl'ud'anka 52 n
Słupsk 33 s
Slutsk 33 x
Smallwood Reservoir 17 f
Smara 36 r
Smederevo 35 e
Smidovich 50 j
Smith 14 l
Smith Bay 13 g
Smithfield 49 o
Smith's Falls 16 l
Smith Sound 91 c
Smithton 77 w
Smolensk 33 z
Smyley Island 82 z
Snag 11 i
Snake 14 j
Snake River (USA) 15 a
Snare River 12 s
Snares Islands 79 p
Snowdrift 12 u
Snowy 78 q
Snowy Mountains 78 q
Soai Dao 65 k
Sobat 42 k
Sobradinho Reservoir 27 r
Sobral 27 f
Soči 54 r
Socompa 28 q
Socorro (Colombia) 23 n
Socorro (USA) 18 j
Socorro (Island) 18 u
Socotra 235 c
Soda Springs 15 n
Söderhamn 33 g
Södertälje 33 f
Sodiri 42 j
Sodo 43 o
Sofala Bay 46 v
Sofia 35 f
Sofiya (Sofia) 35 f
Sofiysk 53 o
Sofiysk (Amur) 59 b
Sogamoso 23 o
Sogeri 71 o
Sognefjörd 33 a
Sogo Nur 57 o
Sohâg 38 i
Sohar 39 z
Soje 49 b
Šokalsky 51 c
Sokcho 58 v
Sokodé 41 a
Sokolo 40 i
Sokoto 41 d
Sokoto (River) 41 c
Sokyrbulak 55 b
Soledade 25 b
Solikamsk 50 w
Sol Iletsk 54 m
Solimões 23 s
Solodniki 54 h
Solok 68 f
Sololo 43 p
Solomon Islands 71/74
Solomon Sea 74 k
Solor Islands 69 v
Solovjevsk 57 i
Solwezi 45 n
Somalia 43
Somerset 16 t
Somerset East 49 n
Somerset Island 13 a
Son 61 t
Sonaly 55 i
Søndrestrømfjord 90 w
Søndre Strømfjord (Fjord) 90 w
Songea 45 u
Songkhla 65 w
Songo 45 s
Sonhat 61 s
Sonoyta 18 f
Sonsonate 22 c
Sonsorol 70 d
Sopo 42 g
Sorabai 60 k
Sordongnokh (Russia) 53 f
Sordongnokh (Russia)) 53 p
Soria 34 e
Sorocaba 29 f
Sorochinsk 54 m
Sorol Atoll 70 j
Sorong 70 p
Soroti 42 y
Söröya 33 i
Sorsk 56 i
Sorsogon 67 i
Sortavala 33 m
Sosna 52 b
Sosnovka 57 f
Sos'va 50 y
Sosyko 54 e
Souanké 41 x
Soubré 40 v
Souk-Ahras 34 w
Sŏul (Seoul) 58 u
Sounda 44 e
Souré 26 m
Sousse 34 y
South Africa 46/48-49
Southampton Island 13 q
Southampton (UK) 34 f
Southampton (USA) 17 n
South Andaman 65 e
South Anyuskiy Mountains 10 b
South Australia 77 b
South Bend 16 s
South Bug 35 k
South Carolina 20 m
South China Sea 66-67
South Dakota 15 g
South East Cape 77 y
Southend 12 w
Southend-on-Sea 32 y
Southern Alps 79 r
Southern Cross 76 f
Southern Indian Lake 12 z
Southern Ocean 87 t
Southern Sierra Madre 19 p
Southern Sporades 35 u
South Georgia Islands 31 y
South Horr 43 o
South Island 79 r
South Magnetic Pole 87 q
South Natuna Islands 68 i
South Orkney Islands 82 g
South Pagai 68 r
South Platte 15 s
South Pole 81 m
South Saskatchewan 15 a
South Scotia Ridge 82 f
South Shetland Islands 82 r
Southwest Cape 79 p
Sovetsk 33 v
Sovetskaya 84 w
Sovetskiy 50 z
Soyapa 18 i
Soyo 44 e
Spaatz Island 82 z
Spain 34/36
Spanish Town 21 o
Spartanburg 20 l
Spassk Dal'niy 58 k
Spearfish 15 e
Spence Bay 13 a
Spencer 15 w
Spencer, Cape 77 r
Spencer Gulf 77 e
Spitsbergen 92 p
Split 35 b
Spokane 14 j
Spratly Islands 66 z
Springbok 48 w
Springdale 17 j
Springfield (Colorado) 15 s
Springfield (Illinois) 16 q
Springfield (Massachusetts) 17 n
Springfield (Missouri) 15 x
Springfield (Ohio) 16 u
Springs 46 q
Springsure 74 r
Spruce Knov 16 w
Squaw Habor 11 n
Sredinnyy Mountains 53 y
Srednekolymsk 53 j
Sredniy 53 w
Sredniy Kalar 52 t
Sretensk 57 k
Srikakulam 64 k
Sri Lanka 64
Sri Mohangarh 60 k
Srinagar 61 a
Stafonovo 50 f
Stakhanov 54 d
Stampriet 48 j
Standerton 49 r
Stanford 15 a
Stanley (Falkland Is.) 31 p
Stanley (N. Dakota) 15 f
Stannakh-Khocho 52 i
Stanovoy Mountains 52 y
Stanthorpe 78 f
Staraja Russa 33 m
Stara Zagora 35 h
Star.Dom 53 d
Stargard 33 s
Staritsa 51 s
Staritsa (Moscow) 50 n
Star. Kayakhnyy 52 b
Star. Kheydzhan 53 s
Staryy Nadym 51 p
Staryy Oskol 54 d
Staten Island 30 z
Statesboro 20 l
Stavanger 33 a
Stavropol 54 t
Stavropolka 55 f
Steamboat 12 p
Steamboat Springs 15 p
Steelpoort 49 e
Steenkool 70 q
Steenstrup Glacier 91 b
Stefansson Island 12 j
Steinhausen 48 i
Steinkier 33 d
Steinkopf 48 w
Stephenville 17 h
Stepnyak 55 i
Sterling 15 r
Sterlitamak 55 a
Stettin (Szczecin) 33 r
Stevens Point 16 d
Stewart (British Colombia) 11 z
Stewart (Yukon Territory) 11 i
Stewart Crossing 11 k
Stewart Island 79 p
Stewart, Mount 74 p
Stewart (River) 11 k
Steynsburg 49 o
Steytlerville 49 n
Stikine 11 z
Stil Bay 48 y
Stirling 32 w
Stockholm 33 g
Stockton 14 u
Stoke-on-Trent 32 w
Stolbovoy 50 i
Stolbovoy (Island) 53 e
Stonehenge 77 i
Stony Rapids 12 v
Stony River 11 c
Store Koldewey 91 u
Storkerson Peninsula 12 i
Storm Bay 77 z
Storm Lake 16 n
Stors, Lake 33 e
Storuman 33 f
Strahan 77 x
Stralsund 33 q
Strand 48 w
Strasbourg 34 j
Strathleven 73 k
Strathmay 73 k
Streaky Bay 77 c
Strelka 53 i
Strelka (Yenisey) 51 w
Strelka-čunja 52 a
Strezhevoy 51 q
Strickland 70 x
Strömsund 33 f
Stryy 35 f
Stung Treng 66 i
Sturge Island 86 v
Sturgeon Bay 16 f
Sturgeon Falls 16 j
Sturt Creek 73 n
Sturt Desert 77 h
Stutterheim 49 p
Stuttgart 34 l
Styr 33 w
Suakin 38 z
Suanhwa 57 w
Suata 23 g
Subotica 35 d
Suceava 35 h
Süchbaatar 57 d
Suchow (Xuzhou) 63 f
Suchow (Suzhou) 63 h
Suchumi 54 r
Suckling, Mount 74 e
Sucre 28 f
Sucuaro 23 q
Sucunduri 25 j
Sucuriu 29 b
Sucurundi 25 j
Sudan 38/42-43
Sudbury 16 i
Sudd 42 j
Suddie 23 k
Sudety 35
Sudocje, Lake 55 o
Sudr 38 j
Sue 42 u
Suez 38 j
Suez Canal 38 j
Suez, Gulf of 38 j
Sugdža 57 e
Suglan 53 a
Sugoy 53 k
Sugun 55 y
Suhai Hu 56 x
Suhait 57 q
Suhsien 63 e
Suide 57 t
Suifenhe 58 j
Suihua 58 g
Suileng 58 g
Suining 62 k
Sui Zian 63 c
Suj 57 r
Sukabumi 68 w
Sukhana 52 h
Sukhinichi 54 b
Sukhona 50 r
Sukkur 60 j
Suknah 37 j
Sulaiman 60 k
Sulaimaniyah (Iraq) 39 e
Sulaimaniyah (Saudi Arabia) 39 t
Sulaiman Range 60 k
Sula Islands 69 k
Sulan Cheer 57 s
Sulawesi (Celébes) 69 t
Suleya 55 b
Suli Hu 56 x
Sullana 24 g
Sullivan 16 p
Sulphur Springs 19 e
Sultan 60 f
Sultanpur 61 s
Sulu Archipelago 67 s
Sulu Sea 67 s
Sulzberger Ice Shelf 81 v
Sum 56 l
Sumaih 42 i
Sumail 60 o
Sumarokovo 51 v
Sumatra 68 f
Sumba 69 s
Sumbawa 69 r
Sumbawa Besar 69 r
Sumbawanga 45 f
Sumbe 44 s
Sümber 57 e
Sumbu 45 e
Sumgait 54 x
Sumprabum 62 r
Sumy 35 l
Sunan 57 n
Suncho Corral 28 u
Sundance 15 d
Sundargarh 61 t
Sunda Strait 68 v
Sunderland 32 y
Sundsvall 33 g
Sungaigerung 68 v
Sungai Ko-lok 65 x
Sungaipenuh 68 f
Sungai Petani 66 q
Sungai Siput Utara 65 x
Sungari 58 i
Sun Kosi 61 i
Suntar 52 g
Sunwu 58 g
Sunyani 40 y
Suolama 52 e
Suon-Tit 52 x
Suordakh 53 c
Superior 15 k
Superior, Lake 15 m
Suphan Buri 66 d
Sur (Oman) 60 q
Sur (Lebanon) 60 q
Surab 60 h
Surabaya 69 o
Surakarta 68 z
Surat (Australia) 74 r
Surat (India) 60 z
Suratgarh 60 m
Surat Thani 65 w
Surgat 51 p
Surigao 67 j
Surin 65 m
Suriname 23/26
Susanville 14 v
Šusenskoje 56 j
Susquehanna 16 y
Sutam 52 y
Sutherland 48 x
Sutlej 60 m
Suttor 74 p
Sutun'ya 52 m
Suva 75 l
Suvorka 57 c
Suwålki 33 v
Suwar 39 b
Suyevatpaul 50 y
Suzun 56 d
Svalbard 92 o
Svartenhuk 90 m
Svartisen 33 e
Sveg 33 e
Svendborg 33 q
Sverdlovsk (Yekaterinburg) 50 y
Sverdrup Channel 89 r
Sverdrup Islands 12 k
Svetlaya 59 a
Svetlovodnaya 58 m
Svetozarevo 35 e
Svobodny 58 h
Swakopmund 48 g
Swan 76 d
Swan Hill 77 v
Swan Island 19 z
Swan River 15 f
Swansea (Australia) 77 z
Swansea (UK) 32 w
Swat 60 m
Swatow 63 r
Swaziland 46/49
Sweden 33
Sweeney Range 83 n
Sweetwater 19 a
Sweetwater (River) 15 o
Swellendam 48 y
Swift Current 15 b
Switzerland 33/34
Syagannakh 53 f
Sydney (Australia) 78 f
Sydney (Canada) 17 h
Sydney Ice Cap 13 d
Syktyvkar 50 t
Sylhet 61 y
Sym 51 u
Sym (River) 51 v
Symya 50 m
Syowa 84 i
Syracuse (USA) 16 y
Syracuse (Sicily) 35 o
Syrdarja (River) 55 s
Syria 38-39/54
Syrian Desert 38 m

Syzran **54** j
Szczecin **33** r
Szczecinek **33** s
Szeged **35** d
Szeping **58** f

T

Tabah **39** c
Tabajara **25** u
Tabarka **34** x
Tabas **39** m
Tabatinga, Serra da **27** p
Tabelbala **36** j
Tablas **67** h
Table Island **13** a
Tableland **72** m
Tabletop **72** x
Tabor **53** i
Tabora **45** f
Tabou **40** v
Tabriz **54** v
Tabuk **38** l
Tacámbaro **19** n
Tachakou **56** r
Tacheng **56** e
Tacloban **67** j
Tacna **25** n
Tacoma **14** h
Tacora **25** o
Tacuarembó **28** z
Tacutu **23** w
Tademait, Hama da **36** m
Tadjourah **43** f
Tadmur **38** m
Taegu **58** v
Taejon **58** u
Tafassâsset, Ténéré du **37** s
Taftan **60** e
Taganrog **54** d
Tagant **36** r
Tagap Ga **62** d
Tagbilaran **67** i
Taghit **36** k
Tagounite **36** h
Taguatinga **27** n
Tagum **67** x
T'agun **56** l
Tagus **34** o
Tahifet **37** q
Tahiti **234** o
Tahlequah **16** n
Tahoua **41** e
Tahrud **60** c
Tai **40** v
Tai'an **63** e
Taibus Qi **57** w
Taichow **58** p
Taichung **63** t
Ta'if **39** o
Taigu **57** u
Tailai **58** e
Tai, Lake **63** g
Taima **38** m
Taimba **51** y
Taimura **51** z
Taimyr **51** l
Taimyr, Lake **51** m
Taimyr Peninsula **51** i
Tainan **63** u
Taiobeiras **29** j
Taipei **63** v
Taiping **65** w
Taitao Peninsula **30** s
Taiwan **63**
Taiwan Strait **63** t
Taiyuan **57** u
Ta'izz **43** g
Tajarhi **37** v
Tajikistan **55** v
Tajito **18** g
Tajo **34** q
Tajsara, Cord. de **28** g
Tajura **37** i
Tak **65** i
Takamatsu **58** y
Takatshwaane **48** m
Ta-Kaw **62** s
Takestan **39** h
Takhta Bazar **55** r
Takijuak Lake **12** g
Takla Makan Desert **56** r
Takpa Shiri **62** b
Takua Pa **65** u
Takum **41** u
Talak **37** p
Talandzha **58** j
Talara **24** g
Talata Mafara **41** e
Talaud Islands **69** l
Talavera de la Reina **34** p
Talawana **72** v
Talbot **76** k
Talca **30** h
Talcahuano **30** g
Talcher **61** u
Taldy-Kurgan **56** n
Talhada, Serra **27** t
Taliabu **69** j
Taliqan **60** k
Talitsa **50** z
Talkeetna **11** f
Tall'Afar **39** c
Tallahassee **20** j
Tall Kalakh **38** l
Tal'menka **56** e
Talodi **42** j
Talon **53** u
Taloyoak (Spence Bay) **13** a
Taloye **52** y
Taltal **28** o
Taluma **52** w
Talwood **78** e
Tamala **76** c
Tamale **40** m
Tamanrasset **37** p
Tamassoumit **36** p
Tamazula **18** z
Tamazunchale **19** p
Tambacounda **40** d
Tambara **49** h
Tambelan Islands **68** j
Tambey **51** b
Tambo (Australia) **74** b
Tambo (Peru) **24** y
Tambo (Peru) **25** n
Tambor **48** f
Tambov **54** f
Tambura **42** u
Tamč **56** k
Tamchaket **36** r
Tamentit **36** l
Tamiahua Lagune **19** q
Tamitatoala **26** v
Tampa **20** k
Tampere **33** j
Tampico **19** p
Tamsagbulag **57** k
Tamshiyaco **24** m
Tamu **62** p
Tamworth **78** e
Tanabe **58** z
Tanacross **11** h
Tanahbala **68** d
Tanahjampea **69** u
Tanahmerah **70** v
Tana (Kenya) **43** q
Tana (Norway) **33** l
Tana (River, Norway) **33** k
Tana, Lake **43** b
Tanama **51** e
Tanami Desert **73** p
Tanana **11** e
Tanbar **77** h
Tanchoj **57** d
Tandil **31** b
Tanega **58** w
Tanezrouft **36** y
Tanezrouft-n-Ahenet **36/37**
Tanga (Tanzania) **45** k
Tanga (Russia) **57** g
Tanganyika, Lake **45** d
Tanggula Shan **61** k
Tanghe **63** c
Tangier **34** p
Tangorin **74** o
Tangra Yumco **61** i
Tangshan **57** y
Tangub **67** u
Tanguiéta **41** b
Tanimbar Islands **70** o
Tanjung **69** e
Tanjungbalai **65** w
Tanjungkarang **68** v
Tanjungpandan **68** x
Tanjungpura **65** v
Tanjungredeb **69** e
Tannu Mountains **56** j
Tano **40** y
Tanot **60** k
Tanout **41** g
Tanque El Revés **18** m
Tanta **38** h
Tantan **36** e
Tanzania **42/45/46**
Tao'an **58** d
Taolañaro **47** q
Taoudenit **36** w
Taourirt **36** k
Taouz **36** j
Taoyüan **63** u
Tapachula **19** t
Tapajós **25** l
Tapaktuan **65** t
Tapanahoni **26** h
Tapauá **25** f
Tapauá (River) **25** e
Tapeta **40** t
Tapirapecó, Serra **23** t
Tapirapua **25** x
Tapis **68** g
Tapsui **50** y
Tapti **60** z
Tapurucuara **25** f
Taquaras **25** s
Taquari **28** m
Tara **55** k
Tara (River) **55** m
Tarabuco **28** g
Taraeuá **25** c
Taragay **55** y
Tarakan **69** e
Taran **51** c
Taranto **35** c
Taranto, Gulf of **35** p
Tarapoto **24** j
Taraq **39** x
Tararua Range **79** v
Tarat **37** s
Tarata **25** o
Tarauacá **25** n
Tarauacá (River) **25** a
Tarbagatay Mountains **56** d
Tarbaj **43** q
Tarbes **34** f
Tarcoola **77** c
Taree **78** g
Tareya **51** j
Tarfaya **36** d
Tarhuna **35** n
Tarialan **57** b
Tariat **57** d
Tariat (Hangayn Mountains) **57** a
Tarif **39** x
Tarija **28** g
Tarim **56** q
Tarim (River) **56** r
Tarim Basin **56** q
Tarim Liuchang **56** t
Tarime **45** g
Tarim Kot **60** h
Tarka La **61** k
Tarkwa **40** z
Tarlac **67** f
Tarnak **60** i
Taro **71** v
Taroom **74** r
Taroudannt **36** g
Tarragona **34** g
Tarso Ahon **37** y
Tartagal **28** h
Tartary, Gulf of **59** b
Tartu **33** k
Tartus **35** z
Tarutung **68** d
Tasauz **55** p
Tashigang **61** l
Tashkent **55** v
Tasik Malaya **68** x
Task **41** h
Taskesken **56** c
Tas-Kumsa **53** a
Tas-Kumyr **55** w
Tasman Bay **79** t
Tasmania **77** w
Tasman Sea **78** v
Tassili-n-Ajjer **37** r
Tassili Oua-n-Hoggar **37** p
Tasso Fragoso **27** o
Tastagol **56** g
Tas-Tumus **52** m
Tastyp **56** h
Tas-Yuryakh **52** f
Tata **36** g
Tatarsk **55** l
Tataurovo **57** e
Tathlina Lake **12** r
Tathlith **39** q
Tatty **55** x
Tatung **57** o
Tatvan **54** s
Tauá **27** s
Taubaté **29** g
Taumarunui **79** v
Taumaturgo **24** z
Taunggyi **62** r
Taupo **79** w
Taupo, Lake **79** w
Tauranga **79** w
Taurus **35** x
Tavda **50** z
Tavda (River) **50** z
Tavira **36** g
Tavoy **65** h
Tavua **75** l
Tawau **69** e
Tawitawi **67** s
Tawitawi Group **67** s
Taxco **19** o
Tayabamba **24** w
Tayebad **60** d
Taygonos, Cape **53** z
Taymylyr **52** j
Tayshet **56** m
Taz **51** f
Taz (River) **51** f
Taz Bay **51** d
Taza **36** j
Tazenakht **36** h
Tazirbu **38** a
Tazovskiy **51** e
Tazzarine **36** i
Tbilisi **54** u
Tchibanga **44** c
Tchien **40** u
Tchin-Tabaradene **41** e
Tcholliré **41** k
Te Anau **79** p
Te Araroa **79** x
Tea Tree **73** r
Tebessa **34** x
Tebicuary **28** y
Teboursouk **34** y
Techiman **40** z
Tecka **30** i
Tecomán **18** y
Tecpan **19** n
Tecuala **18** x
Tedžen **55** p
Teelin **56** m
Tefé **25** f
Tefé (River) **25** e
Tegal **68** y
Tegguidda-n-Tessoum **41** e
Tegina **41** e
Tegucigalpa **22** e
Teguldet **51** v
Tegyulte-Terde **52** z
Tehchow **57** w
Tehran **39** j
Tehuacan **19** q
Tehuantepec **19** r
Tehuantepec, Gulf of **19** r
Tehuantepec, Isthmus of **19** s
Teide, Pico de **36** a
Tekax **19** v
Tekes **56** q
Tekes (River) **56** q
Tekeze **43** c
Tekiliktag **61** e
Tekirdağ **35** h
Teku **69** i
Tela **19** w
Tel-Aviv-Yaffa **38** k
Tele **42** s
Teleckoje, Lake **56** g
Telefomin **70** x
Telegraph Creek **11** y
Telemaco Borba **29** q
Telemba **57** h
Télémélé **40** e
Teles Pires **25** x
Tell Atlas **34** t
Teller **10** m
Telmen Nuur **56** l
Telsen **30** k
Telukbetung
(Bandarlampung) **68** v
Tema **41** n
Temacine **37** d
Tembenchi **51** y
Tembenchi (River) **51** k
Tembo **41** v
Temir **55** b
Temirtau **55** j
Témiscaming **16** j
Temnikov **54** g
Temosachic **18** j
Temple **19** d
Temuco **30** g
Tena **24** j
Tenacatita **18** x
Tenali **64** i
Tenalr **53** h
Tenasserim (River) **65** i
Tenasserim Coast **65** h
Tendaho **43** d
Ten Degree Channel **65** d
Tendelti **42** k
Tendrara **36** k
Ténéré **37** s
Ténéré, Erg du **37** t
Tenerife **36** a
Tenerife (Island) **36** b
Ténès **34** t
Tenchong **62** s
Tengiz, Lake **55** h
Tenkasi **64** s
Tenke **52** g
Tenkeli **53** e
Tenkodogo **40** m
Tennant Creek **73** r
Tennessee (State) **16** r
Tennessee (River) **16** r
Tenosique **19** t
Tensift **36** f
Tenterfield **78** f
Teófilo Otoni **29** j
Tepatitlán **18** z
Tepeji **19** o
Tepi **42** z
Tepic **18** x
Téra **41** a
Terbu **37** w
Terek **54** u
Terensay **55** c
Teresina **27** q
Terezinha **26** j
Termes de Rio Hondo **28** t
Termez **55** t
Termination Island **76** h
Terminos, Laguna de **19** u
Termit, Massif de **41** i
Termit N **41** h
Ternate **69** l
Terney **58** m
Terni **34** m
Ternopol' **35** g
Terrace **14** d
Terrace Bay **48** f
Terra Firma **48** z
Terra Preta **25** j
Terra Haute **16** s
Terror **86** r
Teruel **34** f
Terutao **66** q
Tesig **57** c
Teslin **11** l
Tessalit **36** z
Tessaoua **41** f
Tete **46** g
Tété **42** d
Tetere **52** o
Tetuan **34** p
Tetyushi **54** j
Teuco **28** w
Teulada, Cape **34** x
Tevriz **51** o
Têwo **62** i

Texarkana 19 e
Texas 19 a
Texas City 19 e
Texcoco 19 p
Teya 51 w
Tezejet 55 p
Tezew 33 1
Teziutlán 19 p
Tezpur 62 n
Thabazimbi 49 c
Thailand 62/65/66
Thailand, Gulf of 66 e
Thakhek 66 h
Thallon 78 d
Thamarit 39 x
Thames 32 x
Thamud 39 u
Thane 60 z
Thanh Hoa 62 x
Thanjavur 64 h
Thap Put 65 u
Tharad 60 y
Thar Desert 60 k
Thargomindah 77 j
Tharthar, Lake 39 c
Thasos 35 g
Thaton 65 h
Thatta 60 v
Thayetmyo 65 f
Thebes 38 i
Theda 72 m
The Dalles 14 h
Thedford 15 t
Thelon 12 w
Theodore 74 s
Theodore Roosevelt 26 q
The Pas 15 l
Thermai, Gulf of 35 s
The Sound 33 q
Thessalonika 35 f
Thessaloniki (Thessalonika) 35 f
Thief River Falls 15 i
Thiès 40 b
Thika 45 j
Thimphu 61 k
Thingangyun 65 g
Thionville 34 i
Thiruvananthapurum (Trivandrum) 64 r
Thistle I 77 e
Thlewiaza 12 z
Thollon 77 m
Thomasville 20 j
Thompson 12 y
Thompson Falls 14 l
Thomson 74 n
Thon Buri 65 j
Thorntonia 73 v
Three Hummock Island 77 x
Three Kings Island 79 g
Three Rivers (Australia) 76 f
Three Rivers (USA) 19 c
Thule 91 c
Thunder Bay 16 e
Thurso 32 k
Thurston Island 80 m
Thylungra 77 j
Tiancang 57 n
Tiancheng 57 n
Tianguá 27 e
Tianjin (Tientsin) 57 x
Tianjun 56 z
Tianlin 62 x
Tiaret 36 m
Tibati 41 v
Tibé 40 h
Tibesti 37 y
Tibet 61 g
Tiburon 18 f
Tichitt 36 s
Tichla 36 o
Tichvin 50 n
Ticul 19 v
Tidikelt 37 n
Tidjikja 36 q
Tidra 36 n
Tielongtan 61 d
Tien Shan 56
Tienshui 62 k
Tientsin 57 x
Tierra Blanca 19 q
Tierra Colorada 19 o
Tierra del Fuego 30 w
Tiete 29 d
Tieyon 77 c
Tifariti 36 r
Tifton 20 k
Tigieglo 43 t
Tigil' 53 y
Tigre (Peru) 24 k
Tigre (Venezuela) 23 h
Tigris 39 d
Tiguentourine 37 e
Tijuana 18 c
Tikhoretsk 54 e
Tikrit 39 d
Tiksi 52 m
Tilamuta 69 i
Tilemsés 41 d
Tilemsi 36 z
Tillabéri 41 b
Tillamook 14 g
Tillia 41 d
Tilrhemt 34 u
Timané 28 k
Timan Ridge 50 h
Timaru 79 s
Timbaúba 27 v
Timbedgha 40 g
Timber Creek 73 b
Timbouctou (Timbuktu) 40 l
Timbuktu 40 l
Timeiaouine 37 n
Timétrine 36 y
Timia 37 r
Timimoun 36 m
Timiris, Cap 36 n
Timișoara 35 e
Timmins 16 i
Timón 27 q
Timor 69 x
Timor Leste 69 x
Timor Sea 69 x
Timpton 52 y
Timsah 37 w
Tindouf 36 g
Tiné 42 e
Tinerhir 36 i
Tinfouchy 36 i
Tingo María 24 x
Tingri 61 i
Tinogasta 28 r
Tinos 35 t
Tinrhert Hamada 37 e
Tinsukiä 62 d
Tintina 28 u
Tinyay 53 a
Tin Zaouatene 37 o
Tioman 68 h
Tipperary 73 c
Tiracambu, Serra do 27 a
Tirana 35 d
Tiranë (Tirana) 35 d
Tiraspol 35 j
Tirekh 52 m
Tirekhtyakh 53 c
Tirgelir 53 e
Tirgu Mureș 35 g
Tirich Mir 60 l
Tiroungoulou 42 e
Tiruchchirappalli 64 f
Tirunelveli 64 t
Tiruntán 24 l
Tirupati 64 h
Tisdale 15 e
Tisisat Falls 43 b
Tisza 35 d
Tit 37 p
Titaf 36 m
Tit-Ary 52 l
Titicaca, Lake 25 o
Titule 42 t
Titusville 20 m
Tizimin 19 w
Tizi Ouzou 34 v
Tiznit 36 f
Tlacotalpan 19 r
Tlalamabele 49 b
Tlaltenango 18 z
Tlapacoyan 19 q
Tlaxiaco 19 p
Tlemcen 34 s
Tlisen 37 l
Tmed Bu Haschlscha 37 y
Toamasina 47 r
Toba & Kakar Ranges 60 j
Tobago 23 j
Toba, Lake 68 d
Tobelo 67 y
Tobermorey 73 t
Tobermorey (Australia) 73 x
Tobermorey (Canada) 16 i
Tobi 70 c
Tobo 70 o
Tobol' 55 e
Tobol' (River) 55 d
Tobol'sk 51 n
Tobruk 38 e
Tocache Nuevo 24 w
Tocantínia 26 z
Tocantinópolis 26 z
Tocantins 26 z
Tocopilla 28 c
Tocorpuri 28 e
Tocuyo 23 d
Todertang 43 n
Todos Santos (Bolivia) 25 r
Todos Santos (Mexico) 18 u
Toéssé 40 m
Togiak 11 n
Togliatti 54 k
Togo 41
Togtoh 57 t
Togyz 55 d
Tok 11 h
Tokar 38 z
Tokko 52 u
Tokko (River) 52 v
Tokoroa 79 v
Toktogul 55 x
Toktogul Reservoir 55 w
Tokuma 53 a
Tokushima 58 y
Tōkyō 59 o
Tolbo Nuur 56 i
Toledo (Brazil) 29 o
Toledo (Philippines) 67 i
Toledo (Spain) 34 q
Toledo (USA) 16 u
Toledo Bend Reservoir 20 d
Tolga 37 c
Toli 56 r
Toliara 47 n
Tolka 51 s
Tolka (River) 51 r
Tolo, Gulf of 69 h
Tolon 52 r
Tolstoy, Cape 53 x
Toluca 19 o
Tolybay 55 d
Tom' 56 g
Tomah 16 q
Tomar 55 l
Tomatlán 18 y
Tombador, Serra do 25 x
Tombigbee 20 h
Tombôco 44 f
Tômbua (Porto Alexandre) 48 e
Tomé-Açu 26 m
Tomi 42 o
Tomini, Gulf of 69 g
Tommot 52 y
Tomo 23 q
Tompo 53 b
Tom Price 72 t
Tom Price, Mount 72 t
Tomra 61 i
Tomsk 51 t
Tomtor 53 a
Tomtor (Cherskogo Mountains) 53 e
Tonalá 19 s
Tonatins 23 r
Tondano 67 w
Tonga 42 k
Tonga (State) 235 z
Tongguzbasti 56 q
Tonghae 58 v
Tonghai 62 u
Tongian Islands 67 u
Tongjiang 58 k
Tongliao 58 e
Tongren 62 h
Tongren (Wu Jiang) 62 m
Tongsa Dzong 61 l
Tongshan 63 d
Tongtian 62 e
Tongtianheyan 62 b
Tongulakh 52 j
Tongxin 57 q
Tongyu 58 e
Tongzi 62 l
Toni 42 v
Tonk 61 o
Tonkin, Gulf of 62 y
Tonking 62 w
Tonkling 63 l
Tonle Sap 65 f
Tonnel'nyy 57 h
Tonopah 14 x
Tonsina 11 g
Tontantins 25 c
Tonzang 62 p
Toolebuc 73 v
Toompine 77 j
Toora-Chem 56 l
Toowoomba 78 l
Topeka 15 v
Topia 18 k
Topko 53 p
Top Springs 73 c
Torbat-e-Heidariye 60 d
Torbat-e-Jam 60 d
Torbert, Mount 11 d
Torgo 52 u
Torim 18 h
Torino (Turin) 34 k
Torit 42 y
Torne 33 i
Torngat Mountains 13 y
Tornio 33 i
Toro 28 p
Torodi 41 b
Toro Doum 41 m
Toro, Lago del 30 t
Toronto 16 w
Tororo 42 z
Torrabai 48 f
Torrens Creek 74 o
Torrens, Lake 77 e
Torreón 18 l
Torres Strait 73 k
Tortkuduk 55 k
Tortolas 28 p
Tortosa 34 g
Tortuga 23 f
Torud 39 l
Torun 33 f
Torzhok 50 n
Toson Cengal 56 m
Tostado 28 v
Toteng 48 m
Totma 50 q
Totness 23 z
Totora 28 f
Tottori 58 y
Touba 40 i
Toubkal 36 g
Touboro 41 k
Touggourt 34 w
Tougué 40 e
Toukoto 40 g
Toulépleu 40 u
Toulon 34 j
Toulouse 34 g
Toungoo 62 r
Tours 34 g
Toussoro 42 e
Touws River 48 x
Townsend 15 a
Townsville 74 q
Toxkan 56 o
Toyama 58 z
Toyon-Tirekh 52 m
Tozeur 34 x
Trabzon 54 r
Trail 14 j
Traill Island 91 r
Trang 65 v
Transantarctic Mountains 83 v
Trans Canada Highway 16 c
Transhimalaya 61 g
Trapani 34 z
Traralgon 77 y
Trarza 40 c
Travers 79 t
Treinla-y-Tres 31 f
Trelew 30 l
Trenque Lauquen 31 a
Trenton 16 z
Tres Arroyos 31 b
Três Bocas 25 a
Tres Corações 29 h
Tres Isletas 28 w
Tres Lagos 30 u
Tres Marias 29 g
Tres Marias Islands 18 w
Tres Marias Reservoir 29 g
Tres Puntas, Cabo 30 y
Tres Rios 29 i
Triangulos 19 t
Trichur 64 e
Trier 34 j
Trieste 35 a
Tríkala 35 r
Trincomalee 64 v
Trinidad and Tobago 23 j
Trinidad (Bolivia) 25 s
Trinidad (Colombia) 23 o
Trinidad (Cuba) 20 z
Trinidad (USA) 15 q
Trinity Peninsula 82 t
Tripolis (Greece) 35 s
Tripoli (Lebanon) 38 l
Tripoli (Libya) 37 i
Tripolitania 37 h
Tristan da Cunha 234 y
Trivandrum 64 r
Trobriand or Kiriwina Island 71 r
Trofimovsk 52 l
Troickoje 56 f
Trois Rivières 17 a
Troitsk 55 d
Troitsko-Pechorsk 50 w
Troitskoye 55 a
Trombetas 25 l
Trompsburg 49 n
Tromsö 33 g
Trondheim 33 c
Trondheimsfjord 33 c
Trosh 50 j
Troy 35 u
Troyekurovo 54 e
Troyes 34 i
Troynoy 51 f
Trujillo (Honduras) 20 v
Trujillo (Peru) 24 u
Trujillo (Spain) 34 p
Truro 17 f
Tsaidam Basin 56 w
Tsaochuang 63 f
Tsaratanana Mountains 47 e
Tsau 48 m
Tsavo 45 j
Tsavo National Park 45 k
Tschita 52 r
Tsentral'nyy 56 g
Tsévié 41 n
Tshabong 48 z
Tshane 48 l
Tshele 44 e
Tshikapa 44 k
Tshimbalanga 44 z
Tshofa 45 a
Tsholotsho 49 c
Tshuapa 44 l
Tshwane (Pretoria) 49 p
Tsihombe 47 p
Tsimafana 47 o
Tsimlyanskoye Reservoir 54 g
Tsinan 63 e
Tsingtao 57 z
Tsining 57 u
Tsinling Shan 56 z
Tsiribihina 47 p
Tsiroanomandidy 47 p
Tsitsihar 58 f
Tsoe 49 a
Tsugaru Channel 59 b
Tsumeb 48 j
Tsumkwe 48 k
Tsunyi 62 l
Tsuruga 58 z
Tsushima 58 v
Tswiza 49 f
Tuamotu Archipelago 234 p
Tuan Giao 62 v
Tuangku 65 u
Tuapse 54 r
Tuareg 40 m

Tubaro **29** r
Tubuai Islands **234** o
Tubutama **18** g
Tuckanarra **76** f
Tucson **18** g
Tucumá **26** x
Tucumcari **15** r
Tucupita **23** i
Tucuruf **26** l
Tuekta **56** f
Tueré **26** k
Tugela **49** r
Tugolukovo **54** f
Tuguegarao **67** g
Tugur **53** p
Tugur (River) **53** o
Tukalan **52** a
Tukangbesi Islands **69** v
Tuktoyaktuk **11** l
Tula (Mexico) **19** o
Tula (Russia) **54** d
Tula Mountains **84** m
Tulancingo **19** p
Tulcán **24** i
Tulcea **35** i
Tuli **49** d
Tuli (River) **49** d
Tulia **15** s
Tulihe **57** m
Tulit'a (Fort Norman) **12** b
Tully **74** c
Tulsa **15** v
Tulua **22** y
Tulum **19** w
Tulun **57** a
Tumaco **24** h
Tumanskiy **10** g
Tumany **53** k
Tumat **53** d
Tumatumari **23** x
Tumba, Lake **44** i
Tumbangsamba **69** b
Tumbes **24** g
Tumen **58** i
Tumkur **64** f
Tumu (Ghana) **40** l
Tumu (Libya) **37** v
Tumucumaque, Serra **26** i
Tunduma **45** t
Tunduru **45** v
Tungabhadra Reservoir **64** e
Tungabhadra (River) **64** f
Tungaru **42** k
Tungchuan **62** m
Tungchwan **62** v
Tunghwa **58** f
Tungtin, Lake **63** b
Tungus-Khaya **52** k
Tunguskoye Mountains **51** w
Tuni **61** s
Tunis **34** y
Tunis, Gulf of **34** y
Tunisia **34/37**
Tunja **23** n
Tunor **52** c
Tunuyán **28** q
Tunuyán (River) **28** r
Tunxi **63** f
Tuolba **52** w
Tuostakh **53** c
Tuostakh (River) **53** c
Tupanacca **25** h
Tupelo **19** i
Tupik **52** v
Tupinambarama, Ilha **25** k
Tupiza **28** f
Tupper **12** q
Tupungato **30** i
Tuquan **58** d
Túquerres **24** i
Turabah **39** o
Tura (China) **56** s
Tura (India) **61** y
Tura (Russia) **51** z
Tura (River) **50** y
Turaif **39** a
Turbah **43** g
Turbat **60** s
Turee Creek **72** u
Turgattal **55** e
Turgay **55** e
Turgayskaya Valley **55** e
Turgen **56** i
Türgen (High Altai) **56** x
Turhal **35** m
Turiaçu **27** b
Turin **34** k
Turinsk **50** z
Turkana, Lake **43** n
Turkannakh **52** j
Turkestan **55** u
Turkey **35/54**
Turkey Creek **73** a
Turkmenbashi **54** z
Turkmenistan **55** o
Turks and Caicos Islands **21** s
Turku **33** h
Turneffe Islands **19** w
Túrnovo **35** h
Turnu-Severin **35** l
Turočak **56** g
Turpan **56** u
Turrialba **22** h
Turtkul' **55** q
Turkhan **51** h
Turukhansk **51** h
Turum **42** k
Tuscaloosa **20** h
Tuticorin **64** t
Tutóia **27** d
Tutonchany **51** x
Tutonchany (River) **51** j
Tuul Gol **57** d
Tuvalu **235** y
Tüvsinsiree **57** g
Tuxpan **18** x
Tuxpan (Bay of Campeche) **19** q
Tuxtepec **20** o
Tuxtla Gutiérrez **19** s
Tuyen Quang **62** x
Tuyun **62** y
Tuz, Lake **35** y
Tuz Khurmatu **39** e
Tuzla **35** d
Tvayan **53** x
Tver **50** o
Twee Rivier **48** k
Twee Rivieren **48** y
Twilight Cove **76** k
Twin Falls (Canada) **17** e
Twin Falls (USA) **14** y
Twin Rocks **76** h
Twizel **79** s
Tyler **19** e
Tym **51** s
Tymovskoye **59** d
Tynda **52** w
Tynset **33** d
Tyre (Sour) **38** k
Tyrma **58** k
Tyrrhenian Sea **34** z
Tyry **53** c
Tyubelyakh **53** e
Tyugene **52** l
Tyugyuren **53** e
Tyukalinsk **55** j
Tyukyan **52** h
Tyurnen **50** z
Tyung **52** h
Tyungulyu **52** m
Tzaneen **49** e
Tzekung **62** j
Tzepo **57** x

U

Uainambi **25** b
Uanle Uen **43** u
Uaroo **72** r
Uatum **25** j
Uaupés **25** d
Uaupés (River) **25** c
Ubari **37** u
Ube **58** w
Ubeda **34** q
Uberaba **29** e
Uberaba, Lagoa **28** l
Uberlandia **29** e
Ubon Ratchathani **66** h
Ubundu **45** a
Ucayali **24** l
Uch **60** l
Uchami **51** y
Uchami (River) **51** x
Uchiura Bay **59** c
Uchkuduk **55** r
Uchur **53** n
Uda **53** n
Uda (Vostochny Mountains) **56** m
Uda (Yablon Mountains) **57** g
Udachnyy **52** e
Udaipur **60** z
Udanna **53** f
Uddevaua **33** d
Udd, Lake **33** f
Udine **34** m
Udokan **52** u
Udon Thani **65** l
Udskoye **53** n
Uebonti **69** h
Uele (Russia) **52** g
Uele (Dem. Rep. Congo) **42** r
Uelen **10** l
Uel'kal **10** h
Uengan, Cape **50** m
Ufa **55** a
Ufa (River) **55** a
Ugagli **52** y
Ugalla **45** e
Uganda **42/45**
Ugashik **11** o
Ughelli **41** r
Ugoyan **52** x
Ugulan **53** x
Ugun **52** z
Uige **44** g
Uil **54** m
Uil (River) **55** a
Uinta Mountains **15** n
Uis Mine **48** g
Uitenhage **49** n
Ujjain **61** o
Ujung Pandang **69** s
Ukerewe, Islands **45** f
Ukhta **50** v
Ukiah **14** t
Ukraine **35** j
Ukta **53** g
Ukukit **52** g
Ukwi **48** l
Ulaan Uul **56** m
Ulaanbaatar (Ulan Bator) **57** e
Ulaga **53** a
Ulakhan Botuobuya **52** e
Ulan **56** z
Ulan Bator **57** e
Ulan-Khol **54** v
Ulansuhai Nur **57** s
Ulan-Ude **57** e
Uldz Gol **57** i
Ulety **57** h
Ulgij **56** h
Ulgumdzha **52** h
Ulgumun **52** g
Uliastaj **56** m
Ulindi **45** c
Ul'kan **57** e
Ul'kenkaroy, Lake **55** j
Ullawarra **72** r
Ulovo **53** j
Ulsan **58** v
Ulua **22** d
Ulukuut **53** o
Ulundi **49** s
Ulungur **56** h
Ulungur Hu **56** g
Uluru (Ayers Rock) **77** a
Ulu Tau **55** g
Ul'ya **53** r
Ul'yanovsk **54** i
Ul'yanovskoye **55** k
Umala **25** p
Uman' **35** j
Umbakumba **73** g
Ume **33** g
Umeå **33** h
Umm as Samim **39** y
Umm as Sawin **60** n
Umm el Abid **37** j
Umm Inderaba **42** j
Umm Keddada **42** h
Umm Luj **38** z
Umm Mirdi **38** w
Umm Qasr **39** g
Umm Ruwaba **42** k
Umm Saggat **42** e
Umm Sa'id **39** w
Umm Saiyala **42** k
Umnak **10** y
Umtata **49** q
Umzinto **49** s
Unaí **29** f
Unaizah **39** d
Unalakleet **11** b
Unalaska **10** y
Unare **23** g
Uncia **28** f
Ungava Bay **13** x
Ungava Peninsula **13** u
Unio **27** q
Unio da Vitória **29** q
Unio dos Palmares **27** v
Unimak **10** z
Unimak Pass **10** z
Unini **23** v
Unión **30** l
United Arab Emirates **39**
United States of America **11/14-20**
United States Range **91** g
Unža **50** r
Uoyan **57** n
Upata **23** i
Upemba, Lake **45** b
Upernavik **90** m
Upington **48** y
Upper Egypt **38** v
Upper Red Lake **15** j
Uppsala **33** f
Uqlat al Suqur **39** b
Urad Qianqi **57** s
Ura J'ube **55** u
Urak **53** r
Ural **54** l
Ural'sk **54** k
Urals, South **55** b
Urandangi **73** u
Uranium City **12** u
Uraricoera **23** w
Uraricoera (River) **23** v
Urbano Santos **27** c
Urcos **25** n
Urdakh **53** h
Urdžar **56** d
Uren **50** s
Urengoy **51** e
Urfa **54** r
Urga **55** o
Urgal **58** h
Urgenč **55** p
Urho **56** f
Uriariti **25** x
Uritskiy **55** f
Uritskoye **52** w
Urluk **57** e
Urmia, Lake **54** u
Urru Co **61** j
Uruaçu **29** e
Uruapan **18** z
Urubamba **24** z
Urubamba (River) **24** z
Urucará **25** k
Urucu **25** f
Uruçui **27** p
Uruçui, Serra do **27** o
Uruguaiana **28** z
Uruguay **28-29/31**
Uruguay (River) **28** x
Urumchi **56** t
Urup **59** h
Urusha **52** w
Uruzgan **60** i
Uryung-Khaya **52** f
Usa **50** j
Usakos **48** h
Usal'gin **53** p
Us'-Bel'dir **56** m
Useless Loop **72** p
Ushiarah **39** f
Ushakova **51** e
Ushuaia **30** w
Ushumun **52** y
Usino **71** n
Üsküdar **35** j
Usman **52** x
Usolje Sibirskoje **57** b
Uspenskiy **55** k
Ussuri **58** l
Ussuriysk **58** k
Ust'-Avam **51** j
Ust'-Belaya **10** f
Ust'-Belogolovoye **53** x
Ust'-Charky **53** c
Ust'-Chayka **52** p
Ust'-Džilinda **57** h
Ust'-Ilimpeya **52** b
Ust'-Kada **57** b
Ust'-Kamengorsk **56** e
Ust'Kamo **51** y
Ust'-Kara **50** m
Ust'-Karenga **52** t
Ust' -Karsk **57** k
Ust'-Kata **55** b
Ust'Kolik'yegan **51** q
Ust'Kulom **50** u
Ust'-Kureyka **51** h
Ust'-Kut **52** o
Ust'-Kuyga **53** b
Ust'Ilimsk **52** n
Ust'-Maya **53** n
Ust'-Mil **53** n
Ust'-Muya **52** t
Ust'-Nera **53** e
Ust'Niman **53** n
Ust'Nyukzha **52** v
Ustobe **56** n
Ust-Olenëk **52** h
Ust'Omchug **53** h
Ust'-Ordynskij **52** o
Ust'Ozernoye **51** u
Ust'-Ozilinda **52** s
Ust'Pit **51** w
Ust'-Port **51** g
Ust'-Sopochnoye **53** w
Ust'-Sugoy **53** j
Ust'-Tigil' **53** y
Ust'-Timpton **52** y
Ust'Tsil'ma **50** u
Ust'Tym **51** s
Ust'Tyrma **58** j
Ust'-Un'ya **52** z
Ust-Urt Plateau **54** z
Ust'-Uyskoye **55** e
Ustya **50** r
Usu **56** r
Usumacinta **19** u
Utah **14** z
Utah Lake **15** n
Utica **16** z
Utila **20** v
Utkholok **53** x
Utopia **73** s
Utrecht (Netherlands) **33** n
Utrecht (South Africa) **49** r
Utsunomiya **59** o
Utta **54** l
Utubulak **56** g
Uummannaq **90** m
Uvalde **19** b
Uvinza **45** e
Uvs Nuur **56** j
Uweinat **38** r
Uyandi **53** d
Uyandina **53** e
Uyar **56** k
Uyëdey **53** c
Uyole **45** g
Uyuk **55** v
Uyuni **28** e
Uzbekistan **55** r
Užgorod **35** f
Uzhur **56** h
Užice **35** d
Uzynkair **55** q

V

Vaal **49** q
Vaal Reservoir **49** p
Vaalwater **49** c
Vaasa **33** i
Vác **35** d
Vacaria **29** q
Vach **51** r
Vachs **55** u
Vadodara **60** z
Vaduz **34** k
Vaga **50** r

Vaigač 50 k
Valcheta 30 l
Valdés Peninsula 30 m
Valdez 11 g
Valdivia 30 g
Val-d'Or 16 k
Valdosta 20 k
Vale de Sonho 29 b
Valenç 27 t
Valença do Piauí 27 q
Valence 34 i
Valencia (Spain) 34 s
Valencia (Venezuela) 23 e
Valenciennes 34 h
Valentim, Serra do 27 p
Valentine 15 t
Valera 23 c
Valga 33 k
Valki 35 m
Valladolid (Mexico) 19 w
Valladolid (Spain) 34 d
Valle de la Pascua 23 f
Valle de Santiago 19 n
Valledupar 23 a
Valle Grande 25 s
Vallejo 14 u
Vallenar 28 o
Valletta 35 n
Valley City 15 h
Valleyview 14 k
Valparaíso 28 o
Valsad 60 z
Vals, Cape 70 u
Valuyki 54 d
Vanado Tuerto 28 v
Vanavara 52 n
Van Buren 17 d
Vancouver 14 g
Vancouver Island 14 e
Vandalia 16 r
Van Diemen, Cape 73 b
Van Diemen Gulf 73 c
Vänern, Lake 33 e
Vangaindrano 47 r
Van Gölu 54 t
Vangou 58 l
Vangunu 71 w
Vang Vieng 62 v
Van Horn 18 l
Vanimo 70 x
Van, Lake 54 t
Van Rees Mountains 70 u
Vanrhynsdorp 48 x
Vanrook 73 k
Vansittart Island 13 e
Vantaa 33 j
Vanua Levu 75 m
Vnuatu 75 d
Van Wyksvlei 48 y
Vanzevat 50 z
Vanzhil'kynak 51 s
Varanasi 61 s
Varaždin 35 b
Vargas Guerra 24 j
Varillas 28 c
Varna (Russia) 55 d
Varna (Bulgaria) 35 i
Várzea Grande 25 y
Vasa Barris 27 t
Vasjugan 51 r
Västerås 33 f
Västervik 33 f
Vasyuganye 51 p
Vatican City 34 m
Vatnajökull 32 d
Vättern, Lake 33 e
Vatyna 10 e
Vaughn 15 q
Vaughn (Montana) 15 a
Vaughn (New Mexico) 18 l
Vaupés 25 a
Växjö 33 s
Vayalpad 64 g
Vayvida 51 x
Vegreville 14 m
Vel 50 q
Velasco, Sierra de 28 r
Veles 35 e
Velhas 29 h
Velichayevskoye 54 u
Velikaya 10 f
Velikaya Kema 58 m
Velikije Luki 33 z
Velikiy Ustyug 50 s
Velikovisochnoye 50 i
Vella Lavella 71 u
Vellore 64 g
Velmo 1-oye 51 w
Velsk 50 r
Venada 49 e
Venado Tuerto 31 a
Vendenga 50 t
Venezia (Venice) 34 m
Venezuela 23/26
Venezuela, Gulf of 23 c
Veniaminof, Mount 11 p
Venice 34 m
Venice, Gulf of 34 m
Ventuari 23 s
Ventura 14 w
Vera 28 w
Veracruz 19 q
Vera, Laguna 28 y
Veraval 60 x
Verde, Cape 40 a
Verde Grande 29 i
Verde (River) (Brazil) 29 b
Verde (River) (Mexico) 19 o
Verde (River) (Paraguay) 28 j
Verdigris 15 w
Vereeniging 46 q
Vergemont 73 x
Verissimo-Sarmento 44 k
Verkhneimbatskoye 51 v
Verkhneural'sk 55 c
Verkhnevilyuysk 52 i
Verkhniy Tura 50 x
Verkhnyaya Amga 52 y
Verkhnyaya Taimyra 51 l
Verkhnyaya Toyma 50 s
Verkhoyansk 53 a
Verkhoyanskiy Mountains 52 k
Vermelha, Lagoa 29 p
Vermont 17 n
Vernadsky 82 t
Vernal 15 o
Verona 34 l
Versino-Darasunskij 57 j
Veshenskaya 54 f
Vest Fjord 33 e
Véstia 29 c
Veszprém 35 c
Vetekhtina 52 u
Vetluga 50 s
Vetluzhskiy 50 s
Vetrenyy 53 h
Vey Vozh 50 v
Vezdekhodnaya 52 d
Viacha 25 p
Vicenza 34 m
Vichada 23 q
Vichadero 29 n
Vicksburg 19 h
Victor Harbor 77 s
Victoria (Argentina) 28 w
Victoria (Brazil) 26 j
Victoria (British Columbia) 14 g
Victoria (Chile) 30 h
Victoria (Hong Kong) 63 q
Victoria (Texas) 19 c
Victoria de las Tunas 21 o
Victoria Falls 49 b
Victoria Island 12 h
Victoria, Lake 45 f
Victoria Land 86 t
Victoria (Mount, Burma) 62 o
Victoria, Mount (Fiji) 75 l
Victoria (Mount, Papua New Guinea) 71 o
Victoria Range 76 c
Victoria River Downs 73 b
Victoria (River) 73 b
Victoria (State) 77 w
Victoria Strait 12 l
Victoria West 48 z
Victorica 30 l
Victorino 23 r
Vicuña Mackenna 30 m
Vidim 57 c
Vidin 35 f
Viedma 30 m
Viedma, Lago 30 u
Vienna 35 b
Vien Pou Kha 62 u
Vientiane 62 v
Vietnam 62/66
Vigan 67 f
Vigia 26 m
Vigo 34 a
Vijayawada 64 i
Vikna 33 c
Vila 75 e
Vila Bittencourt 25 b
Vila de Sena 46 u
Vila Velha 29 k
Vila Velha (Serra Lombarda) 26 k
Vilhena 25 v
Viliga-Kushka 53 l
Viljujskoje Reservoir 52 e
Vilkitskogo 92 k
Vilkicky 51 d
Villa Ahumada 18 k
Villa Angela 28 v
Villa Bella 25 r
Villa Delores 28 s
Villaguay 28 x
Villa Guillermina 28 w
Villa Hayes 28 y
Villahermosa 19 t
Villa Iris 30 m
Villa Maria 28 u
Villa Media Agua 28 q
Villa Montes 28 h
Villa Ocampo 18 k
Villa Ojo de Aqua 28 u
Villarrica 30 h
Villa San Martín 28 t
Villa Unión (Argentina) 28 q
Villa Unión (Mexico) 18 x
Villavicencio 23 n
Villazón 28 f
Vilnius 33 w
Vilyuy 52 j
Vilyuy Mountains 52 d
Vilyuysk 52 j
Vindhya Range 61 n
Vineland 16 z
Vinh 62 x
Vinita 16 n
Vinnitsa 35 i
Vinson Massif 81 d
Vioolsdrif 48 v
Virac 67 i
Virawah 60 x
Virden 15 f
Virginia (Minnesota) 15 k
Virginia (State) 16 x
Virginia Range 76 i
Virgin Islands 21 x
Virú 24 x
Virundunagar 64 g
Visalia 14 v
Visby 33 g
Vischarv 55 w
Viscount Melville Sound 12 g
Viseu 27 a
Vishakhapatnam 61 t
Vista Alegre 25 c
Vistula 35 e
Vitebsk 33 z
Viterbo 34 m
Vitichi 28 f
Viti Levu 75 k
Vitim 52 r
Vitim (River) 52 s
Vitim Plateau 52 s
Vitimskiy 52 s
Vitoria 34 d
Vitória 29 k
Vitória da Conquista 29 j
Vivi 51 y
Vivi (River) 51 k
Vivi, Lake 51 k
Vize 51 d
Vizeaino Desert 18 e
Vizianagaram 61 t
Vizinga 50 t
Vladikavkaz 54 u
Vladimir 54 f
Vladivostok 58 k
Vlorë 35 d
Vltava 35 a
Vohidiala 47 r
Volcán Tres Virgenes 18 f
Volga 54 h
Volga Heights 54 h
Volgograd 54 h
Volksrust 49 r
Volochanka 51 j
Volochsk 53 m
Volodarskoye 55 h
Vologda 50 q
Volokolamsk 50 o
Volokon 52 o
Volonga 50 g
Vólos 35 s
Vol'sk 54 i
Volta Lake 41 n
Volta Redonda 29 i
Vorkuta 50 l
Vorob'yeva 52 n
Voronezh 54 e
Vorontsovka 52 s
Vorozheyka 56 h
Vosburg 48 z
Vostochny Mountains 56 l
Vostok I 85 o
Votkinsk 50 v
Vovodo 42 s
Voyeykov Ice Shelf 87 e
Voyvozh 50 w
Vozrozhdeniya 55 p
Vraca 35 f
Vrendenburg 48 w
Vrsac 35 e
Vryburg 49 n
Vryheid 49 s
Vung Tau 66 j
Vyaz'ma 54 b
Vyborg 33 l
Vyčegda 50 v
Vydrino 56 m
Vym' 50 u
Vyngapur 51 q
Vysni Volocek 50 o
Vyzakova 54 i

W

Wa 40 l
Wabag 70 z
Wabash 16 s
Wabowden 15 h
Wabush 17 d
Waco 19 d
Waco-Kungo 44 t
Wad 60 i
Wad Banda 42 i
Waddan 37 k
Waddington, Mount 14 l
Wadi ad Dawasir 39 q
Wadi al Akhdar 38 m
Wadi al Arish 38 j
Wadi al Batin 39 f
Wadi al Ghadaf 39 c
Wadi al Hamd 38 m
Wadi al Hamim 38 c
Wadi al Hayl 39 a
Wadi al Jarin 39 p
Wadi al Jawf 39 r
Wadi al Jiz 39 v
Wadi Allaqi 38 w
Wadi Araba 38 k
Wadi Ar'ar 39 b
Wadi ar Rimah 39 d
Wadi as Sirhan 38 m
Wadi Aswad 39 z
Wadi ath Thartar 39 c
Wadi bin Khawtar 39 x
Wadi Bishah 39 p
Wadi Gabgaba 38 w
Wadi Garara 38 w
Wadi Hadain 38 x
Wadi Halfa 38 v
Wadi Hawran 39 b
Wadi Ibib 38 y
Wadi Masilah 39 u
Wadi Oitbit 39 y
Wadi Oko 38 y
Wadi Qena) 38 j
Wadi Seidna 38 v
Wadi Shihan 39 w
Wadi Tathlith 39 q
Wadi Tibal 39 b
Wad Medani 42 m
Wager Bay 13 b
Wagga Wagga 77 y
Wagin 76 e
Wagon Mound 15 q
Wagontire 14 v
Waigeo 70 c
Waikabubak 69 s
Waikato 79 v
Wailuku 18 q
Waingapu 69 t
Wainwright (Canada) 15 a
Wainwright (USA) 11 b
Waipahu 18 p
Waipara 79 t
Wairoa 79 x
Waitaki 79 s
Wajir 43 q
Waka 42 p
Wakatipu, Lake 79 q
Wakayama 58 z
Wake Island 235 l
Wakkanai 59 c
Walagan 52 x
Wałbrzych 35 b
Walcott 14 e
Waldburg Range 76 e
Walden 15 p
Wales (UK) 32 w
Wales (USA) 10 m
Wales Island 13 c
Walewale 40 m
Walgett 78 c
Walgreen Coast 81 c
Walikale 45 c
Walinga 71 o
Wallabi Group 76 b
Wallace 14 k
Wallal Downs 72 v
Walleroo 77 f
Walla Walla 14 j
Walsenburg 15 q
Walsh 74 b
Walterboro 20 m
Walvis Bay 48 g
Wamba (Nigeria) 41 g
Wamba (River) 44 i
Wamba (Dem. Rep. Congo) 42 v
Wamena 70 v
Wanaaring 78 a
Wanadsor 54 u
Wandels Sea 91 j
Wandingzhen 62 s
Wandoan 78 e
Wanganui 79 u
Wanganui (River) 79 v
Wangaratta 78 o
Wangmo 62 x
Wangqing 58 i
Wang Saphung 65 k
Wanhsien 62 l
Wanning 63 n
Wannoo 76 c
Wanyuan 62 l
Warangal 64 h
Waratah Bay 77 x
Warburton 72 z
Warburton (River) 77 f
Warburton Lake 12 t
Warder 43 u
Wardha 61 q
Warin Chamrap 66 h
Warner Robins 20 k
Waroona 76 d
Warrabri 73 r
Warrawagine 72 v
Warrego 74 p
Warrenton 49 n
Warri 41 r
Warrnambool 77 v
Warsaw 33 u
Warta 33 t
Warwick 74 t
Warzawa (Warsaw) 33 u
Washington DC 16 y
Washington (State) 14 h
Washington, Cape 86 s
Washington Court House 16 u
Washington Land 91 f
Washington, Mount 17 o
Washita 15 v
Watampone 69 t
Waterbury 17 n
Waterford 32 y
Waterloo 16 p

Waterton **16** y
Watertown **15** i
Waterval Boven **49** r
Waterville **17** b
Watsa **42** v
Watsi **44** k
Watsi Kengo **44** k
Watson Lake **12** n
Wau **42** h
Wauchope **73** r
Wau el Kebir **37** x
Wau en Namus **37** y
Wausau (Massachusetts) **17** n
Wausau (Wisconsin) **16** d
Wave Hill **73** c
Wawa (Canada) **16** g
Wawa (Nigeria) **41** c
Waxxari **56** t
Waycross **20** k
Webster City **15** x
Weda **69** m
Weda Bay **69** m
Weda, Tulek **69** m
Weddell Sea **82** y
Wei **62** k
Weichang **57** x
Weifang **58** o
Weimar **33** q
Weinan **62** m
Weining **62** j
Weipa **73** k
Weiser **14** k
Weisse **51** b
Weiya **56** x
Welbourn Hill **77** c
Weldiya **43** d
Welkite **43** c
Welkom **49** o
Wellesley Islands **73** i
Wellington (Australia) **78** d
Wellinglon (Chile) **30** s
Wellington (New Zealand) **79** v
Wellington (South Africa) **48** x
Wells **14** y
Wema **44** l
Wenatchee **14** i
Wenchhow **63** h
Wendo **43** p
Wendover **14** y
Wenlock **74** a
Wenquan **56** q
Wenquan (Tanggula Shan) **61** m
Wenshan **62** w
Wentworth **77** i
Wepener **49** p
Wernecke Mountains **11** k
Weser **33** p
Wessel Islands **73** g
West Bank **38** l
West Cape **79** p
West Cape Howe **76** r
Western Australia **76** f
Western Cordillera **24-25/28**
Western Cordillera (Colombia) **22** y
Western Ghats **60/64**
Western Sahara **36**
Western Sierra Madre **18** k
West Falkland **31** o
West Fayu **71** a
West Frankfort **16** r
Westgate **77** l
West Group **76** h
Westhope **15** f
West Ice Shelf **85** r
Westlock **14** m
Westmar **74** s
West Memphis **16** q
Weslmoreland **73** h
West Palm Beach **20** m
West Papua **70** u
Westport (Ireland) **32** t
Westport (New Zealand) **79** s
West Siberian Sea **51** g
West Virginia **16** v
West Wyalong **78** b
Wetar **69** x
Wetaskiwin **14** m
Wewak **70** z
Wexford **32** v
Weyburn **15** e
Whakatane **79** w
Whale Cove **13** n
Whangarei **79** h
Wheeler Peak (Nevada) **14** y
Wheeler Peak (New Mexico) **15** q
Wheeling **16** w
Whidby Point **77** d
Whim Creek **72** t
White **15** y
White Cliffs **77** i
Whitecourt **14** k
Whitefish Lake **12** v
Whitehorse **11** l
White Island **51** b
White Island **92** d
White, Lake **73** o
White Nile **38/42**
White River (Canada) **16** g
White River (USA) **15** t
White Sea **50** b
White Volta **40** m
Whitewood **73** y
Whitney, Mount **14** w
Whittle, Cape **17** h
Wholdaia Lake **12** w
Whyalla **77** e
Wichita **15** v
Wichita Falls **19** b
Wien (Vienna) **35** b
Wiesbaden **34** j
Wight, Isle of **32** y
Wilcannia **77** i
Wilczek Island **50** l
Wilhelmshaven **33** o
Wilkes Land **87** a
Wilkins Sound **82** w
Wilemstad **23** d
Willeroo **73** c
Williamsburg **16** y
Williams Lake **14** h
Williamson **16** v
Williamsport **16** x
Willis Islands **74** f
Williston (S. Africa) **48** y
Williston (USA) **15** e
Williston Lake **12** p
Willmar **15** j
Willow **11** f
Willowmore **48** z
Willowra **73** q
Wills, Lake **73** n
Wilmington (Carolina) **21** a
Wilmington (Delaware) **16** y
Wilpena **77** g
Wilson Bluff **76** m
Wilson's Promontory **77** y
Wiluna **76** g
Winburg **49** p
Windau **33** h
Windhoek **48** i
Windorah **77** i
Windsor (Canada) **17** j
Windsor (USA) **16** u
Winisk **16** g
Winisk (River) **16** f
Winisk Lake **16** f
Winnemucca **14** w
Winning **72** r
Winnipeg **15** i
Winnipeg, Lake **15** h
Winnipeg (River) **15** j
Winnipeg Beach **15** i
Winnipegosis **15** g
Winnipegosis, Lake **15** g
Winona (Mississippi) **19** i
Winona (Wisconsin) **16** p
Winslow **15** n
Winston Salem **16** w
Winter Garden **20** l
Winterhoek, Groot **48** x
Winton **73** x
Wisconsin **16** q
Wiseman **11** f
Wismar **33** q
Witherspoon, Mount **11** g
Witpülz **48** v
Witrivier **49** f
Witsand **48** x
Wittenberge **33** q
Wittenoom **72** u
Witvlei **48** j
Włocławek **33** t
Woleai Atoll **70** l
Wolf **22** o
Wolf Creek **14** m
Wolf Point **15** d
Wollaston Foreland **91** t
Wollaston Lake **12** w
Wollaston Peninsula **12** f
Wollogorang **73** h
Wollongong **78** f
Wolmaransstad **49** o
Wonarah **73** t
Wondoola **73** w
Wonganoo **76** h
Wŏnju **58** u
Wŏnsan **58** u
Wonthaggi **77** x
Woodgreen **73** r
Woodlark **71** r
Woods, Lake of the **15** k
Woodstock **17** d
Woodward **15** t
Woomera **77** e
Worcester (South Africa) **48** x
Worcester (USA) **17** n
Worland **15** o
Worthington **15** v
Wotu **69** u
Wowan **74** s
Wrangel Island **10** h
Wrangell **11** y
Wrangell Mountains **11** h
Wrath, Cape **32** j
Wright Peak **80** m
Wrigley **12** p
Wrocław **33** t
Wuchow **63** o
Wuda **57** q
Wudao **57** z
Wudil **41** g
Wudinna **77** d
Wudu **62** j
Wuhan **63** d
Wuhsi **63** h
Wuhu **63** f
Wŭjang **61** e
Wu Jiang **62** l
Wukari **41** u
Wular, Lake **61** a
Wuliang Shan **62** t
Wunnummin Lake **16** e
Wuqi **57** r
Würzburg **34** k
Wutunghliao **62** i
Wuwe **62** i
Wuwei **57** p
Wuyiling **58** i
Wuying **58** i
Wuyi Shan **63** r
Wuyo **41** i
Wuyuan **57** s
Wuzhong **57** r
Wyandra **77** k
Wyloo **72** s
Wyndham **73** a
Wynyard **15** d
Wyoming **15** o
Wytheville **16** v

X

Xainza **61** k
Xai-Xai **49** h
Xamavera **48** k
Xambioá **26** z
Xam Nua **62** w
X-Muteba **44** v
Xapuri **25** p
Xar Burd **57** p
Xar Hudak **57** v
Xar Moron **57** z
Xar Moron (Inner Mongolia) **57** t
Xhumo **49** a
Xi **58** c
Xiachengzi **58** i
Xiamen (Amoy) **63** s
Xiangcheng **62** g
Xiang Jiang **63** o
Xiangquan **61** e
Xiangride **62** e
Xi'an (Sian) **62** m
Xianyang **62** l
Xiao'ergou **58** e
Xide **62** i
Xieng Khouang **62** v
Xifeng **62** l
Xi Jiang (Si) **63** o
Xikou **57** m
Xilin Hot **57** w
Xilin Qagan Obo **57** u
Ximiao **57** n
Xin'anjiang **63** g
Xin Barag Youqi **57** k
Xin Barag Zouqi **57** l
Xincai **63** c
Xinge **44** w
Xingsagoinba **62** g
Xingu **26** v
Xingu (River) **26** w
Xinguara **26** y
Xingxingxia **56** y
Xingyi **62** w
Xingzi **63** d
Xinhe **56** q
Xinhuang **62** m
Xinkai **58** d
Xinmin **58** e
Xin Xian **57** u
Xinyang **63** c
Xinyu **63** d
Xinyuan **56** q
Xique-Xique **27** q
Xitieshan **56** y
Xi Ujimqin Qi **57** x
Xiushui **63** d
Xixabangma Feng **61** i
Xixiang **62** l
Xorkol **56** v
Xuancheng **63** g
Xuanwei **62** w
Xueweng **63** u
Xungba **61** f
Xunke **58** h
Xuwen **63** n
Xuyong **62** k

Y

Ya'an **62** i
Yabassi **41** u
Yabelo **43** o
Yablonovy Mountains **57** g
Yabrai Yanchang **57** p
Yacimiento **30** u
Yacuiba **28** h
Yada **51** b
Yadrin **54** h
Yafo **38** k
Yagra **61** f
Yahuma **42** r
Yakeshi **57** l
Yakima **14** i
Yako **40** l
Yakoma **42** r
Yakovlevka **51** g
Yakrik **56** q
Yaku **58** w
Yakutat **11** w
Yakutsk **52** m
Yala (Ghana) **40** m
Yala (Thailand) **65** x
Yalata **77** b
Yali **44** k
Yalifafu **44** l
Yalinga **42** r
Yallalong **76** d
Yalleroi **74** p
Yalong Jiang **62** h
Yalu **58** f
Yalutorovsk **50** z
Yama **53** w
Yamagata **59** o
Yamaguchi **58** w
Yamal Peninsula **51** a
Yambio **42** v
Yamburg **51** d
Yambuya **42** s
Yamma Yamma, Lake **77** h
Yamussoukro **40** w
Yamsk **53** w
Yamuna **61** r
Yamzho Yumco **61** l
Yana **53** b
Yan'an **57** s
Yanbu al Bahr **38** z
Yanchang **63** a
Yandja **44** h
Yandun **56** x
Yangambi **42** s
Yangarey **50** l
Yangbajain **61** k
Yangchow **63** f
Yangchuan **57** v
Yangd Sum **57** w
Yangjiang **63** o
Yangoda **51** h
Yangon (Rangoon) **65** g
Yangtze **63** a
Yanhuqu **61** f
Yankton **15** u
Yanov Stan **51** g
Yanqi Huizu Zizhixian **56** t
Yanrey **72** r
Yanshan **57** x
Yantai **63** h
Yany-Kurgan **55** t
Yanzhou **63** e
Yaoundé **41** v
Yapen **70** t
Yapeyú **28** z
Yap Islands **70** h
Yaptiksale **51** b
Yaqui **18** h
Yaraka **77** j
Yaransk **50** t
Yaransu **54** j
Yarda **37** z
Yarkant **56** o
Yarkino **51** y
Yarlung Zangbo Jiang **61** k
Yarmouth **17** q
Yarongo **51** a
Yarraloola **72** s
Yarraman **78** f
Yarrie **72** v
Yarroto, Lake **51** b
Yartsevo **51** v
Yarumal **22** m
Yashikera **41** c
Yashkino **53** o
Yashkul' **54** h
Yasin **55** x
Yasnaya Polyana **58** m
Yasothon **66** h
Yata (Bolivia) **25** r
Yata (Central African. Rep.) **42** e
Yathkyed Lake **12** y
Yatolema **42** s
Yatsushiro **58** w
Yauri **25** n
Yauyos **24** w
Yavari **24** m
Yavatmal **61** p
Yaxian **62** z
Yazd **39** k
Yazdan **60** d
Ye **65** h
Yebbi Bou **37** y
Yecheng **55** z
Yécora **18** i
Yeggueba **37** v
Yei **42** w
Yekana **42** r
Yekaterinburg **50** y
Yelets **54** d
Yeletskiy **50** m
Yelizavety, Cape **53** r
Yelkhovka **54** k
Yellowknife **12** t
Yellow River (Huang He) **63** f
Yellow Sea **58** r
Yellowstone **15** e
Yelma **76** h
Yeloguy **51** u
Yeloguy (River) **51** t
Yelvertoft **73** u
Yelwa **41** d
Yemen **39/43**
Yenakiyevo **54** d
Yendi **41** a
Yengisar **56** r
Yengisu **56** t
Yenisey **51** u
Yenisey Bay **51** f
Yenisey Mountains **51** w
Yeniseysk **51** w

Yenki **58** i
Yenyuka **52** w
Yeo, Lake **76** j
Yepachic **18** i
Yepoko **51** d
Yeppoon **74** t
Yerbogachen **52** c
Yerema **52** q
Yerevan **54** t
Yermak **55** l
Yermaki **57** f
Yermakovo **51** p
Yermakovo (Yenisey) **51** g
Yermentau **55** k
Yermo **18** l
Yeropol **10** c
Yerupaja **24** w
Yesil' **55** f
Yessey **52** a
Yet **43** s
Yet-Kyuyel' **52** i
Yetti **36** t
Yeva-Liv **50** m
Yeyka **52** c
Yeysk **54** d
Yguazú **29** n
Ygyatta **52** g
Yi'an **58** f
Yichun **58** i
Yidin **62** g
Yidu **63** a
Yimuhe **57** l
Yinchuan **57** q
Yingde **63** o
Yingkow **57** z
Yipinglang **62** u
Yirrkala **73** g
Yirshi **57** l
Yishan **62** y
Yiwu **56** x
Yi Xian **57** z
Yiyang **63** b
Ylas-Yuryakh **52** c
Ynykchanskiy **53** q
Yogyakarta **68** z
Yokadouma **41** x
Yoko **41** w
Yokohama **59** o
Yokosuka **59** o
Yola **41** j
Yolin Mod **58** d
Yolombo **44** m
Yondeng **57** p
Yong'an **63** s
Yongchuang **62** k
Yongden **62** i
Yongding **57** w
Yongren **62** u
Yopal **23** o
Yopurga **55** z
York **32** y
York, Cape (Australia) **73** k
York, Cape (Greenland) **91** b
York Factory **13** o
Yorke Peninsula **77** f
Yorkton **15** f
Yoshkar Ola **54** j
Yŏsu **58** u
Youngstown **16** v
Youshashan **56** v
Youyang **62** m
Yozgat **35** z
Ypé Jhú **28** m
Yreka **14** u
Ystad **33** r
Ytyk-Kyuyel' **53** a
Yuan Jiang **63** a
Yuanjiang **62** u
Yuanlin **57** m
Yuanshanzi **57** n
Yuba City **14** u
Yubi, Cap **36** c
Yucatán Peninsula **20** t
Yucatán, Straits of **20** v
Yuci **63** b
Yudoma **53** q
Yudoma-Krestovskaya **53** q
Yuendumu **73** q
Yueyang **63** c
Yugorenok **53** p
Yugoslavia **35**
Yugo-Tala **53** i
Yu Jiang **62** y
Yukhta **57** e
Yukon **11** b
Yukon Territory **11** j
Yukta **52** c
Yule **72** t
Yuli **56** t
Yulin **63** n
Yulin (Great Wall) **57** s
Yulongxue Shan **62** g
Yuma **18** e
Yun **63** e
Yunhe **63** g
Yunmeng **63** e
Yurga **51** t
Yuribey **51** d
Yurimaguas **24** j
Yurungkax **61** e
Yushu **62** d
Yushu (Manchuria) **58** g
Yuti **28** i
Yutian **61** e
Yutze **57** u
Yu Xian **57** v
Yuzhno-Sakhalinsk **59** d

Z

Zabad **43** f
Zabol **60** e
Zabré **40** m
Zacatecas **18** z
Zacatecoluca **22** c
Zacualtipan **19** p
Zadar **35** b
Zadoi **62** d
Zagazig **38** i
Zaghouan **34** y
Zagora **36** i
Zagreb **35** b
Zagros Mountains **39** i
Zahedan **60** d
Zahksykan **55** h
Zahlé **38** l
Zahran **39** q
Zajsan **56** e
Zajsan, Lake **56** e
Zakamensk **57** c
Zakopane **35** d
Zakouma **42** b
Zalanga **41** h
Zalantun **58** e
Zalari **57** b
Zalim **39** p
Zalingei **42** f
Zambeke **42** t
Zambezi **44** y
Zambia **44-45/46/49**
Zamboanga **67** u
Zamko **41** h
Zamora (Ecuador) **24** h
Zamora (Mexico) **18** z
Zamora (Spain) **34** c
Zamość **33** v
Zamzor **56** m
Zanda **61** d
Zangguy **56** o
Zangla **61** b
Zanjan **54** w
Zanjón **28** q
Zante **35** q
Zanzibar **45** k
Zanzibar Island **45** k
Zapala **30** i
Zaporozh'ye **35** m
Za Qu **62** d
Zarafshan **55** s
Zaragoza **22** m
Zarand **39** m
Zarand (Tehran) **39** i
Zaranj **60** e
Zárate **31** b
Zaraza **23** g
Zard **39** h
Zarghunshar **60** j
Zargun **60** i
Zaria **41** f
Zaris **48** h
Žarma **56** c
Zarqa **38** l
Žarsuat **56** c
Zarumilla **24** g
Zarzis **34** z
Zaskar **61** c
Zastron **49** p
Zatish'ye **53** l
Zavitinsk **58** i
Zawa **56** o
Zayü **62** e
Zealand **33** q
Zebak **55** v
Zedness **36** r
Zeerust **49** b
Zêkog **62** h
Zelenga **54** j
Zella **37** l
Zel'onoje Ozero **52** u
Zémio **42** s
Zemmour **36** r
Zenica **35** c
Zeya **52** y
Zeya (River) **52** z
Zeydan **39** i
Zeyski **52** z
Zhagaly **54** k
Zhag'yab **62** e
Zhaltyr **55** i
Zhamshi **55** l
Zhanay **55** q
Zhangjiaquan **56** x
Zhangpu **63** r
Zhaosu **56** p
Zhaotong **62** j
Zhaoyuan **58** f
Zhari Namco **61** i
Zharkamys **55** a
Zhaxi Co **61** h
Zhecheng **63** d
Zhelezinka **55** l
Zheleznodorozhnyy **50** u
Zhenghe **63** g
Zhenghuang Qi **57** v
Zhenglan Qi **57** w
Zhengzhou (Chengchow) **63** c
Zhenning **62** x
Zhenping **63** a
Zhenyuan **62** u
Zhidoi **62** d
Zhigansk **52** j
Zhirkova **53** k
Zhitomir **35** i
Zhizdra **54** b
Zholymbet **55** j
Zhongba **61** g
Zhongdian **62** g
Zhongshan **85** p
Zhongwei **57** p
Zhongye Islands **67** b
Zhoushan Islands **63** i
Zhoushan Qundao **58** r
Zhungqu **62** j
Zhupanovskiy **53** y
Zichang **62** m
Žigalovo **57** c
Zigazê **61** j
Zighan **38** o
Ziguinchor **40** b
Zihuatenejo **19** n
Zilinan **35** c
Zima **57** b
Zimbabwe **49** s
Zimbabwe (State) **45/46/49**
Zinapécuaro **19** n
Zinder **41** g
Zinga **42** n
Zintan **37** h
Žioj **54** x
Zipaquirá **23** n
Zirō **62** c
Zitácuaro **19** n
Ziway, Lake **43** c
Ziya **57** w
Zlatoust **55** c
Zliten **37** j
Zmiyevka **54** c
Znamenka **35** k
Znojmo **33** s
Zogang **62** e
Zoigê **62** i
Zolotoy **53** c
Zomba **45** u
Zongo **42** o
Zoo Baba **37** v
Zorritos **24** g
Zorzor **40** t
Zouar **37** x
Zouerate **36** q
Zq'gya-Zangbo **61** l
Zrenjanin **35** d
Zubova-Polyana **54** g
Zug **36** p
Zugspitze **34** m
Zumbo **45** r
Zürich **34** k
Zuru **41** e
Zuwarah **34** z
Zuwaylah **37** w
Zvishavane **49** e
Zvolen **35** c
Zyryanka **53** i
Zyryanka (Yenisey Bay) **51** f

a b c d e f g h i j k l m
160° 140° 120° 100° 80° 60° 40° 20° 0°
Greenland
(Kalaallit Nunaat)
(Denmark)
Greenland Sea
Søndrestrømfjord
Nuuk
(Godthåb)
Denmark Strait
ICELAND
Reykjavik
Faeroe Islands (Den.)
Bergen
Davis Strait
Baffin Island
Frobisher Bay
(Iqaluit)
Hudson Strait
Ungava Pen.
Foxe Basin
Boothia Pen.
Southampton I.
Hudson Bay
Victoria I.
Bering Strait
Nome
Alaska
(U.S.A.)
6194 Mc Kinley
Brooks Range
Yukon
Fort Yukon
Fairbanks
Anchorage
Valdez
Seward
Alaska Peninsula
Gulf of Alaska
Inuvik
Dawson
Whitehorse
Juneau
Mackenzie
Coppermine
(Kugluktuk)
Gt. Bear Lake
Port Radium
(Echo Bay)
Yellowknife
Gt. Slave Lake
Peace
Uranium City
Lake Athabasca
Chesterfield Inlet
(Igluligaarjuk)
Churchill
Queen Charlotte Is.
Edmonton
CANADA
Rocky Mountains
Calgary
Regina
Winnipeg
Lake Winnipeg
Schefferville
Labrador
Vancouver I.
Vancouver
Seattle
Columbia
Fraser
Missouri
Thunder Bay
Lake Superior
Duluth
Québec
Montréal
Ottawa
Gulf of St. Lawrence
Newfoundland
St. John's
St. John
Halifax
Portland
Minneapolis
Lake Michigan
Lake Huron
Toronto
Lake Erie
Milwaukee
Chicago
Boston
Salt Lake City
Omaha
UNITED STATES OF AMERICA
(U.S.A.)
Appalachians
New York
Philadelphia
Baltimore
Washington
Denver
St. Louis
San Francisco
Sierra Nevada
Colorado
Oklahoma City
Norfolk
Memphis
Atlanta
Birmingham
Los Angeles
San Diego
Phoenix
El Paso
Rio Grande
Fort Worth
Dallas
Mississippi
Houston
New Orleans
San Antonio
Jacksonville
Corpus Christi
Tampa
Miami
Bermuda
(U. K.)
Azores
(Port.)
ATLANTIC
MEXICO
Monterrey
Gulf of Mexico
Nassau
BAHAMAS
Havana
CUBA
Mérida
León
Guadalajara
Honolulu
Hawaiian Islands
(U.S.A.)
Mexico
5700
Revilla Gigedo Islands
(Mexico)
Belmopan
BELIZE
JAMAICA
Kingston
HAITI
Port-au-Prince
DOMINICAN REP.
Santo Domingo
SAINT KITTS AND NEVIS
ANTIGUA AND BARBUDA
DOMINICA
ST. LUCIA
SAINT VINCENT AND THE GRENADINES
BARBADOS
GRENADA
TRINIDAD AND TOBAGO
GUATEMALA
Guatemala
EL SALVADOR
HONDURAS
Tegucigalpa
Caribbean Sea
NICARAGUA
Managua
COSTA RICA
San José
Barranquilla
Maracaibo
Caracas
Panama
PANAMA
Clipperton (Fr.)
VENEZUELA
Georgetown
GUYANA
Paramaribo
SURINAME
Cayenne
GUIANA
(Fr.)
Medellín
Bogota
COLOMBIA
Orinoco
Cali
PACIFIC
Palmyra
(U.S.A.)
Line Islands
Kiritimati
Equator
Coco
(Costa Rica)
Galapagos Islands
(Ecu.)
ECUADOR
Quito
Guayaquil
6272
Andes
Negro
Amazon
Belem
Manaus
São Luis
Fortaleza
Purus
Tocantins
OCEAN
Chiclayo
PERU
BRAZIL
Recife
Marquesas Is.
Lima
Salvador
Cuzco
Brazilian Plateau
Brasilia
American Samoa
(U.S.A.)
French Polynesia
Tahiti
Tuamotu Archipelago
Lake Titicaca
La Paz
Arequipa
Mato Grosso
Santa Cruz
BOLIVIA
Sucre
Cook Islands
(N.Z.)
OCEAN
Iquique
Belo Horizonte
Mururoa
PARAGUAY
Parana
Tubuai Isalnds
Antofagasta
Salta
Concepción
Rio de Janeiro
São Paulo
6723
San Miguel de Tucumán
Pitcairn I.
(U. K.)
Asunción
Easter I.
(Chile)
Corrientes
Pôrto Alegre
Cordoba
Aconcagua
6959
Santa Fé
Rio Grande
Valparaiso
Mendoza
Rosario
Juan Fernández Islands
(Chile)
Santiago
Buenos Aires
URUGUAY
Montevideo
CHILE
ARGENTINA
Concepción
Mar del Plata
Bahia Blanca
Puerto Montt
Viedma
Comodoro Rivadavia
Falkland Is. /
Islas Malvinas (U. K.)
Stanley
Rio Gallegos
Punta Arenas
Tierra del Fuego
Cape Horn
South Georgia
(U. K.)
Bouvet I.
(Norway)
Tristan da Cunha
(U. K.)
St. Helena
(U. K.)
Ascension
(U. K.)
Glasgow
Belfast
IRELAND
Dublin
Liverpool
UNITED KINGDOM OF GREAT BRITAIN AND NORTHERN IRELAND
London
Amsterdam
Brussels
Paris
FRANCE
Bay of Biskay
Bordeaux
Marseilles
Bilbao
Oporto
PORTUGAL
Lisbon
Madrid
SPAIN
Barcelona
Algiers
Gibraltar
(U. K.)
Tangier
Oran
Rabat
Casablanca
MOROCCO
Atlas Mts.
Marrakesch
4165
Madeira
(Port.)
Canary Is.
(Sp.)
Las Palmas
ALGERIA
El Aaiún
WESTERN SAHARA
Nouâdhibou
MAURITANIA
Nouakchott
MALI
CAPE VERDE
Dakar
SENEGAL
GAMBIA
Bamako
BURKINA FASO
Ouagadougou
GUINEA-BISSAU
GUINEA
Conakry
Freetown
SIERRA LEONE
CÔTE D'IVOIRE
Yamoussoukro
GHANA
TOGO
Monrovia
LIBERIA
Abidjan
Accra
EQUATORIAL GU
SAO TOMÉ AND PRINC
Antarctica
n o p q r s t u v w x y z